TEXTILES

Sept '93 Hannah Worthley .

TEXTILES

SIXTH EDITION

Norma Hollen

Professor Emeritus

Iowa State University

Jane Saddler

Professor Emeritus

Iowa State University

Anna L. Langford

Sara J. Kadolph

Iowa State University

MACMILLAN PUBLISHING COMPANY

New York

PRINTED IN THE UNITED STATES OF AMERICA

Macmillan Publishing Company
866 Third Avenue, New York, New York 10022

Collier Macmillan Canada, Inc.

LIBRARY OF CONGRESS CATALOGING-IN-PUBLICATION DATA

Textiles.

Includes index.
1. Textile industry. 2. Textile fibers. 3. Textile fabrics. I. Hollen, Norma R. II. Hollen, Norma R. Textiles.
TS1446.T47 1988 677 87-7052
ISBN 0-02-367530-6

Printing: 1 2 3 4 5 6 7 Year: 8 9 0 1 2 3 4

ISBN 0-02-367530-6 (Hardcover edition)
ISBN 0-02-946270-3 (International edition)

Preface

This text was written for use in an introductory or single college course in textiles. It provides a broad view of the production and utilization of fabrics with emphasis on consumer values and serviceability.

The text is organized in a logical manner. It deals with textile-related terms, fibers, yarns, fabrication methods, finishes, and care. However, each section of the text is complete enough that any order can be used in presenting the material for study.

The sixth edition of *Textiles* has been revised to include new developments in fibers, yarns, fabrication, and finishes. A new chapter on care has been added because of the importance of understanding how to properly clean and store textile products. The book has been reorganized to present more cohesive units and to keep like material together. For example, man-made fibers and fiber modifications have been combined into one chapter. This new chapter comes before any of the man-made fibers. In addition, all woven chapters are grouped together. The two knit chapters follow the woven fabrics, and the finish chapters have been reorganized. An overview of finishing is presented first, followed by aesthetic finishes, special-purpose finishes, flame-retardant finishes and flammability, and dyeing and printing.

We have tried to maintain the strong points of the book, while correcting the weak points. It is not specific to textiles and clothing majors, but is useful for students who will be working with textile products. Interior design and education students will find much useful information in the book.

The book does not require a physical science background. Technical information is presented to assist the student in understanding the material and to provide background information to those students who have had courses in chemistry.

Illustrations assist in understanding the material. Hence, many illustrations are included in the text. Both photographs and drawings are included to represent actual fabrics and production machinery and to clarify details in the fabric and in production of the fabric.

This text will help students to do the following:

- Use textile terminology correctly.
- Know current laws and labeling requirements that regulate textile distribution.
- Understand how production processes affect the characteristics and cost of textile products.
- Appreciate past developments in textiles and recognize the need for future developments.
- Identify fibers, yarns, and fabrics by analysis and some simple procedures.
- Predict fabric performance based on knowledge of fibers, yarns, fabric constructions, and finishes in conjunction with informative labeling.
- Make wise selections of textile products for specific end uses.

- Care for textile products in a satisfactory manner.
- Develop an interest in textiles that will motivate further study.

A student swatch kit has been developed by Textile Fabric Consultants to use in conjunction with this edition of *Textiles*. It is available through Textile Fabric Consultants, P.O. Box 111431, Nashville, TN 37222.

Acknowledgments

We wish to express our appreciation to the following: Darlene Fratzke of Iowa State University, for her assistance with the care chapter and her comments, in general, about the book; Chuck Greiner of the Front Porch Photography Studio, Story City, Iowa, for his help with the photography; Carolyn Kundel, Ruth Glock, Charles Kim, all of Iowa State University, for their suggestions concerning this edition; and Donna Danielson, Iowa State University, for her illustrations.

We also wish to thank our reviewers for their helpful suggestions: Margaret McBurney, Ashland College; Jane Hooper, Wayne State University; Ann Reed, University of Texas at Austin; Alvertia Quesenberry, Ball State University; Doris Beard, California State University; Billie Collier, Ohio University; Christine Ladish, Purdue University; Lucille Golightly, Memphis State University; Ernestine Reeder, Middle Tennessee State University; and Joan Laughlin, University of Nebraska.

Finally, we would like to thank those members of the textile industry who have provided information, diagrams, and photographs.

N. H.
J. S.
A. L. L.
S. J. K.

Contents

1

Introduction

This chapter begins the detailed study of textiles and the properties they contribute to fabrics, apparel, furnishing, and industrial textiles. A good starting place is the definitions of the component parts of a textile fabric.

Fiber Any substance, natural or man-made, with a high length-to-width ratio and with suitable characteristics for being processed into a fabric.

Yarn An assemblage of fibers, twisted or laid together so as to form a continuous strand that can be made into a textile fabric.

Fabric A planar substance constructed from solutions, fibers, yarns, fabrics, or any combination of these.

Finish Any process used to convert gray goods (unfinished fabric) into finished fabric.

Food, shelter, and clothing are the basic needs of everyone. Most clothing is made from textiles, and shelters are made more comfortable and attractive by the use of textiles. In fact, some shelters are made from textiles. Textiles are used in the production or processing of many things used in day-to-day living, such as food and manufactured goods.

We are surrounded by textiles from birth to death. We walk on and wear textile products; we sit on fabric-covered chairs and sofas; we sleep on and under fabrics; textiles dry us or keep us dry; they keep us warm and protect us from the sun, fire, and infection. Clothing and furnishing textiles are aesthetically pleasing, and they vary in color, design, and texture. They are also available in a variety of price ranges.

The industrial and medical uses of textiles are many and varied. The automotive industry, one of the largest industries in the United States, uses textiles to make tire cords, upholstery, carpeting, head liners, window runners, seat belts, and shoulder harnesses.

Man has traveled to the moon in a 20-layer, $100,000 space suit that has nylon water-cooled underwear. Life is prolonged by replacing wornout parts of the body with woven- or knitted-fabric parts such as polyester arteries and velour heart valves. Disposable garments are worn by doctors and nurses. Bulletproof vests protect police, hunters, and soldiers, and safety belts make automobile travel less dangerous. Three-dimensional, inflatable "buildings" keep out desert heat and Arctic cold.

This text was written for consumers—not average consumers but educated consumers who, when they purchase textile items, want to know *what* to expect in fabric performance and *why* fabrics perform as they do. Textiles are always changing. They change as fashion changes and as the needs of people change. New developments in production processes also cause changes in textiles, as do government standards for safety, environmental quality, and energy conservation. These changes are discussed, but the bulk of the text is devoted to basic information about textile products, with an emphasis on fibers, yarns, fabric construction, and finishes. All of these elements are interdependent and contribute to the beauty, the durability, the care, and the comfort of fabrics.

Much of the terminology used in the text may be new to students and many facts must be memorized. But to understand textiles in a broad aspect one must first learn the basics. The historical development, the basic concepts, and the new developments in textiles are discussed. Production processes are explained briefly. A knowledge of production should give the student a better understanding of, and appreciation for, the textile industry.

In the United States, the textile industry is a tremendous complex. It includes the natural and man-made fiber producers, spinners, weavers, knitters, throwsters, yarn converters, tufters, fiberweb producers and finishers, machinery makers, and many others. More people are employed in the textile industry than in any other manufacturing industry, over 2 million. Textile products valued at over $40 billion are produced each year by systems that are increasingly being directed by computers. In Japan, at the push of a button, an operator can dye wool fabric in over 2,000 color combinations without flaw or error.

The textile industry has developed from an art-and-craft industry perpetuated by guilds in the early centuries, through the Industrial Revolution in the 18th and 19th centuries, when the emphasis was on mechanization and mass production, to the 20th century, with its emphasis on science and technology.

In this century, man-made fibers were developed and modified textured yarns were created. New fabrications and increased production of knits occurred, and many finishes and sophisticated textile production and marketing systems were developed. These developments have been beneficial to consumers. Man-made fibers and durable press finishes have made many items of clothing "easy care." New developments in textiles have also created some problems for consumers, particularly in the selection of apparel and furnishing textiles. So many items look alike. Knitted fabrics look like woven fabrics and vice versa, vinyl and polyurethane films look like leather, fake furs look like real furs, acrylic- and polyester-fiber fabrics look like wool. The traditional cotton fabrics are now often polyester or polyester/cotton blends.

To make textile selection a bit easier for consumers, textile producers and their associations have set standards and established quality-control programs for many textile products. The federal government has passed laws to protect consumers from unfair trade practices, namely, the Wool Products Labeling Act, the Fur Products Labeling Act, the Textile Fiber Products Identification Act, and the Flammable Fabrics Act. The first three laws are "truth-in-fabrics" legislation and, to be beneficial, some knowledge on the part of the consumer about fibers and furs is required. The Flammable Fabrics Act is protective legislation that prohibits the sale of dangerously flammable apparel and furnishing textiles. The Federal Trade Commission issued the Permanent Care Labeling Rule in 1972 and revised it in 1985. Its purpose is to inform the consumer how to care for fabrics and garments.

Emphasis on energy conservation, environmental quality, noise abatement, health, and safety affect the textile industry as well as other industries. The efforts of the textile industry to meet standards set by the federal government affect the consumer indirectly—by raising prices for merchandise or by limiting the choices available. Energy conservation is being achieved by using less water and faster production methods. Nonpollution of streams and air is being achieved by reducing or eliminating the use of water in many finishing processes and by adding equipment to machines to cleanse and purify the water or air before it is emitted. The Occupational Safety and Health Administration has set standards for noise levels and dust and lint levels that make the mills healthier places in which to work. Much progress has been made in providing flame-resistant fibers and finishes in response to the Consumer Products Safety Commission's implementation of the Flammable Fabrics Act. The CPSC also has commissioned the testing of fabrics and finishes for toxicity and carcinogenicity, and it can request that suspect fabrics be removed from the market.

Textile fabrics can be beautiful, durable, comfortable, and easy care. They can satisfy the needs of all people at all times. Knowing how fabrics are created and used will give a better basis for their selection and an understanding of their limitations.

A knowledge of textiles and their production will result in a more informed selection of a textile product for a particular use. A knowledgeable selection will result in a more-satisfied user.

2

Textile Fibers and Their Properties

It is important to understand the factors influenced by fibers because fibers are the basic unit of most fabrics. Fibers contribute to the aesthetic appearance of fabrics; they influence the durability, comfort, and appearance retention of fabrics; they determine, to a large extent, the care required for fabrics; and they influence the cost of fabrics. Successful textile fibers must be readily available, constantly in supply, and inexpensive. They must have sufficient strength, pliability, length, and cohesiveness to be spun into yarns.

Textile fibers have been used to make cloth for the last 4,000 or 5,000 years. Until 1885, when the first man-made fiber was produced commercially, fibers were obtained from only plants and animals. The fibers most commonly used were wool, flax, cotton, and silk. These four natural fibers continue to be used and enjoyed today, although their economic importance relative to all fibers has decreased.

Textile processes—spinning, weaving, knitting, dyeing, and finishing of fabrics—were developed for the natural fibers. Man-made fibers often resemble the natural fibers.

For example, silk has always been a highly prized fiber because of the smooth, lustrous, soft fabrics made from it; it has always been expensive and comparatively scarce. It was, therefore, logical to try to duplicate silk. Rayon (called artificial silk until 1925) was the first man-made fiber. Rayon was produced in filament length until the early 1930s when an enterprising textile worker discovered that the broken and wasted rayon filaments could be used as staple fiber. Acetate and nylon were also introduced as filaments to be used in silk-like fabrics.

Many man-made fibers were developed in the first half of the 20th century, and from that time onward tremendous advances have been made in the man-made fiber industry, primarily modifications of the parent fibers to provide the best combination of properties for specific end uses. The man-made fibers most commonly used today in apparel and home-furnishing fabrics include polyester, nylon, olefin, acrylic, rayon, and acetate.

Fiber Properties

Fiber properties contribute to the properties of a fabric. For example, strong fibers contribute to the durability of fabrics; absorbent fibers are good for skin-contact apparel and for towels and diapers; fibers that are self-extinguishing are good for children's sleepwear and protective clothing.

To analyze a fabric in order to predict its performance, start with the fiber. Knowledge of the fiber's properties will help to anticipate the fiber's contribution to the performance of a fabric and the product made from it. Some contributions of fibers are desirable and some are not. Figure 2–1 illustrates this fact by identifying some of the contributions of a low-absorbency fiber.

Fiber properties are determined by the nature of the *physical structure,* the *chemical composition,* and the *molecular arrangement.*

PHYSICAL STRUCTURE

The physical structure, or morphology, of fibers can be identified by observing the fiber through a light, or electron, microscope. In the text, photomicrographs taken by electron microscopes at

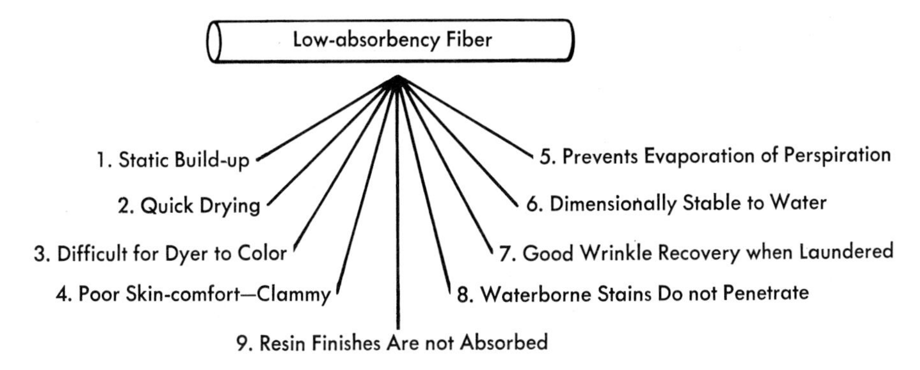

Fig. 2–1 *Properties usually related to low absorbency.*

magnifications of 250–1,000× will be used to clarify details of the fiber's physical structure.

Length. Fibers are sold by the fiber producer as filament, staple, or filament tow. *Filaments* are long continuous fiber strands of indefinite length, measured in yards or meters. They may be either monofilament (one fiber) or multifilament (a number of filaments). Filaments may be smooth or bulked (crimped in some way), as shown in Figure 2–2. Smooth filaments are used to produce silk-like fabrics; bulked filaments are used in more cotton-like or wool-like fabrics.

Staple fibers are measured in inches or centimeters and range in length from 2–46 cm (3/4 of an inch to 18 inches). Staple fibers are shown in Figure 2–3. All the natural fibers except silk are available only in staple form.

The man-made fibers are made into staple form by cutting filament tow into short lengths. *Filament tow* consists of a loose rope or strand of several thousand man-made fibers without a definite twist. Tow is usually crimped after spinning (Figure 2–4).

Diameter, Size, or Denier. Fiber size plays a big part in determining the performance and *hand* of a fabric (how it feels). Large fibers give

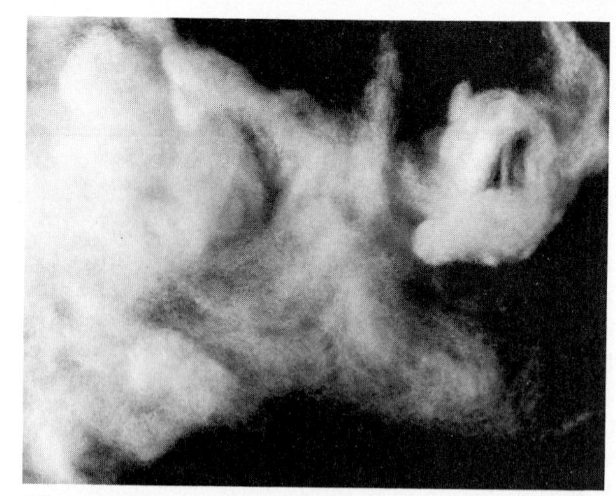

Fig. 2–3 *Man-made staple fiber.*

crispness, roughness, body, and stiffness. Large fibers also resist crushing—a property that is important in carpets, for example. Fine fibers give softness and pliability. Fabrics made with fine fibers will drape more easily.

Natural fibers are subject to growth irregularities and are not uniform in size or development. In natural fibers, fineness is a major factor in determining quality. Fine fibers are of better quality. Fineness is measured in micrometers (a micrometer is 1/1,000 millimeter or 1/25,400 inch).

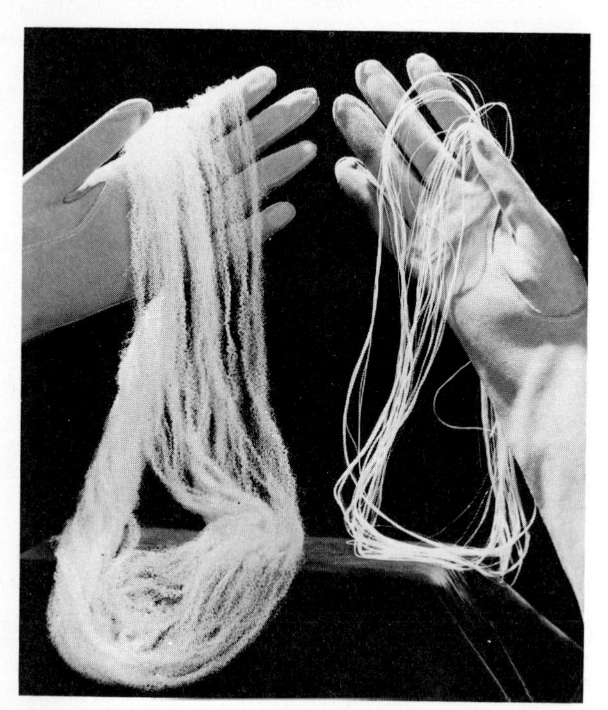

Fig. 2–2 *Man-made filaments: textured-bulk yarn* (left); *smooth-filament yarn* (right).

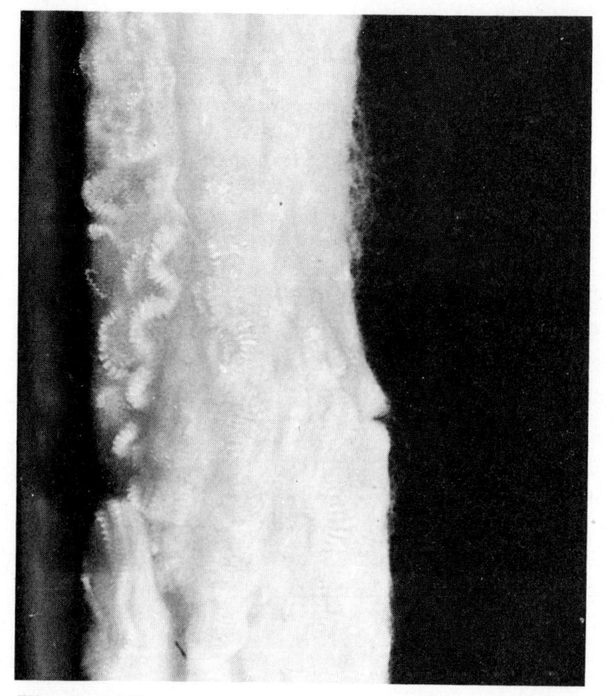

Fig. 2–4 *Filament tow.*

Diameter Range (micrometers)	
Cotton	16–20
Flax	12–16
Wool	10–50
Silk	11–12

In *man-made fibers,* diameter is controlled by the size of the spinneret holes, by stretching or drawing during or after spinning, or by controlling the rate of extrusion of the spinning dope through the spinneret. Man-made fibers can be made uniform in diameter or can be thick-and-thin at regular intervals throughout their length. The fineness of man-made fibers is measured in denier. *Denier* is the weight in grams of 9,000 meters of fiber or yarn. *Tex* is the weight in grams of 1,000 meters of fiber or yarn. Staple fiber is sold by denier and fiber length; filament fiber is sold by the denier of the yarn or tow. Yarn denier can be divided by the number of filaments to give filament denier. For example:

$$\frac{\text{40 denier yarn}}{\text{20 filaments}} = \text{2 denier per filament.}$$

One to 3 denier corresponds to fine cotton, cashmere, or wool; 5 to 8 denier is similar to average cotton, wool, or alpaca; 15 denier corresponds to carpet wool size.

Apparel fibers range from 1 to 7 denier. Fiber of the same denier is not suitable for all uses. Apparel fibers do not make serviceable carpets and carpet fibers do not make serviceable clothing.

Carpet fibers range in denier from 15 to 24. One of the early mistakes made by the carpet industry was that of using clothing fibers for carpets. These fibers were too soft and pliable, and the carpets did not have good crush resistance. Rayon carpet fiber (1953) was the first fiber made especially for carpets, but it is no longer used in carpeting.

Cross-Sectional Shape. Shape is important in luster, bulk, body, texture, and hand or feel of a fabric. Figure 2–5 shows typical cross-sectional shapes. These shapes may be round, dog-bone, triangular, lobal, bean-shaped, flat, or straw-like.

The natural fibers derive their shape from (1) the way the cellulose is built up during plant growth, (2) the shape of the hair follicle and the formation of protein substances in animals, or (3) the shape of the orifice through which the silk fiber is extruded.

The shape of man-made fibers is controlled by the spinneret and the spinning method. The size, shape, luster, length, and other properties of man-made fibers can be varied by changes in the production process.

Surface Contour. Surface contour is defined as the surface of the fiber along its length. Surface contour may be smooth, serrated, striated, or rough. It is important to the luster, hand, texture, and apparent soiling of the fabric. Figure 2–5 shows some of the differences in the surface contours of different fibers.

Crimp. Crimp may be found in textile materials as fiber crimp or fabric crimp. *Fiber crimp* refers to the waves, bends, twists, coils, or curls along the length of the fiber. Fiber crimp increases cohesiveness, resiliency, resistance to abrasion, stretch, bulk, and warmth. Crimp increases absorbency and skin-contact comfort but reduces luster. Inherent crimp occurs wool. Inherent crimp also exists in an undeveloped state in bicomponent man-made fibers where it is developed in the fabric or the completed garment (such as a sweater) by using suitable solvents or heat treatment.

Fabric crimp refers to the bends due to the interlacing or interlooping of yarns in a fabric. When a yarn is unraveled from a fabric, fabric crimp can easily be seen in the yarn. It also may be visible as the yarn is untwisted into fibers.

Fiber Parts. The natural fibers, except for silk, have three distinct parts: an outer covering, called a *cuticle* or skin; an inner area; and a central core that may be hollow. For example, Figure 4–6 and Figure 6–4 show the structural parts of cotton and wool, respectively.

The man-made fibers are not as complex as those of the natural fibers and there are usually just two parts: a skin and a core.

CHEMICAL COMPOSITION AND MOLECULAR ARRANGEMENT

Fibers are classified into groups by their chemical composition. Fibers with similar chemical

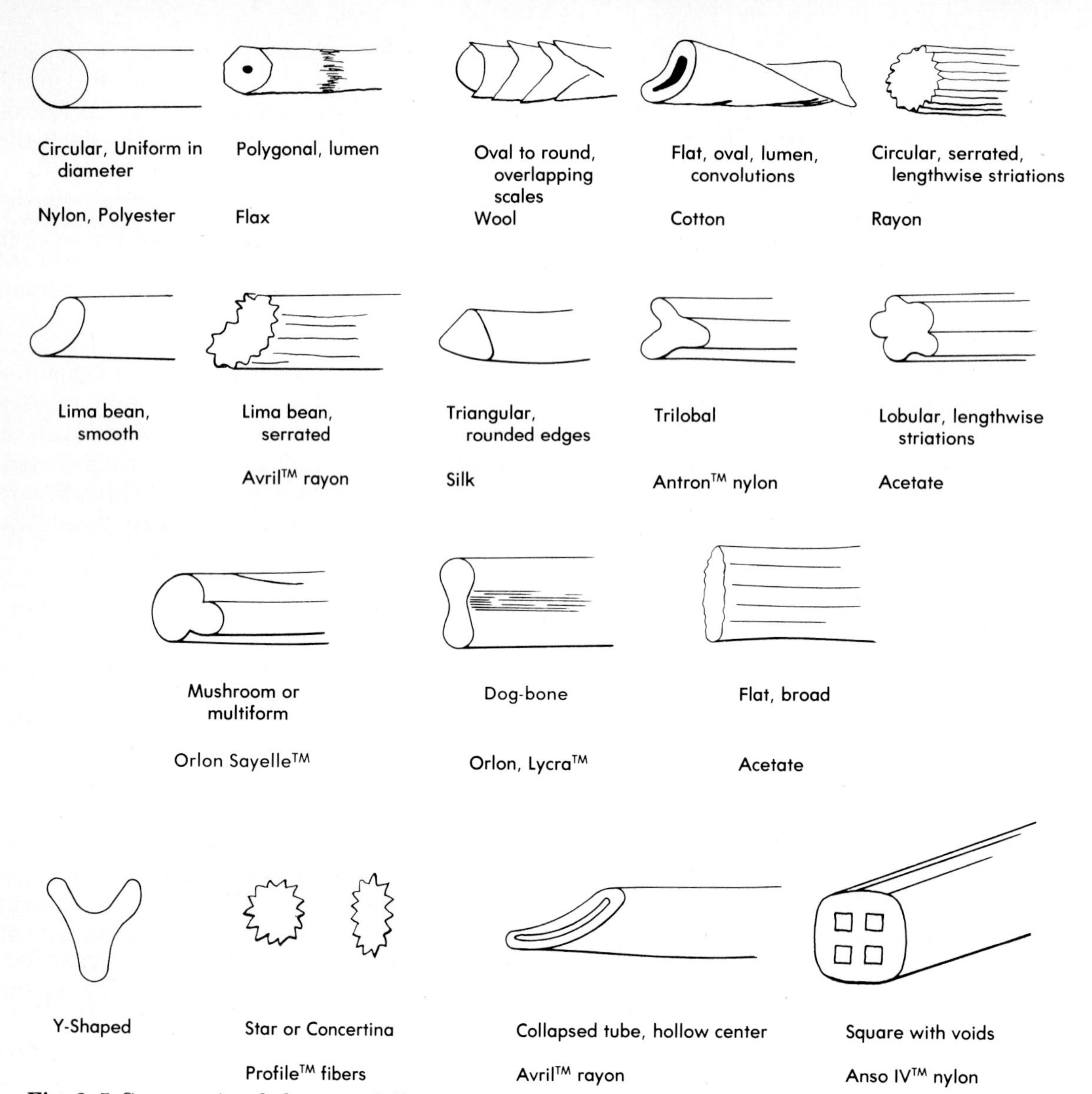

Fig. 2–5 *Cross-sectional shapes and fiber contours.*

compositions are placed in the same generic group. Fibers in one generic group have different properties from fibers in another group.

Fibers are composed of millions of molecular chains. The length of the chains, which varies just as the length of fibers varies, depends on the number of molecules connected in a chain, and it is described as *degree of polymerization.* Polymerization is the process of joining small molecules—monomers—together to form a long chain or a *polymer.* Long chains indicate a high degree of polymerization and a high degree of fiber strength. Molecular chains are not visible to the eye or through the optical microscope.

Molecular chains are sometimes described in terms of weight. The molecular weight is a factor in properties such as fiber strength and extensibility. A fiber of longer chains has a higher strength than a fiber of shorter chains of equal weight; the fiber of longer chains is harder to pull apart than the fiber of shorter chains.

Molecular chains have different configurations in fibers. When molecular chains are nearly parallel to the lengthwise axis of the fi-

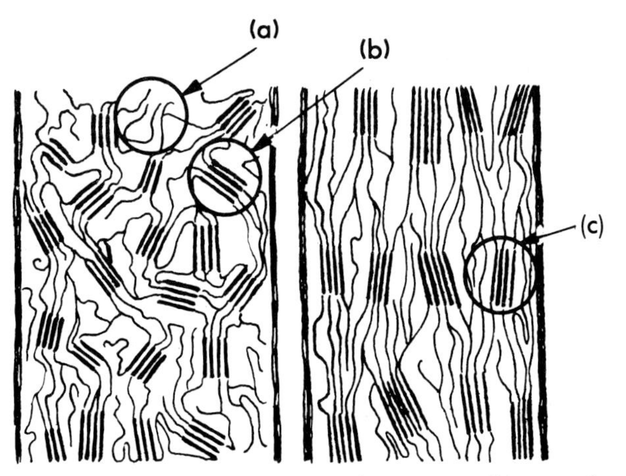

Fig. 2–6 Polymers: (a) amorphous area; (b) crystalline, but not oriented, area; (c) oriented and crystalline area.

ber, they are said to be *oriented;* when they are randomly arranged or disordered, they are said to be *amorphous. Crystalline* is the term used to describe fibers that have molecular chains ordered relative to each other, and usually, but not necessarily, parallel to the lengthwise axis of the fiber (Figure 2–6). Different fibers vary in the proportion of oriented, crystalline, and amorphous regions.

The polymers in man-made fibers are in a random, unoriented state when extruded from the spinneret. *Stretching,* or *drawing,* increases their crystallinity and orients them, reduces their diameter, and packs their molecules together (Figure 2–7). Physical properties of the fiber—such as strength, elongation, moisture absorption, abrasion resistance, and receptivity of the fiber to dyes—are related to the amount of crystallinity and orientation.

Molecular chains are held to one another by *cross links* or by intermolecular forces called *hydrogen bonds* and *van der Waals forces.* The forces are similar to the attraction of a magnet for a piece of iron. The closer the chains are together, the stronger the bonds are. Hydrogen bonding is the attraction of positive hydrogen atoms of one chain for negative oxygen or nitrogen atoms of an adjacent chain. Van der Waals forces are similar but weaker bonds. It is in the crystalline area that hydrogen bonding and van der Waals forces occur. Cross links and intermolecular forces help make crystalline polymers stronger than amorphous polymers.

Fig. 2–7 Before and after drawing the fiber.

Serviceability

Textile serviceability includes the five concepts of aesthetics, durability, comfort, appearance retention, and care. Each concept will be discussed in terms of properties that affect it. For example, the aesthetic properties of luster, drape, texture, and hand, as they relate to apparel and furnishing fabrics, will be defined and discussed.

A note to students: Learn the definitions of the properties. Use the tables as you study each fiber in subsequent chapters to see how that particular fiber compares with other fibers. During this process, evaluate your experience with fabrics made of that fiber and try to understand why they performed the way they did. Once you understand the information, you will remember it and use it now and in future years.

AESTHETIC PROPERTIES

A textile product should be attractive and appropriate in appearance for its end use. Aesthetic properties relate to the way the senses of touch and sight perceive the textile. In evaluating the aesthetics of a textile item, the consumer usually determines whether the item is attractive and appropriate in appearance for its end use.

Luster results from the way light is reflected by a fabric's surface. It is observed by the eye. Shiny or bright fabrics reflect light and are used for certain end uses. Lustrous fabrics reflect a fair amount of light and are used in formal apparel and furnishings. Matte, or dull, fabrics reflect little light and are used most frequently for less-formal looks in apparel and furnishings. Silk fabrics are usually lustrous. Cotton and wool fabrics are usually matte. The luster of man-made fibers can be varied during manufacturing to result in bright, semibright, or matte fibers. Yarn structure, finish, and fabric structure may enhance or decrease the luster of fibers.

Fiber Property Chart

Fiber Property	*Is Due to*	*Contributes to Fabric Property*
Abrasion resistance is the ability of a fiber to withstand the rubbing or abrasion it gets in everyday use.	Tough outer layer, scales, or skin Fiber toughness Flexible molecular chains	Durability Abrasion resistance Resistance to splitting
Absorbency or moisture regain is the percentage of moisture a bone-dry fiber will absorb from the air under standard conditions of temperature and moisture.	Hydroxyl groups Amorphous areas	Comfort, warmth, water repellency, absorbency, static buildup Dyeability, spotting Shrinkage Wrinkle resistance
Aging resistance	Chemical structure	Storing of fabrics
Allergenic potential is the ability to cause some physical reaction, such as skin irritation or watery eyes.	Chemical composition, additives	Comfort
Chemical reactivity is the effect of acids, alkali, oxidizing agents, solvents, or other chemicals.	Polar groups of molecules Chemical composition	Care required in cleaning—bleaching, ability to take acid or alkali finishes
Cohesiveness is the ability of fibers to cling together during spinning.	Crimp or twists, surface properties	Resistance to raveling Resistance to yarn slippage
Cover is the ability to occupy space for concealment or protection.	Crimp, curl, or twist Cross-sectional shape	Warmth in fabric Cost—less fiber needed
Creep is delayed elasticity. Recovers gradually from strain.	Lack of side chains, cross links, strong bonds; poor orientation	Streak dyeing and shiners in fabric
Density—see *Specific gravity.*		
Dimensional stability is the ability to retain a given size and shape through use and care.	Physical structure, chemical structure, coatings	Shrinkage, growth, care, appearance, durability
Drape is the manner in which a fabric falls or hangs over a three-dimensional form.	Fiber size and stiffness	Appearance
Dyeability is the fibers' receptivity to coloration by dyes; dye affinity.	Amorphous areas and dye sites, chemical structure	Aesthetics and colorfastness
Elastic recovery is the ability of fibers to recover from strain. *Elasticity* is another term for elastic recovery.	Chemical and molecular structure: side chains, cross linkages, strong bonds	Processability of fabrics Resiliency Delayed elasticity or creep
Electrical conductivity is the ability to transfer electrical charges.	Chemical structure: polar groups	Poor conductivity causes fabric to cling to the body, electric shocks
Elongation is the ability to be stretched, extended, or lengthened. Varies at different temperatures and when wet or dry.	Fiber crimp Molecular structure: molecular crimp orientation	Increases tear strength Reduces brittleness Provides "give"
Feltability refers to the ability of fibers to mat together.	Scale structure of wool	Fabrics can be made directly from fibers Special care required during washing
Flammability is the ability to ignite and burn.	Chemical composition	Fabrics burn
Flexibility is the ability to bend repeatedly without breaking.	Flexible molecular chain	Stiffness, drape, comfort

Fiber Property Chart *(continued)*

Fiber Property	*Is Due to*	*Contributes to Fabric Property*
Hand is the way a fiber feels: silky, harsh, soft, crisp, dry.	Cross-sectional shape, surface properties, crimp, diameter, length	Hand of fabric
Heat conductivity is the ability to conduct heat away from the body.	Crimp, chemical composition Cross-sectional shape	Warmth
Heat sensitivity is the ability to soften, melt, or shrink when subjected to heat.	Chemical and molecular structure Fewer intermolecular forces and cross links	Determine safe washing and ironing temperatures
Hydrophilic, hygroscopic—see *Absorbency.*		
Loft, or compressional resiliency, is the ability to spring back to original thickness after being compressed.	Fiber crimp Stiffness	Springiness, good cover Resistance to flattening
Luster is the light reflected from a surface. More subdued than shine; light rays are broken up.	Smoothness Fiber length Flat or lobal shape Additives	Luster 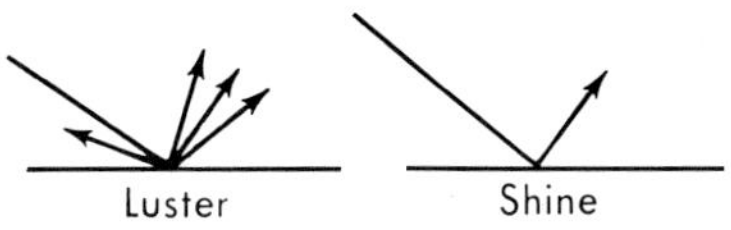 Luster Shine
Mildew resistance	Low absorption	Care during storage
Moth resistance	Molecule has no sulfur	Care during storage
Pilling is the balling up of fiber ends on the surface of fabrics.	Fiber strength High molecular weight	Pilling Unsightly appearance
Resiliency is the ability to return to original shape after bending, twisting, compressing, or a combination of the deformations.	Molecular structure: side chains, cross linkages, strong bonds	Wrinkle recovery, crease retention, appearance, care
Specific gravity and density are measures of the weight of a fiber. Density is the weight in grams per cubic centimeter, and specific gravity is the ratio of the mass of the fiber to an equal volume of water at 4°C.	Molecular weight and structure	Warmth without weight Loftiness—full and light Buoyancy to fabric
Stiffness or rigidity is the opposite of flexibility. It is the resistance to bending or creasing.	Chemical and molecular structure	Body of fabric Resistance to insertion of yarn twist
Strength is defined as the ability to resist stress and is expressed as *tensile strength* (pounds per square inch) or as *tenacity* (grams per denier).	Molecular structure—orientation, crystallinity, degree of polymerization	Durability, tear strength, sagging, pilling Sheerer fabrics possible with stronger fine fibers
Sunlight resistance is the ability to withstand degradation from direct sunlight.	Chemical composition Additives	Durability of curtains and draperies, outdoor furniture, outdoor carpeting
Texture is the nature of the fiber surface.	Physical structure	Luster, appearance
Translucence is the ability of a fiber, yarn, or fabric to allow light to pass through the structure.	Physical and chemical structure	Appearance
Wicking is the ability of a fiber to transfer moisture along its surface.	Chemical and physical composition of outer surface	Makes fabrics comfortable

Drape is the way a fabric falls over a three-dimensional form like a body or table. Fabric may be soft and free-flowing like chiffon, or it may fall in graceful folds like a percale, or it may be stiff and heavy like bridal satin. Fibers influence drape to a degree, but yarns and fabric structure may be more important in determining drape.

Texture describes the nature of the fabric surface. It is identified by both visual and tactile senses. Fabrics may have a smooth or rough texture. Natural fibers tend to give a fabric more texture than man-made fibers because of their inherent variations. Yarns, finishes, and fabric structure greatly affect the texture of a fabric.

Hand is the way a fabric feels to the touch. Fabrics may feel warm or cool, bulky or thin, slick or soft. There are many more descriptive words that may be used. Hand usually is evaluated by feeling a fabric between the fingers and thumb. The hand of fabric may also be evaluated by the way it feels against the skin. Subjective evaluation determines the fabric's acceptability for a particular end use; however, some researchers are developing an objective means of assessing hand.

DURABILITY PROPERTIES

A durable textile product should last an adequate period of time for its end use. Durability properties can be tested in the laboratory, but lab results do not always accurately predict performance during actual use.

Abrasion resistance is the ability of a fabric to withstand the rubbing it gets in use. Abrasion can occur when the fabric is fairly flat—as when the knees of jeans scrape along a cement sidewalk. Edge abrasion can occur when the fabric is folded—as when the bottom of a drapery fabric rubs against a carpet. Flex abrasion can occur when the fabric is moving and bending—as in shoelaces that wear out where they are laced through the shoe. *Flexibility,* the ability to bend repeatedly without breaking, is a very important property that is related to abrasion resistance.

Tenacity, or tensile strength, is the ability of a fabric to withstand a pulling force. Tenacity for a fiber is the force, in grams per denier or tex, required to break the fiber. The tenacity of a wet fiber frequently differs from the tenacity of that same fiber when it is dry. Although the fabric strength depends, to a large degree, on

Abrasion Resistance

Fiber	*Rating*
Nylon	Excellent
Olefin	↓
Polyester	
Spandex	
Flax	
Acrylics	to
Cotton	
Silk	
Wool*	
Rayon	
Acetate	↓
Glass	Poor

*Varies with coarseness of fiber.

Fiber Strength

*Fiber**	*Tenacity (grams/denier)* Dry	Wet
Natural Fibers		
Cotton	3.5–4.0	4.5–5.0
Flax	3.5–5.0	6.5
Silk	4.5	2.8–4.0
Wool	1.5	1.0
Man-Made Fibers		
Acetate	1.2–1.4	1.0–1.3
Acrylic	2.0–3.6	1.8–3.5
Aramid (staple Nomey)	4.0–5.3	4.0–5.0
Flurocarbon	0.8–1.4	same
Glass (multifilament)	9.6	6.7
Modacrylic	1.7–3.5	1.5–2.4
Novoloid	1.5–2.5	1.3–2.3
Nylon 6 (staple)	3.5–7.2	same
Nylon 66 (staple)	2.9–7.2	2.6–5.4
Olefin	3.5–8.0	same
PBI	2.6–3.0	2.1–2.5
Polyester (staple)	2.4–5.5	same
Rayon (Regular)	0.7–2.6	0.4–1.4
Rubber	0.34	same
Saran	1.4–2.4	same
Spandex	0.6–0.9	same
Sulfar	3.0–3.5	same
Vinyon	0.7–1.0	same

*For fibers that are available in several lengths and modifications, the values are for staple fibers with unmodified cross-sections.
Source: Textile World, 1986.

fiber strength, yarn and fabric structure may be varied to yield stronger or weaker fabrics made from the same fibers. Strength may also be measured by how much force it takes to rip the fabric (tearing strength) or to rupture the fabric (bursting strength).

Elongation refers to the degree to which a fiber may be stretched without breaking. It is measured as percent elongation at break. It should be considered in relation to elasticity.

COMFORT PROPERTIES

A textile product should be comfortable as it is worn or used. This is primarily a matter of personal preference and individual perception of comfort under different climatic conditions and degrees of physical activity. Comfort is a complex area and is dependent on characteristics such as absorbency, thermal retention, density, and elongation.

Absorbency is the ability of a fiber to take up moisture from the body or from the environment. It is measured as moisture regain where the moisture in the material is expressed as a percent of the weight of the moisture-free material. Absorbency is also related to static buildup. *Hydrophilic* fibers absorb moisture readily. *Hydrophobic* fibers have little or no absorbency. *Hygroscopic* fibers absorb moisture from the air.

Thermal retention is the ability of a fabric to hold heat. For apparel, it is important for a person to feel comfortably warm in cool weather or comfortably cool in hot weather. A low level of thermal retention is favored in hot weather. This property accounts for the fact that most people dress and use textiles differently in summer and winter weather. Yarn and fabric structure and layering of fabrics are all very important in enhancing this property.

Density is a measure of fiber weight in weight per unit volume. Lighter-weight fibers can be made into thick fabrics that are more comfortable than heavier-weight fibers made into heavy, thick fabrics.

Elongation: Breaking

	% Elongation at Break	
*Fiber**	*Standard***	*Wet*
Natural Fibers		
Cotton	3–7	9.5
Flax	2.0	2.2
Silk	20	30
Wool	25	35
Man-Made Fibers		
Acetate	25	35
Acrylic	20	26
Aramid (Nomex)	22	20
Glass	3.1	2.2
Modacrylic	30–60	same
Nylon (6,6)	16	18
Olefin	10–45	same
PBI	25–30	26–32
Polyester staple	40–45	same
Rayon	15	20
Rubber	500	same
Spandex	400–700	same
Sulfar	35–40	same

Note: A minimum of 10% is desirable for ease in textile processing.

*For fibers that are available in several lengths and modifications, the values are for staple fibers with unmodified cross-sections.
**Standard condition: 65 percent relative humidity, 70°F.

Absorbency

Fiber	*Moisture Regain**
Natural Fibers	
Cotton	7–11
Flax	12
Silk	11
Wool	13–18
Man-Made Fibers	
Acetate	6.0
Acrylic	1.0–2.5
Aramid	6.5
Glass	0.0
Modacrylic	2.0–4.0
Nylon (6,6)	4.0–4.5
Olefin	0.01–0.1
PBI	15
Polyester	0.4–0.8
Rayon	13
Saran	0.1
Spandex	1.3
Sulfar	0.6

*Moisture regain is expressed as a percentage of the moisture-free weight at 70°F and 65 percent relative humidity.

Thermal Properties

Fiber	Melting Point		Softening Sticking Point		Safe Ironing Temperature*	
	°F	°C	°F	°C	°F	°C
Natural Fibers						
Cotton	Does not melt				425	218
Flax	Does not melt				450	232
Silk	Does not melt				300	149
Wool	Does not melt				300	149
Man-Made Fibers						
Acetate	500	230	350–375	184	350	177
Acrylic			430–450	204–254	300	149–176
Aramid	Does not melt—Carbonizes above 700°F (Nomex) or 900°F (Kevlar)				Do not iron	
Glass	2,720		1,560	1,778	Do not iron	
Modacrylic	Does not melt				200–250	93–121
Nylon 6	419–430		340	171	300	149
Nylon 66	480–500		445	229	350	177
Olefin	320–350		285–330	127	150	66
PBI	Does not melt		Decomposes at 860°F			
Polyester PET	482		440–445	238	325	163
Polyester PCDT	478–490		470	254	350	177
Rayon	Does not melt				375	191
Saran	350	177	240	115	Do not iron	
Spandex	446	230	420	175	300	149

*Lowest setting on irons: 185–225°F

APPEARANCE RETENTION PROPERTIES

A textile product shold retain its original appearance during wear and care.

Resiliency is the ability of a fabric to return to its original shape after bending, twisting, or crushing. A common test is to crush a fabric in the hand and watch how it responds when the hand is opened. A resilient fabric will spring back; it is wrinkle resistant if it does not wrinkle easily. It has good wrinkle recovery if it returns to its original look after having been wrinkled.

Dimensional stability can be defined as the ability of a fabric to retain a given size and shape through use and care. Dimensional stability is a desireable characteristic that includes the properties of shrinkage resistance and elastic recovery. *Shrinkage resistance* is the ability of a fabric to retain a given size after care. It is related to the fabric's reaction to water or heat. A fabric that shrinks is smaller after care. In addition to a poor appearance, it will have a poor fit.

Elasticity or *elastic recovery* is the ability of a fabric to return to its original dimensions after elongation. It is measured as the percent of return to its original length. Since recovery varies with the amount of elongation, as well as with the length of time the fabric is stretched, the measurement identifies the percent elongation, or stretch, and the recovery. Fabrics with poor elastic recovery tend to stretch out of shape. Fabrics with good elastic recovery usually exhibit good resiliency.

CARE PROPERTIES

Any treatments that are required to maintain the new look of a textile product during use, cleaning, or storage are referred to as care. Improper care procedures can result in items that are unattractive, not as durable as expected, and uncomfortable. The way fibers react to water, chemicals, and heat in ironing and drying will be discussed in each fiber chapter. Any special requirements of storage will also be discussed.

As students learn these basic properties and

Density and Specific Gravity*

Fibers	Density (g/cc)
Natural Fibers	
Cotton	1.52
Flax	1.52
Silk	1.25
Wool	1.32
Man-Made Fibers	
Acetate	1.32
Acrylic	1.0–2.5
Aramid	1.38–1.44
Flurocarbon	0.8–2.2
Glass	2.48–2.54
Modacrylic	1.35
Novoloid	1.25
Nylon	1.13–1.14
Olefin	0.90–0.91
PBI	1.43
Polyester	1.34–1.39
Rayon	1.48–1.54
Saran	1.70
Spandex	1.20–1.25
Sulfar	1.37

*Ratio of weight of a given volume of fiber to an equal volume of water.

remember the effect of other related properties on each serviceability concept, their understanding of the performance of each fiber will increase. Students are strongly encouraged to study the performance of each fiber for each of these main concepts.

Elastic Recovery

Fiber	% Recovery from 3% Stretch*
Natural Fibers	
Cotton	75
Flax	65
Silk	90
Wool	99
Man-Made Fibers	
Acetate	58 (at 4%)
Acrylic	92
Aramid	100
Glass	100
Modacrylic	88
Nylon	82–89
Olefin	92–98 (at 5%)
Polyester	76
Rayon	80 (at 2%)
Spandex	99 (at 50%)
Sulfar	96 (at 5%)

*Unless otherwise noted.

FIBER PROPERTY CHARTS

The fibers within each generic family have individual differences. These differences are not reflected in the charts shown here, except in a few specific instances. The figures are averages, or medians, and are intended as a general characterization of each generic group. The figures were compiled from the following sources:

Charts: Man-Made Fiber Chart, *Textile World* (August 1986).
Properties of the Man-Made Fibers, *Textile Industries* (1971/1972).
Man-Made Fiber Deskbook, *Modern Textiles* (March 1983).

Bulletins: Textile Fibers and Their Properties, AATCC Council on Technology 1977
Technical Bulletins, Du Pont, Celanese, and Monsanto.
Man-Made Fiber Fact Book, Man-Made Fiber Producers Association, Inc. 1978.

Effect of Acids

Fiber	Effect
Natural Fibers	
Cotton	Harmed
Flax	Harmed
Silk	Harmed by strong mineral acids, resistant to organic acids
Wool	Resistant
Man-Made Fibers	
Acetate	Unaffected by weak acids
Acrylic	Resistant to most
Aramid	Resistant to most
Glass	Resistant
Modacrylic	Resistant to most
Nylon	Harmed, especially nylon 6
Olefin	Resistant
PBI	Resistant
Polyester	Resistant
Rayon	Harmed
Spandex	Resistant
Sulfar	Resistant

Sunlight Resistance

Fiber	*Rating*
Glass	Excellent
Acrylic	↓
Modacrylic	
Polyester	
Sulfar	
Flax	
Cotton	to
Rayon	
PBI	
Triacetate	
Acetate	
Olefin	
Nylon	
Wool	↓
Silk	Poor

Fiber Identification

The procedure for identification of the fiber content of a fabric depends on the nature of the sample, the experience of the analyst, and the facilities available. Because laws require the fiber content of apparel and furnishing textiles to be indicated on the label, the consumer may only need to look for identification labels. If the consumer wishes to confirm or check the information on the label, burning and some simple solubility tests may be used.

Effect of Alkali

Fiber	*Effect*
Natural Fibers	
Cotton	Resistant
Flax	Resistant
Silk	Damaged
Wool	Harmed
Man-Made Fibers	
Acetate	Little effect
Acrylic	Resistant to weak
Aramid	Resistant
Glass	Resistant
Modacrylic	Resistant
Nylon	Resistant
Olefin	Highly resistant
PBI	Resistant to most
Polyester	Degraded by strong
Rayon	Resistant to weak
Spandex	Resistant
Sulfar	Resistant

VISUAL INSPECTION

Visual inspection of a fabric for appearance and hand is always the first step in fiber identification. It is no longer possible to make an identification of the fiber content by the appearance and hand alone because man-made fibers can be made to resemble natural fibers and frequently resemble other man-made fibers. However, observation of the following characteristics is helpful.

1. Length of fiber. Untwist the yarn to determine length. Any fiber can be made in staple length, but not all fibers can be filament. For example, cotton and wool are always staple.

2. Luster or lack of luster.

3. Body, texture, hand—soft-to-hard, rough-to-smooth, warm-to-cool, or stiff-to-flexible.

BURNING TEST

The burning test can be used to identify the general chemical composition, such as cellulose, protein, mineral, or man-made polymers, and thus identify the group to which a fiber belongs. *Blends cannot be identified by the burning test.* If visual inspection is used along with the burning test, fiber identification can be carried further. For example, if the sample is cellulose and also filament, it is rayon; but if it is staple, a positive identification for a specific cellulosic fiber cannot be made.

The following are general directions for the burning test:

1. Ravel out and test several yarns from each direction of the fabric to see if they have the same fiber content. Differences in luster, twist, and color will indicate that there might be two or more generic fibers in the fabric.

2. Hold the yarn horizontally, as shown in Figure 2–8. Use tweezers to protect fingers. Feed the yarns slowly into the edge of the flame from the alcohol lamp and observe what happens. Repeat this several times to check results.

Identification by Burning

Fibers	*When Approaching Flame*	*When in Flame*	*After Removal from Flame*	*Ash*	*Odor*
Cellulose Cotton Flax Rayon	Does not fuse or shrink from flame	Burns	Continues to burn, afterglow	Gray, feathery, smooth edge	Burning paper
Protein Silk Wool	Curls away from flame	Burns slowly	Usually self-extinguishing	Crushable black ash	Burning hair
Acetate	Fuses away from flame	Burns with melting	Continues to burn and melt	Brittle black hard bead	Acrid
Acrylic	Fuses away from flame	Burns with melting	Continues to burn and melt	Brittle black hard bead	—
Glass		Does not burn			
Modacrylic	Fuses away from flame	Burns very slowly with melting	Self-extinguishing, white smoke	Brittle black hard bead	—
Nylon	Fuses and shrinks away from flame	Burns slowly with melting	Usually self-extinguishing	Hard gray or tan bead	Celery-like
Olefin	Fuses and shrinks away from flame	Burns with melting	Usually self-extinguishing	Hard tan bead	—
Polyester	Fuses and shrinks away from flame	Burns slowly with melting; black smoke	Usually self-extinguishing	Hard black bead	Sweetish odor
Saran	Fuses and shrinks away from flame	Burns very slowly with melting	Self-extinguishing	Hard black bead	—
Spandex	Fuses but does not shrink from flame	Burns with melting	Continues to burn with melting	Soft black ash	—

Fig. 2–8 *Fiber identification by the burning test.*

MICROSCOPY

A knowledge of fiber structure, obtained by seeing the fibers through the microscope and observing some of the differences among fibers in each group, is of help in understanding fibers and fabric behavior.

Positive identification of most of the natural fibers can be made by using this test. The man-made fibers are more difficult to identify because some of them look alike and their appearance may be changed by variations in the manufacturing process. Positive identification of the man-made fibers by this means is rather limited. Cross-sectional appearance is helpful if more careful examination is desired.

Longitudinal and cross-sectional photomicrographs of individual fibers are included in the fiber chapters. These may be used for reference when checking unknown fibers.

Solubility Tests

Solvent	*Fiber Solubility*
1. Acetic acid (100%) 20°C	Acetate
2. Acetone, 100%, 20°C	Acetate, modacrylic, vinyon
3. Hydrochloric acid, 20% concentration, 1.096 density, 20°C	Nylon 6, nylon 6,6, vinal
4. Sodium hypochlorite solution (5%) 20°C	Silk and wool (silk dissolves in 70% sulfuric acid at 38°C), azlon
5. Xylene (meta), (100%) 139°C	Olefin and saran (saran dissolves in 1.4 dioxane at 101°C; olefin is not soluble), vinyon
6. Dimethyl formamide, (100%) 90°C	Spandex, modacrylic, acrylic, acetate, vinyon
7. Sulfuric acid, 70% concentration, 38°C	Cotton, flax, rayon, nylon, acetate, silk
8. Cresol (meta), (100%) 139°C	Polyester, nylon, acetate

The following are directions for using the microscope:

1. Clean the lens, slide, and cover glass.

2. Place a drop of distilled water or glycerine on the slide.

3. Untwist a yarn and place the loosened fibers on the slide. Cover with the cover glass and press down to eliminate air bubbles.

4. Place the slide on the stage of the microscope and then focus with low power first. If the fibers have not been well separated, it will be difficult to focus on a single fiber.

5. If a fabric contains two or more fiber types, test each fiber and both warp and filling yarns.

SOLUBILITY TESTS

Solubility tests are used to identify the man-made fibers by generic class and to confirm identification of natural fibers. Two household tests, the alkali test for wool and the acetone test for acetate, are described on pages 54 and 91, respectively.

In using the tests, the specimen should be placed in the liquid in the order listed. The specimen should be stirred for 5 minutes and the effect noted. Fiber, yarns, or small pieces of fabric may be used. The liquids are hazardous and should be handled with care. Chemical laboratory exhaust hoods, gloves, aprons, and goggles should be used.

3

Selection of Textile Products for Consumer Use

Many textiles are used as a part of daily life. Ready-to-wear apparel, over-the-counter fabrics that will be sewn into apparel or furnishing items, and home furnishings are chosen by consumers for their use.

The selection of textile products is a personal decision based on many factors, including current fashion, lifestyle, income, sex, and age. The selection is influenced by aesthetic, psychological, sociological, and economic aspects. The decision to buy a product may be rational or it may be impulsive. This chapter provides a framework to help consumers make rational decisions about textile purchases.

This same framework is very useful for people whose jobs are related to textile products. People employed in merchandising, design, or production—such as retail and wholesale buyers and salespersons, and fabric, garment, and interior designers—can all benefit from a better understanding of textiles and the ways customers use these products. Persons in these jobs can be called pre-selectors. What they select determines what the consumer has available to purchase. Hence, the decisions the pre-selector makes influence, to a large degree, the consumer's choice and ultimate satisfaction.

Before consumers start shopping for a textile product, they will have already decided the item for which they are shopping. They should also think through pertinent factors: Who will be using the item? How will it be used? When will it be used? How long will it need to last? Usually this thinking is in terms of what the item will be used for, or, in other words, the end use.

The end use that is identified affects the subsequent selection. Is the coat to be worn by a college student or for school and play by a rambunctious nine-year-old? Will the fabric be made into a pair of slacks for a teenager or for an elderly person in a nursing home? Are the sheets and bedspread for college use, where they must serve both living and sleeping functions, or are they for the master bedroom of a carefully decorated model home? Is the suit to be worn in the summer for casual wear or will it be worn at work throughout the fall, winter, and spring?

With the end use clearly in mind, additional factors need to be considered. What factors are important in making the item serviceable for its purpose? Factors frequently considered by consumers for items of apparel include the following: price, fit, color, fashion, style, appearance, quality of construction, durability, comfort, and care. Which factors are most important for this item and its end use? Which are least important? What are the rankings of the other factors?

With the end use and a realization of the factors that are important for serviceability of the item in mind, a possible sequence of steps in making the decision might include:

1. Determine a price range or price ceiling for the textile product.

2. Find items that are acceptable by checking available sources, such as local stores and catalogs.

3. Evaluate color, fashion, style, appearance, and quality of construction.

4. Evaluate the serviceability of the textile components of the item.

5. Buy a specific item or decide to do more shopping.

The satisfaction that the consumer receives from the textile material will depend on individual values as well as on the performance of the product. The performance and care of the product depend on the fibers, yarns, fabric construction, and finishes. The manufacturer has decided what combination would be appropriate for the item. The buyer has determined the selection available to the consumer. The final step in the decision-making process is for the consumer to select the item that is most appropriate for his or her own use.

This seems like an involved procedure to follow when shopping for an item, but most of the steps are almost automatic. The choice of where to shop limits the selection. Once in the store, price, fit, color, fashion, style, appearance, and quality of construction can be quickly evaluated. When the selection is narrowed down to two or three items, the consumer can evaluate the serviceability of the textile components of each item. This is where a student of textiles will have more knowledge than a typical customer and will be better able to match end-use requirements to a realistic expected performance of the item.

The serviceability concepts that are used in organizing the material in this book are simple and straightforward. These concepts provide a framework for combining textile facts with personal needs and preferences in a way that will help consumers make wiser decisions about the textile products they purchase and use. As the concepts are combined with the consumer's past and present experiences, they can act as a simple checklist while the purchasing decision is made. The serviceability concepts also provide an excellent framework for the pre-selector. Combined with their experience of what sells and what works for their clients, the serviceability concepts can improve their purchasing decisions.

The *five serviceability concepts* will be used by following through with the examples discussed earlier in this chapter.

1. *Aesthetics.* The bedspread for the model home should be attractive in and of itself; it must also coordinate very well with the décor of the master bedroom and blend well with the overall effect intended for the entire house. The teenager's slacks will need to look good as judged by the individual and his or her peers.

2. *Durability.* The coat for the nine-year-old should be durable enough to withstand hard use. Is it intended to be a summer, fall, spring, or winter jacket? How different are the seasons in that specific location? How quickly is the child growing—does the child tend to wear out clothes or out-grow them? Is the jacket expected to last a season, a year, or several years? The bedspread for college use will be expected to last several years and will be subjected to hard use since dorm beds usually function as sofas.

3. *Comfort.* This factor may be of great importance for a summer suit worn primarily out-of-doors in a hot and humid climate. Comfort also is very important for the person in the nursing home. The fabric must be comfortable next to delicate skin.

4. *Appearance retention.* The suit that will be worn at work needs to resist wrinkles during wear. It should maintain its shape during use. Over the period of time it is worn, it should continue to look professional. The sheets and bedspread for college use will undoubtedly last several years, but how will they look the last year?

5. *Care.* The college student's coat may not need cleaning often. If dry cleaning is required, the coat may be acceptable if it is infrequent. On the other hand, the child's jacket and the teenager's slacks probably should be washable. Slacks used in a nursing home will need to withstand high-temperature washing and will look better after care if made from a wrinkle-resistant fabric.

When selecting among several possibly acceptable items for a specific end use, the serviceability concepts can be posed as questions. See the table, "Serviceability Questions."

To further emphasize the importance of defining the end use for a textile item and to better understand the impact this has on selection, the five serviceability concepts for each suit discussed above have been ranked by a consumer. The most important concept is ranked "1," least important concept is ranked "5." Another consumer might rank these concepts differently

Serviceability Questions

Aesthetics	Is this item attractive and appropriate in appearance for its end use? How does it look and feel?
Durability	Will this item continue to be useable for as long as expected? Or will it wear out sooner than desired?
Comfort	Is this item comfortable enough for use in its purpose? Will it be too warm or too cool? Will it feel good against the skin? Will the comfort change as the fabric wears? Will the comfort be altered by required care procedures?
Appearance retention	Will this item retain its appearance during use and care? Will it resist wrinkles? Will it retain its shape? Do the aspects that make this item attractive have suitable durability, comfort, and care characteristics? Will the item look and feel good as long as it lasts?
Care	Are the treatments required to maintain the new look of this item during use, cleaning, and storage acceptable considering the money and time available? Is the care realistic considering the cost of the item?
Serviceability	What combination of these five concepts is important in making this item useable for its purpose?

Suit

End Use	*Casual: Summer*	*Work: Fall, Winter, and Spring*
Aesthetics	3	1
Durability	5	4
Comfort	1	3
Appearance retention	4	2
Care	2	5

and some concepts may be ranked equally important.

It is most unlikely that a single textile product would be serviceable for both end uses. The needs for each end use are different. Also, another consumer might have ranked the concepts in a different order.

Finally, more than one factor will need to be considered in each end use. Let us look at the suit for work. Aesthetics factors might include a fashion look appropriate for that particular job and company of a color and style becoming to the wearer. Durability requirements would specify how long the suit is expected to be worn—one, two, or more years? Comfort factors include some minimum level of comfort because it is difficult to concentrate if clothes are scratchy or otherwise uncomfortable. Appearance-retention factors might relate to wrinkle resistance. The suit should retain its shape. It should continue to look professional for a reasonable time. Care factors include the method required. Frequently, dry cleaning is required for suits. Care was ranked of lowest priority, so presumably any care method is acceptable.

Now, to return to the shopping process—once several acceptable items have been identified, evaluate the serviceability concepts of aesthetics, durability, comfort, appearance retention, and care. Which item best satisfies the requirements of the end use? Is the cost appropriate? Are all the other selection factors acceptable? Based on this evaluation, make the decision to purchase a specific item.

In summary, end use determines performance requirements. During selection, relate end-use performance requirements to realistic expected performance in the five serviceability areas. Within this framework, textile knowledge can be critical in satisfying needs and wants. The serviceability concepts are defined and discussed in detail in the previous chapter. The concepts are referred to frequently throughout the remainder of the book.

It must be stressed that neither this book nor any course in textiles will give students the answers for any end use. This book will not answer the question: What is the best combination of fiber, yarn, fabrication, and finish for a coat or chair? Consumers make choices based on knowledge of themselves, their needs and expectations, and the use the product will get. If they know about textiles, they can make an intelligent selection. However, with any selection, there will be some good and some not-so-good features. The more knowledge consumers have, the more serviceable the product should be.

Consumers have information available to them on labels, hang tags, and packages. Federal legislation requires that fiber content be stated. With the information in this book, fiber-content information will provide some basis for predicting performance related to durability, comfort, and appearance retention. Aesthetics can be judged by the consumer. Care labels are required by federal legislation. Thus some information is available on which to base rational decisions. How wisely that information is used, or how meaningful that information is to any one person, depends on the person making the choices. In addition, evaluation of other aspects of the product, such as yarn and fabric type, will provide more information on which to base a decision.

4

Cotton

Cotton is the most important apparel fiber throughout the world. A study done by the International Cotton Advisory Committee showed that, in developing countries with a high average temperature, cotton holds 75 percent of the market for apparel. Blends with polyester also are used, but the demand for synthetic fibers is low. On the other hand, in developed countries with a low average temperature, cotton accounts for about 30 percent of the fiber used. In the United States, in 1985, cotton accounted for more than 34 percent of fibers used in apparel.

Cotton has a combination of properties—pleasing appearance, comfort, easy washability, moderate cost, and durability—that make it ideal for warm-weather clothing, active sportswear, work clothes, towels, and sheets. This unique combination of properties has made cotton a standard for people who live in warm and subtropical climates. Even though the man-made fibers have encroached on the markets that were once dominated by 100 percent cotton fabrics, the cotton-look is still maintained, and cotton forms an important part of blended fabrics.

Cotton cloth was used by the people of ancient China, Egypt, India, Mexico, and Peru. Fabrics of cloth from Egypt give some evidence that cotton may have been used there in 12,000 B.C., before flax was known. Cotton spinning and weaving as an industry began in India, and fabrics of good-quality cotton cloth were being produced as early as 1500 B.C. The Pima Indians were growing cotton when the Spaniards came to the New World. One of the items that Columbus took back to Queen Isabella was a hank of cotton yarn.

Cotton was grown in the Southern colonies as soon as they were established. Cotton was planted in Florida in 1556 and in Virginia in the early 1600s. Records show cotton seeds imported from Egypt were planted in Jamestown, Virginia, in 1617. England encouraged cotton production in the colonies and imported the bulk of cotton produced. Throughout the 1600s and 1700s, cotton fibers had to be separated from the cotton seeds by hand. This was a very time-consuming and tedious job; a worker could separate the seeds from the fibers of only one pound of cotton in a day.

With the invention of the saw-tooth cotton gin by Eli Whitney in 1793, the scene began to change. The gin could process 50 pounds of cotton in a day; thus more cotton could be prepared for spinning. Within the next 20 years, a series of spinning and weaving inventions in England mechanized fabric production. The Industrial Revolution was underway! The British cotton-textile trade grew tremendously from 1800–1850.

The other critical factor at this time was the supply of raw cotton. The Southern states were able to meet the greatly increased demand in Britain for raw cotton. The soil and climate of the Southern states were ideal for growing cotton. Small- and medium-sized farms continued to be important, but plantations with large fields and slave labor flourished.

In 1792, 6,000 bales of cotton were produced by the Southern states. Seven years later, with the introduction of the cotton gin, production reached 100,000 bales. The United States was the most important supplier of cotton to the British textile industry for the next 60 years. By 1859, U.S. production was 4.5 million bales of cotton—⅔ of the world production. Cotton was the leading U.S. export.

The picture again changed dramatically with the U.S. Civil War. U.S. cotton production decreased to 200,000 bales in 1864, and Britain looked to other countries to fill her needs. After the war, Western states began producing cotton.

During the time of rapidly expanding cotton production in the Southern states, the New England states were building factories to manufacture yarn and fabric. In 1790, Samuel Slater built the first yarn-spinning mill in Pawtucket, Rhode Island. For the next 150 years, most spinning and weaving of U.S. fabrics took place in the New England states.

After the Civil War, the Southern states began building spinning and weaving mills. Between World War I and World War II, most of the New England mills moved south. Factors important in this change of location included being closer to the supply of cotton, having cheaper power, using less-expensive nonunion labor, and receiving special incentives from state and local governments to encourage companies to set up a mill in their town and state. By 1950, 80 percent of the mills were in the South. In the 1980s, many mills have closed because of increased costs and competition from imported fabrics.

PRODUCTION

Cotton grows in any part of the world where the growing season is long and the climate is temperate to hot with adequate rainfall or irrigation. Cellulose will not form if the temperature is below 70°F. In the United States, cotton is grown from southern South Carolina west to central California and south of that line. Figure 4–1 shows where cotton is currently produced in the United States. Much of the cotton produced in the U.S. is exported.

Upland cotton is the most important kind of cotton grown in the United States. In 1985, 13.7 million bales were produced, with an estimated 5.75 million bales being used by U.S. mills. The rest were exported. Extra-long staple cotton is grown in Arizona, Texas, and New Mexico. The United States produced 148,000 bales of American Pima cotton in 1985 and much of this long-staple cotton was exported.

The United States was the top producer of cotton through 1977. The Soviet Union was the top producer in 1978 and 1980, with the United States leading in 1979 and again in 1981. China took the lead in 1982 and continues to lead in production through 1986.

Factors affecting the U.S. production of cotton include the increasing value of the U.S. dollar compared with other currencies, the increasing imports of cotton apparel and fabrics, the changes in government incentives for growing cotton, and the changes in other countries. One of the most important changes is the emergence of China in world trade with its increased exports and production of cotton and other fibers.

In 1985, world production of cotton was 80.9 million bales. The People's Republic of China produced 28.5 percent of that total; the United States 16.9 percent; and the Soviet Union 15.5

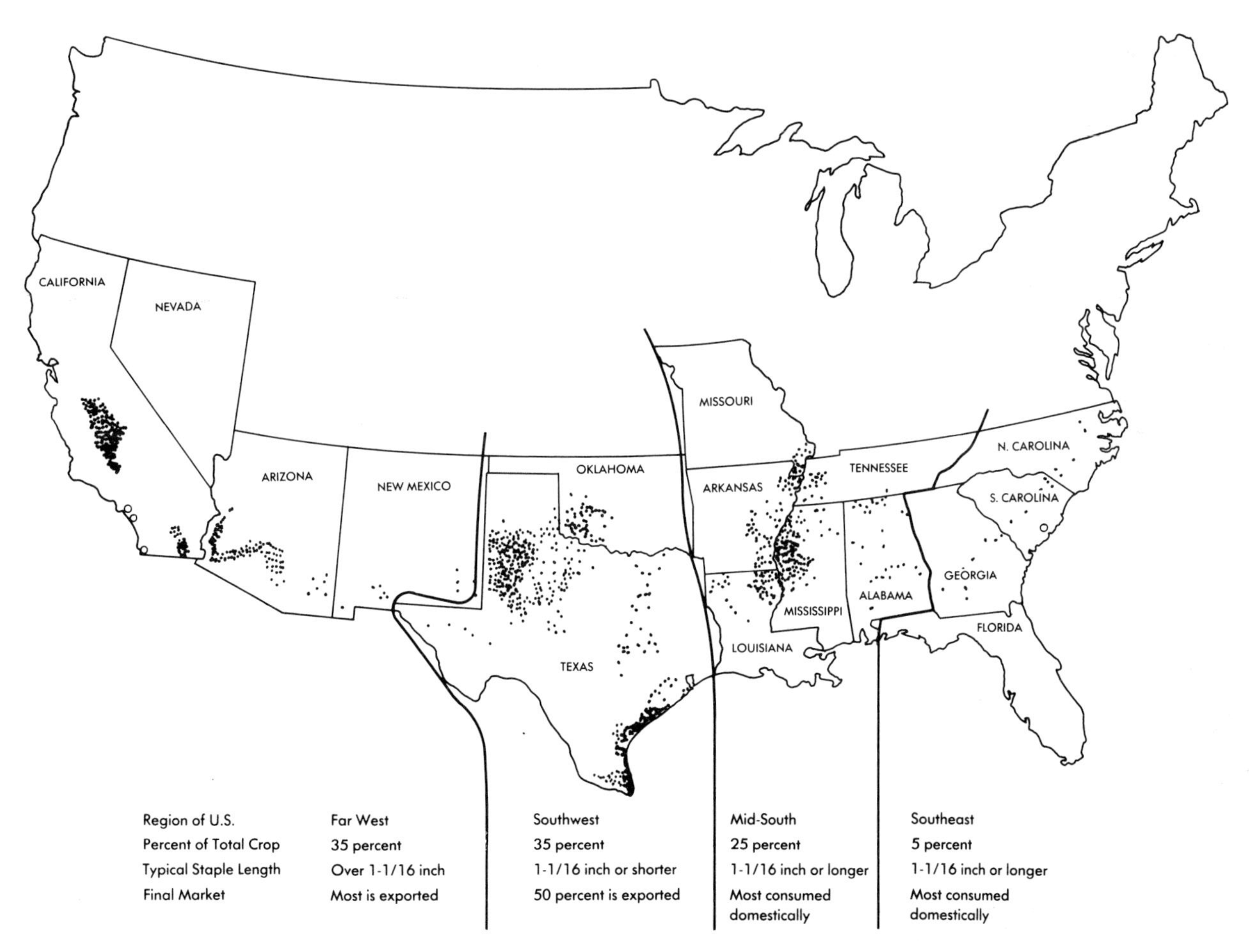

Fig. 4–1 United States Cotton Production, 1984–1985. (Courtesy of National Cotton Council of America.)

percent. India, fourth in production, consistently produces 8–9 percent of the world's production. In 1985, Pakistan, Brazil, Turkey, and Egypt produced over one million bales each, which is 2–4 percent of the world's production.

Cotton grows on bushes 3–6 feet high. The blossom appears, falls off, and the *boll* begins its growth. Inside the boll are seeds from which the fibers grow. When the boll is ripe, it splits open, and the fluffy white fibers stand out like a powder puff (a boll contains seven or eight seeds) (Figure 4–2). Each cotton seed may have as many as 20,000 fibers growing from its surface.

Cotton is picked by machine or by hand (Figure 4–3). Machine-picked cotton contains many immature fibers—an inescapable result of stripping a cotton plant. However, mechanization and weed control have reduced the number of hours required to produce a bale of cotton. After picking, the cotton is taken to a *gin* to remove the fibers from the seed. Figure 4–4 shows a saw gin, in which the whirling saws pick up the fiber and carry it to a knife-like comb, which blocks the seeds and permits the fiber to be carried through.

The fibers, called *lint,* are pressed into bales weighing 480 pounds, ready for sale to a spinning mill. The average yield is 629 pounds per acre. However, Arizona, the state with the largest production per acre, has averaged as much as 863 pounds per acre. The seeds, after ginning, look like the buds of the pussywillow. They are covered with very short fibers—1/8 inch in length—called *linters.* The linters are removed from the seeds and are used to a limited extent as raw material for the making of rayon and acetate. The seeds are crushed to obtain cottonseed oil and meal.

Fig. 4–3 *Cotton field with harvester. (Courtesy of National Cotton Council of America.)*

Fig. 4–2 *Opened cotton boll. (Courtesy of National Cotton Council of America.)*

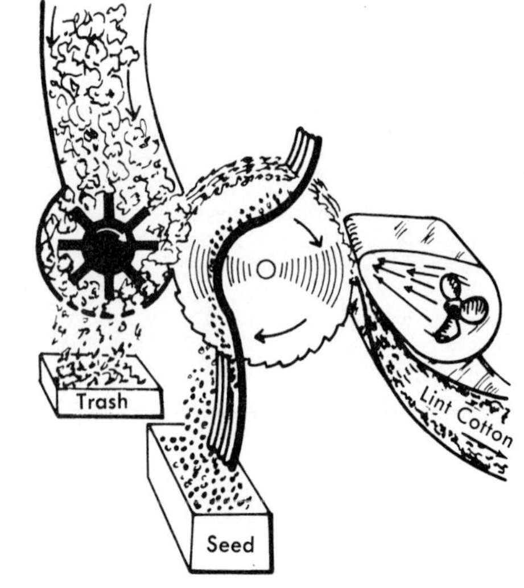

Fig. 4–4 *Cotton gin.*

Fiber Properties

PHYSICAL STRUCTURE

Raw cotton is creamy white in color. The fiber is a single cell, which grows out of the seed as a hollow cylindrical tube over one thousand times as long as it is thick.

The *quality* of cotton depends on its length, strength, fineness, and maturity. Other factors affecting quality are color, leaf residue, and ginning preparation.

Length. Staple length is very important because it affects how the fiber is handled during spinning and it relates to fiber fineness and fiber tensile strength. Longer cotton fibers are finer and make stronger yarns.

Cotton fibers range in length from ½ inch to 2 inches, depending on the variety. There are three groups of cotton that are commercially important:

1. Upland cottons, which are ⅞–1¼ inches in length and were developed from cottons native to Mexico and Central America.

2. Long staple cottons, which are 1⁵⁄₁₆–1½ inches in length and were developed from Egyptian and South American cottons. Different varieties include American Pima, Egyptian, American Egyptian, and Sea Island cottons.

3. Short staple cottons, which are less than ¾ inch in length and are produced primarily in India and eastern Asia.

Long staple fibers are considered to be finer quality because they can be made into softer, smoother, stronger, and more-lustrous fabrics. Because they command a higher price and less is produced than the medium- and short-staple lengths, they are sometimes identified on a label or tag as Pima or Supima. Or they may be referred to as long-staple or extra-long-staple cotton.

Convolutions. *Convolutions,* or ribbon-like twists, characterize the cotton fibers (Figure 4–5). When the fibers mature, the boll opens, the fibers dry out, and the central canal collapses; reverse spirals cause the fibers to twist. The twist forms a natural crimp that enables the fibers to cohere to one another, so that despite its short length, cotton is one of the most spinnable fibers. The convolutions can be a disadvantage, since dirt collects in the twists and must be removed by vigorous washing. Long-staple cotton has about 300 convolutions per inch; short-staple cotton has less than 200.

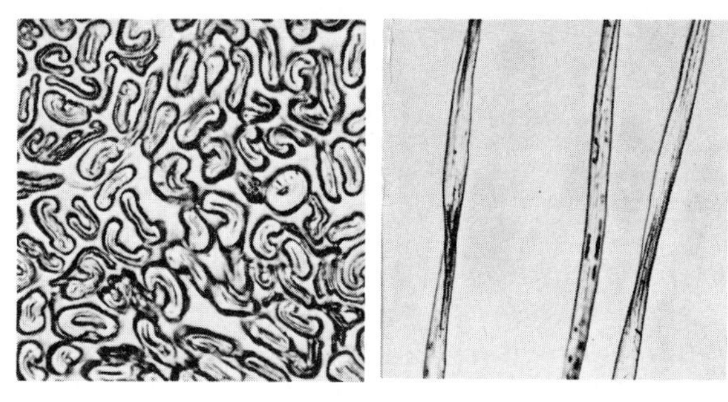

Fig. 4–5 *Photomicrographs of cotton: cross-sectional view at 500×* (left); *longitudinal view 250×* (right). *(Courtesy of E. I. du Pont de Nemours & Company.)*

Fineness. Cotton fibers vary from 16–20 micrometers in diameter. The cross-sectional shape varies with the maturity of the fiber. Immature fibers tend to be U-shaped and the cell wall is thinner; mature fibers are more nearly circular, with a very small central canal. In every cotton boll there are immature fibers. The proportion of immature to mature fibers cause problems in processing, especially in spinning and dyeing. Notice in the photomicrograph of the cross-section the difference in size and shape.

Distinctive Parts. The cotton fiber is made up of a cuticle, primary wall, secondary wall, and lumen (Figure 4–6). The fiber grows to almost full length as a hollow tube before the secondary wall begins to form.

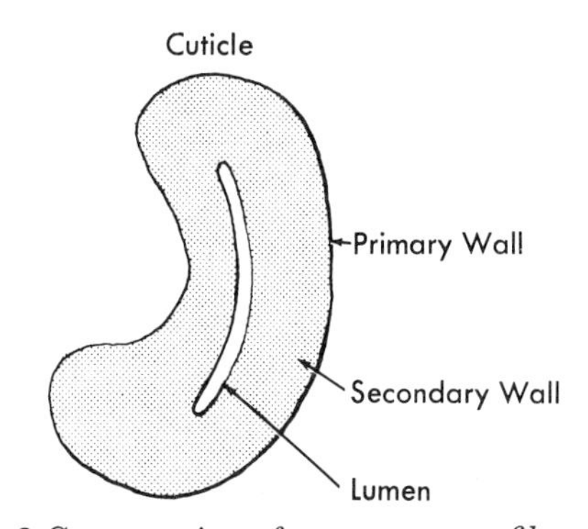

Fig. 4–6 *Cross-section of mature cotton fiber.*

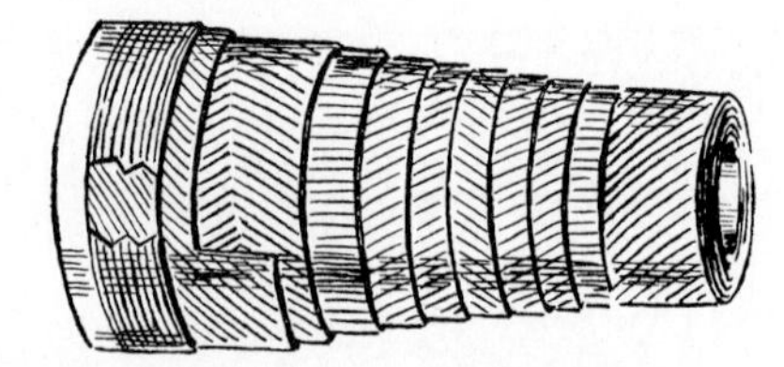

Fig. 4—7 Layers of cellulose (schematic).

The *cuticle* is a wax-like film covering the primary, or outer, wall. The *secondary wall* is made up of layers of cellulose (Figure 4–7).

The layers deposited at night differ in density from those deposited during the day; this causes *growth rings,* which can be seen in the cross-section. The cellulose layers are composed of *fibrils*—bundles of cellulose chains—arranged spirally. At some points the fibrils reverse direction. These *reverse spirals* (Figure 4–8) are an important factor in the convolutions, elastic recovery, and elongation of the fiber. They are also the weak spots, being 15–30 percent weaker than the rest.

Cellulose is deposited daily for 20–30 days until, in the mature fiber, the fiber tube is almost filled.

The *lumen* is the central canal, through which the nourishment travels during growth. When the fiber matures, the dried nutrients in the lumen give the characteristic dark areas that can be seen with the microscope.

Color. Cotton is naturally creamy white. As it ages, it becomes more beige. If it is rained on just before harvest, the fiber is grayer. White fiber is preferred.

Picking and ginning affect the appearance of cotton fibers. Carefully picked cotton is cleaner. Well-ginned cotton is uniform in appearance and white in color. Poorly ginned cotton will have brown flecks in it, called *trash,* such as bits of leaf or stem or dirt. These brown flecks decrease the quality of the fiber. Fabrics made from such fibers are suitable for utility cloths and occasionally are fashionable when a "natural" look is popular.

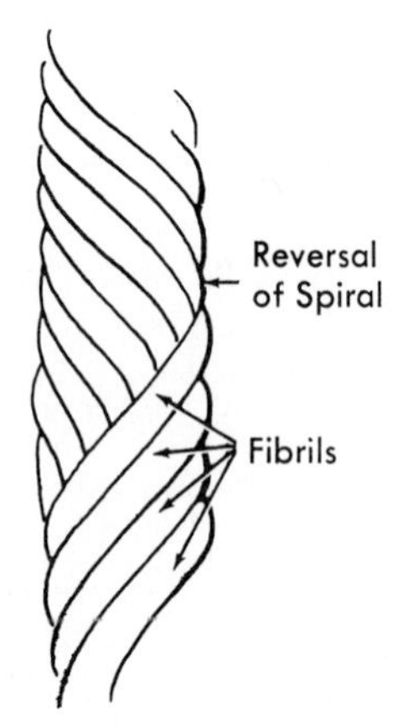

Fig. 4–8 Reverse spirals in cotton fiber.

Grading and classing of cotton is done by hand and by machine. Visual inspection of staple length and color compare the cotton from the bale being graded with standards prepared by the United States Department of Agriculture.

COTTON CLASSIFICATION

Cotton classification is the art of describing quality of cotton in terms of grade and staple length. *Staple length* is identified by using sight and touch and experience to determine the length of a representative bundle of fibers from a bale of cotton. A good cotton classer must be able to tell consistently differences in length of $1/32$ of an inch. Actually, a sample classified as $1\,5/32$ inch will have fibers ranging in length from $1/8$ inch to $1\,5/8$ inch as shown in Figure 4–9.

The grade of cotton is determined by its appearance—more specifically, by its color, leaf residue, and preparation. There are 40 grades for upland cotton. The predominant grade of cotton produced in the United States is strict low-middling cotton. *Strict* in this case means "better than."

Color of cotton is described in terms that range from white to yellow: white, light-spotted, spotted, tinged, and yellow. Color is also described in terms of lightness to darkness: plus, light gray, and gray. This factor of appearance is a combination of grayness and the amount of leaf present in white-cotton grades.

Grading of American Egyptian cotton is based on ten grades. It is yellower in appearance than upland cotton and has a different appearance after ginning because it is done on a different kind of gin and has a higher leaf content.

Cotton is a commodity crop. It is sold by grade and staple length. Strict low-middling cotton is used in mass-produced cotton goods and in cotton/synthetic blends. Better grades of cotton and longer-staple cotton are used in quality shirtings and sheets. Extra-long staple American Egyptian cotton is frequently identified by the terms Pima and Supima because of its higher

Fig. 4–9 *Cotton classed as 1 5/32 inch contains fibers that range in length from less than 1/8 inch up to 1 5/8 inch. (Courtesy of United States Department of Agriculture.)*

quality and price. Pima is seen in women's sweaters, women's panties, and towels.

CHEMICAL COMPOSITION AND MOLECULAR ARRANGEMENT

Cotton, when picked, is about 94 percent cellulose; in finished fabrics it is 99 percent cellulose. Like all cellulose fibers, cotton contains carbon, hydrogen, and oxygen with reactive hydroxyl (OH) groups. Cotton may have as many as 10,000 glucose residues per molecule. The molecular chains are in spiral form.

The basic unit of the cellulose molecule is the *glucose unit.* The glucose unit is made up of the chemical elements carbon, hydrogen, and oxygen.

CHEMICAL NATURE OF CELLULOSE

The chemical reactivity of cellulose is related to the three hydroxyl groups (OH groups) of the glucose unit. These groups react readily with moisture, dyes, and many finishes. Chemicals such as bleaches cause a breakdown of the molecular chain of the cellulose, usually by attacking the oxygen atom and causing a rupture there.

The cellulose molecule is a long, linear chain of glucose units. The length of the chain is a factor in fiber strength.

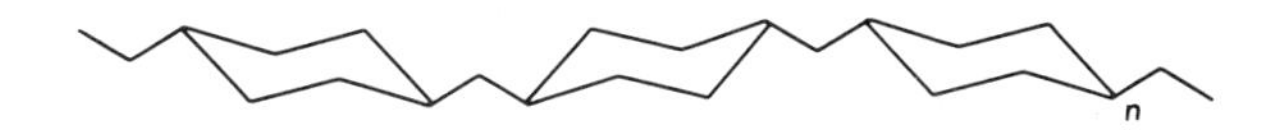

Mercerization. John Mercer, using a cotton cloth to filter a caustic soda (or sodium hydroxide, NaOH) solution, noticed that the cloth was changed during the process. He demonstrated the beneficial effects of caustic soda on cotton and from that time (1844) on, *mercerization* of cotton has been a common finish for yarns and fabrics (see Chapter 32). Mercerization (treating yarns or fabrics with NaOH) causes a physical change in the fiber. In *tension mercerization,* the fabric or yarn being mercerized is under tension. The concentration of the sodium hydroxide solution is high, generally around 20 percent. The sodium hydroxide causes the fiber to swell. Because of the tension during the swelling, the fibers become more rod-like and rounder in cross-section, and the number of convolutions decreases (Figure 4–10).

The following are effects of tension mercerization:

- *Increased strength* to the fibers. The molecular chains are less-spiral in form and are more-oriented in length, making the fibers 30 percent stronger.
- *Increased absorbency* because the fiber swells. This opens up the molecular structure so that more moisture can be absorbed. The moisture regain is 11 percent. Mercerization is done primarily to improve the dyeability of cotton yarns and fabrics.
- *Increased luster* because the fibers become rounder with fewer convolutions and thus reflect more light. Mercerization for luster is done under tension and on long-staple cotton yarns and fabrics.

A second type of mercerization is often referred to as *slack mercerization* because no tension is used. In this process, the fabric or yarn is soaked in a weaker solution of sodium hydroxide than the solution used in tension mercerization. Slack mercerization is done primarily to increase the absorbency and to improve the dyeability of cotton yarns and fabrics.

Liquid Ammonia Finishes. Liquid ammonia is used as an alternative to several preparation finishes, especially mercerization. Liquid ammonia is used on cellulosic-fiber contents and blends with cellulosics. The ammonia swells the fiber, but not to the degree of sodium hydroxide. Fabrics that have had the ammonia treatment have good luster and dyeability. These fabrics do not dye to the same depth as mercerized fabrics. Because the amount of resin needed is less than with mercerized fabrics, ammonia-treated fabrics will have better crease recovery and less loss of strength and abrasion resistance following wrinkle-resistant finishes than mercerized fabrics. Ammonia-treated fabrics are less stiff and harsh than mercerized fabrics. Ammonia treated fabrics have an increase in tensile strength of 40 percent and an increase in elongation of two to three times that of untreated cotton. These fabrics also are less sensitive to thermal degradation.

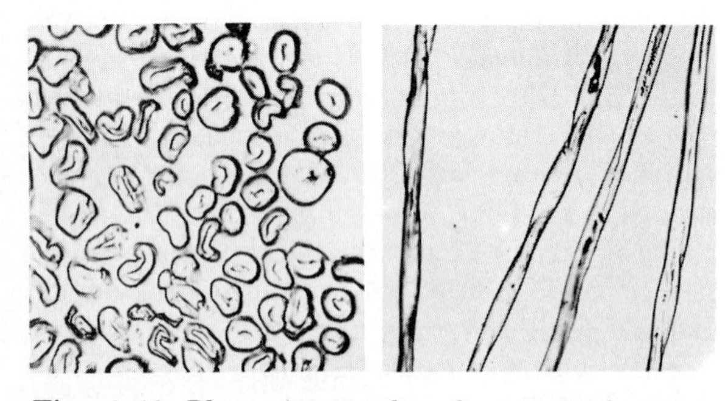

Fig. 4–10 *Photomicrographs of mercerized cotton: cross-sectional view 500×* (left); *longitudinal view 250×* (right). *(Courtesy of E. I. du Pont de Nemours & Company.)*

PROPERTIES

"Once you get a feel for cotton, you won't feel like anything else." This phrase was used to promote cotton in *American Fabrics and Fashions,* Winter 1982. Cotton is a comfortable fiber. Appropriate for year-round use, it is the fiber most preferred for warm-weather clothing, especially where the climate is hot and humid.

Aesthetic. Cotton fabrics certainly have consumer acceptance. Cotton fabrics have a matte appearance. Their low luster is the standard that has been retained with cotton/polyester blends that are increasingly important in apparel and home-furnishing fabrics.

Cotton fabrics with more luster than usual are available. Long-staple cottons make more-lustrous fabrics. Mercerized and ammonia-treated cotton fabrics have a soft, pleasant luster as a result of chemical finishes. Cotton sateen has a luster due to weave structure and possible finishes. Cotton blended with lustrous rayon results in a lustrous cotton-like fabric.

Drape, texture, and hand are affected by choice of yarn size and type, fabric structure, and finish. Soft, sheer batiste; crisp, sheer voile; fine percale; and sturdy denim and corduroy have been popular for years.

Durability. Cotton is a medium strength fiber having a breaking tenacity of 3.5–4.0 g/d. It is stronger when wet. Long-staple cotton makes stronger yarns because there are more points of contact between the fibers when they are twisted together. Because of its higher wet strength, cotton can stand rough handling during laundering.

Abrasion resistance is moderate. Obviously, heavy denim fabrics or corduroy will take longer

Summary of the Performance of Cotton in Apparel Fabrics

AESTHETIC	*ATTRACTIVE*
Luster	Matte, pleasant
Drape	Soft to stiff
Texture	Pleasant
Hand	Smooth to rough
DURABILITY	*MODERATE*
Abrasion resistance	Moderate
Tenacity	Moderate
Elongation	Low
COMFORT	*EXCELLENT*
Absorbancy	Excellent
Thermal retention	Low
APPEARANCE RETENTION	*MODERATE*
Resiliency	Low
Dimensional stability	Moderate
Elastic recovery	Moderate
RECOMMENDED CARE	*MACHINE WASH AND DRY*

to wear through than thinner shirting fabrics. The *elongation* of cotton is low, 3 percent, and it has low elasticity.

Comfort. Cotton makes very comfortable skin-contact fabrics because of its high absorbency and its good heat and electrical conductivity. It is lacking in any surface characteristics that might be irritating to the skin. Cotton has a moisture regain of 7 percent. When cotton becomes wet, the fibers swell and become somewhat plastic. This property makes it possible to give a smooth, flat finish to cotton fabrics when they are ironed, and makes high-count woven fabrics water repellent. However, as cotton fabrics absorb more moisture, they feel wet or clammy—think of dish or bath towels.

Cotton is a comfortable fiber to wear in hot and humid weather. The fibers absorb moisture and feel good against the skin in high humidity. The fiber ends in the spun yarn hold the fabric slightly off the skin for greater comfort. Moisture passes freely through the fabric, thus aiding evaporation and cooling. Cotton is very comfortable for wear all year—heavier fabrics such as denim and corduroy are usually cotton and are perennial favorites for fall and winter wear in some areas. However, in cold rain or snow, cotton may be too absorbent to be comfortable.

Appearance Retention. Overall appearance retention is moderate. Cotton has very low resiliency. The hydrogen bonds holding the molecular chains together are weak, and when fabrics are bent or crushed, particularly in the presence of moisture, the chains move freely to new positions. When pressure is removed, there are no forces within the fibers to pull the chains back to their original positions so the fabrics stay wrinkled. Creases can be ironed in easily, and wrinkles can be pressed out easily, but wrinkling while wearing remains a problem.

Unless cotton fabrics are given a durable-press finish, or blended with polyester and given a durable-press finish, they wrinkle easily during both wear and care. The wrinkled look is very popular at times—wrinkles are even added during finishing so they will remain. However, at other times, fashion dictates a neat, nonwrinkled look.

All-cotton denim and all-cotton knits will shrink unless they have been given a durable-press finish or a shrinkage-resistant finish. Untreated cottons shrink less when washed in cool water and drip dried; they shrink more when washed in hot water and dried in a hot drier. When they are worn again, they tend to recover some of their original dimensions—think of your experiences with cotton-denim jeans.

All-cotton fabrics and garments that have been given a wrinkle-resistant or durable-press finish or have been treated for shrinkage generally should not shrink noticeably. However, you may need to be a little more careful of handwoven cotton fabrics or those imported from India and China, unless specific information about shrinkage is available on the label.

Elastic recovery is moderate. Cotton recovers 75 percent from 2–5 percent stretch. In other words, cotton tends to stay stretched out in areas of stress, such as in the elbow or knee areas of garments.

Care. Cotton requires no special care during washing and drying. White cottons can be washed in hot water. Colored cottons will retain their color better if washed in warm water instead of hot water. If the garments are not heavily soiled, cold water will clean them adequately. Cotton releases all types of soil readily.

Chlorine bleach may be used on cottons if the directions are followed; bleaching should be considered a spot-removal method and not be routinely used with every load of wash. Excessive bleaching weakens cellulosic fibers.

Less wrinkling will occur in the dryer if the cotton garments are removed when they are dry and not left in the dryer longer than necessary. Ironing is not as common at present. For many cotton clothes, a steam iron is adequate to remove wrinkles. However, some all-cotton clothes respond better to pressing while damp with a hot and dry iron. These cottons will iron nicely, but they handle differently than blends do. Polyester/cotton fabrics need to be ironed at a lower temperature to avoid melting the polyester.

Cottons should be stored clean and dry. If they are damp, mildew can form. Mildew first appears as little black dots, but it can actually eat through the fabric, causing holes if enough time elapses. If the clothing merely smells of mildew, it can be laundered or bleached and it will be fine. But if the mildew has progressed to visible spots, they may or may not be removable. More-extensive damage cannot be corrected.

Cotton is harmed by acids. Fruit and fruit-juice stains should be promptly treated with cold water before they set and become even more difficult to remove. Cotton is not greatly harmed by alkalis. Cotton can be washed with strong detergents and under proper conditions it will withstand chlorine bleaches. Cotton is resistant to organic solvents so that it can be safely dry-cleaned.

Cotton oxidizes in sunlight, which causes white and pastel cottons to yellow and all cotton to degrade. Some dyes are especially sensitive to sunlight and when used in window-treatment fabrics the dyed areas disintegrate.

Cotton is not thermoplastic. It can safely be ironed at high temperatures. Cotton burns readily.

USES

Cotton is the most important apparel fiber in the United States. In 1983, it accounted for 37.5 percent of apparel fabrics. (Polyester was a close second with 35 percent of apparel fabrics.) Cotton had been the most widely used fiber for apparel, home furnishing, and industrial uses for over 125 years, but in 1973, 1974, and 1975, both cotton and polyester were used almost equally. By 1976, polyester was used more widely than cotton when all uses were combined.

A report in *The Cotton Situation,* November 1985, by the United States Department of Agriculture, stated:

Years from now, the U.S. textile industry may look back on 1985/86 as the season when cotton finally turned the corner, once and for all, in its competition with man-made fibers. "Natural" blend pants and shirts of 60 percent cotton and 40 percent polyester are increasingly available at retail stores. Use of heavyweight, 100-percent denim is rising, and the cotton content of other apparel products, ranging from socks and underwear to women's wear, is increasing.

A shift in consumer preferences toward natural fibers, improved technology for using cotton, a fashion trend toward heavyweight denim, and a decline in the cotton/polyester price ratio, combined with a perception that cotton will remain relatively less expensive, underlie the rise in cotton's market share. The influence of trends in fashion and consumer preferences cannot be quantified, but cotton's share of domestic consumption began rising in 1980. . . . Retailers indicate that shifts in consumer tastes are causing these changes." (pp. 4–5)

Other people closely associated with cotton agree that cotton's share of the market is rising, but they think discussing competition between fibers is an outdated approach. Blends are too important a part of the market to be ignored. The increased usage of cotton will come from (1) greater use of 100 percent cotton in apparel; (2) increased availability of cotton-rich blends (those with 60 percent or more cotton) in apparel; and (3) more cotton used with other fibers to replace 100 percent man-made fiber fabrics, especially in bottom-weight fabrics.

Looking at cotton as a part of the total fiber market, cotton accounts for the following percentages of fiber used in these broad end-use categories in 1985:

- 40 percent of apparel
- 19 percent of home furnishings
- 16 percent of industrial uses

Major End Uses of Cotton—1985

Percent of Cotton's Share of Market		Thousand Bales*	Percent of Market Share Held by Cotton
40%	*APPAREL*		
	Men's and boys' trousers and shorts	854	67%
	Men's and boys' shirts	558	53
	Men's and boys' underwear	373	71
	Women's and misses slacks and jeans	369	51
	Women's and misses blouses and shirts	163	32
	Women's and misses dresses	83	20
	Men's and boys' gloves and mittens	82	84
	Men's and boys linings and pockets	79	23
	Girls' and children's slacks and jeans	67	55
	Girls' and children's hosiery	57	39
	Girls' and children's blouses and shirts	50	44
	Men's hosiery	49	22
19%	*HOME FURNISHINGS*		
	Towels and wash cloths	685	93%
	Sheets and pillowcases	406	48
	Drapes, upholstery, and slipcovers	387	28
	Bedspreads	90	44
	Tablecloths, napkins, and placements	50	52
	Rugs and carpets	38	1
	Comforters and quilts	29	43
	Blankets	27	14
	Curtains	27	14
16%	*INDUSTRIAL AND OTHER CONSUMER PRODUCTS*		
	Medical supplies	143	52%
	Retail piece goods	138	33
	Thread, industrial	88	30
	Tarpaulins	70	54
	Abrasives	54	87
27	Overall market share		

*480-lb bales.
Source: Cotton Counts Its Customers, 1986, National Cotton Council.

Focusing on cotton alone, of the 2,980 million pounds of cotton consumed in the United States in 1984, the following percentages were used:

- 57 percent of cotton went into apparel
- 27 percent of cotton went into home furnishings
- 13 percent of cotton went into industrial uses
- 3 percent of cotton went into exports

Details of specific end uses within those broad categories are given in the table, "Major End Uses of Cotton—1985."

The greatest amount of cotton is used for apparel. All-cotton fabrics are used where comfort is of primary importance and appearance retention is not as important—or where a more-casual plissé fabric or crinkle cotton is acceptable. Cotton blended with polyester in durable-press fabrics is much easier to find on the market, both in ready-to-wear apparel and in over-the-counter fabrics. It is possible to find cotton-rich blends if you look for them—60 percent or 70 percent cotton blended with polyester and given a durable-press finish. Most blends retain the pleasant appearance of cotton, have the same or increased durability, are less comfortable in conditions of extreme heat and humidity or high physical activity, and have better appearance retention during wear in comparison with 100 percent cotton fabrics. However, removal of oily soil is a greater problem with blends.

The statistics on end use do not distinguish between all-cotton items and cotton blends. What is in your wardrobe? What blend levels are present? Do you have all-cotton garments? Do you prefer all cotton for some uses? The last column in the table, "Major End Uses of Cotton—1985," gives some indication of how important cotton and cotton blends are in the specific end uses.

Cotton, Inc., is the organization that promotes the use of cotton by consumers. This organization wants people to recognize the seal of cotton trademark. They also have been promoting the use of Natural Blend® fabrics and apparel—items that have at least 60 percent cotton in them.

Imports of cotton textiles have increased over recent years until in 1984 they totaled 3.1 million bales, or 37 percent, of domestic cotton consumption. Half of the imports are cotton apparel, 30 percent of imports are cotton fabrics—Hong Kong is the largest supplier. Korea, Taiwan, and China are all important suppliers. What have your experiences been with the performance of these imported garments? Can you make generalized comparisons about the quality or performance differences, if any, between garments imported from different countries?

Cotton is also a very important home-furnishing fabric. Towels are mostly cotton—softness, absorbency, wide range of colors, and washability are important in this end use. Durability is increased in the base fabric, as well as in the selvages and hems by blending polyester with the cotton. However, the loops of the towel are all cotton so that maximum absorbency is retained.

Sheets and pillowcases are mostly blends of cotton with polyester. All-cotton sheets are available. Blend levels and counts vary a great deal. Muslin and percale sheets are common, and flannelette sheets are available in the fall and winter. Spring- and fall-weight blankets made of cotton are also on the market. Cotton bedspreads are available in a variety of weights.

Fig. 4–11 *Cotton®* *seal* (top) *for fabrics and apparel made of 100 percent cotton and Natural Blend® seal* (bottom) *for fabrics and apparel made of at least 60 percent cotton.*

Drapes, curtains, upholstery fabrics, and slipcovers are made of cotton, as well as of polyester/cotton blends. Many heavyweight cotton upholstery fabrics are very attractive and very durable. They are comfortable and easy to spot clean. They retain their appearance well. Resiliency is not a problem with heavyweight fabrics that are stretched over the furniture frame.

Medical supplies are frequently cotton. Since cotton can be autoclaved (heated to a high temperature to sanitize it), it is very important in hospitals. Absorbency, washability, and low static build-up are also important properties in hospital uses.

Industrial uses include abrasives, book bindings, luggage and handbags, shoes and slippers, tobacco cloth, woven wiping cloths, and wall-covering fabrics.

5

Flax and Other Natural Cellulosic Fibers

All plants are fibrous. The fiber bundles of plants give strength and pliability to their stems, leaves, and roots. Natural cellulosic textile fibers are obtained from plants where the fibers can be readily and economically separated from the rest of the plant. They can be classified according to the portion of the plant from which they are removed (see the following chart).

Natural Cellulosic Fibers

Seed Hairs	*Bast Fibers*	*Leaf Fibers*
Cotton	Flax	Piña
Kapok	Ramie	Abaca
Coir	Hemp	Sisal
	Jute	Hennequin
	Kenaf	

Cotton, discussed in the previous chapter, is an example of a *seed fiber.* Flax is an example of a *bast fiber,* a fiber that is obtained from the stem of the plant. Sisal is an example of a *leaf fiber.*

These fibers differ in physical structure but are alike in chemical composition. The arrangement of the molecular chains in fibers, although similar, varies in orientation and length, so that performance characteristics related to polymer orientation and length will differ. Fabrics made from these fibers will thus have different appearances and hand but will react to chemicals in essentially the same way and will require essentially the same care. Properties common to all cellulosic fibers are summarized in the chart on page 37.

In this chapter, natural cellulosic fibers that account for a relatively small amount of U.S. textile fiber consumption will be discussed. Many of these fibers have found little use in the United States; nevertheless, a discussion of these fibers has a place in an introductory textiles course because some of these fibers are imported into the United States and others may be encountered during travel. Although many of these fibers are not of great importance to the U.S. economy, these fibers may be significant to the economy of the countries where they are produced. There are, of course, many other natural cellulosic fibers that will not be discussed because of their extremely limited use.

Bast Fibers

Bast fibers come from the stem of the plant. Hand labor is often required to process bast fibers, so that production of these fibers has flourished in countries where labor is cheap. Complete mechanization in the production of bast fibers has yet to be achieved. Harvesting is done by pulling or cutting the plants. Flax is usually pulled either by hand or by machine. After harvesting, the seeds are removed from the plants.

Since the fiber extends into the root, harvesting is done by pulling up the plant or cutting it close to the ground to keep fiber length as long as possible. After harvesting, the seeds are removed by pulling the plant through a machine in a process called *rippling.*

Bast fibers lie in bundles in the stem of the plant just under the outer covering or bark. They are sealed together by a substance composed of pectins, waxes, and gums. To loosen the fibers so that they can be removed from the stalk, the pectin must be decomposed by a process called *retting* (bacterial rotting). There are some individual fiber differences in the process, but the major steps are the same. Retting can be done in the fields (dew retting); in ponds or pools (pool retting); in tanks (tank retting), where the temperature and bacterial count can be carefully controlled; or in chemical retting, in which chemicals such as sodium hydroxide are used. Chemical retting is a much faster process than any other method. However, extra care must be taken or irreversible damage can occur to the fiber.

After the plants have been rinsed and dried, the woody portion is removed by breaking the outer covering, or *scutching,* in which the stalks are passed between fluted metal rollers. Most of the fibers are separated from one another, and the short fibers are removed by *hackling,* or combing. Figure 5–1 shows flax at different stages of processing.

The processes of spinning, weaving, and finishing cause further separation of the fibers. With most bast fibers, length and fineness dimensions are not clearly definable. The primary fibers are bound together in fiber bundles and never completely separate into individual fibers. These fibers, as they are commonly used, are

Properties Common to All Cellulose Fibers

Properties	*Importance to Consumer*
Good absorbency	Comfortable for summer wear Good for towels, diapers, handkerchiefs, and active sportswear
Good conductor of heat	Sheer fabrics cool for summer wear
Ability to withstand high temperature	Fabrics can be boiled or autoclaved to make relatively germ free. No special precautions in ironing
Low resiliency	Fabrics wrinkle badly unless finished for recovery
Lacks loft. Packs well into compact yarns	Tight, high-count fabrics can be made Makes wind-resistant fabrics
Good conductor of electricity	Does not build up static
High density (1.5±)	Fabrics are heavier than comparable fabrics of other fiber content
Harmed by mineral acids, but little affected by organic acids	Fruit stains should be removed immediately from a garment to prevent setting
Attacked by mildew	Store clean items under dry conditions
Resistant to moths, but may be damaged by crickets and silverfish	
Flammability	Cellulose fibers ignite quickly, burn freely, and have an afterglow and gray, feathery ash. Filmy or loosely constructed garments should not be worn near an open flame
Moderate resistance to sunlight	Draperies should be lined

made up of many primary fibers. It is this characteristic of fiber bundles that give bast-fiber fabrics their characteristic thick-and-thin yarns.

FLAX

Flax is one of the oldest textile fibers. Fragments of linen fabric were found in the prehistoric lake dwellings in Switzerland; linen mummy cloths, more than 3,000 years old, were found in Egyptian tombs. The linen industry flourished in Europe until the 18th century. With the invention of power spinning, cotton replaced flax as the most important and widely used fiber.

Flax is a prestige fiber as a result of its limited production and relatively high cost. The term *linen* refers to cloth made from flax. This term is, however, often misused today in referring to fabrics that look like linen—fabrics that have thick-and-thin yarns and are fairly heavy or crisp. The term *Irish linen* always refers to fabrics made from flax. (The former use of flax in sheets, tablecloths, and towels has given us the term *linens* to describe textile items—for example, bed linens and table linens.)

The unique and desirable characteristics of flax are its body, strength, and thick-and-thin fiber bundles, which give texture to fabrics. The main limitations of flax are low resiliency and lack of elasticity. Most dress and suiting linens are given wrinkle-resistant finishes.

Major producers of flax include the Soviet Union, Belgium, Ireland, and New Zealand.

Fig. 5–1 Flax fiber at different stages of processing.

STRUCTURE

The primary fiber of flax averages 0.5–2.15 inches in length and a few micrometers in diameter. However, as stated earlier, these primary fibers are bound together in fiber bundles.

Flax fibers can be identified under the microscope by crosswise markings called *nodes* or *joints* (Figure 5–2). The markings on flax have been attributed to cracks or breaks during harvesting, or to irregularity in growth. The fibers may appear slightly swollen at the nodes and resemble somewhat the joints in a stalk of corn. The fibers have a small, central canal similar to the lumen in cotton. The cross-section (Figure 5–2) is several-sided, or polygonal, with rounded edges.

Flax fibers are grayish in color when dew retted, and yellowish in color when water retted. Flax has a more highly oriented molecular

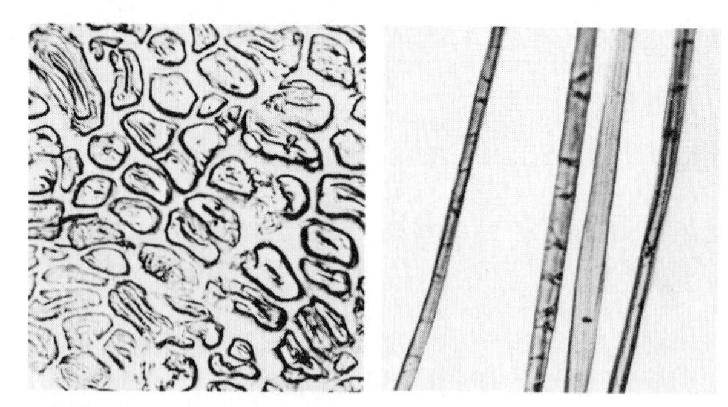

Fig. 5–2 Photomicrographs of flax: cross-sectional view (left); *longitudinal view* (right). *(Courtesy of E. I. du Pont de Nemours & Company.)*

structure than cotton, and is, therefore, stronger than the cotton fiber.

Short flax fibers are called *tow,* and the long, combed, better-quality fibers are called *line.* Line fibers are ready for spinning into yarn. The short tow fibers must be carded to prepare them for spinning into yarns that are used in less-expensive fabrics.

CHEMICAL COMPOSITION AND MOLECULAR ARRANGEMENT

Flax is similar to cotton in its chemical composition. The major differences between the two fibers are that flax has a higher degree of poly-

Summary of the Performance of Flax in Apparel Fabrics

AESTHETICS	*EXCELLENT*
Luster	High
Texture	Thick and thin
Hand	Stiff
DURABILITY	*GOOD*
Abrasion resistance	Good
Tenacity	Good
Elongation	Low
COMFORT	*HIGH*
Absorbency	High
Thermal retention	Good
APPEARANCE RETENTION	POOR
Resiliency	Poor
Dimensional stability	Adequate
Elastic recovery	Low
RECOMMENDED CARE	*DRY CLEAN OR MACHINE WASH*

merization (the cellulose polymer is longer) and a greater degree of orientation and crystallinity.

PROPERTIES

Aesthetic. Flax has a high, natural luster that is broken up by the fiber bundles. The fiber bundles give an irregular appearance to yarns made from flax. This irregular appearance is part of the charm of linen fabrics. The luster of flax can be increased by pounding linen fabrics with wooden hammers, a process called *beetling*. This finish produces flat yarns with good luster. The effect is not permanent unless a resin is also used.

Because flax has a higher degree of orientation and crystallinity and the fiber diameter is larger than cotton, the resulting fabrics are stiffer in drape and harsher in hand.

Durability. Flax is strong for a natural fiber. It has a breaking tenacity of 5.5 g/d when dry and 6.5 g/d when wet. Before synthetic fibers were invented, linen thread was used to sew shoes. Flax has very low elongation of approximately 3 percent. Elasticity is poor, with a 65 percent recovery at an elongation of only 2 percent. Flax also is a stiff fiber. With poor elongation, elasticity, and stiffness, fabrics of flax should not be folded repeatedly in the same place. With repeated folding, the fabric will break. Also, avoid pressing folds when ironing linen to minimize the stress on the fibers at the fold line. Flax has good abrasion resistance for a natural fiber. The good abrasion resistance is related to the fiber's high orientation and crystallinity.

Comfort. Flax has a high moisture regain of 12 percent, and it is a good conductor of electricity. Hence, static is no problem. Flax is also a good conductor of heat, so it makes an excellent fabric for warm-weather wear. Flax has a high specific gravity of 1.54, which is the same as cotton.

Care. Flax is resistant to alkalis and organic solvents. It is also resistant to high temperatures. Linen fabrics can be dry-cleaned or washed without special care and bleached with chlorine bleaches. Linen fabrics have very low resiliency and require ironing after washing. They are more resistant to sunlight than cotton.

Crease-resistant finishes can be used on linen, but the resins usually decrease strength and abrasion resistance. The wrinkling characteristics of flax are responsible for the strong high-fashion image of linen fabrics. Linen fabrics must be stored dry, otherwise mildew will become a problem.

Uses. Flax is used primarily for fashion fabrics in both apparel and home furnishings due to its high price. Flax is common in warm-weather professional or high-fashion apparel and upholstery, table linens, and window-treatment fabrics.

Identification Tests. Flax burns readily in a manner very similar to cotton. An easy way to differentiate between these two cellulosic fibers is to study their fiber length. Cotton is seldom over 2.5 inches in length; flax is almost always longer than that. Flax is also soluble in strong acids.

RAMIE

Ramie is also known as rhea or grasscloth. It has been used for several thousand years in China. It is a tall shrub from the nettle family that requires a hot, humid climate. Ramie is a fast-growing plant that can be harvested several times a year. It has been grown in the Everglades and Gulf Coast regions of the United States, but it is not currently produced in those areas.

Ramie fibers must be separated from the plant stalk by *decortication*. In this process, the bark and woody portion of the plant stem are separated from the ramie fiber. Because this process required a lot of hand labor, ramie was not commercially important until less-labor-intensive ways of decorticating were developed. Now that relatively inexpensive ways of decorticating ramie are available, ramie is becoming a commercially important fiber. Ramie found a place in the U.S. market because it was a nonquotaed fiber in the Multi-Fiber Agreement regulating the amount of specific fibers that can be imported. For this reason, many items of ramie combined with another fiber have appeared in the United States in recent years. Ramie is produced in the People's Republic of China, the Philippines, and Brazil.

Ramie is long, lustrous, and fine. It has an absorbency similar to flax. Ramie has a density

Fig. 5–3 Photomicrographs of ramie: cross-sectional view (left); *longitudinal view* (right). *(Courtesy of Donna Danielson.)*

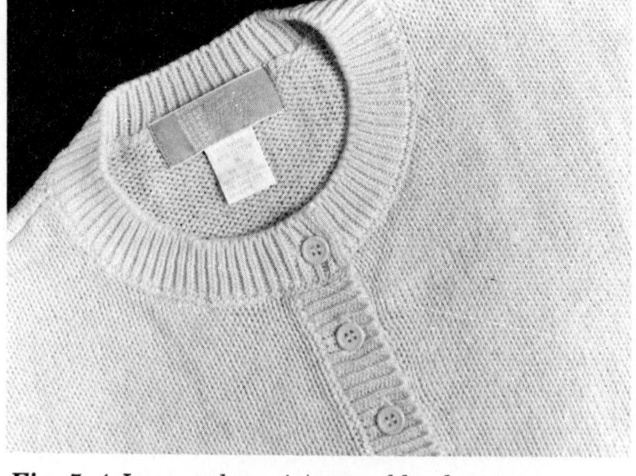

Fig. 5–4 Imported ramie/cotton blend sweater.

of 1.56, which is just slightly greater than that of flax.

When seen under the microscope, ramie is very similar to flax fiber (see Figure 5–3). It is pure white. It is one of the strongest fibers known and its strength increases when it is wet. It has silk-like luster. Ramie also has a very high resistance to rotting and mildew.

Ramie has some disadvantages. It is stiff and brittle, owing to the high crystallinity of its molecular structure. Consequently it lacks resiliency and is low in elasticity.

Ramie is resistant to shrinkage. It has a low elongation potential and will break if folded repeatedly in the same spot. Ramie has good absorbency, but will not dye as well as cotton. It has poor resiliency and should have a durable-press finish. Ramie is resistant to insects and microorganisms.

Ramie is used in a wide variety of imported apparel items including sweaters, shirts, blouses, and suitings (see Figure 5–4). It is often found in blends, particularly with cotton. It is also used in table linens, ropes, twines, nets, and industrial uses. Major problems experienced with ramie include its poor elasticity and color retention.

HEMP

The history of *hemp* is as old as that of flax. Hemp resembles flax; however, because hemp lacks the fineness of better-quality flax, it has never been able to compete with flax for clothing. Some varieties of hemp, though, are very difficult to distinguish from flax.

The high strength of hemp makes it particularly suitable for twine, cordage, and thread. Hemp is not very pliable or elastic. It does not rot readily when exposed to water. Hemp was commercially important up to the end of World War II. After World War II, the demand for hemp declined because many other natural and synthetic fibers took over its end uses.

JUTE

Jute was known as a fiber in Biblical times and probably was the fiber in sack cloth. Jute is one of the cheapest textile fibers, and it is second in production to cotton of all the natural cellulosic fibers. It is grown throughout Asia, chiefly in India and Bangladesh. The primary fibers in the fiber bundle are short and brittle, making jute one of the weakest of the cellulosic fibers.

Jute is creamy white to brown in color. It is soft, lustrous, and pliable when first removed from the stalk. But, on exposure to water, it turns brown, weak, and brittle. Jute has poor elasticity and elongation.

The greatest part of jute production goes into sugar and coffee bagging; it is also used for carpet backing, rope, cordage, and twine. Olefin is a strong competitor in these end uses. Because jute is losing its market, other uses for it are being investigated by jute-producing countries. For example, Bangladesh is investigating jute as a reinforcing fiber in resins to create preformed low-cost housing.

Burlap is used as a fashion fabric for decorative home furnishings. Chemical finishes can be used to overcome the natural odor and the stiffness of the fiber. Jute has low sunlight resistance and poor colorfastness. It is brittle and subject to splitting and snagging. It also deteriorates quickly when exposed to water.

KENAF

Kenaf is a soft bast fiber from the kenaf plant. The fiber is light yellow to gray, long in length, and harder and more lustrous than jute. It is used for twine, cordage, and other purposes similar to jute. Kenaf is produced in Central Asia, India, Africa, and some Central American countries. Kenaf is being investigated as a source of paper fiber.

Seed Fibers

Seed fibers are those from the seed pod of the plant. By far the most important seed fiber is cotton. In this section, some minor seed fibers will be discussed briefly.

In the production of seed fibers, the first step is separation of the fiber from the seed. With some seed fibers, the seed also is used in producing oil and feed for animals. After the seed and fiber have been separated, the fibers may be carded to get the fibers in a parallel arrangement for production of yarns or used as bundles of fibers for fiberfill.

KAPOK

Kapok is obtained from the seed of the Java kapok or silk cotton tree. The fiber is very lightweight and soft. Kapok is hollow and very buoyant. With use, kapok has a tendency to break down into a powder. The fiber is difficult to spin into yarns so it is used primarily as fiberfill for personal flotation devices and for pillow and upholstery padding.

COIR

Coir is the fiber obtained from the fibrous mass between the outer shell and the husk of the coconut. The fibers are removed by soaking the husk in saline water for several months. Coir is a very stiff, cinnamon-brown fiber. It has good abrasion, water, and weather resistance. It is used for floor mats and outdoor carpeting. Sri Lanka is the major producer of coir fiber.

Leaf Fibers

Leaf fibers are those fibers obtained from the leaf of the plant. Most leaf fibers are long and fairly stiff. In processing, the leaf is cut from the plant and the fiber is split, or pulled, from the leaf. Most of these fibers have poor dye affinity and are used in their natural color.

PIÑA

Piña is obtained from the leaves of the pineapple plant. The fiber is soft, lustrous, and white or ivory. The fiber is highly susceptible to acids and enzymes, so any acid stains should be rinsed out immediately and enzymatic presoaks should be avoided. Hand washing is recommended for piña. The fiber is used to produce lightweight sheer fabrics that are fairly stiff. These fabrics are often embroidered and used for formal wear in the Philippines. Piña is also used to make mats, bags, table linens, and clothing (see Figure 5–5).

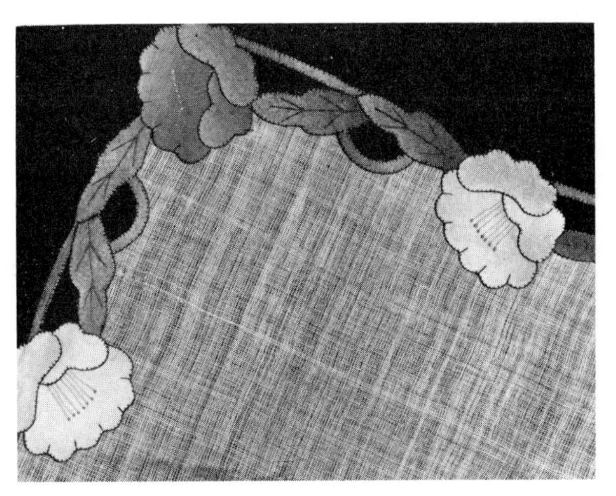

Fig. 5–5 *Piña place mat.*

ABACA

Abaca is obtained from a member of the banana-tree family. Abaca fibers are coarse and very long; some may reach a length of 15 feet. Abaca is off-white to brown in color. The fiber is strong, durable, and flexible. It is used for ropes, cordage, floor mats, table linens, and clothing. It is produced in Central America and the Philippines. Abaca is sometimes referred to as Manila hemp even though it is not a true hemp (see Figure 5–6).

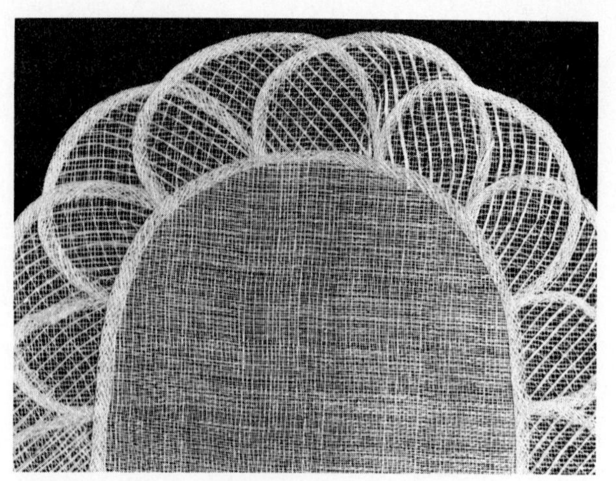

Fig. 5–6 *Abaca place mat.*

SISAL AND HENEQUEN

Sisal and *henequen* are closely related plants. They are grown in Africa, Central America, and the West Indies. Both fibers are smooth, straight, and yellow. They are used for better grades of rope, twine, and brush bristles. However, since both fibers are degraded by salt water, they are not used in maritime ropes. In addition to these end uses, sisal may be substituted for horsehair in upholstery.

6

Wool and Other Animal-Hair Fibers

Natural protein fibers are of animal origin: Wool and specialty wools are the hair and fur of animals and silk is the secretion of the silkworm. The natural protein fibers are prestige fibers today. Silk, vicuña, cashmere, and camel's hair have always been in this category. Wool is still the most widely used protein fiber, but it is no longer as readily available as it was through the 1960s. Wool and animal-hair fibers will be discussed in this chapter. Silk will be discussed in Chapter 7.

Protein fibers are composed of various amino acids that have been formed in nature into polypeptide chains with high molecular weight. They contain the elements carbon, hydrogen, oxygen, and nitrogen. Wool, in addition, contains sulfur. Protein fibers are amphoteric, having both acidic and basic reactive groups. The protein of wool is keratin, whereas that of silk is fibroin.

The following is a simple formula for an amino acid:

```
            R
            |
            CH        O
          /    \    //
     NH2         C
                  \
                   OH
amino group       carboxyl group
  (basic)            (acidic)
```

Protein fibers have some properties in common because of their chemical composition. These properties are important because they indicate the care required for the fabrics. Silk and wool have some different properties because their physical and molecular structures are different (see the following chart).

Properties Common to All Protein Fibers

Properties	*Importance to Consumer*
Resiliency	Resist wrinkling. Wrinkles hang out between wearings. Fabrics tend to hold their shape during wear.
Hygroscopic	Comfortable in cool, damp climate. Moisture prevents brittleness in carpets.
Weaker when wet	Handle carefully during washing. Wool loses about 40 percent of its strength and silk loses about 15 percent.
Specific gravity	Fabrics feel lighter than cellulosics of the same thickness.
Harmed by alkali	Use neutral or slightly alkaline soap or detergent. Perspiration weakens the fiber.
Harmed by oxidizing agents	Chlorine bleaches damage fiber so should not be used. Sunlight causes white fabrics to turn yellowish.
Harmed by dry heat	Wool becomes harsh and brittle and scorches easily with dry heat. Use steam! White silk and wool turn yellow.
Flame resistance	Do not burn readily; are self-extinguishing; have odor of burning hair; and form a black, crushable ash.

Wool

Wool was one of the first fibers to be spun into yarns and woven into cloth. Wool and flax were the most widely used textile fibers when fibers were spun by hand before the Industrial Revolution.

Now, many people consider both wool and silk to be luxury fibers. Designers continue to use these fibers extensively in their collections. The consumer is most likely to have a wool sweater, suit, or coat. The high initial cost of wool products and the cost of their care have led many customers to classify wool garments as investment clothing. These factors have encouraged the substitution of acrylic, polyester, or blends of wool with these fibers in many end-use products.

In the 1960s, when man-made fibers were used in increasing amounts in sweaters, blankets, carpeting, and in many kinds of outerwear, wool was promoted as Nature's Wonder Fiber by the Wool Bureau. This is an apt description of wool. Wool has a combination of properties that are unequaled by any man-made fiber; namely, ability to be shaped by heat and moisture, ability to absorb moisture in vapor form without feeling wet, comfortable warmth in cold weather, initial water repellency, feltability, and flame retardance.

Sheep were probably the first animals domesticated by man. The covering of primitive sheep consisted of two parts: a long, hairy outercoat that was used primarily for rugs and felt, and a light, downy undercoat that was very desirable for clothing. The fleece of present-day domesticated sheep is primarily the soft undercoat. It is thought that cross breeding sheep to increase the amount of undercoat began about A.D. 100. By A.D. 1400, the Spanish had developed the Merino sheep, whose fleece contains no hair or kemp fiber. Kemp is a coarse, brittle, dead-white fiber found in the fleece of primitive sheep and still found in the wools of all breeds of sheep except the Merino.

Sheep were not known to the American Indian, before the Spanish brought sheep to the southwest in the 16th century. The sheep of the Navajo are descendants of an unimproved long-haired breed introduced during the 16th or 17th century.

Sheep raising on the Atlantic seaboard began in the Jamestown, Virginia, colony in 1609 and in the Massachusetts settlements in 1630. From these centers, the sheep-raising industry spread rapidly because it was vital to the welfare of the colonists. In 1643, twenty families of wool combers and carders emigrated from England, settled in the Massachusetts Bay colony, and produced and finished wool fabric.

Most textiles were imported until 1650, when the movement towards self-sufficiency in fabric production became strong. Following the Civil War, the opening of free grazing lands west of the Mississippi prepared the way for sheep raising on a larger scale than was possible on Eastern farms. There were 50 million range sheep by 1884; that year was the high point of sheep production in the United States. During the 1950s and 1960s, the sheep population was around 32 million. In the 1980s, it has further declined to 9 million.

PRODUCTION

In 1983–84, Australia produced 25 percent of the world's wool supply. It was followed by the Soviet Union (16 percent), New Zealand (12 percent), China (7 percent), and Argentina (6 percent). The United States ranked tenth, producing only 1.6 percent of the world supply.

Fig. 6–1 *Merino sheep. (Courtesy of Australian Wool Corp.)*

Merino sheep produce the most valuable wool (Fig. 6–1). About 43 percent of merino wool comes from Australia. Good quality ewes produce 15 pounds of wool per fleece, while rams produce 20 pounds. Australian Merino wool is 3–5 inches long and very fine.

Fine wool is produced in the United States by four breeds of sheep: Delaine-Merino, Rambouillet, Debouillet, and Targhee (Fig. 6–2). More than half of this fine wool is produced in Texas and California. It is 2½–5 inches long.

Sheep are raised in every state of the United States, with the exception of Hawaii, but most

Fig. 6–2 *Rambouillet ewes. (Courtesy of the Texas Department of Agriculture. Photographer: Dan Morrison.)*

sheep are raised in Texas (18 percent) and California (11 percent). Wyoming, Colorado, South Dakota, and seven other western states are important producers (43 percent). The remaining 28 percent of sheep are raised in the East, South, and Midwest.

The greatest share of U.S. wool production is of medium-grade wools. These fibers have a larger diameter than the fine wools and a greater variation in length, from 1½–6 inches. There are 15 breeds of sheep commonly found in the United States. The breeds vary tremendously in appearance, type of wool produced, and other uses—such as meat, fat, and milk.

Sheep are generally sheared once a year in the spring. The fleece is removed with power shears that look like large barber's shears. A good shearer can handle 100–225 sheep per day. An expert can shear a sheep in less than 5 minutes. Shearers start in the Southern states in February and work northward, finishing in June.

The fleece is removed with long, smooth strokes, beginning at the legs and belly. A good shearer will leave the fleece in one piece. After shearing, the fleece is folded together and put in bags to be shipped to market.

At this point the wool is called *raw wool,* or *grease wool,* and it contains impurities such as sand and dirt, grease and dried sweat (called *suint*), which account for 30–70 percent of the weight of the fleece. Once these impurities are removed, the wool is called *clean,* or *scoured, wool.* The grease is a valuable byproduct; in its purified state, it is known as *lanolin* and is used in manufacturing face cream, cosmetics, soaps, and ointments.

Pulled wool is obtained from animals that are sold for meat. The pelts are washed and brushed, then treated chemically to loosen the fibers.

Grading and sorting are two marketing operations that put wools of like character together. In *grading,* the whole fleece is judged for fineness and length. Each fleece contains more than one quality of wool. In *sorting,* the individual fleece is pulled apart into sections of different-quality fibers. The best-quality wool comes from the sides, shoulders, and back, while the poorest wool comes from the lower legs.

Different qualities of wool are required for a particular product. For example, fine wool may be used in a lightweight worsted fabric while a coarse wool could be used in a bulky sweater. The end use of the product determines the grade of wool required.

Quality of apparel wool is based on fineness and length and does *not* necessarily imply durability because fine fibers are not as durable as coarse fibers. Fineness, color, crimp, strength, length, and elasticity are wool-fiber characteristics that vary with the breed of the sheep.

TYPES AND KINDS

Many different qualities of wool are available for the production of yarns and fabrics. As mentioned before, wool from Merino sheep is a very high-quality wool. Breeds of sheep produce fibers with different characteristics. Labels on wool garments almost never give information about the breed of sheep; the fiber is simply identified as wool. Wool is the fleece of the sheep. The term *wool* legally includes fiber from the Angora goat, Cashmere goat, camel, alpaca, llama, and vicuña. Wool comes from several sources:

- Sheared wool—from live sheep
- Pulled wool—from the pelts of meat-type sheep
- Recycled wool—from worn clothing and cutters' scraps

Wool is often blended with less-expensive fibers to reduce the cost of the fabric or to extend its use. Congress passed the Wool Products Labeling Act in 1939 (amended in 1980) to protect consumers as well as producers, manufacturers, and distributors from the unrevealed presence of substitutes and mixtures and to inform the consumer of the source of the wool fiber. The law requires that the label must give the fiber content in terms of percent and also give the source. The act does *not* state anything about the quality. The consumer must rely on feel and texture to determine quality.

The terms that appear on the label of a garment made of wool fiber are defined by the Federal Trade Commission as follows:

1. Wool—new wool or wool fibers reclaimed from knit scraps, broken thread, and noils. (Noils are the short fibers that are combed out in the making of worsted yarns.)

2. Recycled wool—scraps of new woven or felted fabrics are *garnetted* (shredded) back to the fibrous state and used again in the manufacture of woolens. Wool (shoddy) from old clothing and rags are cleaned and sorted and shredded into fibers. Recycled wool is often blended with new wool before being respun. It is usually used in interlinings and mackinaw-type fabrics that are thick and boardy.

Recycled wool is important in the textile industry. However, these fibers lose some of the desirable properties of new wool during the wool-garnetting process. Some fibers are broken by the mechanical action and/or wear. The fibers are not as resilient, strong, or durable as new wool, yet the fabrics made from them perform very well.

The term *virgin wool* on a label does *not* necessarily mean good quality. The term is not defined by the law but has been defined by the Federal Trade Commission as wool that has never been processed in any way. This eliminates knit clips and broken threads from being labeled as virgin wool.

Lamb's wool comes from animals less than 7 months old and is finer and softer because it is the first shearing and the fiber has only one cut end; the other end is the natural tip (Figure 6–3). The term lamb's wool usually identifies it on a label.

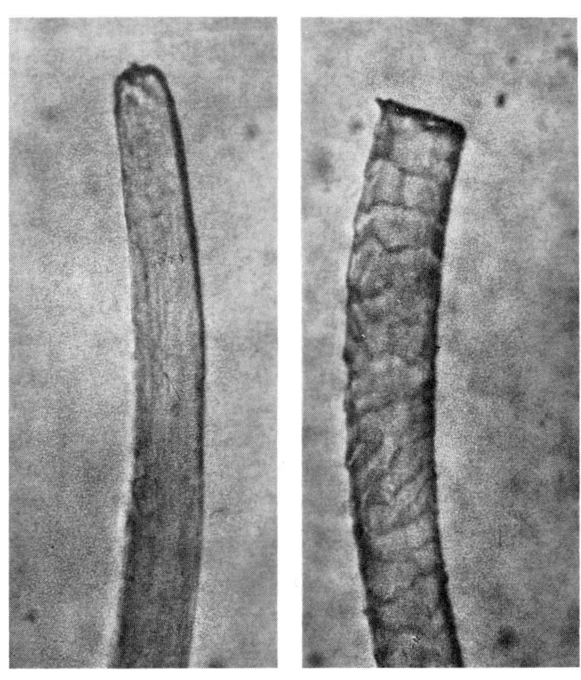

Fig. 6–3 *Lamb's wool fiber: natural tip* (left); *cut tip* (right).

PHYSICAL STRUCTURE

Length. The length of Merino wool fibers ranges from 1½ to 5 inches, depending on the kind of animal and the length of time between shearings. Long, fine wool fibers, used for worsted yarns and fabrics, have an average length of 2½ inches. The shorter fibers, which average 1½ inches in length, are used in woolen fabrics. Certain breeds of sheep produce coarse, long wools that measure from 5–15 inches in length. These long wools are used in specialty fabrics and hand weaving.

The diameter of wool fiber varies from 10–50 micrometers. Merino lamb's wool may average 15 micrometers in diameter. The wool fiber is made up of a cuticle, cortex, and medulla (Figure 6–4).

Medulla. When it is present, the *medulla* is a honeycomb-like core containing air spaces that increase the insulating power of the fiber. It appears as a dark area when seen through the microscope, but is usually absent in fine wools.

Cortex. The *cortex* is the main part of the fiber. It is made up of long, flattened, cigar-shaped cells with a nucleus near the center. In natural-colored wools, the cortical cells contain *melanin,* a colored pigment.

The cortical cells on the two sides of the wool fiber have somewhat different chemical and physical properties; they react differently to moisture and temperature. They are responsible for the three-dimensional crimp, which is

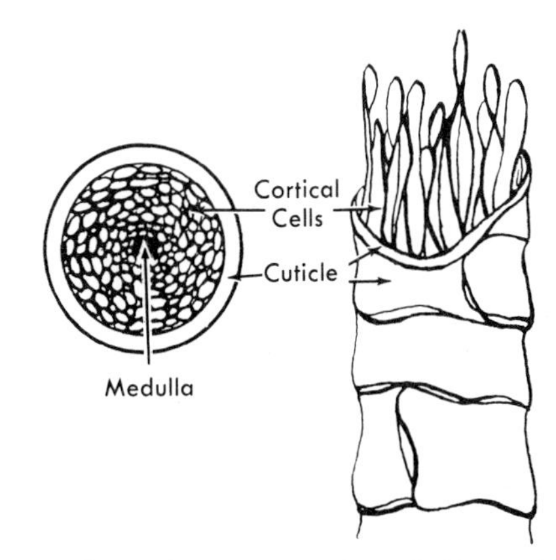

Fig. 6–4 *Physical structure of wool fibers. (Courtesy of Werner von Bergen from* Industrial and Engineering Chemistry, *September 1952; reprinted by permission.)*

unique to the wool fiber. Figure 6–5 shows the crimp in wool fiber. Fine Merino wool may have as many as 30 crimps per inch. Lower-quality wools may have as few crimps as 1–5 per inch. The irregular lengthwise waviness gives wool fabrics three very important properties: cohesiveness, elasticity, and loft. Crimp helps individual fibers cling together in a yarn; this cohesiveness increases the strength of the yarn over what one would expect from a low-strength fiber. Elasticity is increased—crimp makes the fiber act like a spring. As force is exerted on the fiber, the fiber first straightens out from its naturally wavy state to a flat state, without any damage to the fiber. Once the force is released, the wool fiber gradually will return to its crimped position. Crimp also is an important factor in the loft that wool fabrics exhibit. Because of the crimp of the fibers, yarns and fabrics made from wool are lofty or bulky and retain this loftiness throughout use. Two-dimensional crimp, or waviness, could be diagrammed like this: ∿∿∿. Figure 6–6, shows wool with crimp having this appearance.

Actually, the crimp in wool is three dimensional. The fiber itself turns even as it goes up and down in waves. Another way of saying this is that the fiber twists around its axis. This is drawn in Figure 6–6. Remember that the cortical cells on the two different sides of the fiber react differently to heat and moisture. Because wool has these two different parts, it is called a *natural bicomponent fiber*. To illustrate this bicomponent nature, think how a wool fiber will react to water. One side of the fiber will swell more than the other side; this will cause a decrease in the natural crimp of the fiber. When the fiber dries, the crimp will return.

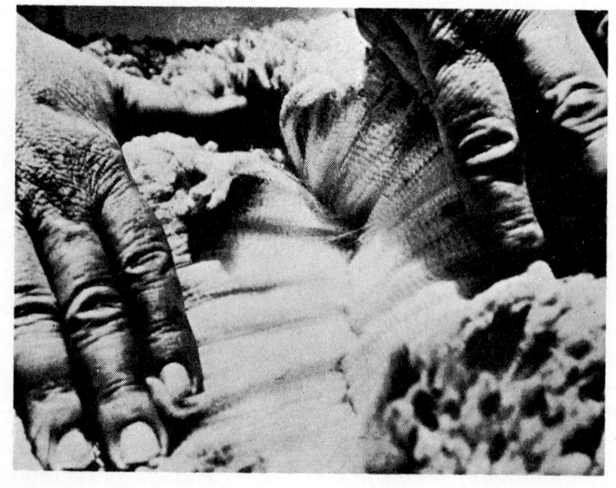

Fig. 6–5 *Natural crimp in wool fiber.*

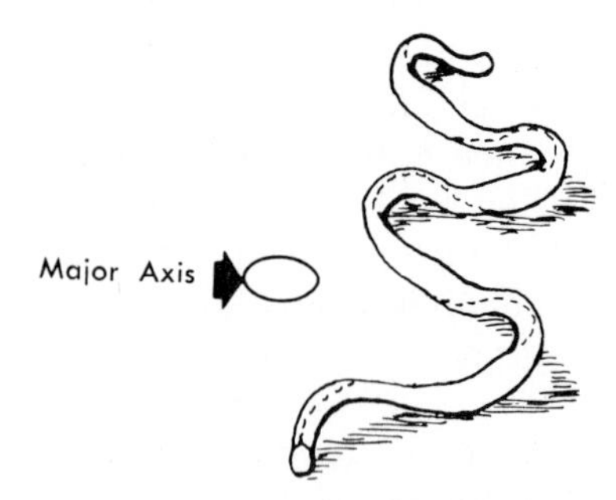

Fig. 6–6 *Three-dimensional crimp of wool fiber. (From G. E. Hopkins,* Wool As An Apparel Fiber. *Copyright 1953 by Holt, Rinehart and Winston, Inc.)*

Wool has been described as a giant molecular coil spring with outstanding resiliency. This *resiliency* is excellent when the fiber is dry and poor when it is wet. If dry fabric is crushed in the hand, it tends to spring back to its original shape when the hand is opened. The wool fiber can be stretched to as much as 30 percent of its original length. When stress is applied, the waves and bends of the fiber straighten out, and when stress is removed, the fibers recover their original length. Recovery takes place more slowly when the fabric is dry. Steam, humidity, and water hasten recovery. This is why a wool garment will lose its wrinkles more rapidly when it is hung over a bathtub of steamy water.

Cuticle. The cuticle is made up of an epicuticle and a horny, nonfibrous layer of scales. The *epicuticle* is a thin, nonprotein membrane that covers the scales. This layer gives water repellency to the fiber, but is easily damaged by mechanical treatment. In fine wools, the *scales* completely encircle the shaft and each scale overlaps the bottom of the preceding scale like parts of a telescope. In medium and coarse wools, the scale arrangement resembles shingles on a roof or scales on a fish (Figure 6–7). The free edges of the scales project outward and point toward the tip of the fiber. They cause skin irritation for some people. The scale covering gives wool its abrasion resistance and felting property.

Felting, a unique and important property of wool, is based on the scale structure of the fiber. Under mechanical action, such as agitation, friction, and pressure in the presence of heat and moisture, the wool fiber tends to move rootward and the edges of the scales interlock, thus pre-

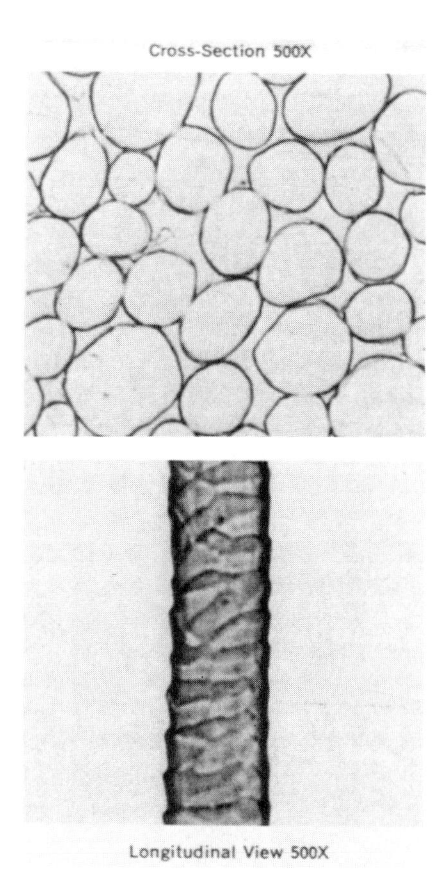

Fig. 6–7 *Photomicrographs of wool. (Courtesy of American Association of Textile Chemists and Colorists.)*

venting the fiber from returning to its original position in the fabric. The result is the shrinkage, or felting, of the cloth.

Movement of the fibers is speeded up and felting occurs more rapidly under extreme or severe conditions. A wool garment can be shrunk down to half its original size. Lamb's wool will felt more readily than other wool. In soft, fluffy fabrics the fibers are not firmly held in position and are free to move, so these fabrics are more susceptible to felting than are the firmly woven worsteds. The felting property is an advantage in making felt fabric directly from fibers without spinning or weaving. The felting property is a disadvantage because it makes the laundering of wool more difficult. Treatments to prevent felting shrinkage (see Chapter 34) are based on the principle of smoothing off the rough edges of the scales.

Fulling, or *milling,* is a cloth-finishing process in which the cloth is washed in a thick soap solution and squeezed by wooden rollers to shrink the cloth and close up the weave by bringing the yarns closer together. After fulling, the cloth has more body and cover, as shown in Figure 6–8. The shrinking is dependent on the action of heat and moisture on the molecule structure and also on the scale structure. *Fulling, the process, should not be confused with felting, the property.*

Fig. 6–8 *Wool cloth before* (left) *and after* (right) *fulling.*

CHEMICAL COMPOSITION AND MOLECULAR ARRANGEMENT

Wool fiber is a protein called *keratin*. It is the same protein that is found in human hair, fingernails, horns, and hooves. Keratin is made up of carbon, hydrogen, oxygen, nitrogen, and sulfur. These combine to form over 17 different amino acids. Five amino acids are shown in Figure 6–9. The individual wool molecule consists of flexible molecular chains held together by natural crosslinks—cystine (or sulfur) linkages and salt bridges.

Figure 6–9 somewhat resembles a ladder, with the crosslinks analogous to the cross bars of the ladder. This simple structure can be useful in understanding some of the properties of wool. Imagine a ladder made of plastic that is pulled askew. When wool is pulled, its inherent tendency is to recover is original shape; the crosslinks are very important in this recovery. However, if the crosslinks are damaged, the structure is destroyed and recovery cannot occur.

A more-realistic model of the structure of wool molecules would have this ladder-like structure alternating with a helical structure. About 40 percent of the chains are in a spiral formation, with hydrogen bonding occurring between the closer parts. The ladder-like formation occurs where cystine crosslinks are or where other bulky amino acids mean the chains cannot pack closely together. The spiral formation works like a spring and is also important in the resilience, elongation, and elastic recovery of wool fibers. Figure 6–10 shows the helical structure of wool.

The cystine linkage is the most important part of the molecule. Any chemical, such as alkali, that damages this linkage can destroy the entire structure. In controlled reactions, the linkage can be broken and then reformed. Minor modifications of the cystine linkage that result from ironing and steaming have a beneficial effect; those from careless washing and exposure to light have a detrimental effect.

Shaping of Wool Fabrics. Wool fabrics can be shaped by heat and moisture—a definite aid in tailoring. Puckers can be pressed out; excess fabric can be eased and then pressed flat or rounded as desired. Pleats can be pressed into wool cloth with heat, steam, and pressure, but they will not last through washing.

Hydrogen bonds are broken by moisture and heat so the wool structure can be re-shaped by the mechanical action of the iron or press. Simultaneously, the heat dries the wool and new hydrogen bonds are formed in the wool structure as the water escapes as steam. The new hydrogen bonds maintain the wool in the new shape so long as the humidity is low. In high humidity or if the wool is dampened with water, the new hydrogen bonds are broken and the molecular structure reverts to its former shape. This is why garments shaped by ironing lose their creases or flatness and show relaxation shrinkage on wetting.— "What Happens When Setting Wool," Textile Industries, *130 (October 1966): 344.*

Permanent "set" can be achieved in much the same way and by chemicals similar to those used in the permanent waving of hair. The Si-Ro-Set finish, developed in Australia, uses the chemical ammonium thyglycollate. The fabric is sprayed or soaked with the chemical and then set as pleats or the like by steaming or steam pressing for a required period of time. During setting, the cystine linkage splits between the two sulfur atoms and new linkages are formed.

Wool fabrics are *subject to shrinkage*. The somewhat amorphous molecular structure of wool permits water molecules to penetrate, and,

Summary of the Performance of Wool in Apparel Fabrics

AESTHETIC	*VARIABLE*
Luster	Matte
DURABILITY	*HIGH*
Abrasion resistance	Moderate
Tenacity	Low
Elongation	High
COMFORT	*HIGH*
Absorbency	High
Thermal retention	High
APPEARANCE RETENTION	*HIGH*
Resiliency	High
Dimensional stability	Low
Elastic recovery	Excellent
RECOMMENDED CARE	*DRY CLEAN*

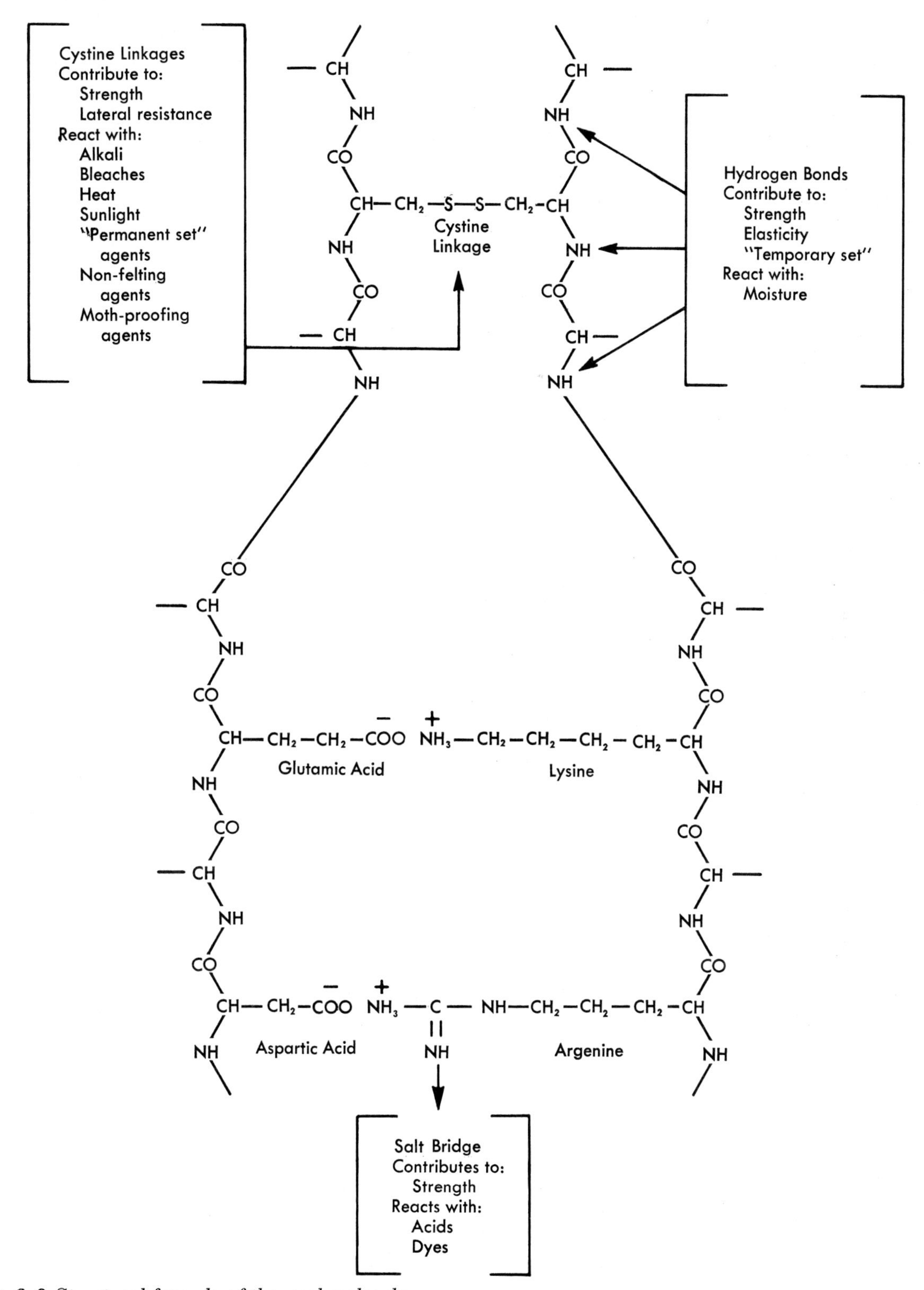

Fig. 6–9 Structural formula of the wool molecule.

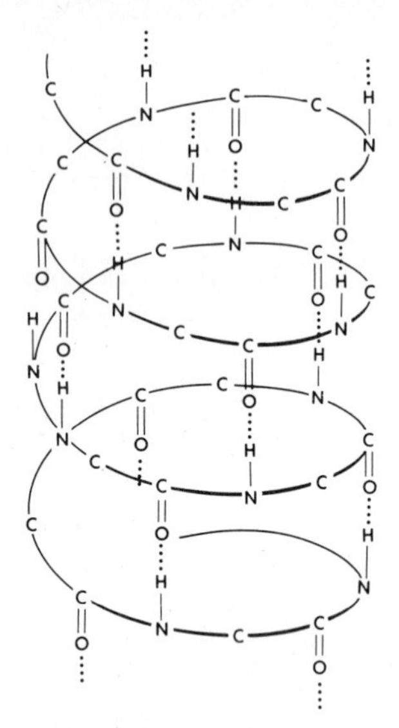

Fig. 6–10 Helical arrangement of the wool molecule. (Courtesy of International Wool Secretariat.)

when they do, the wool fiber swells and the molecular chains can be easily deformed.

Another finish that minimizes the effect of moisture in laundering is the Wurlan finish, developed by the Western Utilization Research and Development Laboratory at Albany, California. A polyamide-type solution, which forms a microscopic film on the surface of the scales and masks them, is applied.

PROPERTIES

One slogan the Wool Bureau used to promote wool in 1985 was "Wool . . . Good Looks That Last." Wool garments look good, wear well, and are comfortable.

Aesthetic. Wool, because of its physical structure, contributes loft and body to fabrics. Wool sweaters and men's and women's suit fabrics are the standard "looks" by which similar man-made fiber fabrics are measured. (See Figure 6–11.)

Wool has a matte appearance. Fibers are sometimes blended with wool from sheep that produce longer fibers or with specialty hair fibers such as mohair to make more-lustrous fabrics.

Drape, texture, and hand can be varied by choice of yarn structure, fabric structure, and finish. Sheer-wool scarves, medium-weight drapey printed-wool challis, medium-weight suit fabrics for summer, heavyweight suit fabrics for winter, and very heavy coat fabrics demonstrate the spectrum of possibilities. No wonder designers love to work with wool!

Durability. Wool fabrics are very durable. They have moderate abrasion resistance due to the scale structure of the fiber and its flexibility. The flexibility of wool is excellent. Wool fibers can be bent back on themselves 20,000 times without breaking, as compared to 3,000 times for cotton and 75 times for rayon. Atmospheric moisture helps wool to retain its flexibility. Wool carpets, for example, become brittle if the air is too dry. The crimp and scale structure of wool fibers make them very cohesive so they cling together to make strong yarns.

Wool fibers have a low tenacity. When they are pulled lengthwise until they break, they exhibit a tenacity of 1.5 g/d dry and 1.0 g/d wet. The durability of wool is the result of the excellent elongation (25 percent) and elastic recovery

Fig. 6–11 Man's business suit made from wool. (Courtesy of Gant Clothing.)

(99 percent) of the fibers. When stress is put on the fabric, the crimped fibers elongate, then the molecular chains unfold. When stress is removed, the crosslinks pull the fibers back almost to their original positions. The combination of these properties, excellent flexibility, elongation, and elastic recovery, results in wool fabrics that can be used and enjoyed for many years.

Comfort. Wool is more hygroscopic than any other fiber. It has a moisture regain of from 13–18 percent under standard conditions. All animal fibers are superior to other fibers in that they absorb moisture without surface wetting; they are *hygroscopic*. This phenomenon has long been recognized as a major factor in understanding why garments made from protein fibers are so comfortable to wear. Hygroscopic fibers minimize sudden temperature changes at the skin. This is illustrated by the difference in warmth between an all-polyester suit and an all-wool suit. In the winter, when people go from a dry indoor atmosphere into the damp outdoor air, the wool fibers absorb moisture and generate heat, protecting the wearers from the cold.

Outdoor-sports enthusiasts have long recognized the superior comfort provided by wool. After a period of activity, no sudden chilling occurs when exercise stops because the wool fiber absorbs moisture and releases heat. The process of the wool slowly drying and the moisture evaporating into the air occurs slowly enough that the wearer is more comfortable than in any other fiber.

Wool fibers are initially water repellent. In a light rain or snow, the water will run off or remain on the fabric surface. In a heavier rain, wool will absorb a lot of moisture without feeling wet. Eventually wool will absorb enough moisture so that it will feel wet and heavy.

Wool is a poor conductor of heat so that warmth from the body is not dissipated readily. Wool's excellent resiliency is important in providing warmth. The wool fibers can recover from crushing and the fabrics will remain porous and capable of incorporating much air. "Still" air is one of the best insulators because it keeps body heat close to the body.

Coarse-wool fabrics are often irritating to the skin. Some people are allergic to the chemical components of wool and itch or break out in a rash or sneeze when they wear or handle wool. Fabrics that seem scratchy may be a low-quality wool. Feel them and try them on before purchasing. Wear them layered with a smooth fabric underneath them or try some better-quality fabrics.

Wool has a medium density (1.32 g/cc). People often associate heavy fabrics with wool since it is used in fall and winter wear when the additional warmth given by heavy fabrics is desirable. Lightweight wools are very comfortable in the changeable temperatures of spring and early fall.

One way to compare fiber densities is to think of blankets. A winter blanket of wool is heavy and warm. An equally thick blanket of cotton would be even heavier (cotton has a higher density), but not as warm. A winter blanket of acrylic would be lighter in weight (acrylic has a lower density than either). Personal preferences, room temperature, and humidity need to be considered before deciding which fiber content would be more comfortable.

Appearance Retention. Wool is a very resilient fiber. It resists wrinkling and recovers well from wrinkles. It wrinkles more readily when wet. Wool maintains its shape fairly well during normal use. Women's slacks and straight skirts should be lined to prevent baggy knees or saggy seats. Men's worsted suiting fabrics are designed to retain their shape well.

When wool fabrics are dry cleaned, they retain their size and shape well. When wool sweaters are hand washed, they need to be cared for properly to avoid shrinking. Follow care instructions for washable woolens.

Wool has an excellent elastic recovery—99 percent at 2 percent elongation. Even at 20 percent elongation, recovery is 63 percent. Recovery is excellent from the stresses of normal usage. The fabric regains its shape best when allowed to rest at least 24 hours between wearings.

Care. Wool does not soil readily, and the removal of soil from wool is relatively simple. Grease and oils do not spot wool fabrics as readily as they do fabrics made of other fibers. (Wool in its natural state is about 25 percent grease). Wool garments do not need to be washed or dry cleaned after every wearing. They do not wrinkle very much. They can be spot treated. Layer them with a washable shirt to decrease odor pickup.

Use a good brush on collars and the inside of cuffs after each wearing. A firm, soft brush not only removes dust but also gently lifts the fibers back to their natural springiness. Damp fabrics should be allowed to dry before brushing. Garments should have a period of rest between wearings to recover from deformations. Baggy elbows and skirt seats will become less baggy as the garment rests. Hanging the garment over a tub of hot, steamy water or spraying a fine mist of water on the cloth will speed up recovery.

Wool is very weak when it is wet. Its wet tenacity is 1.0 g/d, a third lower than its relatively low dry strength. Wet elongation increases to 35 percent before breaking. Resiliency and elastic recovery decrease when wool is wet. The redeeming properties of dry wool that make it durable in spite of its low tenacity do not operate when it is wet. Wet wool is weak; handle it very gently.

Dry cleaning is the recommended method of caring for wool garments. Dry cleaning minimizes the potential problems that may occur during hand or machine washing. The cost of dry cleaning may be viewed as a disadvantage and needs to be evaluated on an individual basis. The dry cleaner removes spots and stains, does minor repairs, cleans the garment, steam presses it, and can also store it. For many people, the convenience of these services is worth the cost. Incorrect care procedures can be disastrous and costly; the garment can be ruined so it can no longer be worn.

Follow care instructions. Dry cleaning is always appropriate for wool garments. Knit sweaters can be dry cleaned or they can be hand washed if correct procedures are followed. When hand washing, warm water that is comfortable to the hand is recommended. Avoid agitation; squeeze gently. Support the garment, especially if it is knit, so it does not stretch unnecessarily. Fabrics and garments, woven or knit, that are labeled machine washable are usually fiber blends or have been given a special finish so they can be laundered safely. Follow any special instructions given. These special instructions usually include warm or lukewarm water and a gentle cycle for a short period of time, with line drying or flat drying.

Chlorine bleach will damage wool. Verify this by putting a small piece of wool in fresh chlorine bleach (bleach gets weaker with age). What happens? The wool dissolves! This can be used as a simple test to identify protein fibers. Chlorine bleach is an oxidizing agent. Wool is also very sensitive to the action of alkali. The wool reacts to the alkali by turning yellow, then becoming slick and jelly-like, and finally going into solution. If the fabric is a blend, the wool in the blend will disintegrate, leaving only the other fibers.

Avoid using the dryer for wool clothing or blankets. If care instructions state to machine dry, use a low-temperature setting and remove the garment while it is still slightly damp to the touch, letting it finish drying flat.

Following these care procedures will prevent extreme shrinkage. Under conditions of heat, moisture, agitation, and pressure, wool will felt—the fibers will shorten and their scales will lock together in an irreversible process. If an adult-sized sweater is carelessly machine washed for an average length of time with regular agitation and then dried in a hot dryer, it will be child-sized and very stiff when it is removed.

Use care in ironing wool apparel. Wool becomes weak and harsh at elevated temperatures, and it scorches readily. Wool fabrics should always be pressed with moist heat.

If wool fabrics become shiny from pressure, sponge with a 5 percent solution of white vinegar. Then steam, and the fibers will swell and become fluffier. If surface fibers wear off, use fine sandpaper to restore the nap. Fine sandpaper will also remove *light* scorch.

Wool is attacked by moth larvae and other insects. The most effective way to prevent moth damage to wool is to alter the molecular structure of the fiber. The mothproofing process consists of chemically breaking the cystine linkage (—S—S—) and reforming it as $-S-CH_2-S-$ (see Chapter 34).

Unless mothproofed, wool fabrics should be stored so that they will not be accessible to moths. Moth larvae will also eat, but not digest, any fiber that is blended with wool. Wool fabrics should be cleaned before storage and should be stored with moth crystals in a closed container.

Wool burns very slowly and is self-extinguishing. It is normally regarded as flame-resistant. When used for curtains, carpets, and upholstery in trains, planes, ships, hotels, and other public buildings, wool is often given a flame-retardant finish.

When wool burns, it gives off only moderate

amounts of smoke and carbon monoxide and is a minor impediment to evacuation from the site of a fire as compared to other materials that are likely to be present. The temporary resistance to burning of wool may be increased by the use of phosphates or borates. Durable fire-retardant effects can be obtained by the use of small amounts of a variety of protective compounds with little detectable change in the physical or chemical behavior of the wool fiber.

USES

Only a small amount of wool is used in the United States. Domestic consumption of wool was 195 million pounds, or 1.7 percent of all fiber used in the United States in 1985. A greater amount of wool is imported than is produced domestically.

The majority of wool (72.8 percent) is used in apparel. Home furnishings account for 15.4 percent, industrial uses 6.7 percent, and exports 5 percent. Wool accounts for 3.3 percent of all fibers used for apparel.

The most important use of wool is for adult apparel: coats, jackets, suits, dresses, skirts, and slacks made from woven fabrics of varying weights; and suits, dresses, skirts, and sweaters made from knitted fabrics. In the early 1980s, there was a trend towards greater consumption of women's suits containing wool. Manufacturers tried to develop a greater variety of women's suiting fabrics to meet this demand.

Uses of Wool in 1985

Percent	*End Use*	*Million Pounds*
72.8	Apparel	
	Top weight	2.7
	Bottom weight	105.7
	Underwear and nightwear	3.9
	Sweaters	16.8
	Retail piece goods	2.9
	Socks	2.7
	Hand-knitting yarns	4.3
15.4	Home Furnishing	
	Carpets	11.9
	Upholstery	13.6
	Blankets	4.1
6.7	Industrial	
	Felts	12.1

Source: Textile Organon, 57 (9), Sept. 1986.

Performance standards are stricter for menswear than for womenswear. Wool suits perform very well and look great. They fit well because they can be shaped through tailoring. The fabrics hang well. The suits last a long time. They are very comfortable under a variety of conditions and retain their good looks during wear and care. Suits are typically dry cleaned to retain their best looks and because of the shaping components.

Blends of different synthetic fibers with wool for suiting materials are increasingly important. They result in fabrics that are more appropriate in warmer conditions. Polyester is the most important fiber used. Varying percentages of wool and polyester are used. Other fibers may also be used; sometimes two or three synthetics are blended with wool.

Wool is important in knitwear—especially sweaters. Wool is also used in retail piece goods, socks, and hand-knitting yarns.

In 1964, the Wool Bureau adopted the Woolmark as a symbol of quality to be used on all merchandise that meets the Wool Bureau's specifications for quality. In 1970—recognizing the increasing use of blends—the Wool Bureau adopted a Woolblend mark for blends with at least 60 percent wool. The Woolmark and Woolblend symbols are shown in Figure 6–12.

In the home-furnishing area, the major use of wool is in carpets and rugs. Many woven Axminister and Wilton rugs with Persian-type designs are made from wool. More contemporary looks are also available. Most are imported, although some are made in the United States. Wool rugs are more expensive than nylon carpets, but people who prefer them like the pat-

Fig. 6–12 *Woolmark® and Woolblend® symbols of quality. (Courtesy of the Wool Bureau, Inc.)*

terns and appreciate the color, texture, and appearance of wool. Wool rugs account for a very small share of the rug market. Wool is also found in upholstery fabrics and blankets.

In industrial uses, wool is important in felts. These are used under heavy machinery to help decrease noise or for a variety of other uses.

Specialty Hair Fibers

Most specialty wools are obtained from the goat, rabbit, and camel families.

Specialty wools are available in smaller quantities than sheep's wool so they are usually more expensive. Like all natural fibers, wools vary in quality.

Specialty wool fibers are of two kinds: the coarse long outerhair and the soft, fine undercoat. Coarse fibers are used for interlinings, upholstery, and some coatings; the very fine fibers are used in luxury coatings, sweaters, shawls, suits, and dress fabrics.

MOHAIR

Mohair is the hair fiber of the Angora goat. In 1985, 41.4 million pounds of mohair were produced worldwide. South Africa produced the most (38 percent), followed by Turkey (29 percent) and the United States (24 percent). Texas is the major producer in the United States. Most U.S. mohair (80 percent of it) is exported to the United Kingdom, with smaller amounts going to South Africa and Italy. The goats (Figure 6–13) are sheared twice a year, in the early fall and early spring. The fiber length is

- 4–6 inches for half a year
- 8–12 inches for a full year

Mohair fibers have a circular cross-section. Scales on the surface are scarcely visible and the cortical cells show through as lengthwise striations. There are some air ducts between the cells that give mohair its lightness and fluffiness. Few of the fibers have a medulla.

Mohair is one of the most resilient fibers and has none of the crimp found in sheep's wool, giving it a silk-like luster and a smoother surface that is more resistant to dust than wool. Mohair has fewer scales than wool, so mohair fibers are smoother than wool fibers. (Figure 6–14). Mohair is very strong and has good affinity for dye. The washed fleece is a lustrous white.

Chemical properties are the same as those of wool. Mohair makes a better novelty *loop* yarn than wool or the other specialty hair fibers.

Uses of Mohair

- Upholstery and draperies
- Men's suitings
- Bouclé coatings for women
- Pile fabrics; embossed and curled like fur
- Laces
- Wigs and hairpieces
- Oriental-type rugs

Figure 6–15 is the quality symbol used on all mohair products that meet performance standards established by the Mohair Council.

QIVIUT

Qiviut, a rare and luxurious fiber, is the underwool of the domesticated musk ox. See Fig. 6–16. Successful domestication projects have been conducted in Alaska. A large musk ox will provide 6 pounds of wool each year. The fiber can be used just as it comes from the animal, for it is protected from debris by the long guard hairs and has a low lanolin content. The fleece is not shorn but is shed naturally and is removed from the

Goat Family	*Camel Family*	*Others*
Angora goat—mohair	Camel's hair	Angora rabbit—angora
Cashmere goat—cashmere	Llama	Fur fibers
	Alpaca	Musk ox—qiviut
	Vicuña	
	Guanaco	

Fig. 6–13 Angora goats produce mohair fibers. (Courtesy of the Mohair Council of America.)

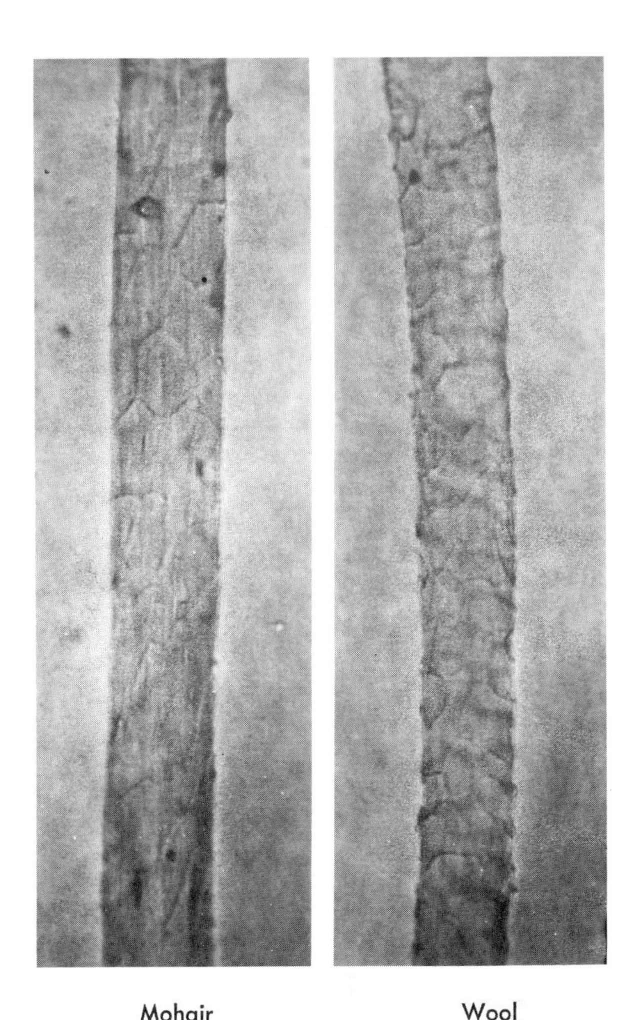

Fig. 6–14 Microscopic view of mohair and wool. (Courtesy of U.S.D.A., Livestock Division, Wool and Mohair Lab.)

Fig. 6–15 Mohair symbol of quality.

guard hairs as soon as it becomes visible. In 1984, qiviut was $125 an ounce.

Eskimo women hand knit with qiviut. Their first products were lacy scarves with designs taken from the Eskimo artifacts, and each pattern is identified with a particular village.

ANGORA

Angora is the hair on the angora rabbit, which is raised in France and in small amounts in the United States. See Figure 6–17. Each rabbit produces only a few ounces of fiber, which is very fine, fluffy, soft, slippery, and fairly long. It is pure white in color.

CAMEL'S HAIR

Camel's hair is obtained from the two-hump Bactrian camel. These camels are found in the area of Asia from Turkey east to China and north to Siberia. Camel's hair is said to have the

Fig. 6–16 The musk ox of Alaska produces qiviut fiber. (Courtesy of Fairbanks Convention and Visitor Bureau.)

Fig. 6–17 The Angora rabbit produces a soft, white fiber. (Courtesy of Mary Goodwin.)

best insulation of any of the wool fibers, since it keeps the camel comfortable under extreme conditions of temperature during a day's journey through the cold mountain passes and the hot valleys. The hair is collected by a "trailer" who follows the camel caravan and picks up the hair as it is shed and places it in a basket carried by the last camel. The trailer also gathers the hair in the morning, at the spot where the camels lay down for the night. A camel produces about 5 pounds of hair a year.

Because the camel's hair gives warmth without weight, the finer fibers are much prized for clothing fabrics. They are often used in blends with sheep's wool, which is dyed the tan color of camel's hair.

There are so many qualities of cashmere and camel's hair that care should be taken by consumers to determine the quality of fiber they are buying. The best way to judge the quality is by the feel. Consumers should be guided by the reputation of the manufacturer or retailer.

CASHMERE

Cashmere comes from a small goat raised in Kashmir, China, Tibet, and Mongolia. The fibers vary in color from white to gray to brownish gray. The goat has an outercoat of long, coarse hair and an innercoat of down. The hair is combed by hand from the animal during the molting season, and care is taken to separate the coarse hair from the fine fibers. Only a small part of the fleece is the very fine fiber, probably not more than one-quarter pound per goat. Cashmere is used in high-quality apparel, especially women's sweaters and coatings. Fabrics are warm, buttery in hand, and have beautiful draping characteristics. Cashmere is more sensitive to chemicals than wool.

LLAMA AND ALPACA

Llama and *alpaca* are domesticated animals of the South American branch of the camel family. The fiber is 8–12 inches in length and is noted for its softness, fineness, and luster. The natural colors are white, light fawn, light brown, dark brown, gray, black, and piebald.

VICUÑA AND GUANACO

Vicuña and *guanaco* are wild animals of the South American camel family. They are very rare, and the animals must be killed to obtain the fiber. The governments of countries where vicuña and guanaco are found have limited the number of animals that can be harvested each year in order to protect the herds from extinction. Vicuña is the softest, finest, rarest, and most expensive of all textile fibers. The fiber is short, very lustrous, and a light-cinnamon color.

Azlon

Man-made protein fibers, *azlon,* are made by dissolving and resolidifying protein substances from animal or grain sources. They are no longer made in the United States. In the 1940s and 1950s, Aralac, made from milk casein, and Vicara, made from the zein of corn, were produced, but these fibers were not successful because they were too weak to be used alone and too expensive to compete with other blending fibers, particularly rayon and acetate.

Chinon is an azlon fiber that is presently manufactured in Japan and imported to the United States. It is made from milk casein combined with acrylic resins. Its characteristics are silk-like; it is used in scarves, ties, blouses, and sweaters. It is also used in the pharmaceutical industry.

7

Silk

Silk is a natural protein fiber. It is similar to wool in that it is composed of amino acids arranged in a polypeptide chain. Silk is produced by the larvae of a moth, while wool is produced by animals. All protein fibers have some general characteristics in common, such as their sensitivity to oxidizing agents and alkalis and their weaker strength when wet. Some characteristics differ significantly between the two fibers, such as the natural luster and uncrimped nature of silk.

Silk culture, according to Chinese legend, began in 2640 B.C. when a Chinese Empress Hsi Ling Shi became interested in silkworms and learned how to reel the silk and make it into fabric. It was through her efforts that China developed a silk industry that was monopolized for 3,000 years. Silk culture spread to Korea and Japan, westward to India and Persia, and then finally to Spain, France, and Italy. Silk fabrics, imported from China, were coveted by other countries; in India, the fabrics were often picked out and rewoven into looser fabrics or combined with linen to provide more yardage from the same amount of silk filament. Major producers of silk are Japan, Thailand, and the People's Republic of China.

Silk is universally accepted as a luxury fiber. The International Silk Association of the United States emphasizes the uniqueness of silk by its slogan "Only silk is silk." Silk has a unique combination of properties not possessed by any other fiber:

- "Dry" tactile hand
- Natural luster
- Good moisture absorption
- Lively suppleness and draping qualities
- High strength

The beauty and hand of silk and its high cost are probably responsible for the man-made fiber industry. In the early days, there was no scarcity of other natural fibers and thus no need to try to duplicate them. Silk is a solid fiber with a simple physical structure. It is this physical nature of silk that some modifications of man-made fibers (rayon and acetate) and synthetic fibers (nylon and polyester) attempt to duplicate. Man-made fibers with a triangular cross-section are the most successful.

PRODUCTION

Sericulture is the name given to the production of cultivated silk, which begins with the silk moth laying eggs on specially prepared paper. The cultivated silkworm is *Bombyx mori.* When the eggs hatch, the caterpillars, or larvae, are fed fresh, young mulberry leaves. After about 35 days and 4 moltings, the silkworms are approximately 10,000 times heavier than when born and ready to begin spinning a cocoon, or chrysalis case. A straw frame is placed on the tray and the silkworm starts to spin the cocoon by moving its head in a figure-eight (see Figure 7–1). The silkworm produces silk in two glands and forces the liquid silk through openings, called *spinnerets,* in its head. The two strands of silk are coated with a water soluble, protective gum called *sericin.* When the silk comes in contact with the air, it solidifies. In 2 or 3 days, the silkworm has spun approximately 1 mile of filament and has completely surrounded itself in a cocoon. The silkworm then begins to change into a chrysalis and then into a moth. Usually the silkworm is killed (stifled) with heat before it reaches the moth stage. If the silkworm is allowed to reach the moth stage, it will be used for breeding additional silkworms. The moth secretes a fluid that dissolves the silk at one end of the cocoon, permitting the moth to crawl out. Cocoons from which the moth has emerged cannot be used for filament silk yarns and the silk from these cocoons is not as valuable.

To obtain filament silk from the cocoon after the silkworm has been stifled, the cocoons are brushed to find the outside ends of the filaments.

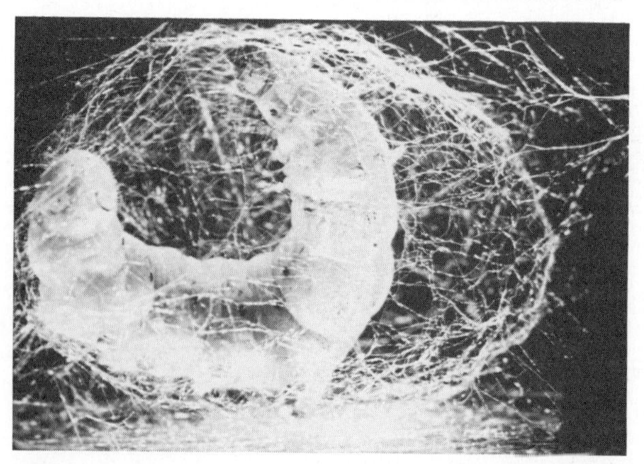

Fig. 7–1 *Silk caterpillar spinning silk fibers to form cocoon. (Courtesy of Stock, Boston. © Cary Wolinsky, 1984.)*

Several filaments are gathered together and wound onto a reel. This process is referred to as *reeling,* and it is performed in a manufacturing plant called a *filature.* Each cocoon yields approximately 1,000 yards of silk filament. This is *raw silk,* or *silk-in-the-gum.* Several filaments are combined to form a yarn. The operators in the filature must be careful to join the fibers so that the diameter of the reeled silk remains uniform in size. Uniformly reeled filament silk is the most valuable (see Figure 7–2).

As the fibers are combined and wrapped onto the reel, twist can be added to hold the filaments together. Adding twist is referred to as *throwing,* and the resulting yarn is called a *thrown yarn.* There are several types of thrown yarns. The type of yarn and amount of twist relate to the type of fabric desired. The simplest type of thrown yarn is a singles. In a *singles,* three to eight filaments are twisted together to form a yarn. Commonly used for filling yarns in many silk fabrics, singles may have two to three twists per inch.

Much usable silk is not reeled because long filaments cannot be taken from a damaged cocoon. Cocoons where the filament broke or where the moth was allowed to mature, and silk from the inner portions of the cocoon yield *silk noils,* or *silk waste.* This silk is degummed and spun as any other staple fiber or blended with another staple fiber and spun into a yarn. Spun silk is less expensive than filament silk.

Wild silk production is not controlled as it is for cultivated silk. Although many species of wild silkworms produce wild silk, the two most common species are *Antheraea mylitta* and *Antheraea pernyi.* The silkworms feed on oak and cherry leaves and produce fibers that are much less uniform in texture and color. The fiber may be brown, yellow, orange, or green, with brown the most common color for wild silk. Since the cocoons are harvested after the moth has matured, the silk cannot be reeled and must be used as spun silk. Wild silk is also referred to as *tussah silk.* It is coarser, darker, and cannot be bleached. Hence, white and light colors are not available in tussah silk.

Duppioni silk results when two silkworms spin their cocoons together. The yarn is irregular in diameter, with a thick-thin appearance, and used in linen-like silk fabrics for apparel.

Fig. 7–2 *Reeling of silk. (Courtesy of Stock, Boston. Photographer: Ira Kirschenbaum.)*

FIBER PROPERTIES

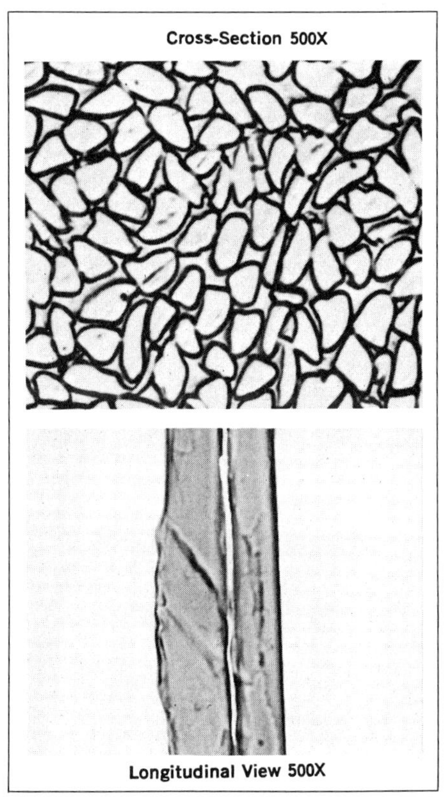

Fig. 7–3 *Photomicrographs of silk fiber: cross-sectional view* (top); *longitudinal view* (bottom). *(Courtesy of American Association of Textile Chemists and Colorists.)*

Summary of the Performance of Silk in Apparel Fabrics

AESTHETIC	*VARIABLE*
Luster	Beautiful and soft
DURABILITY	*HIGH*
Abrasion resistance	Moderate
Tenacity	High for natural fibers
Elongation	Moderate
COMFORT	*HIGH*
Absorbency	High
Thermal retention	Good
APPEARANCE RETENTION	*MODERATE*
Resiliency	Moderate
Dimensional stability	High
Elastic recovery	Moderate
RECOMMENDED CARE	*DRY CLEAN*

PHYSICAL STRUCTURE

Silk is a natural continuous-filament fiber. It is a solid fiber, smooth but irregular in diameter along its shaft. The filaments are triangular in cross-section with rounded corners (Figure 7–3). Silk fibers are very fine—1.25 denier/filament. Tussah silk may have slight striations along the longitudinal length of the fiber.

CHEMICAL COMPOSITION AND MOLECULAR STRUCTURE

The protein in silk is *fibroin,* which contains 15 amino acids in polypeptide chains. Silk has reactive amino (NH_2) and carboxyl (COOH) groups. Silk has no crosslinkages and no bulky side chains. The molecular chains are not folded as in wool, but are almost fully extended and packed closely together. Thus silk is highly oriented, which gives the fiber its strength. As with all fibers, there are some amorphous areas between the crystalline areas, giving silk its elasticity.

PROPERTIES

Aesthetic. Silk can be dyed and printed in brilliant colors. It is adaptable to a variety of fabrication methods. Thus it is available in a wide variety of fabric types. Because of cultivated silk's smooth but slightly irregular surface and triangular cross-section, the luster of this fiber is soft with an occasional sparkle. It is this luster that has been the model for many man-made fibers. Fabrics made of cultivated silk usually have a smooth appearance and a luxurious hand.

Tussah silk has a duller luster because of its coarser size, less-regular surface, and presence of sericin. Fabrics made of tussah silk have a more-pronounced texture.

In filament form, silk does not have good covering power. Before the development of strong synthetic fibers, silk was the only strong filament, and silk fabrics were often treated with metallic salts such as tin, a process called *weighting,* to give the fabric better drape, covering power, and dye absorption. Silk has *scroop,* a natural rustle, which can be increased by treating with an organic acid such as acetic or tartaric acid.

Durability. Silk has moderate abrasion resistance. Because of its end uses and cost, silk seldom receives harsh abrasion.

Silk is one of the strongest natural fibers, with a tenacity ranging from 3.5–5.0 g/d dry. It may lose up to 20 percent of its strength when wet. Its strength is excellent in relationship to its fineness.

Silk has a breaking elongation of 20 percent. It is not as elastic as wool because there are no crosslinkages to pull back the molecular chains. When silk is elongated by 2 percent, its elasticity is only 90 percent. Thus, when silk is stretched a small amount, it does not return to its original length, but remains slightly stretched. When silk is used in apparel, this poor elasticity may be seen in baggy elbows in blouses.

Comfort. Silk has good absorbency with a moisture regain of 11 percent. Silk is a poor conductor of electricity; thus problems may develop with static cling. Silk fabrics are comfortable in summer in skin-contact apparel. Silk, like wool, is a poor conductor of heat so that silk scarves and raw-silk suitings are comfortably warm in the winter. The weight of a fabric is important in heat conductivity—sheer fabrics, possible with filament silk, will be cool whereas heavy-suiting fabrics will be warm.

Silk is smooth and soft and thus not irritating to the skin.

The density of silk is 1.25 g/cc, which gives strength and light weight to silk products. Weighted silk is not as durable as regular silk, and it wrinkles more readily. In 1932, the Federal Trade Commission ruled that anything labled pure silk or pure dye silk could contain no more than 15 percent weighting for black and 10 percent weighting for all other colors. Anything exceeding these levels is weighted silk. At present, very little silk is weighted.

Appearance Retention. Silk has moderate resistance to wrinkling. This is related to silk's elastic recovery. Because the fiber does not recover well from elongation, it does not resist wrinkling as well as some other fibers.

Silk fibers do not shrink. Because the molecular chains are not easily distorted, silk swells a small amount when wet. Fabrics made from true crepe yarns will shrink if laundered, but this is caused by the yarn structure, not the fiber content.

Care. Dry-cleaning solvents do not damage silk. In fact, dry cleaning often is recommended for silk items because of yarn structures, dyes that have poor fastness to water or laundering, or garment- or fabric-construction methods. Some silk items can be laundered in a mild detergent solution with gentle agitation. Since silk may lose up to 20 percent of it strength when wet, care should be taken with wet silks to avoid adding any unnecessary stress. Silk items should be pressed after laundering. Pure dye silks should be ironed damp with a press cloth. Wild silks should be dry cleaned and ironed dry to avoid losing sericin, which gives the fabric its body.

Silks may water-spot easily so care should be taken to avoid this problem. Before hand or machine washing, test in an obscure place of the item to make sure the dye does not water-spot.

Silk can be damaged and yellowed by strong soaps or detergents and high temperatures. Chlorine bleaches should be avoided. However, bleaches of hydrogen-peroxide and sodium-perborate are safe to use if the directions are followed carefully.

Silk is resistant to dilute mineral acids and organic acids, but it is damaged by strong alkaline solutions. A crepe-like surface effect may be created by the shrinking action of some acids.

Silk is weakened by exposure to sunlight and perspiration. Many dyes used to color silk are damaged by sunlight and perspiration. Silks tend to yellow with age and exposure to sunlight and chlorine bleach. Furnishing fabrics of silk should be protected from direct exposure to sunlight.

Silks may be attacked by insects, especially carpet beetles. Care should be taken when storing silks to be sure they are clean because soil may attract insects that do not normally attack silk.

Weighted silks deteriorate even under good storage conditions and are especially likely to break at folds. Historic items often exhibit a condition known as *shattered silk,* where the weighted silk is disintegrating. The process cannot be reversed.

Identification Tests. Silk burns like other protein fibers. Silk is soluble in sodium hydroxide; however, it dissolves more slowly than wool.

USES

Silk is used primarily in apparel and home-furnishing items because of its appearance and cost. Silk is extremely versatile and can be used to create a variety of fabrics from sheer, gossamer chiffons to heavy, beautiful brocades and velvets. Because of silk's absorbency, it is appropriate for warm-weather wear. Because of its low heat conductivity, it is also appropriate for cold-weather wear. In furnishings, silk is often blended with other fibers to add a soft luster to the furnishing fabric. Silk blends are often used in window-treatment and upholstery fabrics. Occasionally, beautiful and expensive handmade rugs will be made of silk.

8

Introduction to Man-Made Fibers

In 1664, Robert Hooke suggested that if a proper liquid were squeezed through a small aperture and allowed to congeal, a fiber like that of the silkworm might be produced. In 1889, the first successful fiber was made from a solution of cellulose by a Frenchman, Count de Chardonnet. In 1910, rayon fibers were commercially produced in the United States, and acetate was produced in 1924. By 1939, the first noncellulosic, or synthetic, fiber—nylon—was made. Since that time, many more generic fibers and modifications or variants of these generic fibers have appeared on the market. The increasing number of new fiber names appearing on labels can create a great deal of confusion for the consumer.

Man-Made Fibers Generic Names

Cellulosic	*Noncellulosic or Synthetic*		*Mineral*
Acetate	Acrylic	Nytril*	Glass
(Triacetate*)	Anidex*	Olefin	Metallic
Rayon	Aramid	PBI	
	Azlon*	Polyester	
	(Lastrile*)	Rubber	
	Modacrylic	Saran	
	Novoloid	Spandex	
	Nylon	Sulfar	
		Vinal*	
		Vinyon*	

*Not produced in the United States.

Legislation and Generic Names

In 1958, Congress passed legislation to regulate labeling of textiles in order to protect the consumer through the enforcement of ethical practices and to protect the producer from unfair competition resulting from the unrevealed presence of substitute materials in textile products. This law, the Textile Fiber Product Identification Law, covers *all* fibers except those already covered by the Wool Products Labeling Act, with certain other exceptions.

Although the law was passed in 1958, it did not become effective until 1960. During this interval, the Federal Trade Commission held hearings in regard to inequalities or injustices that the law might cause. Then it established rules and regulations to be observed in enforcing the law. The following list of man-made fiber generic names was established by the Federal Trade Commission in cooperation with the fiber producers. A *generic name* is the name of a family of fibers all having similar chemical composition. (Definitions of these generic names are included with the discussions of each fiber.)

The following information, in English, is required on the label for most textile items.

1. The percent of each natural or man-made fiber present must be listed in the order of predominance by weight. The percent listed must be correct within a tolerance of 3 percent. What that means is that if the label states a fiber content of 50 pecent cotton, the minimum can be no less than 47 percent and the maximum can be no more than 53 percent.

If a fiber or fibers represent less than 5 percent by weight, the fiber cannot be named unless it has a clearly established and definite functional significance. In those cases where the fiber has a definite function, the generic name, percentage by weight, and functional significance must be listed. For example, a garment that has a small amount of spandex, may have a label that reads "96% Nylon, 4% Spandex for elasticity."

2. The name of the manufacturer or the company's registered number such as WPL or RN. In many cases, the company's registered number is listed with the letters and the number. (Trademarks may serve as identification, but they are not required information. Often a trademark is listed with the generic fiber name.)

3. The first time a trademark appears in the required information, it must appear in immediate conjunction with the generic name and in type or lettering of equal size and conspicuousness. When the trademark is used elsewhere on the label, the generic name shall accompany it in legible and conspicuous type the first time it appears.

4. The name of the country where the product was processed or manufactured, such as "Made in USA." If the item was manufactured in the United States from imported fabric, the label might read "Made in USA of imported fabric."

Trade Names

An experimental fiber may be given a *trade name* (trademark), which distinguishes the fiber from other fibers of the same generic family that are made and sold by other producers. A producer may adopt a single trade name, word, or symbol, which may be used to cover all (or a large group) of the fibers made by that company. For example, "Orlon" is no longer used to designate a single acrylic fiber made by Du Pont, but is a broad descriptive name covering a family of related Du Pont acrylic fibers, each of which is sold to the manufacturer by type number. Trade names are also usually protected by a quality-control program.

The fiber producer must assume all the responsibility for promoting the fiber. The company must sell not only to its customers, the manufacturers, and retailers, but to the customer's customer—the consumer.

Fiber Spinning

It took many years to develop the first spinning solutions and devise spinnerets to convert the solutions into filaments. The first solutions were made by treating cellulose so it would dissolve in certain substances. It was not until the 1920s and 1930s that we first learned how to build long-chain molecules from simple substances.

All man-made fiber spinning processes are based on these three general steps.

1. Preparing a viscous or syrupy dope.
2. Extruding the dope through a spinneret to form a fiber.
3. Solidifying the fiber by coagulation, evaporation, or cooling.

The *raw material* may be a natural product such as cellulose or protein, or it may be chemicals that are synthesized into resins. These raw materials are made into solutions by dissolving them with chemicals or by melting. The solution is referred to as the *spinning solution,* or *dope.*

Extrusion is a very important part of the spinning process. It consists of forcing or pumping the spinning solution through the tiny holes of a spinneret.

A *spinneret* is a small thimble-like nozzle (Figure 8–1). Rayon is spun through a spinneret that is made of platinum—one of the few metals that will withstand the action of acids and alkalies. Acetate and other fibers are extruded through stainless-steel spinnerets. Spinnerets are costly—as much as $1,000 each—and new developments are closely guarded secrets. The making of the tiny holes is the critical part of the process. Fine hair-like instruments or laser beams are used. Ordinarily the holes are round, but many other shapes are used for special fiber types (see page 8).

Each hole in the spinneret forms one fiber. *Filament fibers* are spun from spinnerets with 350 holes or less. Together these fibers make a filament yarn. *Filament tow* is an untwisted rope of thousands of fibers. This rope is made by putting together the fibers from 100 or more spinnerets, each of which may have as many as three thousand holes (Figure 8–2). This large rope of fibers is crimped and is then ready to be made into staple by cutting to the desired length. (See Chapter 19 for methods of breaking filament tow into staple.)

Spinning Methods. Spinning is done by five different methods. These methods are compared briefly in Figure 8–3. Details of the methods are given in later chapters.

Man-made fibers are produced to satisfy a market or supply a special need. The first man-made fibers made it possible for the consumer to

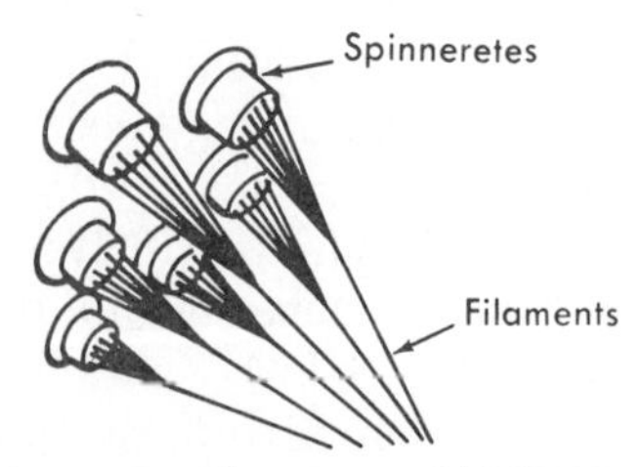

Fig. 8–1 *Spinnerets. (Courtesy of AVTEX Fibers, Inc.)*

Fig. 8–2 Collecting fibers from several spinnerets to make a rope, called filament tow, which will be cut into staple fiber.

have silk-like fabrics at low cost. The synthetics gave the consumer fabrics with improved properties unlike any natural fiber fabric.

The production program for a new fiber is long and expensive, and millions of dollars are invested before any profit can be realized. First, a research program is planned to develop the new fiber. Then a pilot plant is built to scale-up laboratory procedures to commercial production. This pilot plant may produce as much as 5 million pounds of fiber, which is used to test and evaluate the fiber and to determine and evaluate end uses. When the fiber is ready, a commercial plant is built.

A patent on the process gives the producer 17 years of exclusive right to the use of the process—time to recover the initial cost and make a profit. The price per pound during this time is high, but it drops later. The patent owner can license other producers to use the process. Continuing research and developmental programs correct any problems that arise and produce new fiber types modified for special end uses.

Common Fiber Modifications

One advantage of the man-made fibers is that each step of the production process can be precisely controlled to "tailor," or modify, the parent fiber. These modifications are the result of a producer's continuing research program to correct any limitations, explore the potential of its fibers, and develop properties that will give greater versatility in the end uses of the fibers.

The *parent fiber* is the fiber in its simplest form. It is often sold as a "commodity fiber" by generic name only, without benefit of a trade name. The parent fiber has been called by the following names: regular, basic, standard, conventional, or first-generation fiber.

Modifications of the parent fiber are usually sold under a brand or trade name. Modifications are also referred to as types, variants, or second-generation fibers.

The following are fiber modifications of the second generation:

1. Modification of fiber shape:
Cross-section, thick and thin, hollow

2. Modification of molecular structure and crystallinity:
High tenacity, low pilling, low elongation

3. Additives to polymer or fiber solution:
Cross dye, antistatic, sunlight resistance, fire retardant

4. Modifications of spinning procedures:
Crimp, fiberfill.

Complex modifications have been engineered to combine two polymers as separate entities within a single fiber or yarn. These have been referred to as third-generation fibers.

1. Bicomponent fibers

2. Blended filament yarns

Fig. 8–3 Methods of spinning man-made fibers. (First three drawings courtesy of Man-Made Fiber Producers Association.)

Wet Spinning: Acrylic, Rayon, Spandex

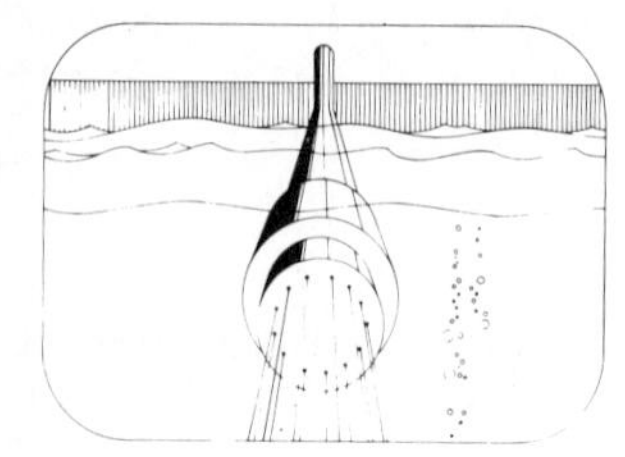

1. Raw material is dissolved by chemicals.
2. Fiber is spun into chemical bath.
3. Fiber solidifies when coagulated by bath.

Oldest process
Most complex
Weak fibers until dry
Washing, bleaching, etc., required before use

Dry Spinning:
Acetate, Acrylic, Modacrylic, Spandex, Triacetate, Vinyon

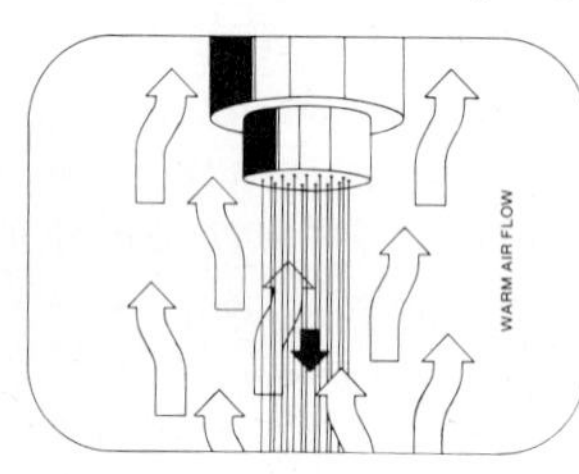

1. Resin solids are dissolved by solvent.
2. Fiber is spun into warm air.
3. Fiber solidifies by evaporation of the solvent.

Direct process
Solvent required
Solvent recovery required
No washing, etc., required

Melt Spinning: Nylon, Olefin, Polyester, Saran

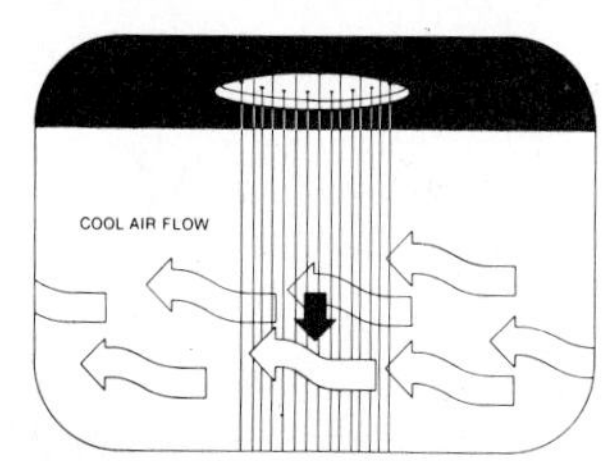

1. Resin solids are melted in autoclave.
2. Fiber is spun out into the air.
3. Fiber solidifies on cooling.

Least expensive
Direct process
High spinning speeds
No solvent, washing, etc., required
Fibers shaped like spinneret hole

Dispersion or Emulsion Spinning: Polytetrafluoroethylene

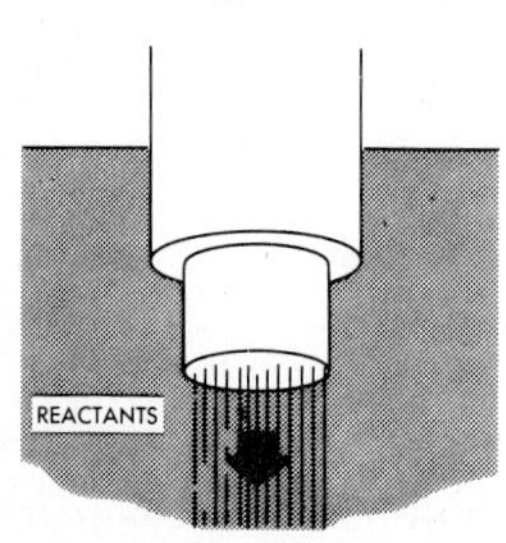

1. Polymer is dispersed as fine particles in a carrier.
2. Dispersed polymer is extruded through a spinneret and coalesced by heating.
3. Carrier is removed by heating or dissolving.

Expensive
Used only for those fibers that are insoluble
Carrier required

Reaction Spinning: Spandex

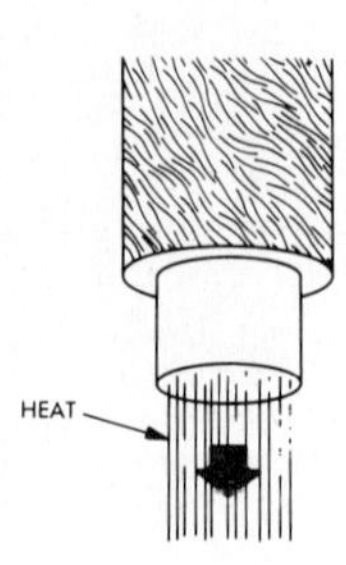

1. Monomers are placed in solution.
2. Polymerization occurs during extrusion through spinneret system of reactants.
3. Solvents may be used to control fiber size.

Simple recovery of solvents
Less expensive than dry spinning for spandex
Difficult to get uniform, light fibers

Modification of Shape

NONROUND FIBER TYPES

Changing the cross-sectional shape is the easiest way to alter the mechanical and aesthetic properties of a fiber. This is usually done by changing the shape of the spinneret hole to produce the fiber shapes desired. All kinds of shapes are possible: flat, trilobal, quadralobal, pentalobal, triskelion, cruciform, clover leaf, and alphabet shapes such as Y and T (see page 8).

The *flat shape* was one of the first variations produced. "Crystal" acetate and "sparkling" nylon were ribbon-like fibers that were extruded through a long, narrow spinneret hole. Flat fibers tend to catch and reflect light much as a mirror does, so fabrics have a glint or sparkle. The *trilobal shape* has been widely used in both nylon and polyester fibers (Figure 8–4). It is spun through a spinneret with three triangularly arranged slits. The trade name Antron designates selected round, trilobal, and pentalobal fibers made by Du Pont. The following are some of the advantages of trilobal shape:

- Beautiful silk-like hand (depending on end-use requirements)
- Subtle opacity
- Soil-hiding (cloaks dirt)
- Built-in bulk without weight
- Moisture-heightened wicking action
- Silk-like sheen and color
- Crush resistance in heavy deniers
- Gives good textured crimp.

Fig. 8–4 Stereoscan photograph of trilobal nylon. (Courtesy of E. I. du Pont de Nemours & Company.)

Cadon is a *triskelion*-shaped fiber (a three-sided configuration similar to a boat propeller) carpet nylon by Monsanto. Trevira® is a *pentalobal* polyester fiber produced by Hoechst (see Figure 8–5). Encron 8 is *octolobal*. "Touch" nylon has a *Y-shaped* cross-section. All these fibers have characteristics similar to the trilobal fibers. Dacron Type 83 is a *cruciform*-shaped staple fiber (shaped like a cross) with a crisper, more cotton-like hand. It has been optically whitened. It has been suggested that the cruciform shape might be the perfect shape for a fiber.

THICK-AND-THIN FIBER TYPES

Thick-and-thin fiber types have variations in diameter along their length as a result of uneven drawing or stretching after spinning. When woven into cloth, these yarns give the effect of a duppioni silk fabric or give a linen-like texture. The thick areas, or nubs, will dye a deeper color to create interesting tone-on-tone color effects. Barré in knits can be eliminated by thick-and-thin yarns. Many surface textures are possible by changing the size and length of the nubs or slubs.

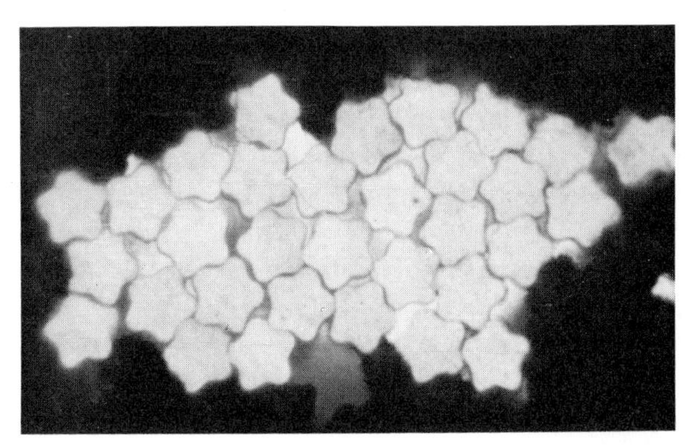

Fig. 8–5 Trevira® polyester pentalobal cross-section 312×. (Courtesy of Hoechst Fibers Industries.)

HOLLOW, OR MULTICELLULAR, FIBER TYPES

The hair or fur of many animals contains air cells that provide insulation in cold weather. The feathers of birds are hollow to give them buoyancy. Similar air cells and hollow filaments are possible in man-made fibers by the use of gas-forming compounds added to the spinning solution, by air injection at the jet face as the fiber is forming, or by the shape of the spinneret holes.

Hollow melt-spun fibers can be formed by pyrolizing a portion of the polymer flowing to the spinneret to form gas and then extruding the bubble containing polymer as hollow filaments. The spinneret hole can be shaped to produce hollow fibers. When extruded through a C-shaped hole, the fiber closes immediately. Other spinneret holes spin the fiber as two halves that immediately close to make the hollow fiber (see Figure 8–6). Examples of trade names of hollow fibers include Hollofil and Quallofil by du Pont.

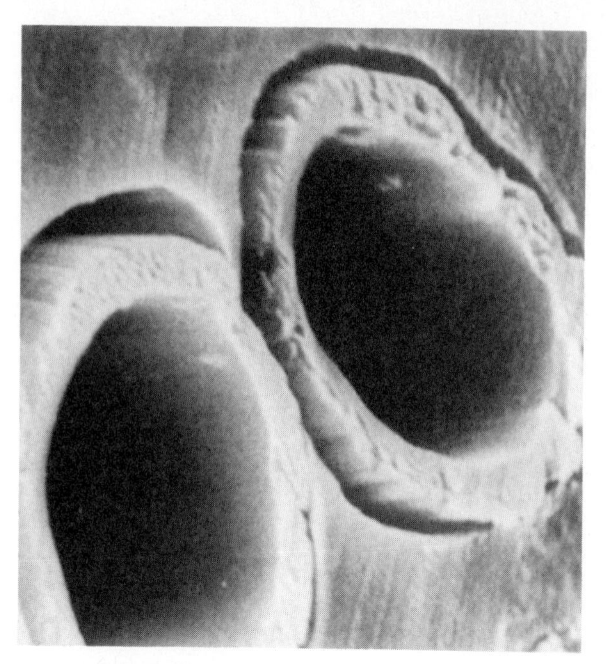

Fig. 8–6 *Cross-section of Hollofil polyester. (Courtesy of E. I. du Pont de Nemours & Company.)*

Modifications of Molecular Structure and Crystallinity

HIGH-TENACITY FIBER TYPES

Stretching a fiber changes its stress/strain curve, which is the basis of tenacity. Fiber strength is increased by (1) drawing or stretching the fiber to align or orient the molecules, thus strengthening the intermolecular forces, and/or by (2) chemical modification of the fiber polymer to increase the degree of polymerization.

High-tenacity rayon results from the rate of coagulation, the spinning speed, and/or modifiers added to the spinning solution. The spinning speed for *regular rayon* is quite high and the coagulation and regeneration occur almost simultaneously. This results in the formation of an oriented skin and amorphous core. In high-tenacity rayon, the spinning speed is reduced and the zinc sulfate content of the bath is increased, so there is an increase in the proportion of skin and a decrease in the core to the point where the core may disappear and the fiber will be an all-skin structure. The increase in orientation of an all-skin fiber increases tensile strength. Stretching is done by passing the fibers around two Godet wheels, one of which rotates faster than the other. After the high-tenacity fiber emerges from the coagulating bath, it is then given a high degree of stretch in a hot-water or dilute-acid bath. This stretching increases orientation and strength. Figure 8–7 is a photomicrograph of a high-tenacity all-skin rayon. Amino compounds may be added to the spinning solution to increase the viscosity and produce a higher degree of polymerization.

High-strength rayons are suitable for blending with nylons and polyesters. They are widely used for industrial purposes such as in conveyor belts and tire cord. Comiso is a high-strength apparel fiber made by Beaunit. Tyrex, Dynacor, and Suprenka are trade names for tire cord.

High-Tenacity Synthetic Fibers. Molecular chain length can be varied in the melt-spun fibers at the polymer stage by changes in time, temperature, pressure, and chemicals. Long molecules are harder to pull apart than short molecules. Hot drawing of polyester and cold drawing of nylon align the molecules in such a way that the intermolecular forces are strengthened. Some of the high-tenacity (low-elongation)

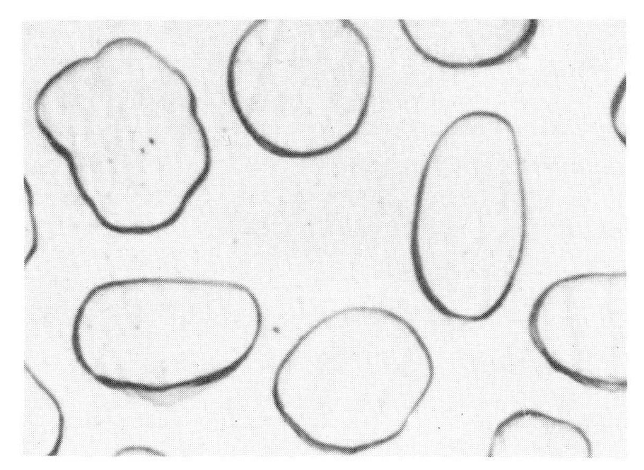

Fig. 8–7 High-tenacity rayon fiber. (Courtesy of AVTEX Fibers, Inc.)

nylons and polyesters were designed to strengthen cotton blends for durable press. Polyester's stress/strain curve more closely matches that of cotton, so they "pull together" to give greater durability.

High-tenacity polyester is now used in tire cord. Nylon has always had its widest use in industrial items such as tires for cars, trucks, and planes. Aramid fibers are the strongest and toughest fibers that have ever been made.

LOW-PILLING FIBER TYPES

Low-pilling fiber types are engineered to reduce the flex life by reducing the molecular weight as measured in terms of intrinsic viscosity. When flex-abrasion resistance is reduced, the fiber balls (pills) break off almost as soon as they are formed and the fabric retains its attractive appearance. These low-pilling fibers are not as strong as other types but are durable enough for apparel uses and are particularly suited to soft knitting yarns. (Review the discussion of molecular weight in Chapter 2.)

Dacron Type 35 is a higher-modulus, low-pilling staple fiber for blending with cotton. Type 65 is an extremely low-pilling, basic-dyeing staple for knitted garments; Dacron 107-W and Trevira Type 350 are optically brightened, pill-resistant staple for cotton/polyester blends in underwear.

BINDER STAPLE

Binder staple is a semidull, crimped polyester with a very low melting point. (Melting point relates to molecular structure.) It was designed to develop a thermoplastic bond with other fibers under heat and pressure. It sticks at 165°F and will shrink 55–75 percent at 200°F; Type 450 Fortrel is a fiber of this type.

LOW-ELONGATION FIBER TYPES

Low-elongation fiber types are designed as reinforcing fibers to increase the strength and abrasion resistance of cotton and cellulosic fabrics. The low elongation results from changing the balance of tenacity and extension. High-tenacity fibers have lower elongation properties. End uses are mainly in work clothing—items that get hard wear. Kodel 421 is a low-elongation fiber for use in blends with cotton.

Additives to the Polymer, or Spinning, Solution

DELUSTERING

The basic fiber is usually a *bright* fiber. It reflects light from its surface. To deluster a fiber, titanium dioxide—a white pigment—is added to the spinning solution before the fiber is extruded. In some cases, the titanium dioxide can be mixed in at an earlier stage, while the resin polymer is being formed. The degree of luster can be controlled by varying the amount of delusterant, producing dull or semidull fibers. Figure 8–8 shows three cones of yarn of different lusters.

Delustered fibers can be identified under the microscope by what appear to be peppery black spots (Figure 8–8). The particles of pigment absorb light or prevent reflection of light. Absorbed light causes degradation, or "tendering," of the fiber. For this reason, bright fibers that reflect light suffer less light damage and are better for use in curtains and draperies. The initial strength of a delustered fiber is less than that of a bright fiber. Rayon, for example, is 3–5 percent weaker when it is delustered.

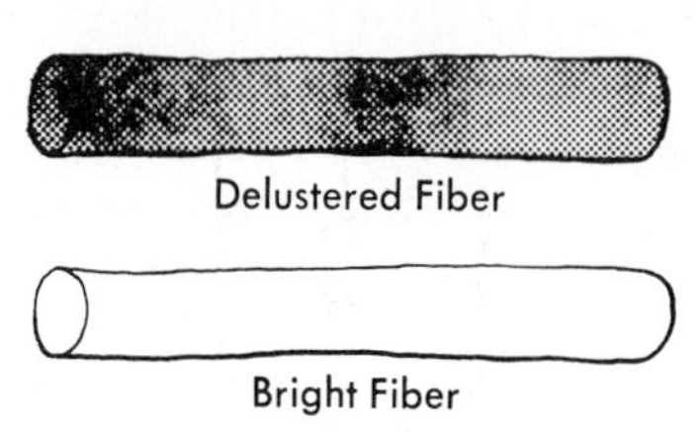

Fig. 8–8 (Top) *Rayon yarns. (Courtesy of AVTEX Fibers, Inc.)* (Bottom) *Fibers as they would look under a microscope.*

SOLUTION DYEING OR MASS PIGMENTATION

Solution dyeing, or mass pigmentation, was developed in response to the gas-fading of many dyes used to dye acetate. *Solution dyeing* is the addition of colored pigments or dyes to the spinning solution. These fibers are referred to as *solution dyed, mass pigmented, dope dyed, spun dyed,* or *producer colored.* If the color is added before the fiber hardens, the term *gel dyeing* may be used. Solution dyeing offers the potential of providing color permanence that is not obtainable in any other way. The lightfastness and washfastness are unchanged for the life of the item. Because the color is uniformly distributed throughout the fiber, crocking and other color changes with use are not problems.

Because of the difficulty in obtaining a truly black dye that has reasonable colorfastness properties, black pigments are usually the first ones to be used. Other colors beyond black are produced as suitable colorfast pigments are developed. Solution-dyed fibers cost more per pound than uncolored fibers. This difference is offset later by the cost of yarn or piece (fabric) dyeing. The solution-dyed fibers are used in all kinds of end uses, such as upholstery, window-treatment fabrics, and apparel. One disadvantage of the solution-dyed fibers is that the manufacturer must carry a large inventory to be able to fill orders quickly. The manufacturer is also less able to adjust to fashion changes in color, because it is not possible to strip color from these fibers and redye them.

WHITENERS AND BRIGHTENERS

Whiteners and *brighteners* are added to the spinning solution to make whiter fibers or fibers that resist yellowing. The additive used is an optical bleach or fluorescent dye that causes a whiter light to be reflected from the cloth. These whiteners are permanent to washing and dry cleaning. They are an advantage in many items because they eliminate the necessity for bleaching.

CROSS-DYEABLE FIBER TYPES

Cross-dyeable fiber types are very different from solution-dyed fibers. Solution-dyed fibers have colored pigment added to the spinning solution, so they are colored as they emerge from the spinneret.

Cross-dyeable fiber types are made by incorporating dye-accepting chemicals into the molecular structure. Some of the parent fibers are nondyeable or have poor acceptance of certain classes of dyes; the cross-dyeable types were developed to correct this limitation. Cross-dyeable types are white when spun, but will react with dyes later when the fabric is to be colored.

Two or more of these cross-dyeable fiber types can be used in a fabric that is then immersed in a suitable dye-bath mixture, and each fiber type reacts to pick up a different color. As many as five colors have been achieved in one dye bath. Designs may be heather, tone-on-tone, floral, or geometric, depending on the arrangement of the fiber types within the fabric. Solution-dyed fibers may be combined with cross-dyeable fiber types to increase color possibilities. Some of the cross-dyeable fiber types are basic dyeable, acid dyeable, disperse dyeable, acid-dye resist basic dyeable, and nondyeable.

ANTISTATIC FIBER TYPES

Static is a result of the flow of electrons. Fibers conduct electricity according to how readily electrons move in them. If static builds up in a fiber so that it has an excess of electrons, it is negatively charged and it will be attracted to something that is positively charged—something that has a deficiency of electrons. This attraction is illustrated by the way clothing clings to the body. Water will dissipate static. Because the heat-sensitive fibers, especially the synthetics, have such low water absorbency, static charges will build up rapidly during cold, dry weather, and they are slow to dissipate. If the fibers can be made wettable, the static charges will dissipate quickly and there will be no annoying static buildup.

For the person whose clothes are clinging, an immediate but temporary solution is that of using a wet sponge or paper towel and rubbing it over the garment or slip to drain away the static. The benefits will last much longer if one of the fabric softeners is used in place of plain water. Finishes and sprays are applied to the surface of the fiber and are lost during washing and wear.

The antistatic fiber types give durable protection because the fiber is made wettable by incorporating an antistatic compound—a chemical conductor—in the fiber so that it becomes an integral part of the fiber. The compound is added to the fiber-polymer raw material so that it is evenly distributed throughout the fiber-dope or spinning solution. It changes the fiber's hydrophobic nature to a hydrophilic one and raises the moisture regain so that static is dissipated more quickly. The moisture content of the air in the home should be kept high enough to provide moisture for absorption. Even "bone-dry" cotton will build up static. Static control is also achieved by incorporating a conductive filament into the filament (Figure 8–9). In the chart, Antistatic Fiber Variants, trade names and end uses are listed.

Fig. 8–9 *Antistatic polyester. (Courtesy of E. I. du Pont de Nemours & Company.)*

The soil-release benefits of the antistatic fiber types has been outstanding. The antistatic fibers retard soiling by minimizing the attraction and retention of dirt particles, and the opacity and luster in the yarn have soil-hiding properties. Soil redisposition in laundry is dramatically reduced. Oily stains, even motor oils, are released easily.

Antistatic Fiber Variants

Parent Fiber	*Trademark*	*Fiber Modification*	*End Use*	*Producer*
Nylon	Antron III	Three filaments of carbon-black core surrounded by sheath of nylon	Apparel Carpets	du Pont
Nylon	X-Stat	Seven silver-coated nylon filaments	Carpets	
Nylon	Ultron	Conjugate spun 95 percent nylon 6,6 and 5 percent nylon/carbon-black polymer stripe	Carpets	Monsanto
Nylon	Bodyfree		Apparel	Allied Chemical
Nylon	Enkalure	Fine denier–anticling	Apparel	American Enka
Polyester	Dacron III (Figure 8–9)	Polymeric conductive core	Carpets	du Pont

SUNLIGHT-RESISTANT TYPES

Ultraviolet light is the source of fiber degeneration as well as color fading. When ultraviolet light is absorbed, the damage results from an oxidation-reduction reaction between the radiant energy and the fiber or fiber dye. Stabilizers such as nitrogenous compounds may be added to the fiber to increase their light resistance. These stabilizers must be carefully selected for the fiber and the dye. Estron SLR is a sunlight-resistant acetate fiber. Delustered fibers are more sensitive to sunlight than bright fibers.

FLAME-RESISTANT FIBER TYPES

Flame-resistant fiber types give better protection to consumers than do topical flame-retardant finishes (see Chapter 35).

The man-made fibers that are always flame resistant are aramid, novolid, modacrylic, glass, PBI, saran, sulfar, and vinyon. Other man-made fibers can be modified by changing their polymer structure or by adding water-insoluble compounds to the spinning solution. These fiber modifications make the fibers inherently flame resistant. The fibers vary in their resistance to flame.

The following chart lists the naturally fire-resistant fibers produced in the United States.

Modifications of Spinning Procedure

When producers started to make staple fiber, mechanical crimping was done to broken filaments and later to filament tow to make the fibers more cohesive and thus easier to spin into yarns. Other techniques were developed to give permanent crimp to rayon and acetate and to provide bulk or stretch to all fibers—filaments as well as staple.

Crimping of fibers is important in many end uses for cover and loft in bulky knits, blankets, carpets, battings for quilted items, pillows, and the like, and for stretch and economy in hosiery and sportswear. One of the first crimping techniques was developed for use with rayon.

Viscose rayon fiber with latent or potential crimp is produced by coagulating the fiber in a bath of lower acid and higher salt concentration. A skin forms around the fiber and then bursts. A thinner skin forms over the rupture. The crimp develops when the fiber is immersed in water. Avicron is a latent-crimp rayon fiber made by American Viscose. It is a heavy-denier novelty filament fiber used in pile fabrics (Figures 8–10 and 8–11).

Crimped polyester fibers are used for fiberfill

Flame-Resistant Fibers*

Fiber Name	*Trade Name*	*Producer*
Aramid	Kevlar, Nomex	E. I. du Pont de Nemours
Glass	Star Rov, Stran Vitron, Vitron-Strand	Manville Corp.
Glass	Fiberglas Beta	Owens-Corning Fiberglas Corp.
Glass	LEX, TEXO	PPG Industries, Inc.
Modacrylic	SEF	Monsanto Chemicals Co., Fibers Div.
PBI	Arazole	Celion Carbon Fibers
Saran		Ametek, Inc.
Sulfar		Albany International
Sulfar		Johnson Filament
Sulfar	Ryton	Phillips Fiber Corp.
Sulfar	PPS	Shakespere Monofilament

**Textile World,* August, 1986.

Fig. 8–10 *Cross-section of Avicron rayon. Note the difference in thickness of the skin on the two sides. (Courtesy of AVTEX Fibers, Inc.)*

batting for pillows, furniture, carpets, sleeping bags, and quilted apparel. The fiberfill used for furniture is made into bats that are needle-punched to prevent lumping. Helically (spiral) crimped fibers are produced by cooling one side of the fiber faster than the other side as the melt-spun fiber is extruded. This uneven cooling causes a curl to form in the fiber. The same effect can be achieved by heating one side of the fiber during the stretching or drawing process. This helical crimp has more springiness than the conventional mechanical sawtooth crimp, and these fibers are used where high levels of compressional resistance and recovery are needed.

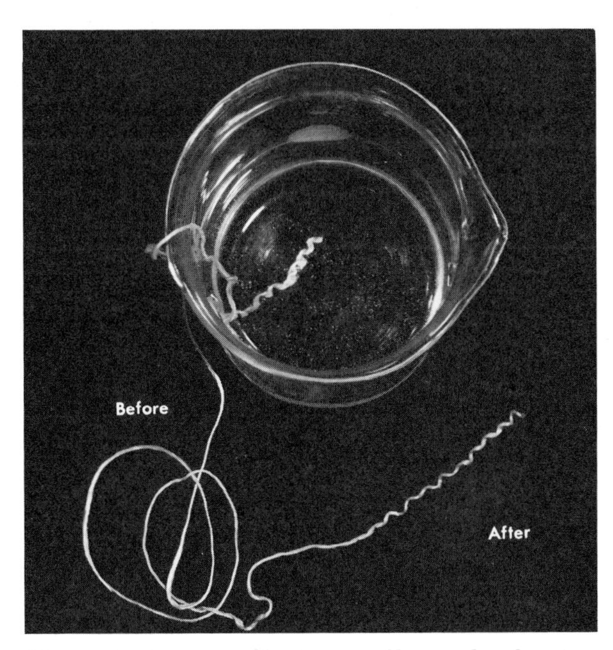

Fig. 8–11 *Avicron self-crimping fiber curls when immersed in water.*

Bulky yarns were developed by throwsters to make stretch fabrics about 1950 (see Chapter 18). In the 1970s, fiber producers started texturizing yarns. Undrawn and partially drawn yarns are sold for texturizing.

Third-Generation Fiber Types

BICOMPONENT FIBERS

A *bicomponent fiber* is a fiber consisting of two polymers that are chemically different, physically different, or both. If the two components would fall into two different generic classes, the term *bicomponent bigeneric* may be used. Bicomponent fibers may be of several types. In the first type, the fibers are spun with the two polymers side-by-side, called *bilateral.* In the second type, one polymer is surrounded by another in an arrangement called a *core sheath.* In the third type, short fibrils of one polymer are imbedded in another polymer called a *matrix fibril* (see Figure 8–12).

The original discovery that the two sides of a fiber can react differently when wet was made during studies of wool in 1886. In 1953, it was discovered that the difference in reaction was the result of the bicomponent nature of wool, which results from a difference in growth rate and in chemical composition. To produce a bicomponent rayon, two viscose solutions—one aged and one unaged—are spun through spinneret holes, each of which is separated into two halves by a divider through the center (Figure 8–12). The resulting fiber has a side-by-side, or bilateral, arrangement. This fiber is straight as it is spun but will crimp when immersed in hot or cold water and will retain its crimp after it dries. Crimp potential of this kind is referred to as *latent,* or *inherent, crimp.* When the latent crimp of a bicomponent, bilateral fiber is devel-

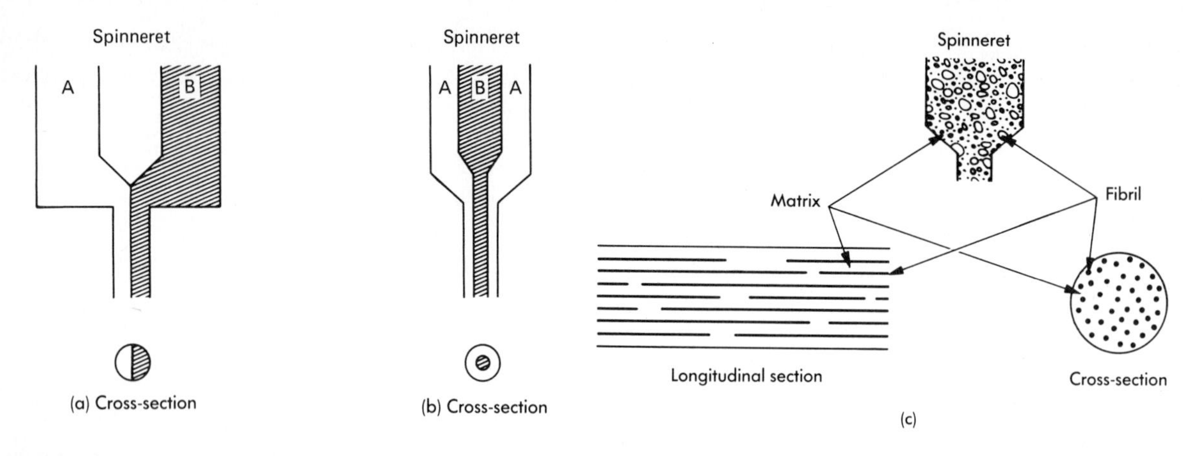

Fig. 8–12 Bicomponent fiber structure: (a) bilateral; (b) core-sheath; (c) matrix-fibril.

oped by heat and/or moisture, it is a three-dimensional helical crimp with the shorter component on the inside. Helical crimp (or curl) gives more bulk and stretch than other types of crimp.

The pipe-in-pipe procedure is another way to spin bicomponent fibers. In this case a sheath-core structure is formed, one component forming the sheath and the other the core (Figure 8–12).

Conjugate spinning was used in 1959 to produce Orlon 21 (Figure 8–13), the first acrylic bicomponent bilateral structure that would respond to wetting and drying in the same manner as the wool fiber. The fiber is spun straight and made into a garment such as a sweater, which is then exposed to heat; one side of the fiber shrinks and the fiber takes on a helical crimp. The reaction of the fibers to water occurs during laundering. As the fiber gets wet, one side swells and the fiber uncrimps. As the crimp relaxes, the sweater increases in size. The crimp will return as the sweater dries and it will regain its original size if properly handled. The sweater should not be drip-dried or placed on a towel to dry because the weight of the water and the resistance of the towel will prevent the sweater from regaining its original size. The right way to dry the sweater is to either machine dry it at low temperatures or place it on a smooth, flat surface and "bunch-it-in" to help the crimp recover. This Orlon bicomponent fiber is called a *reversible crimp fiber* because of the uncrimping and crimp recovery action during the laundry process. When Orlon 21 is used in quality products, the trade name Sayelle is used on the label.

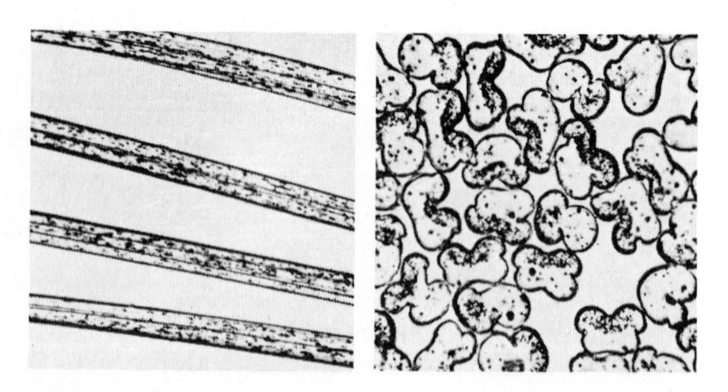

Fig. 8–13 Photomicrographs of semidull Orlon Sayelle: longitudinal (left) *and cross-sectional* (right) *views. (Courtesy of E. I. du Pont de Nemours & Company.)*

Other bicomponent acrylic fibers are Orlon Type 27, Orlon Type 33 for carpets, Acrilan Type 45 for carpets, Creslan Type 68-B, and Creslan Type 83 CF-5 carpet fiber. Civona is a bicomponent Orlon that was created for worsted spun-machine knitting yarns having a fine, Merino wool–like handle. Bi-loft is a bicomponent acrylic made by Monsanto for apparel.

The bicomponent fibers have been classified as latent-crimp-water (viscose) and latent-crimp-heat types. Nylon bicomponents are of the latter type. Nylon was the first of the synthetics to be made as a bicomponent fiber. Cantrece was the trade name adopted by du Pont for hosiery made of its monofilament-bicomponent fiber. Cantrece I was introduced in 1963, but it lacked the elastic recovery needed to prevent bagging at the knees. After a period of research, Cantrece II was introduced successfully. Like the other synthetic bicomponent fibers (Types 880, 881, 882, and 890 are used in Cantrece products), the two sides of the fiber differ in their reaction to heat. One side will shrink, causing the fiber to curl helically.

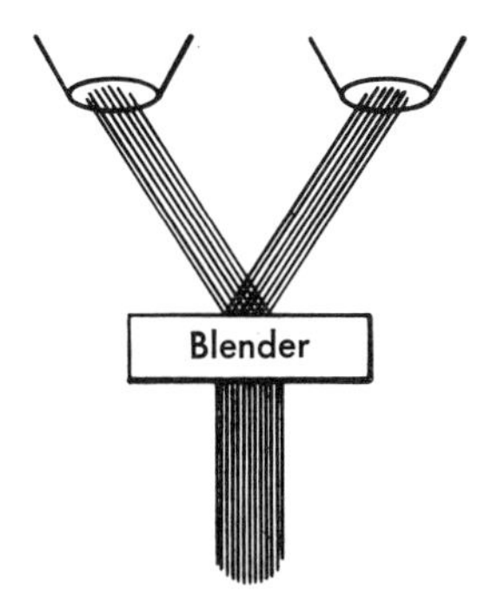

Fig. 8–14 *Blended filament yarn. Different fibers are spun, then blended to form a yarn.*

Bicomponent polyester and olefin fibers have been developed for use in the clothing, upholstery, and carpet fields. Cordelan is a vinal/vinyon bicomponent-bigeneric flame-resistant fiber. It is made of three polymers: polyvinyl chloride, polyvinyl alcohol, and a copolymer of PVC/PVA. It is a matrix fibril–type fiber.

BLENDED FILAMENT YARNS

Blended filament yarn differs from bicomponent and bicomponent-bigeneric fibers in that the blending takes place after the fibers are spun (Figure 8–14). This is a less-complex combination and can be made from a wider range of materials; the combinations can be tested very quickly in fabric form.

Lanese is Celanese's trademark for its acetate/polyester core-bulked filament yarn used in apparel and home-furnishing fabrics (Figure 8–15). Creslan 67A, 67AB, and 710 are blends of monocomponent and bicomponent Creslan acrylic staple fibers. SEF modacrylic types SX6 and SX7 are producer blends of 65 percent SEF modacrylic and Spectran polyester.

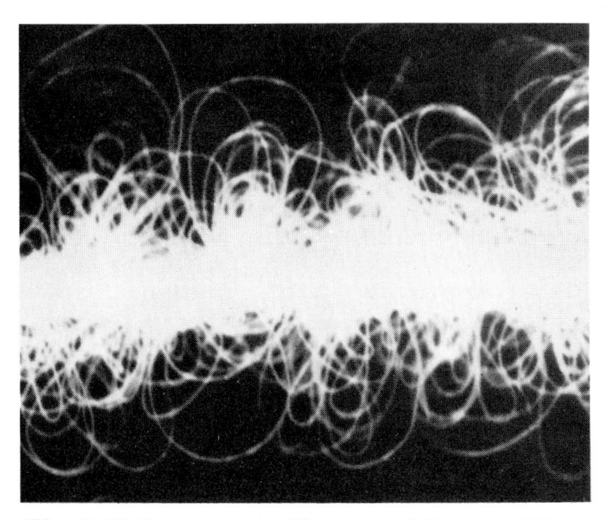

Fig. 8–15 *Lanese yarn. (Courtesy of Celanese Fibers, Inc.)*

ROTOFIL, OR FACIATED, YARNS

The purpose of these yarns is to give better texture and hand to fabrics. The yarns are combinations of coarse filaments for strength and fine broken filaments for softness. They are made by a combination of crimping and twisting. Trevira Dawn by Hoechst Fibers Industries is a polyester rotofil yarn.

Carpet Fibers

Carpeting is an end use in which a large volume of fiber or yarn is needed. Carpeting also requires a specific combination of fiber properties for satisfactory appearance and performance. The ideal carpet is durable, resilient (shows no traffic pattern), resistant to soil, and is easily cleaned. The major portion of a carpet consists of the fiber that forms the face. Fibers with high abrasion resistance make the most durable carpets, but the useful life of the carpet is also dependent on appearance retention—shedding, fading, "walked-down" areas, pilling, and soil and static resistance. Some carpets may need to be replaced before they are worn out. Wool was, at one time, the standard carpet fiber; but the amount of carpet wool produced throughout the world has declined, whereas the need for carpet fiber has increased tremendously. At present, nylon is the most widely used carpet fiber.

Abrasion Resistance. Abrasion resistance is the major factor in wear performance or durability. Durability has been defined as the time required to wear out the face fibers. The thickness of the yarn tufts, the denseness of the pile, and the kind of fiber are major factors in durability. In resistance to abrasion, the carpet fibers are rated as follows: nylon (unexcelled); polyester and olefins (very good); acrylics, modacrylics, and wool (good).

Comparison of Wool and Man-Made Carpet Fibers

Fiber Characteristic	*Wool*	*Man-Made*
Fiber diameter	Coarse–blends of various wools	15–18 denier or blend of various deniers
Fiber length	Staple	Staple or filament blend of various lengths
Crimp	3D Crimp	Sawtooth crimp, 3D crimp, bicomponent, textured filament
Cross-section	Oval	Round, trilobal, multilobal, square with voids, 5 pointed star
Resiliency and resistance to crushing	Good	Medium to excellent depending on fiber
Resistance to abrasion	Good	Good to excellent
Resistance to water-borne stains	Poor	Good to excellent
Resistance to oily stains	Good	Poor
Fire retardancy	Good	Modified fiber or topical finish
Static resistance	Good	Poor to good depending on fiber

Compressional Resiliency. Compressional resiliency is the tendency of a carpet fiber to spring back to its original height after being bent or otherwise deformed. This is particularly important in areas of heavy traffic or under the crushing force of the legs of heavy furniture. Nylon is unexcelled in overall recovery. Wool and the acrylics and polyesters are very satisfactory and have better immediate recovery. The cellulose fibers have poor recovery. As finer denier fibers were used to achieve a softer hand in the late 1970s, the autoclave heat setting process was developed so that these finer yarns performed with comparable resiliency.

Carpet fibers are larger in diameter than apparel fibers. This larger size gives the fiber more resistance to bending or crushing. A combination of different deniers is often used in a carpet.

Soiling. Soiling is a strongly fiber-dependent property. It may be real or "apparent." Soil retention is a function of fiber cross-sectonal shape. "Apparent" soiling is a function of fiber color and optical properties, such as transparency or opacity, as well as fiber shape. Smooth circular fibers retain the minimum quantity of soil. However, when circular fibers are transparent, as nylon is, the soil shows through and the circular shape tends to magnify it, so the "apparent" soil seems much worse.

Fibers can be made more opaque by changing the cross-sectional shape from circular to nonround. Light is reflected by the angles of the indented surfaces of nonround fibers, giving them greater opacity. For this reason (as well as bulk), the trilobal and Y-shaped fibers are often used (Figure 8–16). However, in areas of heavy,

Fig. 8–16 *Trilobal carpet fiber. (Courtesy of E. I. du Pont de Nemours & Company.)*

oily soil, circular fibers may be better, as they have fewer crevices where soil can be deposited and from which it must be removed. The delustering agent, titanium dioxide, will increase the opacity of a fiber and thus reduce the apparent soiling, but it makes the carpet look dull and chalky and it affects dyeing properties. Nylon fibers have now been spun in such a way as to leave voids in the fiber that scatter light, thus giving the same effect as the delusterant (Figure 8–17). Fibers such as Stainmaster by du Pont can be modified to be soil resistant.

U.S. Man-Made Fiber Capacity in Millions of Pounds in 1985*

Acetate and rayon	558
Acrylic	631
Glass	1394[1]
Nylon	2343
Olefin	1249
Polyester	3341

*Source: *Chemical and Engineering News,* June 9, 1986, page 38.
[1]Figure from 1984; 1985 information was not available.

Static. Static is one of the annoying problems associated with carpets in terms of comfort (static shock) and soiling. Some fibers generate more static than others. Cut pile generates more static than loop pile. Some carpet backings are better conductors than others. Conductive carbon black can be added to the latex adhesive to reduce carpet static. The shoes worn by the individual have an effect on the amount of static produced. Shoes with leather soles and rubber heels generate more static than others.

Metallic fibers can be used in a limited way to control static. Brunsmet, a stainless steel fiber from 2 to 3 inches long, can be mixed throughout any kind of spun yarn to make the yarn a good conductor. Only one or two fibers per tuft will carry the static from the face fiber to the backing. So far, this kind of carpet yarn has been used in places in which static is a special problem, such as hospitals and rooms where sensitive computer equipment is kept. Zefstat, an acrylic or nylon spun yarn by Badische, contains an especially treated strip of aluminum blended in so that it is invisible. As little as 2 percent aluminum will dissipate static as fast as it is generated. In addition, synthetic and man-made fibers can be modified to be antistatic, as explained earlier in the chapter.

Fig. 8–17 *Nylon fiber with voids. (Courtesy of E. I. du Pont de Nemours & Company.)*

Man-Made Fiber Capacity

In 1928, man-made fibers accounted for 5 percent of textile-fiber consumption in the United States; in 1986, man-made fibers comprise approximately 75 percent of U.S. textile consumption. See Figure 8–18, which compares domestic consumption of man-made fiber, cotton, and wool.

Man-Made versus Natural Fibers

A comparison of natural and man-made fibers is made in the following chart (page 80).

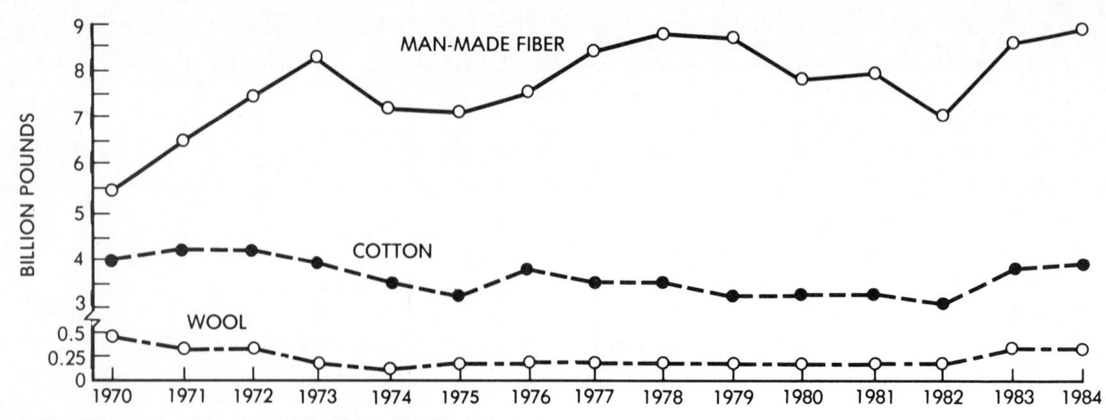

Fig. 8–18 *Domestic Consumption: Man-Made Fiber, Cotton, and Wool. (Courtesy of* Textile Organon, *March, 1985.)*

Comparison of Natural and Man-Made Fibers

Natural	*Man-Made*
Produced seasonally and stored until used	Continuous production
Vary in quality because they are affected by weather, nutrients, insects, or disease	Uniform in quality
Lack uniformity	Uniform or made purposely nonuniform
Physical structure depends on natural growth of plant or animal	Physical structure depends on fiber-spinning processes and after treatments
Chemical composition and molecular structure depend on natural growth	Chemical composition and molecular structure depend on starting materials
Properties are inherent	Properties are inherent
Properties conferred on fabrics can be changed by yarn and fabric finishes	Properties of fibers can be changed by varying spinning solutions and spinning conditions
	Properties conferred on fabrics can be changed by fabric finishes
Only silk is available in filament	Fibers can be any length
Less versatile	Versatile, changes can be made more quickly
Fibers are absorbent	Most (rayon and acetate are exceptions) have low absorbency
Not heat sensitive	Most (rayon is the exception) are heat sensitive)
Require fabric finish to be heat-set	Most (rayon and acetate are exceptions) can be heat-set
Research, development, and promotion done by trade organizations	Research, development, and promotion done by individual companies as well as by trade organizations

9

Rayon: A Man-Made Cellulosic Fiber

Rayon is a man-made cellulosic fiber in which the starting material, wood pulp or cotton linters, is physically changed. Rayon, the first man-made fiber, was developed before scientists knew much about molecular chains—how they are built up in nature or how they can be built up in the laboratory. The developers of rayon were trying to make artificial silk. Three methods of manufacturing rayon were developed in Europe in the 1880s and 1890s.

Frederick Schoenbein discovered in 1846 that cellulose would dissolve in a mixture of ether and alcohol if it were first treated with nitric acid, but the resulting fiber was highly explosive.

In 1884, Count Hilaire de Chardonnet, in France, made the first successful rayon by changing the nitrocellulosic fiber back to cellulose. This process was dangerous and difficult, and the nitrocellulosic process has not been used anywhere in the world since 1949.

In 1890, Louis Despeissis discovered that cellulose would dissolve in a cuprammonium solution, and in 1919 J. P. Bemberg made a commercially successful cuprammonium rayon. It was produced in the United States from 1926 until 1976. In 1892 in England, Cross, Bevan, and Beadle developed the viscose method. The viscose method is the only process currently used in the United States.

Commercial production of viscose rayon in the United States started in 1911 by the American Viscose Co. (Avtex Fibers). The fiber was sold as artificial silk until the name "rayon" was adopted in 1924. Viscose filament fiber, the first form of the fiber to be made, was a very bright, lustrous fiber. Because it had low strength, this fiber was used in the crosswise direction of the cloth; the lengthwise yarns were of silk, a fiber strong enough to withstand the tension of the loom. The double Godet wheel, invented in 1926, stretched the filaments and gave them strength. Delustering agents, which were added to the spinning solution, made it possible to have dull as well as bright fibers. Solution-dyed fibers came later.

In 1932, machinery was designed especially for making staple fiber. Large spinnerets with ten times as many holes were used and the fibers from several spinnerets were collected as a rope called *tow,* which was then crimped and cut.

The first uses of rayon were in crepe and linen-like apparel fabrics. The high twist that was required to make the crepe yarn reduced the bright luster of the fibers. "Transparent velvet" (made in France), sharkskin, tweed, challis, and chiffon were other fabrics made from these first rayons.

The physical properties of rayon remained about the same until 1940, when high-tenacity rayon for tires was developed. It proved to be superior to cotton for that use, and by 1957 cotton had disappeared from the tire-cord market. After high-tenacity tire cord and heavy-denier carpet fiber were developed, 65 percent of the rayon produced went into industrial and home-furnishing uses and less went into apparel.

Continued research and development led to what has been considered the greatest technological breakthrough in rayon—high-wet-modulus rayon. Production in the United States started in 1955. This modified fiber made it possible for rayon to be used for washable fabrics, dresses, sheets, towels, and also in blends with cotton. High-wet-modulus rayon stimulated a resurgence in the use of rayon in apparel.

High-wet-modulus rayon is frequently referred to as HWM rayon to distinguish it from regular or viscose rayon. In fact, HWM rayon is a viscose rayon, but in common usage viscose rayon refers to the weaker fiber. HWM rayon is also called high-performance (HP) rayon, or polynosic rayon. Polynosic has recently been adopted as a trade name by a U.S. company—this adds to the confusion of names because polynosic is used as a generic name for HWM rayon in Europe.

In 1960, 12 companies were producing rayon in the United States; in 1982, there were four producers, only one of which produced filament rayon.

Major producers of rayon in the United States include Avtex Fibers, Inc., BASF Corporation, Courtaulds North American, Inc., and North American Rayon Corporation. It is estimated that the output of rayon will not be increased because of the high cost of replacement machinery. Rayon is no longer the inexpensive fiber it once was—now it is generally comparable in price to cotton.

PRODUCTION

In the production of rayon, purified cellulose is chemically converted to a viscous solution that is pumped through spinnerets into a bath that

Spinning Processes for Viscose Rayon

Regular or Standard		*High Wet Modulus*
1. Blotter-like sheets of purified cellulose		1. Blotter-like sheets of purified cellulose
2. Steeped in caustic soda 3. Liquid squeezed out by rollers	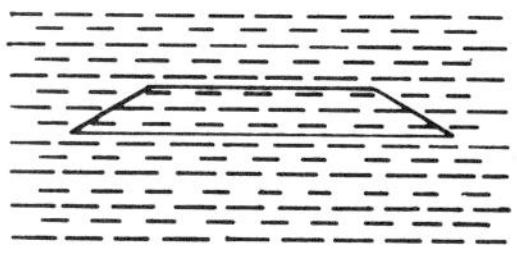	2. Steeped in weaker caustic soda 3. Liquid squeezed out by rollers
4. Shredder crumbles sheets to alkali crumbs 5. Crumbs aged 50 hours 6. Crumbs treated with carbon disulfide to form cellulose xanthate, 32 percent CS_2	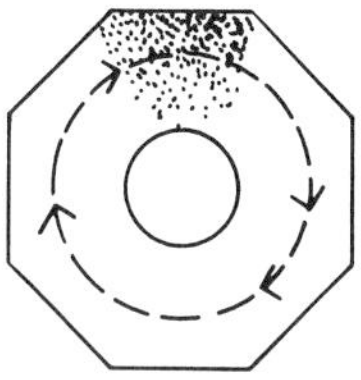	4. Shredder crumbles sheets to alkali crumbs 5. No aging 6. Crumbs treated with carbon disulfide to form cellulose xanthate, 39–50 percent CS_2
7. Crumbs mixed with caustic soda to form viscose solution	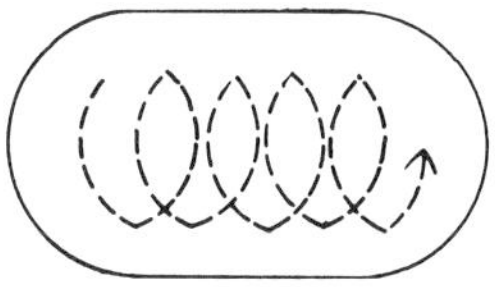	7. Crumbs mixed with 2.8 percent sodium hydroxide to form viscose solution
8. Solution aged 4–5 days 9. Solution filtered	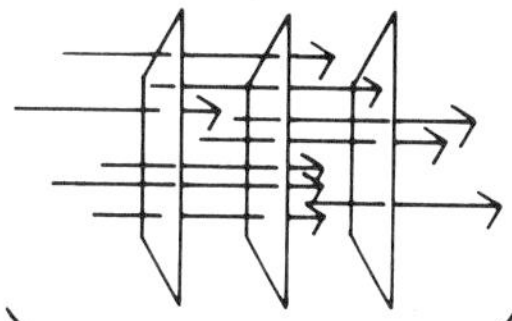	8. No aging 9. Solution filtered
10. Pumped to spinneret and extruded into sulfuric acid bath	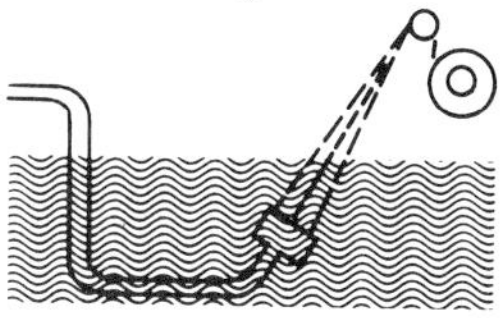	10. Pumped to spinneret and extruded into acid bath
10 percent H_2SO_4 16–24 percent Na_2SO_4 1–2 percent $ZnSO_4$	Spinning bath	1 percent H_2SO_4 4–6 percent Na_2SO_4
120 meters/minute	Spinning speed	20–30 meters/minute
45–50°C	Spinning-bath temperature	25–35°C
25 percent	Filaments stretched	150–600 percent

changes it back to solid 100 percent cellulose filaments. This is done by the wet-spinning process (see page 68). In the following chart, the processes for making regular and high-wet-modulus rayon are described. The differences in the spinning process produce fibers with different properties. In the high-wet-modulus process, the maximum chain length and fibril structure are maintained as much as possible.

A 1984 article in a technical journal stated that experiments were being made with solvent spinning of rayon. It was expected to go into commercial production by 1989 for industrial end uses.

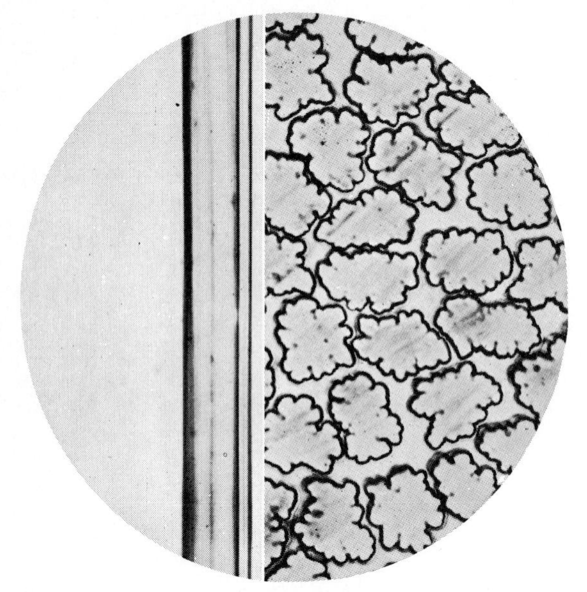

Fig. 9–1 Photomicrographs of viscose rayon: longitudinal (left) *and cross-sectional* (right) *views. (Courtesy of E. I. du Pont de Nemours & Company.)*

PHYSICAL STRUCTURE

Regular viscose is characterized by lengthwise lines called *striations*. The cross-section is a *serrated* circular shape (Figure 9–1). The shape of the fiber results from the presence of zinc sulfate in the spinning bath and from the liquid lost from the fiber during coagulation. The indented shape is an advantage in dye absorption because of an increase in surface area.

High-wet-modulus rayon that is spun into a bath with less zinc sulfate has a rounder cross-section. Figure 9–2 shows the difference in cross-sectional shapes of regular and high-wet-modulus rayon.

Filament rayon yarns have from 80–980 filaments per yarn and vary from 40–5,000 denier. Staple fibers and tow have a range of 1.5–15 denier, the 15 denier being carpet fiber or a heavy-use fiber. Staple fibers are usually crimped mechanically or by chemical means (Figure 9–3).

Rayon fibers are naturally very bright. This was one of the limitations to the use of the early fibers because bright filaments make very lustrous fabrics that are limited in use to dressy or luxury-type garments. The addition of delustering pigments (see page 71) remedied this problem. Pigment colors can be added to the fiber-spinning solution to make dull or colored fibers.

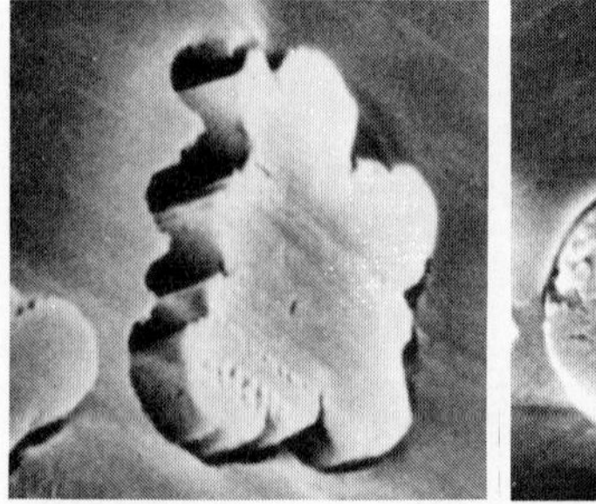

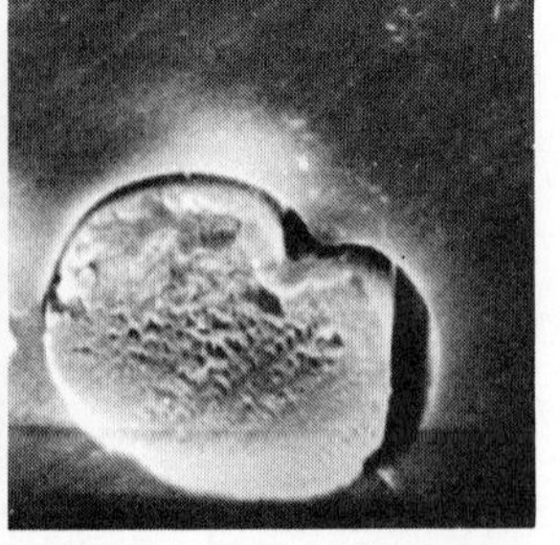

Fig. 9–2 Stereoscan photograph of Fibro, regular rayon (left) *and Vincel, high-wet-modulus rayon* (right). *(Courtesy of* Modern Textiles *magazine.)*

CHEMICAL COMPOSITION AND MOLECULAR STRUCTURE

Rayon—a manufactured fiber composed of regenerated cellulose, as well as manufactured fibers composed of regenerated cellulose in which substituents have replaced not more than 15 percent of the hydrogens of the hydroxyl groups.—Federal Trade Commission.

Rayon is 100 percent cellulose and has the same chemical composition as natural cellulose. The molecular structure of rayon is also the same as cotton and flax except that the rayon molecular chains are shorter and are not as crystalline. The breakdown of the cellulose occurs when the alkali cellulose and the viscose solution are aged. In regular rayon, the breakdown of the chains is quite severe. When the solution is spun into the acid bath, regeneration and coagulation

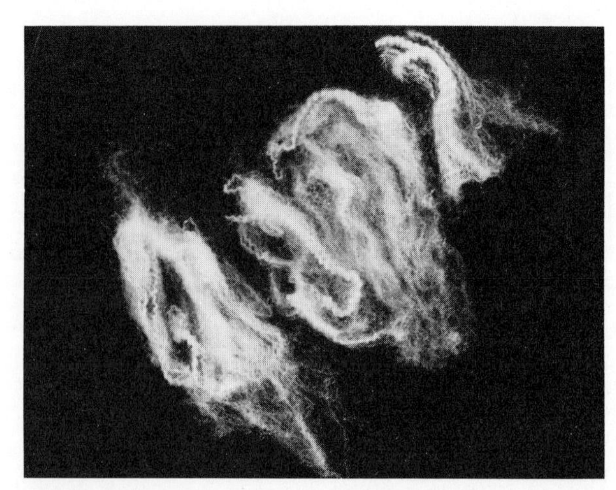

Fig. 9–3 Crimp in viscose rayon staple.

Comparison of Cotton, Regular Rayon, and High-Wet-Modulus Rayon

Properties	*Cotton*	*Regular Rayon*	*High-Wet-Modulus Rayon*
Fibrils	Yes	No	Yes
Molecular chain length	10,000	300–450	450–750
Swelling in water, percent	6	26	18
Average stiffness	57–60	6–50	28–75
Tenacity, grams/denier			
Dry	4.0	2.0	4.5
Wet	5.0	1.0	3.0
Breaking elongation, percent	12	11	30

take place very rapidly. Stretching aligns the molecules to give strength to the filaments.

In high-wet-modulus rayon, since the aging is eliminated, the molecular chains are not shortened as much. Because the acid bath is less concentrated, there is slower regeneration and coagulation so that more stretch and thus more orientation of the molecules can be made. HWM rayon retains its microfibrilar structure. This means its performance is more similar to cotton than to that of regular rayon. The following comparison chart shows the similarity between cotton and HWM rayon.

PROPERTIES

Rayon fibers are highly absorbent, soft, comfortable, easy to dye, and versatile. Fabrics made of these fibers have good drapability. Rayon fibers are used in apparel, home furnishings, medical/surgical products, and nonwovens.

Aesthetic. Since the luster, fiber length, and diameter of the fiber can be controlled, rayon can be made into cotton-like, linen-like, wool-like, and silk-like fabrics. As a blending fiber, rayon can be given much the same physical characteristics as the other fiber in the blend. If it is chosen instead of cotton or to blend with cotton, rayon can give the look of mercerized long-staple cotton to a fabric.

Durability. Regular rayon is not a very strong fiber and it loses about 50 percent of its strength when wet. The breaking tenacity is 0.7–2.6 g/d. Rayon has a breaking elongation of 15 percent dry and 20 percent wet and has the lowest elastic recovery of any fiber. All of these factors are the result of the amorphous regions in the fiber. Water enters the amorphous areas very readily,

Summary of the Performance of Rayon in Apparel Fabrics

	Regular Rayon	*HWM Rayon*
AESTHETIC	*VARIABLE*	*VARIABLE*
DURABILITY	*LOW*	*MODERATE*
Abrasion resistance	Low	Moderate
Tenacity	Low	Moderate
Elongation	Moderate	Low
COMFORT	*EXCELLENT*	*EXCELLENT*
Absorbency	High	Excellent
Thermal retention	Low	Low
APPEARANCE RETENTION	*LOW*	*MODERATE*
Resiliency	Low	Low
Dimensional stability	Low	Moderate
Elastic recovery	Low	Moderate
RECOMMENDED CARE	*DRY CLEAN*	*MACHINE WASH*

causing the molecular chains to separate as the fiber swells. This breaks the hydrogen bonds and permits distortion of the chains. When water is removed, new hydrogen bonds form, but in the distorted state.

HWM rayon has a more-crystalline and oriented structure so that the dry fiber is relatively strong. It has a breaking tenacity of 2.5–5.5 g/d, a breaking elongation of 6.5 percent dry and 7 percent wet, and an elastic recovery that is greater than that of cotton.

Comfort. Both types of rayon make very comfortable fabrics. They are absorbent, having a moisture regain of 13 percent. This eliminates any static. They are smooth and soft.

Appearance Retention. The resiliency of both rayons is low. This can be improved in HWM rayon fabrics by adding a durable-press finish. However, the finish may decrease strength and abrasion resistance. Dimensional stability of regular rayon is low. Fabrics may shrink or stretch. The fiber is very weak when wet and has low elastic recovery. Performance of HWM rayon is better—it exhibits moderate dimensional stability that can be improved by shrinkage-control finishes. The fiber is not likely to stretch out of shape and elastic recovery is moderate.

Care. Regular rayon fabrics have limited washability because of the low strength of the fibers when wet (0.7–1.8 g/d). Unless resin-treated, rayon fabrics have a tendency to shrink progressively. This shrinkage cannot be controlled by Sanforization. Regular rayon fabrics generally should be dry cleaned.

HWM rayon fabrics have greater washability. (Wet breaking tenacity of 1.8–4.0 g/d.) They have stability equal to cotton and strength equal to or better than cotton; they can be mercerized and Sanforized, and they wrinkle less than regular rayon in washing and drying.

The chemical properties of rayon are like those of cotton and other cellulosic fibers. They are harmed by acids, resistant to dilute alkalis, and are not affected by organic solvents; thus they can be safely dry cleaned. Rayon is attacked by silverfish and mildew.

Rayon is not greatly harmed by sunlight. It is not thermoplastic and thus can withstand a fairly high temperature for pressing. Rayon burns readily, like cotton.

USES

Rayon is the sixth most important fiber in the United States. Quantities produced have been decreasing; 607 million pounds in 1970, 461 million pounds in 1981, and 355 million pounds in 1982. Filament rayon has been decreasing in importance.

Uses of Rayon in 1984	*Million Pounds*
Woven fabrics	177.7
Nonwovens	126.3
Other uses	18.8
Circular and flat knits	4.3
Flocking	2.7
Carpet-face yarns	0.3
	331.8

Presently, rayon is mostly used in woven fabrics. In 1986, more rayon was seen in apparel, in both all-rayon fabrics as well as in blends with other fibers. Antique-satin drapery fabrics in a blend of rayon and acetate continue to be a classic fabric in interior decoration.

The second most important use of rayon is in nonwoven fabrics, where absorbancy is important. Items include industrial wipes; medical supplies, including bandages; diapers; sanitary napkins; and tampons. These disposable products are biodegradable.

TYPES AND KINDS

Over the years, the uses of rayon have changed. Technology has developed so rapidly that, of the 160 types of rayon made now, only five of those types were made in 1976!

Types and Kinds of Rayon

Rayon	*Trademark*	*Producer*	*Uses*
Staple fiber	Fibro	Courtaulds	Apparel and home furnishings
	Fibrenka	BASF	
Filament, staple, tow	Narco	North American Rayon	Apparel and home furnishings
	Avtex Rayon	Avtex	
Solution dyed	Coloray	Courtaulds	Home furnishings and industrial
	Jetspun	BASF	
	Kolorbon	BASF	
	Skybloom	BASF	
Acid dyeable	Enkrome	BASF	Apparel and home furnishings
	Fibro DD	Courtaulds	
Varied cross-section	Enkaire	BASF	Apparel
	Viloft	Courtaulds	
Intermediate or high tenacity	Hi-Narco	North American Rayon	Industrial
	Super-Narco	North American Rayon	
	I. T.	BASF	
	Aviloc (adhesive treated)	Avtex	
	Fibro HT	Courtaulds	
High wet modulus	Avril	Avtex	Apparel and home furnishings
	Avril II	Avtex	
	Avril III	Avtex	
	Avril Prima	Avtex	
	Polynosic	BASF	
	Vincel	Courtaulds	
	Zantrel	BASF	
Optically brightened	Super White	BASF	All
Flame retardant	Durvil	Avtex	Apparel
High absorbency	Absorbit	BASF	Sanitary supplies
	Avsorb	Avtex	Sanitary supplies
Adhesive-treated yarns	Beau-grip	North American Rayon	Industrial
Rayon/cotton	Cotron	Avtex	All
Hollow filament	ViLoft		
Self-crimping	Avicron	American Viscose	Home furnishings

10

Acetate: The First Heat-Sensitive Fiber

Acetate was the second man-made fiber produced in the United States: production began in 1924. Acetate originated in Europe, using a technique to produce a spinning solution for a silk-like fiber. The early experiments were not successful because the treated cellulose was only soluble in an expensive, highly toxic solvent. It was later discovered that, with further treatment, a nontoxic, less-expensive solvent could be used. The Dreyfus brothers, who were experimenting with acetate in Switzerland, went to England during World War I and perfected the acetate "dope" as a varnish for airplane wings. After the war, they perfected the process of making acetate fibers.

More problems had to be solved with the acetate process than with the rayon processes, in which, unlike the acetate process, the treated cellulose was changed back to 100 percent cellulose. The acetate fiber is a different chemical compound. In the original acetate or primary acetate there were no hydroxyl groups; in the modified or secondary acetate there were only a few hydroxyl groups. Secondary acetate is usually referred to as acetate to distinguish it from triacetate, which has essentially no hydroxyl groups. Thus the fibers could not be dyed with any existing dyes. Disperse dyes were developed especially for acetate and triacetate.

Acetate had better properties than rayon for use in silk-like fabrics. It had natural body, which made it good for blends with rayon in staple form for wool-like fabrics.

When a new fiber comes on the market, problems often arise. Because acetate was the first thermoplastic or heat-sensitive fiber, consumers were confronted with fabrics that melted under a hot iron. This was a long time before durable-press fabrics and homemakers were accustomed to ironing all apparel. The problem was further confused because manufacturers introduced and named acetate as a kind of rayon.

Another problem with acetate was fume fading—a condition in which certain disperse dyes changed color (blue to pink, green to brown, gray to pink) as a result of atmospheric fumes. Solution dyeing was developed to correct this problem in 1951. This process is now possible for all man-made fibers. In 1955, an inhibitor was developed that gave greatly improved protection to the dyes under all conditions that cause fading. However, fume fading can still be a problem.

Acetate fabrics have a luxurious feel and appearance as well as excellent drapability. They are economical.

PRODUCTION

In 1960, six companies were producing acetate; in 1986 there were two: Celanese and Eastman. The basic steps in the manufacturing process are listed in the following chart.

Manufacturing Process

Acetate
1. Purified cellulose from wood pulp or cotton linters
2. Mixed with glacial acetic acid, acetic anhydride, and a catalyst
3. Aged 20 hours—partial hydrolysis occurs
4. Precipitated as acid-resin flakes
5. Flakes dissolved in acetone
6. Solution is filtered
7. Spinning solution extruded in column of warm air. Solvent recovered (see Fig. 10–1)
8. Filaments are stretched a bit and wound onto beams, cones, or bobbins ready for use

Triacetate was produced by Celanese until the end of 1986, when their last triacetate plant was closed. Some triacetate is imported into the United States, so it is important for consumers to be aware that triacetate is a thermoplastic

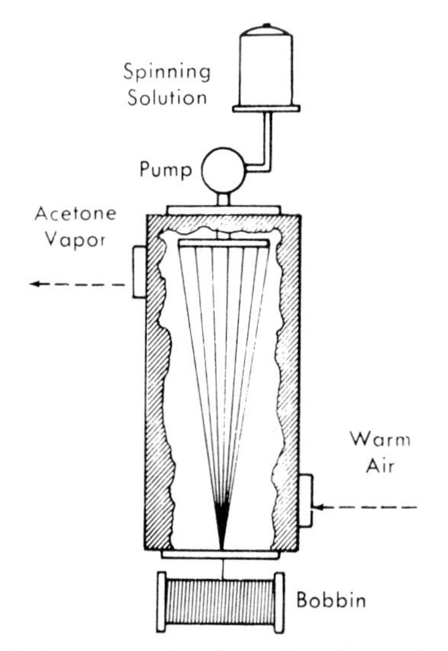

Fig. 10–1 *Acetate spinning chamber. (Courtesy of Tennessee Eastman Co.)*

fiber. It can be heat set for resiliency and dimensional stability. Thus it is a machine-washable fiber.

PHYSICAL STRUCTURE

Acetate is produced as staple or filament. Much more filament is produced because of its silk-like end use. Staple fibers are crimped and usually blended with other fibers. The cross-section of acetate is lobular or flower-petal shaped. (Lobular shape is characteristic of silk-like fibers.) The shape results from the evaporation of the solvent as the fiber solidifies in spinning. Notice in Figure 10–2 that one of the lobes shows up as a false lumen.

The cross-sectional shape can be varied. Y-shaped fibers have been produced for fiberfill for pillows and battings; flat filaments have been produced to give glitter to fabrics.

CHEMICAL COMPOSITION AND MOLECULAR ARRANGEMENT

Acetate—a manufactured fiber in which the fiber-forming substance is cellulose acetate. Where not less than 92 percent of the hydroxyl groups are acetylated, the term triacetate may be used as a generic description of the fiber.—Federal Trade Commission.

H OH
H
OH H
—O— —O—
H
H
O
CH_2OH
Glucose

→

H CH_2COOH
H
OH H
—O— —O—
H
H
O
CH_2CH_2COOH
Acetate

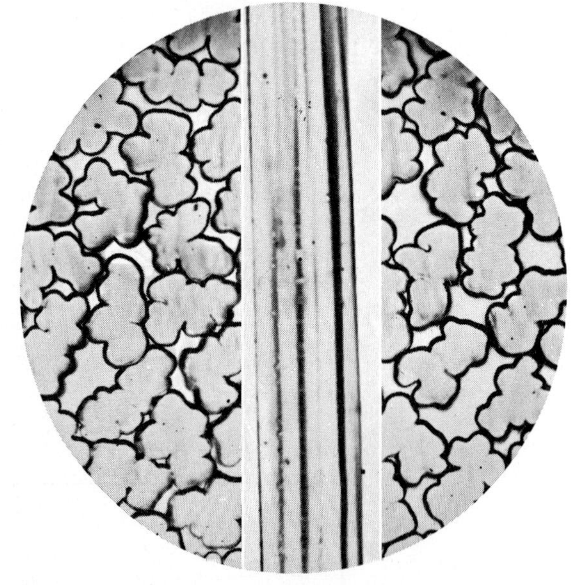

Fig. 10–2 Photomicrographs of acetate fiber: longitudinal and cross-sectional views. (Courtesy of E. I. du Pont de Nemours & Company.)

Acetate is an ester of cellulose and therefore has a different chemical structure than rayon or cotton. In acetate, two of the hydroxyl groups have been replaced by acetyl groups. The bulky acetyl groups tend to keep the molecules apart so they do not pack into regions of regularity (crystalline areas). There is less attraction between the molecular chains as a result of a lack of hydrogen bonding. Water molecules do not penetrate as readily, which accounts for the lower absorbency of acetate. The changed chemical structure also explains the different dye affinity of acetate. Acetate is thermoplastic.

PROPERTIES

Acetate has a combination of properties that make it a valuable textile fiber. It is low in cost and has natural body to give it good draping qualities.

Summary of the Performance of Acetate in Apparel Fabrics

AESTHETIC	*EXCELLENT*
Luster	High
Drape	High
Texture	Smooth
Hand	Smooth
DURABILITY	*LOW*
Abrasion resistance	Low
Tenacity	Low
Elongation	Moderate
COMFORT	*MODERATE*
Absorbency	Moderate
Thermal retention	Moderate
APPEARANCE RETENTION	*LOW*
Resiliency	Low
Dimensional stability	Moderate
Elastic recovery	Low
RECOMMENDED CARE	*DRY CLEAN*

Aesthetics. Acetate has been promoted as the beauty fiber. It is widely used in satins, brocades, and taffetas in which luster, body, and beauty of fabric are more important than durability or ease of care. Acetate has, and keeps, a good white color. This is one of its advantages over silk, which yellows readily.

Durability. Acetate is a weak fiber having a breaking tenacity of 1.2–1.4 g/d. It loses some strength when wet. Other weak fibers have some compensating factor, such as good elastic recovery in wool or spandex, but acetate does not. Acetate has a breaking elongation of 25 percent. Acetate also has poor resistance to abrasion. A small percentage of nylon is often combined with acetate to make a stronger fabric.

Comfort. Acetate has a moisture regain of 6.0 percent and is subject to static buildup.

Appearance Retention. Acetate fabrics are not very resilient. They wrinkle as they are worn. When the fabrics are washed, they often develop wrinkles that are difficult to remove. Acetate has moderate dimensional stability. The fibers are weaker when wet and can be shrunk by too much heat. Elastic recovery is low, 58 percent.

Care. Acetate should be dry cleaned unless other care procedures are recommended on the label of the garment. Acetate is resistant to weak acids and to alkalis. It can be bleached with hypochlorite or peroxide bleaches. Acetate is soluble in acetone. Acetate cannot be heat-set at a temperature high enough to give permanent shape to fabrics or to insure that embossing is durable.

Acetate is thermoplastic and heat sensitive; it becomes sticky at 177–191°C (350–375°F) and melts at 230°C (446°F). Figure 10–3 shows two fabrics, one acetate and the other triacetate, which were pressed with an iron set at cotton setting. The acetate fabric softened and shrank, whereas the triacetate fabric showed only a slight imprint of the iron. Triacetate has a higher melting point than acetate.

Acetate has better sunlight resistance than silk or nylon but less than the cellulose fibers. It is resistant to moths, mildew, and bacteria.

Fiber Identification. The *acetone test* is a specific identification test for acetate. None of the other fibers will dissolve in acetone. Figure 10–4 shows a procedure for testing the acetate content of a fabric. Use a dropper bottle, glass rod, watch glass, and cleaning tissue. Test individual yarns first. The presence of other fibers, in blends or combinations with acetate, can be determined. The structure will not disintegrate if

Fig. 10–3 *Effect of heat on (1.) triacetate and (2.) acetate*

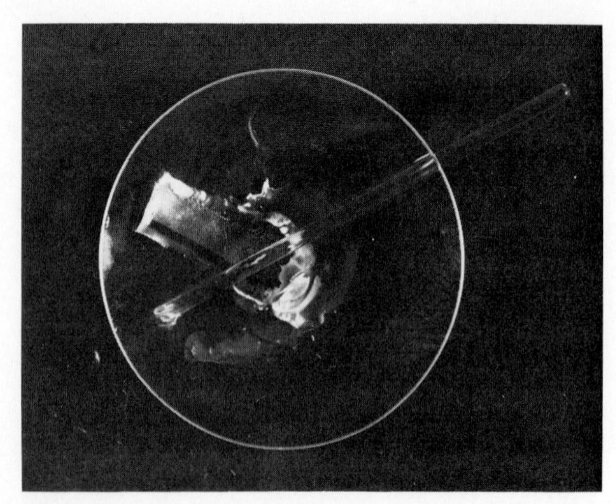

Fig. 10–4 Acetone test for identification of acetate fiber.

only a small amount of acetate is present, but it will feel sticky and will stiffen permanently when the solvent evaporates.

The burning test will also identify acetate. Acetate gives off an acetic—vinegar-like—odor that is specific for acetate. It burns freely, melts, and then decomposes to a black char. Fire-resistant fiber types have been developed.

Comparison With Rayon. Rayon and acetate are the two oldest man-made fibers and have been produced in large quantities, filling a very important need for less-expensive fibers in the textile industry. They lack the easy care, resilience, and strength of the synthetics and have had difficulty competing in uses where these characteristics are important. Rayon and acetate have some similarities because they are made from the same raw material, cellulose. The manufacturing processes differ, so the fibers have many individual characteristics and uses. Some of these are listed in the following table.

USES

Acetate is a minor fiber in terms of usage. Approximately 185 million pounds of acetate were produced in 1984. Americans used about the same amount of wool. Acetate is used in both apparel and home furnishings, but not in industrial products.

The most important use of acetate is in lining fabrics. The aesthetics of acetate—its luster, hand, and body—and its relatively low cost, make it appropriate for this use. However, since acetate is not a durable fiber, the fabric must be carefully selected for the end use or the consumer will be dissatisfied.

The second important use of acetate is in robes and loungewear. It is frequently seen in brushed-tricot and fleece fabrics. In these uses

Comparison of Rayon and Acetate

Rayon	*Acetate*
Differences	
Wet spun	Dry spun
Regenerated cellulose	Chemical derivative of cellulose
Serrated cross-section	Lobular cross-section
More staple produced	More filament produced
Scorches	Melts
High absorbency	Fair absorbency
No static	Static
Not soluble in acetone	Soluble in acetone
Industrial uses—tires	Very few industrial uses
Not used for fiberfill	Used for fiberfill
Color may crock or bleed	Color may fume fade
Mildews	Resists mildew
Moderate cost	Low cost
Similarities	
Low strength	Low strength
Low abrasion resistance	Low abrasion resistance
Chlorine bleaches can be used	Chlorine bleaches can be used
Flammable	Flammable

in the knit structure, the fiber is washable and performs well.

The third important use for acetate is in drapery fabrics. Antique-satin fabrics made of blends of acetate and rayon are very common. They come in an amazingly wide assortment of colors—nearly any décor can be matched. Usually most of the acetate is on the back side of the fabric—the side that faces the window and the sun. Sunlight-resistant variants that can increase the lifetime of the drapery fabrics have been developed. Most of the front side of the fabric is made of rayon yarns and has some textural interest from slub yarns.

The fourth important use of acetate is in fabrics for more-formal wear, such as dresses and blouses. Taffeta, moiré taffeta, satin, and brocade are very common and popular fabrics.

Other important uses of acetate fabrics include bedspreads and quilts, satin sheets, fabrics sold for home sewing, and ribbons.

TYPES AND KINDS

Types of acetate are solution dyed, flame retardant, sunlight resistant, fiberfill, textured filament, modified cross-section, and thick-and-thin slub-like filament.

Types and Kinds of Acetate

Acetate	*Trademarks*	*Producer*
Regular	Celanese Acetate	Celanese
	Estron	Eastman
Solution dyed	Celaperm	Celanese
	Chromspun	Eastman
Modified cross-section	Celafil	Celanese
	Celacloud (fiberfill)	Celanese
Textured or crimpable	Celacrimp	Celanese
	Celara	Celanese
	Loftura	Eastman
Combinations		
Acetate/polyester core bulked yarn	Lanese	Celanese
Sunlight and weathering resistant	SLR	Eastman

11

Nylon: The First Synthetic Fiber

Synthetic Fibers

Synthetic fibers are made by putting together simple chemical compounds (monomers) to make a complex chemical compound (polymers). They are also called chemical, or noncellulosic, man-made fibers. The fibers differ in the elements used, the way they are put together as polymers, and the method of spinning used. The synthetic fibers include polyamide, polyacrylic, polyester, polyolefin, polyurethane, and polyvinyl. The synthetic fibers have many properties in common that are listed in the following chart.

COMMON PROPERTIES

Heat Sensitivity. All man-made fibers, except rayon, are heat sensitive. *Heat resistance* is the resistance of a fiber to heat exposure. The term *heat sensitivity* is used with specific meaning for fibers that soften or melt with heat; those that scorch or decompose are described as being heat resistant. Heat sensitivity is important in use and care of fabrics as well as in manufacturing processes. Heat is encountered in washing, ironing, and dry cleaning during use, and in dyeing, scouring, singeing, and other fabric-finishing processes.

The fibers differ in their level of heat resistance. This difference is reflected in the table of safe-ironing temperatures in Chapter 2. The speed of ironing has been found by research to average about 40 inches per minute. This means that in normal ironing the fabric never gets as hot as the sole plate of the iron. If the iron is slowed down or allowed to stand in one spot, the heat will build up. If heat-sensitive fabrics get too hot, the yarns will soften and pressure from the iron will flatten them (Figure 11–1). This flattening will be permanent. Flattening of the

Properties Common to Synthetic Fibers

Properties	*Importance to Consumers*
Heat sensitive	If iron is too hot, fabric will shrink and then melt. Hole melting from cigarettes. Pleats, creases, and so forth can be heat-set in fabrics. Fabric can be stabilized by heat setting. Yarns can be textured for bulk. Fur-like fabrics can be produced.
Resistant to most chemicals	Can be used in laboratory and work clothing where chemicals are used.
Resistant to moths and fungi	Storage is no problem. Useful in sandbags, fishlines, tenting.
Low moisture absorbency	Products dry quickly. Resist waterborne stains. Stains can be sponged off. Lack comfort in humid weather. Increases possibility of static. Water does not cause shrinkage. Difficult to dye.
Oleophilic	Oil and grease absorbed into the fiber must be removed by dry-cleaning agents.
Electrostatic	Clothes cling to wearer. May cause sparks that can cause explosions or fires. Shocks in cold, dry weather are unpleasant.
Abrasion resistance good to excellent (acrylics lowest)	Good appearance retained longer because holes and worn places do not appear as soon. Color does not wear off as fast.
Strength good to excellent	Strongest fibers make good ropes, belts, and women's hosiery. Resist breaking under stress.
Resilience excellent	Easy-care apparel, packable for travel. Less wrinkling during wear.
Sunlight resistance good to excellent (nylon modified to improve resistance)	Webbing for outdoor furniture. Indoor/outdoor carpet. Curtains and draperies. Flags.
Flame resistance	Varies from poor to excellent. Check individual fibers.
Density or specific gravity	Varies as a group but tend to the lightweight.
Pilling	May occur in staple-length fibers.

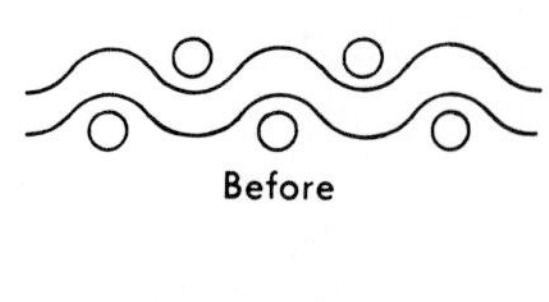

Fig. 11–1 *Heat and pressure cause permanent flattening of the yarn (glazing).*

surface is called *glazing*. Garment alteration is difficult in heat-sensitive fabrics because creases are hard to press in or out. Fullness cannot be shrunk out of the top of a sleeve or other parts of the garment, so patterns have to be adjusted to remove some of the fullness in areas where fullness is usually controlled by shrinkage.

Heat Setting. *Heat setting* is a factory process that uses heat to stabilize yarns or fabrics made of heat-sensitive fibers. The yarn or fabric is heated to bring it almost to the melting point specific for the fiber being heat set. This temperature range is from 375–445°F. The fiber molecules move freely, dissipating stresses within the fiber. The fabric is kept under tension until it cools, to prevent shrinkage. After cooling, the fabric or yarn will be stable to any heat lower than that at which it was set, but changes can be brought about by higher temperatures. Heat setting may be done at any stage of finishing, depending on the level of heat resistance of the fiber and other qualities. Figure 11–2 illustrates heat setting of flat fabric. Heat setting is both an advantage and disadvantage to the consumer.

Pilling. The strength of fibers is a basic factor in the problem of fabric pilling. *Pilling* is the formation of bunches or balls on the surface of the fabric and occurs on fabrics that have free-fiber ends when the ends get tangled by rubbing. The pills often break off before the garment be-

Fig. 11–2 *Heat-setting nylon fabric.*

Heating Setting

Advantages	*Disadvantages*
Embossed designs are permanent. Pleats and shape are permanent. Size is stabilized. Pile is crush resistant. Knits do not need to be blocked. Clothing resists wrinkling during wear.	"Set" creases and wrinkles are hard to remove in ironing or in garment alteration. Care must be taken in washing or ironing to prevent the formation of set wrinkles.

comes unsightly, but with nylon and the polyesters the fibers are so strong that few pills break off and they accumulate on the fabric's surface. Pills are of two kinds: lint and fabric. *Lint pills* are more unsightly, because they contain not only fibers from the garment but also fibers picked up in the wash water or through contact with other garments and even through static attraction.

The best single treatment to prevent pilling is *singeing*. Singeing is a necessity for polyester/cotton and polyester/worsted blends. Singeing consists of running the fabric between two gas flames or two hot plates so it will be singed on both sides at one pass. It should be done very rapidly. The ends of the polyester fibers melt and shrink into the core of the yarn, making it harder for the fibers to work to the surface and form pills. The tips of the fibers look like match heads under the microscope. Singeing should be after dyeing because these fused ends will take a deeper dye.

The construction of the fabric is an important factor in the prevention of pilling. Close weave, high-yarn twist or plied yarns, and longer-staple fibers are recommended. Resin finishes of cotton and fulling of wool are finishes that help prevent pilling.

Static Electricity. Static electricity is generated by the friction of a fabric when it is rubbed against itself or other objects. If the electrical charge is not removed, it builds up on the surface. When the fabric comes in contact with a good conductor, a shock, or transfer, occurs. This transfer may produce sparks that, in a gaseous atmosphere, can cause explosions. Static electricity is always a hazard in such places as dry-cleaning plants and operating rooms. Operating-room personnel are forbidden to wear nylon or polyester uniforms because of the danger. Static tends to build up more rapidly in dry, cold regions. Other problems involving static include

1. Soil and lint cling to the surface of the fabric and dark colors become very unsightly. Brushing simply increases the problem.
2. Dust and dirt are attracted to curtains.
3. Fabrics cling to the machinery at the factory and make cutting and handling very difficult. Static is responsible for increased defects and makes a higher percentage of seconds.
4. Clothes cling to the wearer and cause discomfort and an unsightly appearance. Temporary relief can be obtained by the wearer if a damp sponge or paper towel is wiped across the surface to drain away the static. More-permanent relief can be obtained by the use of fabric softeners. These are effective when used as directed.

Antistatic finishes are applied to many of the fabrics at the factory, but they frequently wash out or come out in dry cleaning.

Oily Stains. Fibers that have low moisture absorption usually have an affinity for oils and greases. They are *oleophilic.* These stains are very difficult to remove and require prespotting with a concentrated liquid soap or a dry-cleaning solvent.

Nylon

Nylon was the first synthetic fiber and the first fiber conceived in the United States. The discovery of nylon was not planned, but resulted from a fundamental research program by Wallace Carothers that was designed to extend basic knowledge of the way in which small molecules are united to form giant molecules or polymers.

In 1928, the du Pont Company decided to establish a fundamental research program. If anything was discovered, it would be good for the company—a means of diversification. The slogan of du Pont is "Better Things for Better Living." Du Pont hired Dr. Carothers, who had done research on high polymers, to direct a team of scientists. These people created many kinds of polymers, starting with single molecules and building them up into long molecular chains. One of Carothers' assistants noticed that when a glass rod was taken out of one of the polyester stills, the solution adhering to it stretched out into a solid filament. The filament could be stretched even further and it did not go back to its original length. This stimulated the group to concentrate on textile fibers. The polyester they were working on had too low a melting point; the decision was made to concentrate on developing polyamides.

By 1939, du Pont was making a polyamide fiber—nylon 6,6—in a pilot plant. Nylon 6,6 was introduced to the public in women's hosiery where it was an instant success. The term *nylon* was chosen for the fiber. It had no special meaning but had a nice textile sound like cotton and rayon. (At the time there were no laws specifying generic names for fibers. Acetate was still considered a kind of rayon.)

Nylon was called the *Miracle Fiber* for several years. The first thermoplastic fiber ever used, it had a combination of properties unlike any natural or man-made fiber in use in the 1940s. It was stronger and more resistant to abrasion than any fiber; it had excellent elasticity; it could be heat set, and permanent pleats became a reality. For the first time, gossamer-sheer, frilly lingerie was durable and machine washable. Nylon's high strength, light weight, and resistance to sea water made it suitable for ropes, cords, sails, and the like.

As nylon entered more end-use markets, its disadvantages became apparent—static buildup, poor hand, lack of comfort in skin-contact apparel fabrics, and low resistance to sunlight in curtains. But fortunately, as each problem appeared, more was learned about fibers, and ways were found to overcome the disadvantages.

In 1960, five firms in the United States were producing nylon. In 1983, there were nineteen firms in the United States producing nylon: three produced only nylon 6,6; nine produced only nylon 6; and seven produced both nylon 6 and nylon 6,6. Four companies in the last group produced one or two additional nylons as well, including nylon 6,12; nylon 11; or nylon 12.

PRODUCTION

Polyamides are made from various substances. The numbers after nylon indicate the number of

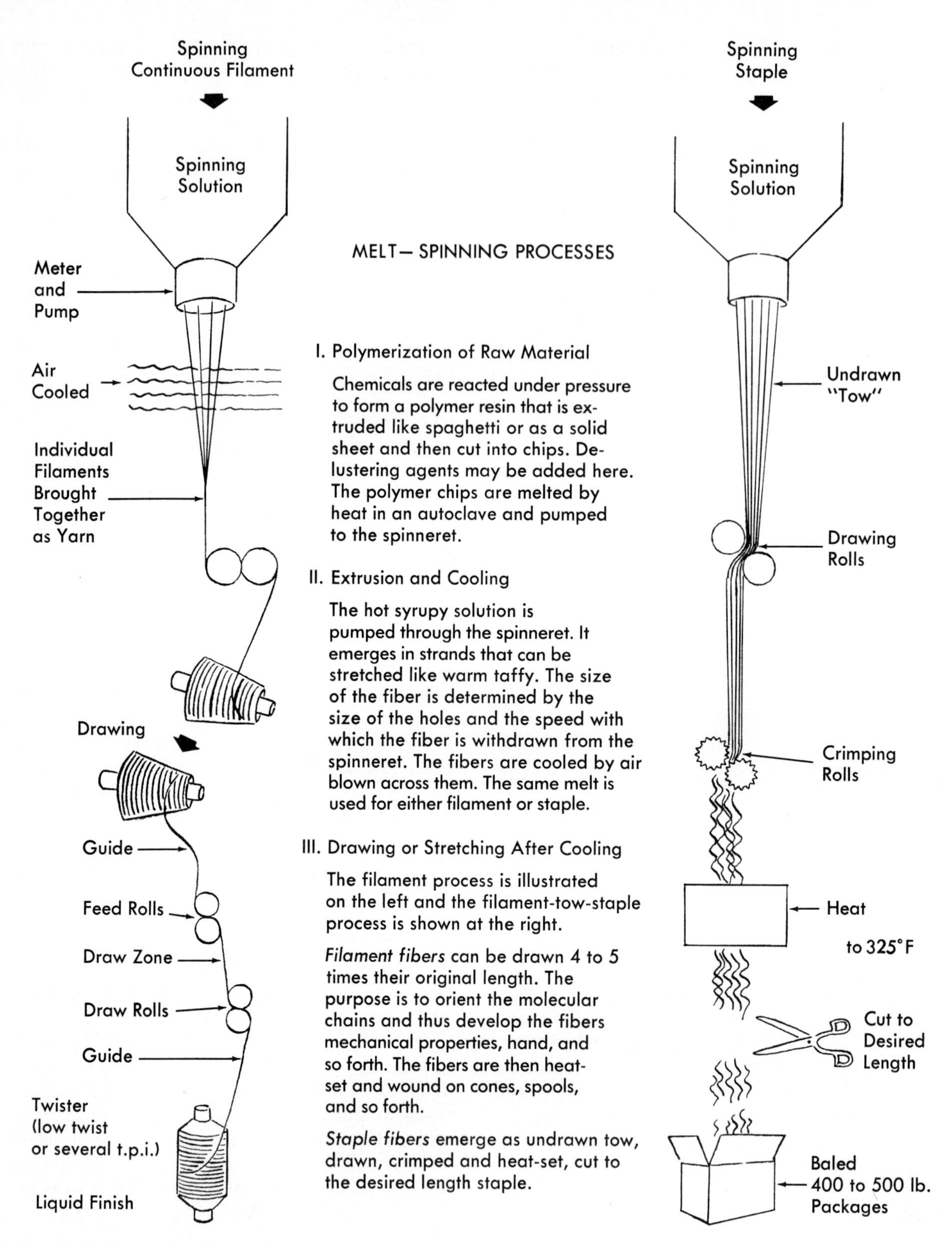

Fig. 11–3 Chart of melt-spinning processes.

carbon atoms in the starting materials. Nylon 6,6 is made from hexamethylene diamine, which has six carbon atoms and adipic acid, which has six carbon atoms.

While nylon 6,6 was being developed in the United States, scientists in Germany were working on nylon 6, to which they gave the trade name of Perlon. It is made from a single substance, caprolactam, which has six carbons. Allied and American Enka both began producing nylon 6 in the United States in 1954. In the United States during the 1980s, two-thirds of the nylon produced was nylon 6,6 while the remaining one-third was nylon 6.

Melt Spinning. Nylon is melt spun; this process was developed by du Pont. The basic steps in the melt-spinning process, for both filament and staple fiber made from filaments, are shown in Figure 11–3. *Melt spinning* is essentially a simple process. It can be demonstrated by a laboratory experiment that is fun to do. A flame, a pair of tweezers, and a piece of nylon are all that are needed. Heat the fabric until quite a little melt has formed, then quickly draw out the melt with tweezers as shown in Figure 11–4.

Commercial melt spinning consists of forcing nylon melt through the holes of the stainless-steel plate of a heated spinneret. The fiber cools in contact with the air, solidifies, and is wound on a bobbin. Figure 11–5 shows commercial spinning of nylon as it is extruded through the spinneret into cool air.

The chain-like molecules of the fiber are in an amorphous, or disordered, arrangement and the filament fiber must be *drawn* to develop the desirable properties of the fiber, such as strength, pliability, toughness, and elasticity. Nylon is cold drawn. Drawing aligns the molecules, placing them parallel to one another and bringing them closer together so they are more crystalline and oriented. The fiber is also reduced in size. The amount of draw varies with intended use. The draw ratio determines the decrease in fiber size and the increase in strength.

PHYSICAL STRUCTURE

Nylon is made as multifilaments, monofilaments, staple, and tow in a wide range of deniers and staple lengths. They are produced as bright, semidull, and dull lusters. They vary in degree of polymerization (D. P.) and thus in strength. They are available as partially drawn or completely finished filaments.

Regular nylon has a round cross-section and is perfectly uniform throughout the filament (Figure 11–6). Under the microscope, the fibers look like fine glass rods. They are transparent unless they have been delustered or solution dyed.

At first, the uniformity of nylon filaments was a distinct advantage over the natural fibers—especially silk. Silk hosiery often had rings caused by thicker areas in yarns, which detracted from their beauty. However, the perfect uniformity of nylon produced woven fabrics with

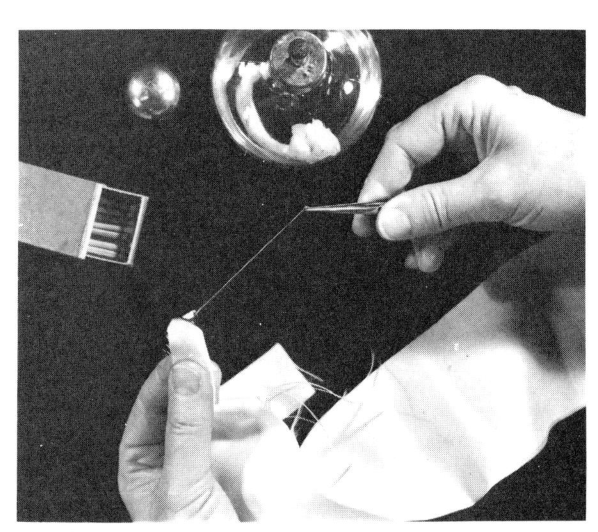

Fig. 11–4 *Spinning a melt-spun fiber by hand.*

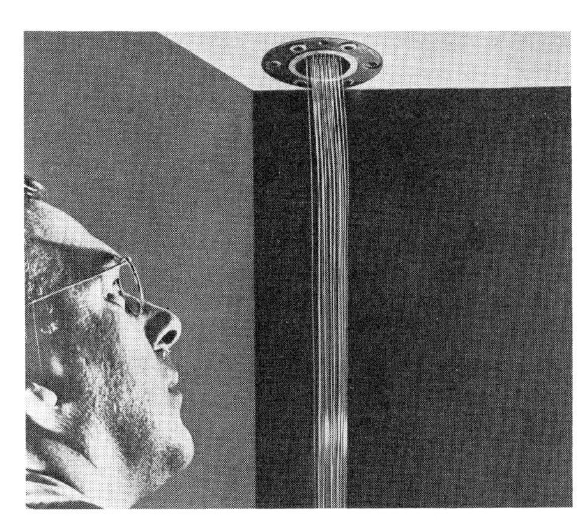

Fig. 11–5 *Spinning nylon fiber. (Courtesy of E. I. du Pont de Nemours & Company.)*

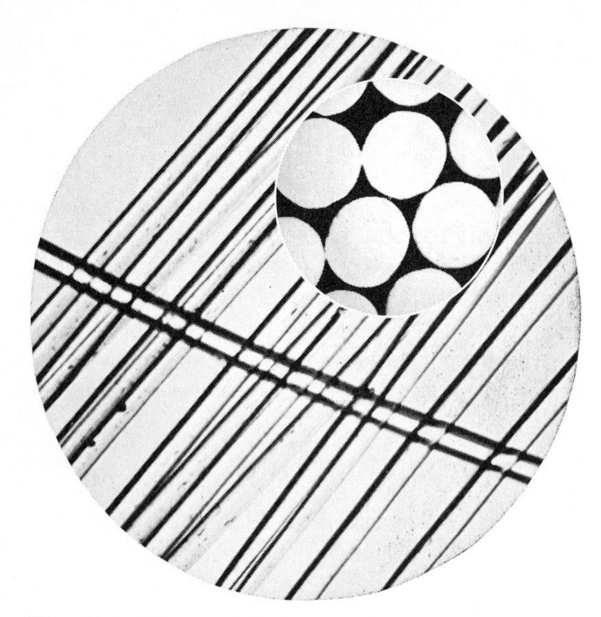

Fig. 11–6 *Photomicrographs of nylon fiber: longitudinal and cross-sectional (inset) views. (Courtesy of E. I. du Pont de Nemours & Company.)*

a dead feel. They lacked the liveliness of silk. This condition was corrected by changing the shape of the spinneret holes in 1959. Trilobal fibers give a silk-like hand to nylon fabrics (see Fig. 11–7). Qiana is a certification mark used by du Pont for trilobal fibers. In nylon carpets, trilobal fibers and square fibers with voids give good soil-hiding characteristics.

CHEMICAL COMPOSITION AND MOLECULAR ARRANGEMENT

Nylon—a manufactured fiber in which the fiber-forming substance is any long-chain, synthetic polyamide in which less than 85 percent of the amide linkages $\left[-\underset{\underset{O}{\|}}{C}-NH- \right]$ *are attached directly to two aromatic rings.—Federal Trade Commission.*

Fig. 11–7 *Photomicrograph of trilobal nylon. (Courtesy of E. I. du Pont de Nemours & Company.)*

The various nylons are all polyamides with recurring amide groups. They all contain the elements carbon, oxygen, nitrogen, and hydrogen. They differ in their chemical arrangement and this accounts for slight differences in properties.

The molecular chains of nylon vary in length. They are long, straight chains with no side chains or crosslinkages. Cold drawing aligns the chains so that they are oriented with the lengthwise direction and are highly crystalline. High-tenacity filaments have a longer chain length than regular nylon. Staple fibers are not cold drawn after spinning and thus have fewer crystallites. They have lower tenacities than filaments.

Nylon has a protein-like molecule and is related chemically to the protein fibers silk and wool. Both have amino dye sites that are important in the reaction of acid dyes. Nylon possesses far fewer dye sites than wool.

PROPERTIES

Aesthetic. Nylon has been very successful in hosiery and in knitted-filament fabrics such as

Summary of the Performance of Nylon in Apparel Fabrics

AESTHETIC	*VARIABLE*
DURABILITY	*EXCELLENT*
Abrasion resistance	Excellent
Tenacity	Excellent
Elongation	High
COMFORT	*LOW*
Absorbency	Low
Thermal retention	Moderate
APPEARANCE RETENTION	*HIGH*
Resiliency	High
Dimensional stability	High
Elastic recovery	Excellent
RECOMMENDED CARE	*MACHINE WASH*

Comparison of Nylon 6,6 and Nylon 6

Nylon 6,6	*Nylon 6*
Made of hexamethylene diamine and adipic acid	Made of captolactam
$\left[\overset{O}{\overset{\|}{C}}(CH_2)_4\overset{O}{\overset{\|}{C}}NH(CH_2)_6NH \right]_n$	$\left[NH(CH_2)_5\overset{O}{\overset{\|}{C}} \right]_n$
Advantages	*Advantages*
Heat setting 205°C (401°F) Pleats and creases, can be heat set at higher temperatures Softening point 250°C (482°F) Difficult to dye	Heat setting 150°C (302°F) Softening point 220°C (428°F) Better dye affinity than nylon 6,6; takes deeper shades Softer hand Greater elasticity, elastic recovery, and fatigue resistance Better weathering properties, including better sunlight resistance

tricot and jersey because of its smoothness, light weight, and high strength. The luster of nylon can be selected for the end use—it can be lustrous, semilustrous, or dull. Trilobal nylons have a pleasant luster.

The drape of fabrics made from nylon can be varied, depending largely on the yarn size and fabric structure selected. High-drape fabrics are found in sheer-knit overlays for nightgowns and in sheer-woven overlays in formals. Stiff fabrics are found in taffetas for formal wear or parkas. Very stiff fabrics include webbing for luggage handles and seat belts. These also vary in filament size.

Smooth textures are frequently found. These too can be varied by using spun yarns or by changing the knit or woven structure. The hand frequently associated with nylon fabrics is smooth because of the filament yarn and flat tricot-knit construction. Textured-yarn fabrics are bulkier.

Durability. Nylon has outstanding durability. High-tenacity fibers are used in seat belts, tire cords, ballistic cloth, and other industrial uses. Regular-tenacity fibers are used in apparel.

Tenacity (Grams Per Denier)	*Nylon 6,6*	*Nylon 6*
High-tenacity filament	5.9–9.8	6.5–9.0
Regular-tenacity filament	2.3–6.0	4.0–7.2
Staple	2.9–7.2	3.5–7.2
Bulked-continuous filament		2.0–4.0

High-tenacity fibers are stronger, but they have lower elongation than regular-tenacity fibers. During production, the high-tenacity fibers are drawn out more than the regular-tenacity fibers; they are more crystalline and oriented.

Breaking Elongation (Percent)	*Nylon 6,6*	*Nylon 6*
High-tenacity filament	15–28	16–20
Regular-tenacity filament	25–65	17–45
Staple	16–75	30–90
Bulked-continuous filament		30–50

In addition to excellent strength and high elongation, nylon has excellent abrasion resistance. Nylon carpet fibers outwear all other fibers.

This combination of properties make nylon the fiber for women's hosiery. No other fiber has been able to compete with nylon in pantyhose. The sheer, almost transparent, fiber is flattering. The fiber is more durable in wear than any other fiber for its sheerness. Filament hosiery develops runs because the fine yarns break and the knit loop is no longer secure. Very sheer hosiery made for evening wear is less durable than the coarser yarn that is intended for daily use.

The high elongation and excellent elastic recovery of nylon account for nylon's outstanding performance in hosiery. Hosiery is subjected to a relatively high degree of elongation; nylon recovers better after high elongation than other fibers do. As it is worn, nylon hosiery recovers its original shape at the knees and ankles, instead of bagging. Another factor that helps it retain its shape during wear is that the shape of hosiery can be pre-set by steaming. The stocking is knit in a tube, then shaped to fit the leg by heat setting the thermoplastic yarns.

Nylon is used for lining fabrics in coats or jackets. A number of years ago, acetate lining fabrics were the most commonly used. They were a poor choice because acetate has low strength and low abrasion resistance. Now nylon linings are available in raincoats, jackets, and winter coats. These linings are more durable.

Nylon is not very durable as a curtain or drapery fabric because it is weakened by the sun.

Comfort. Nylon has low absorbency. Even though its moisture regain is the highest of the synthetic fibers, 4.0–4.5 percent, nylon is not as comfortable a fiber to wear as the natural fibers.

Moisture Regain	*Nylon 6,6*	*Nylon 6*
At 70°F, 65 percent relative humidity	4.0–4.5	2.8–5.0
At 70°F, 95 percent relative humidity	6.1–8.0	3.5–8.5

Filament nylon was used in men's woven sports shirts in 1950. The smooth, straight fibers were packed together very compactly in yarns to minimize the transparency of the fiber. This transparency was especially noticeable in light shades when the shirt was wet from perspiration. The yarn and fabric structure resulted in shirts that felt like a plastic sheet wrapped around the body. As the man perspired, the fiber did not absorb the moisture nor did the dense fabric let any moisture escape. The humidity between the skin and the fabric rose, and the person perspired more and became even more uncomfortable. The shirts were especially uncomfortable in warm, humid weather.

Later, fabric structures were made with woven-in open spaces that permitted ventilation. Those were in dress shirts instead of sports shirts so the men were less active. The fabric structure resulted in a more-comfortable shirt.

Because of this early and inappropriate use, nylon got a bad image. On the basis of this experience, fiber producers began to develop fabric quality-control programs through which they could exercise control over the final product and thus protect the image of their fibers. Today, textured and spun yarns used in knit fabrics result in more-comfortable shirts.

Jersey-knit fabrics and tricot-knit fabrics made from nylon are more comfortable than woven-nylon fabrics because the additional air spaces within the fabric structure allow heat and moisture to escape more readily.

The very factors that make nylon uncomfortable under one set of conditions make it very comfortable under a different set. Nylon is widely used for wind-resistant jackets and parkas. The smooth, straight fibers pack closely together into yarns that can be woven into a compact fabric with very little space for wind to penetrate.

Another disadvantage of low absorbency is the development of static electricity by friction at times of low humidity. This disadvantage can be overcome by use of antistatic-type nylon fibers, by antistatic finishes, and by blending with high-absorbency, low-static fibers.

Nylon is very widely used in pantyhose and women's panties. Because of the low absorbency of nylon, doctors recommend that an absorbent crotch panel be used to decrease vaginal irritation and infection. Frequently, cotton is the fiber used for the panel.

Appearance Retention. Nylon fabrics are highly resilient because they are thermoplastic. They resist wrinkling during wear because they have been heat set. The same process can be used to make permanent pleats, creases, and embossed designs that last for the life of the garment. Shrinkage resistance is also high because the heat setting and the low-absorbancy fiber are not affected by water.

Elastic recovery is excellent. Nylon recovers fully from 8 percent stretch. No other fiber does as well. At 16 percent elongation, it recovers 91 percent immediately. This property makes nylon an excellent fiber for hosiery, tights, ski pants, and swim suits.

Elastic Recovery	Percent Recovery	At Percent Elongation
High-tenacity	89	3
filament	99–100	2–8
Regular-tenacity	88	3
filament	98–100	1–10
Staple	82	3
	100	2

Nylon does not wrinkle much in use, it is stable, and it has excellent elastic recovery—so it retains its appearance very well during wear.

Care. Nylon introduced the concept of "easy-care" garments. In addition to retaining their appearance and shape during wear, garments made from nylon fabrics retain their appearance and shape during care.

The wet strength of nylon is 80–90 percent of its dry strength. Wet elongation increases slightly. Little swelling occurs in wet fabric made of nylon 6,6. This is in marked contrast to cellulosic fibers. Nylon 6 swells 13–14 percent, cotton swells 40–45 percent and viscose rayon swells 80–110 percent.

To minimize wrinkling, use warm wash water and gentle agitation and spin cycles. Hot water may cause wrinkling in some fabric constructions. Wrinkles set by hot wash water can be permanent. Hot water will remove greasy and oily stains when necessary. Usually the additional wrinkling that occurs in the wash can be pressed out without any problem.

Nylon is a "color scavenger." White and light-colored nylon fabrics pick up color or dirt that is in the wash water. A red sock that loses color into the wash water of a load of whites will turn the white nylon fabrics a dingy pink-gray. This extra color may be so difficult to remove that you may have to resort to color remover. Discolored nylon and grayed or yellowed nylon can be avoided by following correct laundry procedures.

Since nylon has low absorbency, it is quick drying. Travelers find this convenient because items can be hand washed at night and ready to wear the next morning. Because the fabrics are so quick drying, they need to be dried a short time. Do not overdry the fabrics. They can be put in with the rest of the clothes at the beginning of the cycle and removed when dry, or they can be put in near the end of the cycle, or they can be dried on a line.

Dryer temperatures should be warm or low. Avoid using the hot setting on commercial gas dryers. Figure 11–8 is the melted and fused result of a nylon garment dried in an overheated gas dryer with socks of a different fiber content.

Nylon does have problems with static, particularly when the air is not humid, so a fabric softener may be used in the washer or dryer. Nylon should be ironed at a low temperature setting—270–300°F. Home-ironing temperatures are not high enough to press seams, creases, and pleats permanently in home-sewn garments or to press out wrinkles acquired in washing. Using too hot an iron will cause glazing, then melting.

The chemical resistance of nylon is generally good. Nylon has excellent resistance to alkali and chlorine bleaches but is damaged by strong acids. Soot from smoke in industrial cities contains sulfur, which on damp days combines with atmospheric moisture to form an acid that has been responsible for epidemics of runs in stockings. Certain acids, when printed on the fabric, will cause shrinkage that creates a puckered damask effect. Nylon will dissolve in formic acid and phenol.

Nylon is resistant to moths and fungi.

Nylon has low resistance to sunlight. Better resistance is achieved in curtain fabrics by using bright rather than delustered fibers, which absorb rather than reflect light.

Identification. The burning test is one way to identify nylon. Untreated nylon does not flash burn and does not readily support the spread of the flame after the ignition sources are removed.

Fig. 11–8 *The melted and fused remains of nylon garments dried in an overheated gas dryer.*

When exposed to a flame, nylon fuses and draws away from the flame before it will ignite. When it burns, the nylon fibers melt and drip and some of the flame is carried down with the drip. The odor is celery-like, and white smoke is given off. In untreated nylon, the melt will harden as a tan bead. A black bead forms when dyes are present, and certain finishes will increase the flammability.

USES

Nylon is the third most widely used fiber in the United States. It follows polyester and cotton in pounds used and is far ahead of all other fibers. The highest production of nylon occurred in 1979. Since then, total pounds produced have decreased.

Fiber	*Million Pounds of Fiber Used (1985)*	*Percent*
Polyester	3,351	30.1
Cotton	2,947	26.5
Nylon	2,379	21.4
Olefin and vinyon	1,239	11.1
Acrylic	644	5.8
Rayon and acetate	558	5.0
Total	11,118	99.9

One estimate for 1986 states that the uses of nylon were as follows:

Home furnishings	60 percent
Apparel	21 percent
Industrial	19 percent

The single most important use of nylon is for carpets. Over 85 percent of the face fibers on carpets in the United States in 1986 were nylon. Tufted carpets are an excellent end use for nylon because of its aesthetic appearance, durability, appearance retention, and ability to be cleaned in place. The combination of nylon fiber and the tufting process resulted in relatively low-cost carpeting that has brought about the widespread use of carpeting in both residential and commercial buildings.

In 1980, 30 percent of nylon went into filament yarns for carpets. Another 23 percent was used in staple-carpet yarns. Thus 53 percent of all nylon produced in the United States was used for carpets. The other 2 percent in home furnishing was used for upholstery fabrics.

The second important use of nylon is for apparel. Lingerie fabrics are an end use for which nylon is the predominant fiber. The knit structure is open enough so the fabrics are comfortable most of the time. The fabrics are attractive and durable; they retain their appearance well and are easy care. Panties, bras, nightgowns, pajamas, and lightweight robes are frequently made from nylon.

Women's sheer hosiery is an important end use of nylon. How frequently they are simply called "nylons"! No other fiber has the combination of properties that make it so ideal for that use. The very sheer hosiery is often 12–15 denier instead of the once standard 30 denier yarn or monofilament. Sheers give the look that is wanted, but they are less durable. Hosiery yarns may be monofilament or multifilament-stretch nylon. They may be plain or textured.

Short socks or knee-high socks are sometimes made from nylon. More frequently they are nylon blends with cotton or acrylic, with the nylon adding strength and stretch.

Active sportswear where comfort stretch is important—leotards, tights, swim suits, and ski wear—is another end use for nylon. Other apparel fabrics are used for blouses and dresses. Nylon-taffeta windbreakers and parkas are commonly seen in cooler weather. Lining fabrics, especially for jackets and coats, are sometimes made of nylon.

Industrial uses for nylon are varied. Within this group, the most important use of nylon is for tire cord. In 1981, 11 percent of the nylon produced was used for this purpose. Nylon is facing stiff competition in this specialized market. It captured the market from high-tenacity rayon, but now may lose the market to polyester, aramid, and/or steel.

Although nylon is strong and abrasion resistant, with high elongation and high elasticity, it has a tendency to "flat spot." Flat spotting occurs when a car has been stationary for some time and a flattened place forms on the tire. The car will have a bumpy ride for the first mile or so until the tire recovers from flattening. With the advent of belted-radial tires and the availability of heat-resistant aramid and steel, the market is changing again. The nylon or polyester fibers that are used in the cord of radial tires go rim to rim over the curve of the tire.

Types and Kinds of Nylon

Cross-Section	Dyeability	Crimp or Textured	Others
Round	Acid dyeable	Mechanical crimp	Antistatic
Heart-shaped	Cationic dyeable	Crimp-set	Soil hiding
Y-shaped	Disperse dyeable	Producer textured	Bicomponent
8-shaped	Deep dye	Undrawn	Faciated
Delta	Solution dye	Partially drawn	Thick and thin
Trilobal	Heather	Steam crimped	Antimicrobial
Triskelion	Optically whitened	Bulked continuous filament	Sunlight resistant
Trinode		Latent crimp	Flame resistant
Pentagonal			Delustered
Hollow			High tenacity
			Crosslinked

Car interiors are another example of the varied uses for nylon. The average car uses 25 pounds of fiber, most of which is nylon. Upholstery fabric (called body cloth), carpet for the interior, trunk lining, door and visor trims, head liners on the interior of the car roof, and seat-belt webbing are all nylon fabrics of one kind or another. In addition, clutch pads, brake linings, and yarns to reinforce radiator hoses and other hoses are needed. This again is a very competitive market. Polyester is gaining importance. Research is being done to determine the fiber and fabric structure that is most appropriate for air bags.

Additional industrial uses include the following: parachute fabric, cords and harnesses, glider-tow ropes, ropes and cordage, conveyor belts, fishing nets, mail bags, and webbings.

The category of industrial uses also includes consumer uses and sporting goods. Consumer uses include umbrellas, clotheslines, toothbrush bristles, hair-brush bristles, paint brushes, and luggage. A popular nylon fabric for soft-sided luggage is a 430-denier woven-oxford canvas. In 1986, nylon was used for almost three-fourths of all soft-sided luggage.

Nylon is important in sporting goods. It is used for tents, sleeping bags, spinnaker sails, fishing lines and nets, racket strings, back packs, and duffle bags.

TYPES AND KINDS OF NYLON

It has been said that as soon as a new need arose, a new type of nylon was produced to fill the need. This has led to a large number of types of nylon that are identified by trademarks.

Nearly 200 variants of nylon were listed in the 1983–1984 *Textile Industries* "Man-made Fiber Variant Chart." Differences in luster, denier, mechanical crimp, or staple length were not considered sufficient to define a variant in that list. Over half of the variants of nylon were made by du Pont. Only 8 of the 19 producers of nylon were included in the list. The types and kinds of nylon fibers are too numerous to list as was done for rayon and acetate. The lists on this page illustrate many modifications of nylon.

Some Trademarks and Producers

Nylon 6,6		Nylon 6	
Trade Names	Producer	Trade Names	Producer
Antron, Cantrece, Cordura	du Pont	Anso, Caprolan, Captiva, Hydrofil	Allied
Ultron, Wear-Dated	Monsanto	Natural Touch	BASF Fibers
		Zefran	
		Zefsport	
		Zeftron	
		Shareeen	Courtaulds

12

Polyester

The polyester polymers were part of the high-polymer research program of Wallace Carothers in the early 1930s. When work on the polyesters was discontinued by du Pont in favor of the more-promising nylon fiber, research on polyesters continued in England, and the first polyester fiber, Terylene, was produced there under a patent that controlled the production rights for the world. In 1946, du Pont purchased the exclusive right to produce polyesters in the United States. The du Pont fiber was given the trade name Dacron—a name that is commonly mispronounced. The correct pronunciation is "day'kron."

Polyester was introduced to Americans at a press conference in 1951 where a man's suit was displayed. This suit was still presentable after being worn continuously for 67 days without pressing. It had been dunked in a swimming pool twice and it had been washed by machine. But it had not been pressed. The outstanding resiliency of polyester, whether dry or wet, coupled with its outstanding dimensional stability after heat setting made it an instant favorite.

Dacron was first produced commercially in 1953. In 1958, Kodel, a different kind of polyester, was introduced by Eastman Kodak Company. In 1960, four companies were producing polyester; in 1986, there were 15 producers, including du Pont, BASF Fibers, Celanese, Hoechst, and Avtex Fibers.

Polyester is the most widely used synthetic fiber. Polyester is sometimes referred to as the "workhorse" fiber of the industry. The filament form of the fiber has been said to be the most versatile fiber, and the staple form has been called the "big mixer" because it can be blended with so many other fibers, contributing its good properties to the blend without destroying the desirable properties of the other fiber. Its versatility in blending is one of the unique advantages of polyester.

By the time the polyesters were synthesized, much had been learned about high polymers and about the structure of fibers. Many of the problems of production had been solved—for example, controlled luster and strength, spinning methods, making of tow for staple fibers, and crimping of staple. Continuing research is being done on heat setting, high-temperature dyeing, and static control. Man-made fibers were being promoted vigorously by their trade names. The generic names nylon, rayon, acetate, and acrylic had been agreed on. When the polyesters were introduced, they were backed by quality-control programs that limited the use of a trade name to those products that met standards set by the fiber producers. Consumers readily accepted polyesters.

The polyesters have probably undergone more research and developmental work than any other fiber. The polymer is "endlessly engineerable," and many physical and chemical variations are possible. These modified fibers are designed to improve the original polyester in areas where it has shown either a deficiency or a limitation in its use. One of the important physical changes has been that of changing from the standard round shape to a trilobal cross-section that gives the fiber silk-like properties. A chemical modification, high-tenacity staple, was developed for use in durable-press fabrics. The strength of the polyester reinforces the cotton fibers, which are weakened by the finishing process. Current research is focused on developing a more "natural" polyester—polyester with a hand and absorbency more like the natural fibers.

The properties of polyester that make it the most widely used man-made fiber are listed in the following chart.

PRODUCTION

Polyester is made by reacting dicarboxylic acid with dihydric alcohol. The fibers are melt spun by a process that is very similar to that used to make nylon. The polyester fibers are hot drawn (nylon is cold drawn) to orient the molecules and make significant improvements in strength and elongation, and especially in the stress/strain properties. As the polyester fibers, like the nylons, have the ability to retain the shape of the spinneret hole, modifications in cross-sectional shape are possible.

Figure 12–1 is a diagram of the production of polyester staple fiber. The diagram includes several steps with chips, or small pieces of hardened polymer. Many manufacturers eliminate the chip stage and extrude the polymer directly into fiber.

PHYSICAL STRUCTURE

Polyester fibers are produced in many types—filament yarns, staple fibers, and tow. Filaments

Properties of Polyester

Properties	*Importance to Consumers*
Resilient—wet and dry	Easy-care apparel, home furnishings, packable garments
Dimensional stability	Machine washable
Resistant to sunlight degradation	Good for curtains and draperies
Durable, abrasion resistant	Industrial uses, sewing thread, good for work clothes
Aesthetics superior to nylon	Blends well with natural or other man-made fibers, good silk-like filaments

are high tenacity or regular, bright or delustered, white or solution dyed. Staple fibers are available in deniers from 1.5–10 and are delustered. They may be regular, low pilling, or high tenacity.

Regular polyester fibers, when seen under the microscope, are so much like nylon that identification is difficult. The smooth rod-like fibers have a circular cross-section (Figure 12–2). The fibers are not as transparent as the nylon fibers. They are white, so they normally do not need to be bleached. However, whiter types of polyester fibers have been produced by the additon of optical whiteners (fluorescent compounds) to the fiber-spinning solution. The pitted appearance of the Dacron fiber is caused by the delusterant that was added to the spinning solution.

A variety of cross-sectional shapes are produced: round, trilobal, octolobal, oval, hollow, voided, hexalobal, and pentalobal (star-shaped).

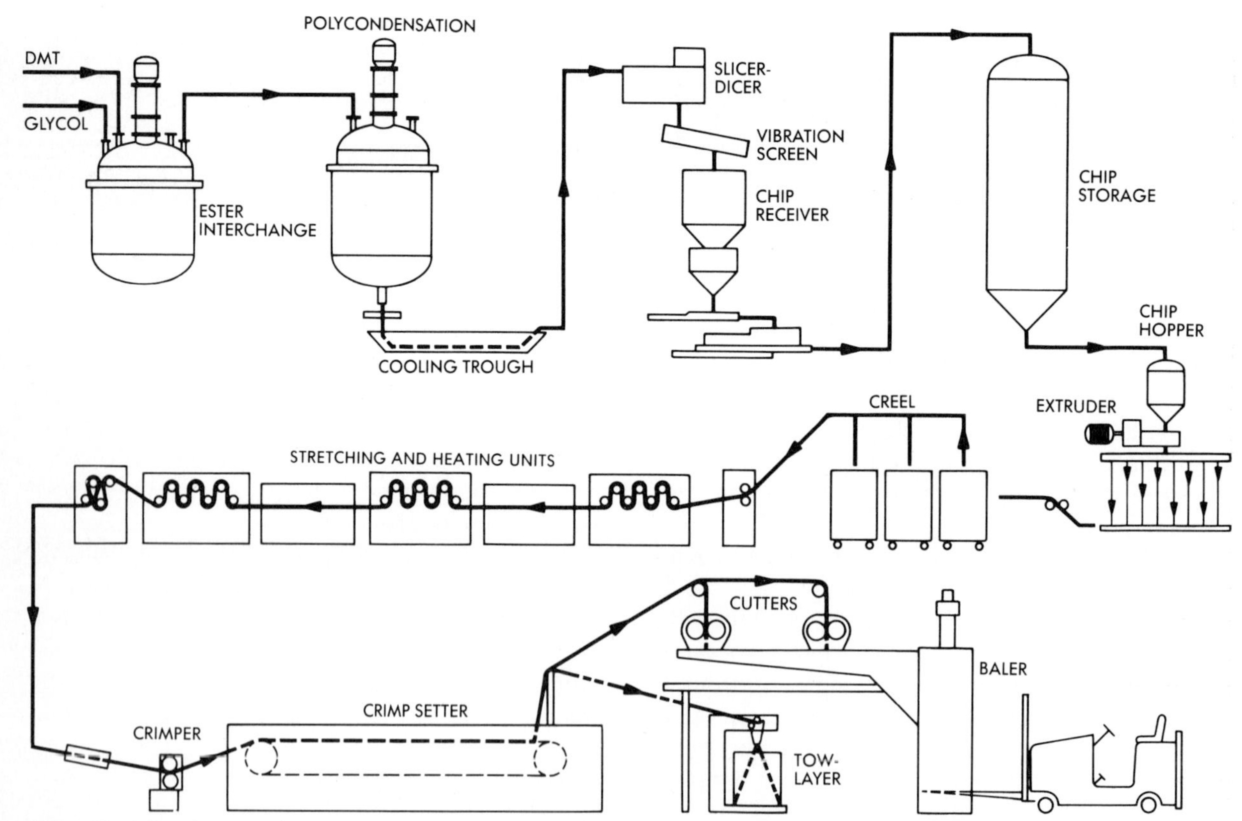

Fig. 12–1 Production diagram for polyester staple fibers. (Courtesy of Hoechst Fibers Industries.)

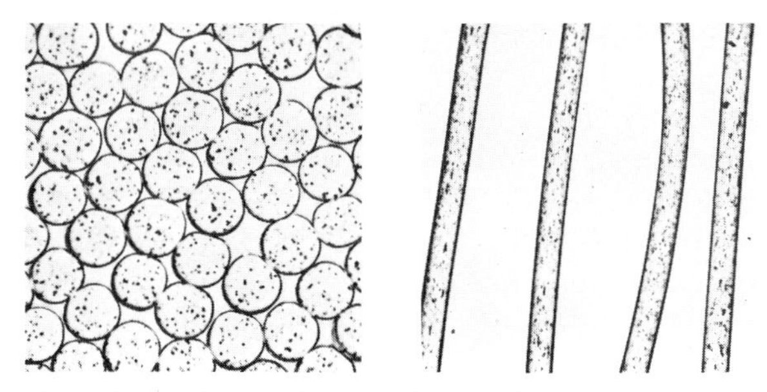

Fig. 12–2 *Photomicrographs of Dacron polyester: cross-section of regular-delustered Dacron 500×* (left); *longitudinal view 250×* (right). *(Courtesy of E. I. du Pont de Nemours & Company.)*

CHEMICAL COMPOSITION AND MOLECULAR ARRANGEMENT

Polyester fibers—manufactured fibers in which the fiber-forming substance is any long-chain synthetic polymer composed of at least 85 percent by weight of an ester of a substituted aromatic carboxylic acid, including but not restricted to substituted terephthalate units, $p(-R-O-\overset{\underset{\|}{O}}{C}-C_6H_4-\overset{\underset{\|}{O}}{C}-O-)$, *and para substituted hydroxybenzoate units,* $p(-R-O-C_6H_4-\overset{\underset{\|}{O}}{C}-O-)$. *– Federal Trade Commission.*

Polyester fibers are made from two kinds of terephthalate polymers. The original fibers Terylene and Dacron were spun from polyethylene terephthalate (abbreviated PET). In 1958, Eastman Chemical Products, Inc., introduced a new type of polyester, Kodel, which is spun from 1,4 cyclohexylene-dimethylene terephthalate, commonly known as PCDT. The differences are listed in the following table.

One should not assume from this chart that only Kodel II is low pilling. The PET spinning solutions may be homopolymers or copolymers. The copolymers are pill-resistant, lower-strength staple fibers used primarily in knits and carpets.

Polyester fibers have straight molecular chains that are packed closely together and are well oriented with very strong hydrogen bonds.

PROPERTIES

Polyester fibers are outstanding in their wet and dry resiliency. Because of polyester, ironing has almost been eliminated from apparel and bed and table linens, although many people still do touch-up pressing.

Aesthetic. Polyester fibers accommodate themselves in blends so that a natural-fiber look and

Comparison of PET and PCDT Fibers

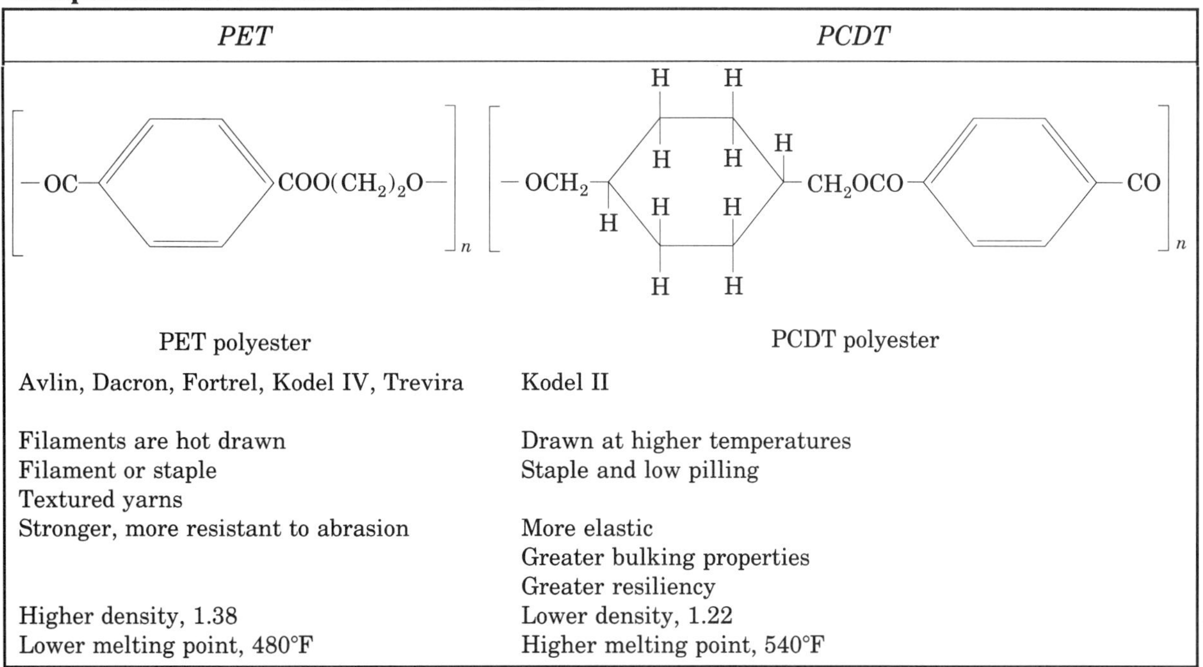

PET	*PCDT*
$[-OC-C_6H_4-COO(CH_2)_2O-]_n$	$[-OCH_2-C_6H_{10}-CH_2OCO-C_6H_4-CO]_n$
PET polyester	PCDT polyester
Avlin, Dacron, Fortrel, Kodel IV, Trevira	Kodel II
Filaments are hot drawn	Drawn at higher temperatures
Filament or staple	Staple and low pilling
Textured yarns	
Stronger, more resistant to abrasion	More elastic
	Greater bulking properties
	Greater resiliency
Higher density, 1.38	Lower density, 1.22
Lower melting point, 480°F	Higher melting point, 540°F

Summary of the Performance of Polyester in Apparel Fabrics

AESTHETIC	*VARIABLE*
DURABILITY	*EXCELLENT*
Abrasion resistance	Excellent
Tenacity	Excellent
Elongation	High
COMFORT	*LOW*
Absorbency	Low
Thermal retention	Moderate
APPEARANCE RETENTION	*HIGH*
Resiliency	Excellent
Dimensional stability	High
Elastic recovery	High
RECOMMENDED CARE	*MACHINE WASH*

texture are maintained with the advantage of easy care. They are very widely used in blends with cotton for shirts, slacks, and skirts. The appearance of these fabrics is like cotton; their appearance retention during both wear and care strongly proclaims the influence of polyester.

Thick-and-thin yarns of polyester and rayon give a linen-look to summer-weight blouse and suit fabrics. Wool-like fabrics are found in both summer-weight and winter-weight men's suiting fabrics.

Silk-like polyesters have been very satisfactory in appearance and hand. The trilobal polyester fibers were developed as the result of a study by du Pont to find a man-made filament that would have the aesthetic properties of silk. The study, made in cooperation with a silk-finishing company, began by investigating the effect of silk-finishing processes on the aesthetic properties of silk fabrics, since silk seemed to acquire added richness in the fabric form.

In silk fabric, sericin (gum) makes up about 30 percent of the weight. The boil-off finishing process removes the sericin and creates a looser, more-mobile fabric structure. If the fabric is in a relaxed state while the sericin is being removed, the warp yarns take on a high degree of weave crimp. This crimp and the looser fabric structure together create the liveliness and suppleness of silk. The suppleness has been compared to the action of the coil-spring "Slinky" toy. The properties are quite different when the boil-off is done under tension. The weave crimp is much less, and the response of the fabric is more like that of a flat spring; thus the supple nature is lost. This helps to explain the difference between qualities of silk fabric.

The results of the silk fabric study indicated that the unique properties of silk—liveliness, suppleness, and drape of the fabric; dry "tactile" hand; and good covering power of the yarns—are the result of (1) the triangular-like shape of the silk fiber; (2) the fine denier per filament; (3) the loose, bulky yarn and fabric structure; and (4) a highly crimped fabric structure.

The process was then applied to polyesters. The fibers were spun with a trilobal shape and made into fabrics that were processed by a silk-finishing treatment. The polyesters were particularly suited to this study. They are unique because they can be treated with a caustic soda to dissolve away the surface, leaving a thinner fiber, yarn, or fabric without changing the fiber basically.

Man-made fibers are normally processed under tension by a continuous method rather than by a batch method. Because of the results of the du Pont research study of silk, the trilobal fabrics are processed in a completely relaxed condition. Finishing starts with a heat-setting treatment to stabilize the fabric to controlled width, remove any wrinkles, and impart resistance to wrinkling. The next step is a very important caustic-soda (alkali) treatment, which dissolves away a controlled amount of the fiber. This step is similar to the degumming of silk and it gives the fabric structure greater mobility. All remaining finishes are done with the fabric completely relaxed to get maximum weave-crimp. (Antron nylon is finished in the same way, except that there is no caustic treatment.) Figure 12–3 shows the effect of the alkali treatment on a fabric made of a circular-cross-section polyester.

Durability. The abrasion resistance and strength of polyesters are excellent, and the wet strength is comparable to the dry strength. The high strength is developed by hot-drawing, or stretching, to develop crystallinity and also by increasing the molecular weight. The breaking tenacity of polyester is varied depending on the end use.

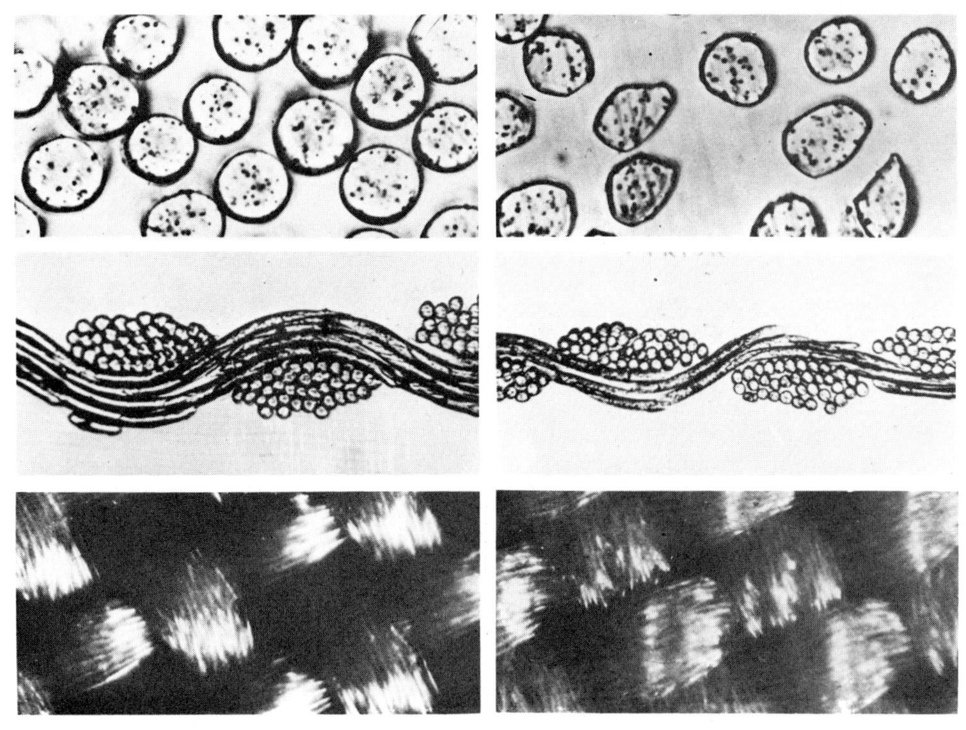

Fig. 12–3 Photomicrographs showing effect of heat-caustic treatment. Original fabric on left; fabric after treatment on right. Dacron polyester fiber cross-section 1,000× (top); *fabric cross-section 200×* (center); *fabric surface 50×* (bottom). *(Courtesy of E. I. du Pont de Nemours & Company.)*

Type of Polyester Fiber	*g/d*	*Typical End Use*
High-tenacity filament	6.8–9.5	Tire cord, industrial uses
Regular-tenacity filament	2.8–5.6	Apparel and home furnishings
High-tenacity staple	5.8–7.0	Durable-press apparel
Regular-tenacity staple	2.4–5.5	Apparel and home furnishings

The stronger fibers have been stretched more, so their elongation is lower than the weaker fibers. This is particularly dramatic in the case of partially oriented filament fibers. These are sold to manufacturers who will stretch them more during the production of textured yarns. Their tenacity is 2.0—2.5 g/d. These filament fibers are lower in strength than the staple fibers! Yet their elongation far exceeds that of the other fibers. Their elongation is 120–150 percent! They can be thought of as being partially manufactured fibers until the texturing is completed.

Type of Polyester Fiber	*Percent Elongation*
High-tenacity filament	9–27
Regular-tenacity filament	18–42
High-tenacity staple	24–28
Regular-tenacity staple	40–45

Comfort. Absorbency is quite low for the polyesters, ranging from 0.4–0.8 percent moisture regain. Poor absorbency lowers the comfort factor of skin-contact apparel.

Woven fabrics made from round polyester fibers can be very uncomfortable to wear in warm, humid weather or to wear when the person is perspiring. Moisture does not escape easily from between the skin and the fabric, and the fabric feels slick and clammy.

To increase the comfort of a garment, select a loose-fitting garment design, a thin and somewhat open fabric design, spun rather than filament yarns, trilobal rather than round fibers, and finishes that absorb, or wick, moisture. The soil-release finishes have improved the wicking characteristics of the polyesters, thus improving the breathability and comfort of the fabrics. Additional work on finishes and chemical or fiber modifications is being done in an effort to increase the comfort of polyester.

Blends of polyester/cotton are more comfortable to wear in humid weather than are 100 percent polyester fabrics. Blends of cotton/polyester, where cotton accounts for 60–90 percent of the fabric, are even more comfortable in humid weather. Since cotton is so absorbent, it absorbs most of the moisture. The rest of the moisture is wicked along the outer surface of the polyester fibers to the fabric surface where it evaporates. Polyester is resilient when it is wet, so the fabric does not matt down. Polyester is light in weight and dries quickly.

Polyester exhibits moderate thermal retention. It is generally not as comfortable as wool or acrylic for cold-weather wear. Blends with wool are very successful in increasing its comfort, particularly in men's suits. A lot of work has been done engineering polyester for fiberfill. Fiber modifications—including hollow fibers, binder staple, and crimped fibers—are performing very well.

Polyesters are more *electrostatic* than the other fibers in the heat-sensitive group. Static is characteristic of fibers that have low absorbency. Static is very annoying when it causes clothes to cling to each other. Static is a definite disadvantage because lint is attracted to the surface of fabrics and it is difficult to keep dark-colored fabrics looking neat. Curtains soil more rapidly. New fabrics usually have an antistatic finish, but it is often removed by washing or dry cleaning. The fabric softeners as a laundry aid are good antistatic agents. Temporary relief from static can be gained by running a damp sponge over a garment or by using an antistatic spray. Density of most polyester fibers is 1.38. Hollow variants for fiberfill are less dense.

Appearance Retention. *Resiliency* relates to tensile-work recovery and refers to the extent and manner of recovery from deformation. The following table indicates that polyester has a high recovery when the elongation is low, an important factor in the suiting market. Only small deformations are involved in the wrinkling of a suit, and Dacron recovers better than nylon under those conditions. The recovery behavior of Dacron is similar to that of wool at the higher elongations, which helps explain the compatability of polyester and wool blends. Nylon exhibits better recovery at the higher elongations, so it performs better in garments that are subject to greater elongation—hosiery, for example.

Tensile Recovery from Elongation of:

Fiber	*1%*	*3%*	*5%*	*15%*
Polyester 56 (regular)	91	76	63	40
Nylon 200 (regular)	81	88	86	77

Source: E. I. du Pont de Nemours & Company, *Technical Bulletin X-142* (September 1961).

The other polyesters are similar to Dacron in their wet- and dry-wrinkle recovery. This has given them an advantage over wool in tropical suitings, since wool has poor wrinkle recovery when wet. Under conditions of high atmospheric humidity and body perspiration, polyester suits do not shrink and are very resistant to wrinkling. However, when polyester garments do acquire wear wrinkles, as often happens at the waist of a garment where body heat and moisture "set" the wrinkles, pressing is necessary to remove them.

Resiliency and quick drying make the polyesters especially good for fiberfill batts in quilted fabrics—for example, quilts, bedspreads, parkas, and robes.

To summarize, the resiliency of polyester is excellent; it resists wrinkles and, when wrinkled, it recovers well whether wet or dry. Elastic recovery is high for typical apparel items. The dimensional stability of polyester is high. When properly heat set, it retains its size. It can be permanently creased or pleated satisfactorily.

Pilling was a severe problem with fabrics made from the unmodified polyesters. Pilling changes the appearance of fabrics, making them look shabby before they are worn out. Polyester fabrics did not pill more than wool fabrics, but the pills held on and did not break off as they did with wool. Pilling is a problem with all smooth, round fibers of high strength. Low-pilling fiber types have been developed to minimize the problem and to make them more suitable to use in blends with wool and cellulose and in napped and pile fabrics. The finishing process of singeing also helps control pilling of polyester/

cotton blends. Singeing must be done carefully or the fiber ends will appear darker.

Care. Polyester has revolutionized the way Americans care for everyday clothing. White cottons were washed in hot water; colored cottons were washed in warm water. They were dried on a line or in a dryer at a high temperature. Cottons were ironed damp with a hot iron to remove wrinkles. Some people starched their cottons for additional luster and crispness. Clothing was bought large and expected to shrink—especially knit items. Once an item had been worn all day, it looked like it—it was a mass of wrinkles! Wool suits and coats were dry cleaned.

The revolution in clothing care occurred because of the dimensional stability heat setting produces in nylon and polyester fabrics, both knit and woven. Equally important, it occurred because of the advent of durable-press fabrics, notably polyester/cotton blends. During the 1950s and 1960s, cycles on washing machines changed, and special instructions were developed on handling the new durable-press clothing to minimize wrinkling. Now care instructions for polyester/cotton durable-press fabrics can be summarized as follows: Wash in warm water; dry with medium heat in a dryer and remove promptly when the cycle is over; hang; touch-up press with a steam iron.

The excellent abrasion resistance and tenacity, and the high elongation of polyester are unaffected by water; polyester remains the same wet as dry. The low absorbancy of polyester (0.4 percent) means that it resists water-borne stains and is quick to dry. The excellent resiliency of polyester keeps it looking good during wear and minimizes wrinkling during care so only light pressing is required to remove what wrinkles have occurred. Shrinkage resistance is high—indeed, because of heat setting, dimensional stability is excellent.

Warm water is generally recommended to minimize wrinkling of polyester or polyester-blend fabrics. However, hot water (120–140°F) may be needed to remove greasy or oily stains or built-up body soil because polyester is oleophilic, or oil absorbent; polyester has a tendency to retain oily soil. Perhaps one of the most familiar examples of this is "ring around the collar." With polyester shirts or polyester/cotton blends, the soil usually responds to pretreatment, then laundering. If, however, these means are not enough to remove the soil, and it starts to gray the collar in that area, using hot water should help.

Besides holding on to oily soil, another way this can affect laundry results is that soil in the wash water can redeposit on clothing and make it dingy. While polyester is not the color scavenger that nylon is, white polyester will dull or gray if washed with colored or heavily soiled garments. Soil-release finishes applied to fabrics can also make a difference in how well garments clean. The effect of these finishes decreases after many washings.

Another adverse problem of polyester apparel is a tendency to exhibit bacterial odor. Whether or not this is a problem depends on a variety of factors: the person, the fabric, and the laundry procedures. Apparently this is a problem when soil has built up on the fabric; bacteria grow there and an odor results. Use of hot water wash; laundry agents such as borax, which minimizes odor; or bleach to remove the soil buildup and kill the bacteria may minimize the problem. Several detergents were marketed in 1985 and 1986 in response to this problem.

Polyester fibers are generally resistant to both acids and alkalis and can be bleached with either chlorine or oxygen bleaches. This is very important because the largest single use of polyester is in blends with cotton for durable-press fabrics. Polyester fibers are resistant to biological attack and to sunlight damage. The polyester fibers are the most important filaments for sheer curtains.

Polyesters are thermoplastic. They must be heat-set to obtain stability and permanent pleats in garments. Washing in warm water followed by tumble drying is recommended, but polyesters may be safely washed in hot water to remove greasy and oily stains or to remove body oils. Hot water may cause fabrics to wrinkle more and may cause color loss.

The heat properties of the polyesters are used to advantage in the production of fiberfill for pillows, quilts, and linings. The fiber is flattened on one side or made asymmetrical while it is softened by heat, and it will then take on a tight spiral curl of outstanding springiness. Fiberfill can be made of a blend of fiber deniers to give different levels of support for pillows. Lumpiness in pillows can be prevented by spot-welding the fibers to each other by running hot needles through the pillow bat. The snowmobile suits in Figure 12–4 have unquilted nylon shells over Kodel polyester fiberfill insulation.

Identification. The polyesters, like nylon, withdraw from the flame before igniting, so they do not flash burn. They also melt and drip and the flame is carried down with the drip. A black bead forms when the melt hardens. The polyesters can be distinguished from the nylons by odor and smoke. The polyesters have an aromatic odor and produce a heavy black smoke that contains pieces of soot. The fabric must burn briskly before black smoke and soot are evolved, so an adequate sample must be tested to make the identification positive.

USES

Polyester is the most widely used fiber in the United States. Its use has steadily increased since its introduction in 1954. Peak mill consumption of polyester was 3,800 million pounds in 1979. Statistics on pounds produced and percent of U.S. mill consumption of polyester are shown in the table.

Fig. 12–4 Snowmobile suits with Eastman Kodel polyester fiberfill. (Courtesy of the Rowland Company, Inc.)

Year	*Million Pounds*	*Percent of Mill Consumption*
1965		4.0
1970	1,600	17.0
1975	3,100	30.0
1980	3,500	31.0
1985	3,351	30.1

Cotton had been the most widely used fiber, but in 1973, 1974, and 1975, cotton and polyester were used almost equally. By 1976, polyester began to be used more than cotton.

Polyester still takes second place to cotton for apparel uses. Of the 4.2 billion pounds of fibers processed for apparel fabrics in 1985, cotton, polyester, and acrylic accounted for almost 88 percent of all apparel fabrics!

Fiber	*Percent of Apparel Fabric Fibers Used*
Cotton	40.2
Polyester	33.3
Staple	19.2
Filament	14.1
Acrylic	14.2
Other	12.3

Another way of looking at the importance of polyester is seeing the market share it has captured in various markets (see Figure 12–5).

What are the most important uses of polyester? Using data on domestic shipments of yarn in 1984, staple polyester accounted for 64 percent (1,992 million pounds) and filament polyester accounted for 36 percent (1,145 million pounds) of fiber.

Spun Yarns of Polyester Are Used For:

Broadwoven fabrics	1,039.1 million pounds
Fleece and other knit fabrics	283.0
Fiberfill	276.1
Nonwoven fabrics	177.2
Carpet-face yarns	131.4
Pile fabrics	18.4
Blankets	5.5
Other uses	61.3
Total	1,992.0 million pounds

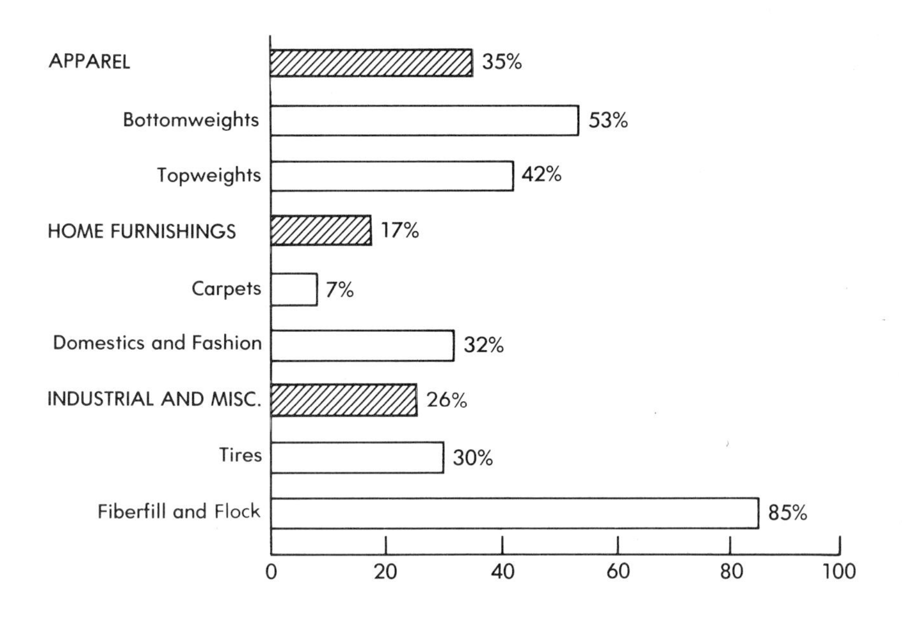

Fig. 12–5 Polyester share of major markets, 1983. (American Fabrics and Fashions, *1965. Number 132. Copyright © 1987 Bobbin International, Inc. All Rights Reserved.)*

Filament Yarns of Polyester Are Used For:

Broadwoven fabrics	392.2 million pounds
Knit fabrics	312.0
Tire cord	174.4
Narrow webs	38.6
Other rubber	32.7
Rope, cordage, and fishing line	15.7
Other uses	179.4
Total	1,145.0 million pounds

Feeder and partially oriented yarns account for 675.5 million pounds of polyester.

The most important use of polyester is in woven fabrics. Frequently, spun yarns blended with cotton are seen. Polyester filament yarns may be used in one or both directions of the fabric. Most of the woven fabrics are polyester/cotton blends made into durable-press fabrics. For topweight fabrics—shirts and blouses—65 percent polyester/35 percent cotton blends are commonly seen. For bottomweight fabrics—slacks, skirts, and suits—50 percent polyester/50 percent cotton is seen. These blended fabrics are attractive, durable, comfortable (except in very hot and humid conditions), retain their appearance well, and are easy care. Their excellent performance has resulted in their widespread use and continued popularity.

The first use of staple polyester was in tropical suitings for men's summer suits. The suits were light in weight and machine washable—something unique in men's clothing. The very low absorbency of the polyester fibers limited the comfort factors of these first garments, a disadvantage that was overcome by blending polyester with cotton and/or wool. In 1977, staple polyester began to be widely used in heavier cotton-like fabrics such as denim or gabardine.

Woven fabrics are very important in apparel. They are also important in home-furnishing uses—polyester and polyester blends are widely used in sheets, blankets, bedspreads, curtains that match bedspreads, mattress ticking, and tablecloths. They are being used in more upholstery fabrics.

The filaments are also used in sheer curtains,

where the excellent light resistance of the fibers and the fineness of denier made them particularly suitable for ninon and marquisette.

The second important use of polyester is in knitted fabrics. Slightly more filament yarns than spun yarns are used. Polyester as well as polyester/cotton blend yarns are used. Knit fabrics of polyester can look sporty or sophisticated. They wear well, are comfortable in well-chosen fabric structures and garment designs, retain their appearance well, and are easy care.

The first use of polyester filaments was in knit shirts for men and blouses for women. The use of filament polyester increased tremendously when the "set" textured yarns were developed and used in double-knit and woven fabrics for suits, dresses, jackets, and lightweight coats in the 1970s. Both smooth and textured filaments have wide use in career apparel such as uniforms for nurses and waitresses.

The third important use of polyester is in a specialized area—fiberfill. Used in pillows, comforters, bedspreads, other quilted household and apparel fabrics, and winter jackets, polyester has captured the major share (85 percent) of the market. Other fiberfill would be made from down, feathers, or other fibers—mainly cotton or acetate. The polyester used for fiberfill is engineered for resiliency and maintaining loft. The durability, comfort, and easy care of polyester also make it appropriate for this end use. Fiberfill is not visible during use—but poor performance shows up in lumpy fabrics or pillows.

Nonwoven fabrics are the fourth important use of polyester. Sewn-in interfacings, fusible interfacings, pillow covers, and mattress interlinings are examples of uses for nonwoven polyester fabrics. They are used where the durability of rayon is inadequate and where absorbency is not needed. They compete with olefin in many industrial uses.

Tire cord is the fifth important use of polyester. Polyester has taken a large share of the new tire market away from nylon because polyester tires do not "flat spot" like nylon tires do.

Polyester accounts for about 7 percent of carpets that are produced. When they were introduced, polyester carpets had a softer hand than most nylon carpets. Since then, nylon carpet fibers have been made in finer deniers to give them a softer hand. Performance of polyester carpets has not been quite as high as nylon carpets under most conditions of use. The first polyester carpets suffered from a "walked-down" look after a period of wear in heavy-traffic areas. This now has been corrected by autoclave heat setting the fibers.

Polyester is chosen for many other consumer

Variants of Polyester

Cross-Section	*Dyeability*	*Crimp or Textured*	*Tenacity*
Round Trilobal Triangular Trilateral Pentalobal Scalloped oval Octolobal Heptalobal Hollow	Disperse dyeable Cationic dyeable Solution dyed Optically whitened Deep dye Extra bright Bright heather Dark heather	Producer textured Partially oriented Undrawn filament	Regular tenacity Intermediate tenacity High tenacity High elongation Mid-modulus High modulus

Shrinkage	*Other*
High shrinkage Normal shrinkage Low shrinkage Heat stabilized Chemically stabilized, adhesive activated	Pill resistant Homopolymer Copolymer Bicomponent Bigeneric Polished high luster Binder fiber Soft luster

Polyesters for Specialized Uses

Producer	*Trade Name*	*Use*
Allied	A.C.E.	Tire cord
du Pont	Hollofil	Fiberfill and insulating fibers
	Reemay	Spunbonded nonwoven fabrics
	Sontara	Spunlaced nonwoven fabrics
Celanese	Angelette	
	ESP	
	Ceylon	
	Comfort Fiber	Staple fiber for apparel uses
	Loftguard	Staple fiber for industrial uses
	Polar Guard	
	Lambda	Filament yarn with spun-yarn characteristics
	Serene	
	Superba	
	Spunese	
	Wondercrepe	

BASF Fiber Polyester Trade Names

Trade Name	*Filament or Spun*	*Special Characteristics or Uses*
Crepesoft	F	Textured, inherent crepe effect in appearance and hand
Encron 8	F	Textured, octagonal cross-section for antiglitter effect
Enkadrain	F	Industrial matting
Enkasonic	F	Industrial sound-rated material
Golden Glow	F	Bright triangular cross-section for soft luster
Golden Touch	F	Textured, fine denier per filament (1.5) for soft hand
Golden Touch-Corduroy		
Golden Touch-Suede		
Letha-Suede	F	Textured, soft hand
Matte Touch	F	Octagonal cross-section
Natural Touch	S	Natural luster with soil hiding, antistatic properties
Plyloc	F	Torque-free yarn
Polyextra	F	Producer textured yarn
Silky Touch	F	Trilobal bright fiber with soft luster and hand
Spunloc	F	
Stablienka	F	Monofilament yarn in nonwoven fabric used for civil-engineering fabrics
Strialine	F	Thick-and-thin yarns—heavy areas dye deeper
Ultra Touch	F	Soft hand

and industrial uses: pile fabrics, tents, ropes, cording, fishing line, cover stock for disposable diapers, garden hoses, sails, seat belts, filters, fabrics used in road building, seed and fertilizer bags; artificial arteries, veins, and hearts; and sewing threads. Research is being done to increase the ways polyester is used industrially.

TYPES AND KINDS

Slightly over 200 variants of polyester were listed in the 1983–1984 *Textile Industries* "Manmade Fiber Variant Chart." Ten of the 16 companies producing polyester were included in the list. Du Pont listed 74 variants, Celanese listed 41, and Hoechst and Eastman each listed around 25.

Some of the more-commonly seen trade names for polyester fibers, and the companies that produce them, include: Avlin by Avtex Fibers, Fortrel by Celanese, Dacron by du Pont, Kodel by Eastman, and Trevira by Hoechst. The following chart lists the terminology used to describe the types and kinds of polyester. Each company has a large variety of specific fibers or yarns that combine one or more of these many variables.

13

Olefin Fibers

Many attempts were made to polymerize ethylene in the 1920s. Ethylene was polymerized and used as an important plastic during World War II, but filaments made from it did not have sufficient strength or a high enough melting point for use in textile fibers. In 1954, Karl Ziegler in Germany developed a process in which the melting point of polymerized ethylene filaments was raised but it still was too low for apparel fibers. Polyethylene fibers were used in some industrial end uses. In Italy, Giulio Natta worked with polypropylene and was successful in making linear polymers of high molecular weight that proved to be suitable for textile applications. By 1957, Italy was producing olefin fibers; U.S. production of olefin fibers started in 1960 with the production of Herculon by the Hercules Powder Company, Inc.

Olefin fibers have a combination of properties that make them good for home furnishings, apparel that does not need ironing, and industrial uses. Olefin fibers are strong and resistant to abrasion, inexpensive, chemically inert, and thermoplastic but static resistant. There were 15 major producers of polypropylene in the United States in 1986.

PRODUCTION

The term *olefin* is derived from the Latin *oleum,* meaning oil. (Oleum is also the root of the word *oleomargarine.*)

Two processes are used to produce olefin. The high-pressure system polymerizes ethylene gas in an autoclave at 200°C (392°F) under a pressure of 10 tons per square inch. This system is used to produce polyethylene. Polyethylene is not as commonly used as a fiber as polypropylene. The low-pressure system polymerizes the propylene gas at a lower temperature with a catalyst and hydrocarbon solvent. The low-pressure system is less expensive and produces a polymer (polyethylene) more suitable for textile uses. The extrusion process is similar to that of nylon and polyester. Olefin is melt spun into water or cool air and cold drawn to six times its spun length. Olefins differ from polyester and nylon in that the solution crystallizes very rapidly (undrawn fibers are crystalline), so that the spinning conditions and after treatments greatly affect the fiber properties. Olefin is an inexpensive fiber. The low price of olefin, coupled with its properties, explains the widespread use of olefin in industrial uses, household furnishings, and, to a limited degree, apparel. Polyethylene is used in industrial end uses. Polypropylene is used in home furnishing, apparel, and industrial end uses. Olefin is one synthetic fiber with a growing production. See the following table.

Production of Synthetic Fibers (millions of lb)*

Fibers	1985	1980	1975	(peak year)
Acrylic	631	779	525	1981
Nylon	2,343	2,358	1,857	1983
Olefin	1,249	748	497	still growing
Polyester	3,341	3,989	2,995	1980

**Source: Chemical and Engineering News,* June 9, 1986, page 38.

PHYSICAL STRUCTURE

Olefins are produced as monofilament, multifilament, staple fiber, and tow with variable tenacities. The fibers are colorless, usually round in cross-section, and have a somewhat waxy feel (Figure 13–1). The fibers may have an irregular cross-section for specialized end uses.

CHEMICAL COMPOSITION AND MOLECULAR ARRANGEMENT

Olefin fibers—manufactured fibers in which the fiber-forming substance is any long chain synthetic polymer composed of at least 85 percent by weight of ethylene, propylene, or other olefin units except amorphous (noncrystalline) polyolefins qualifying . . . as rubber.—Federal Trade Commission.

The process used to obtain polypropylene fibers was a remarkable development. Polypropylene is a three-dimensional structure with a backbone of carbon atoms and methyl groups standing out from the chain. Natta observed that three configurations could be developed when propylene was polymerized and that when all the methyl groups were on one side of the chain, the molecular chains could pack together and crystallize. Natta developed the process in which polymerization would take place in this manner, and, together with Ziegler, received the Nobel Prize in 1963 for his achievement.

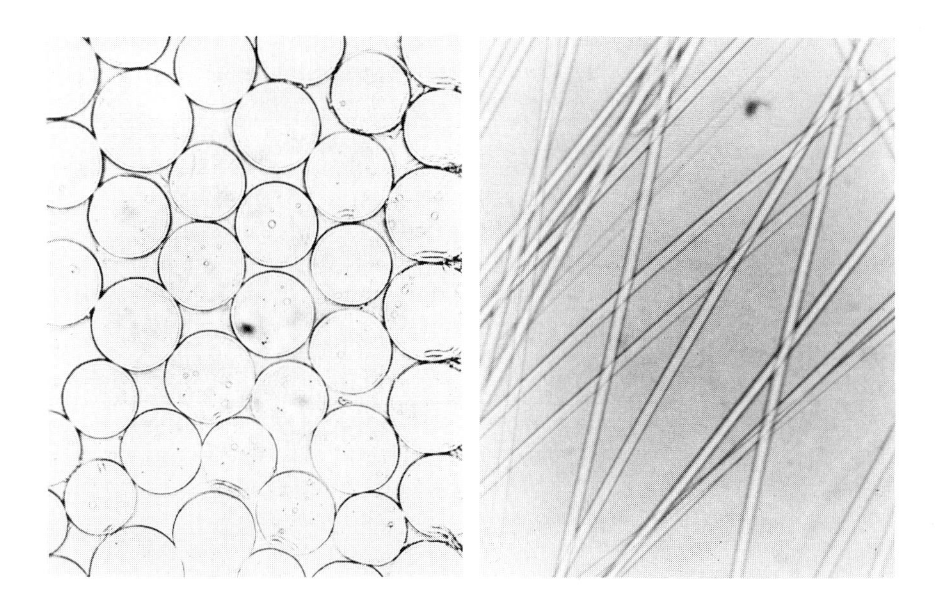

Fig. 13–1 Photomicrographs showing cross-sectional (left) *and longitudinal* (right) *views of Herculon® olefin. (Courtesy of Hercules, Inc.)*

Karl Ziegler's work on catalysts to polymerize ethylene and Giulio Natta's discovery of sterio-specific polymerization made it possible to obtain high-molecular-weight crystalline polypropylene polymers. *Steriospecific polymerization* means that all the molecules are specifically arranged in space so that all the methyl groups have the same location and there are no polar groups. Natta called this phenomenon *isotactic.* In the atactic form, the methyl groups are randomly oriented, resulting in an amorphous polymer that would qualify as rubber.

```
        H       H       H
        |       |       |
    H   C   H   C   H   C
  \ | / | \ | / | \ | / | \
    C   H   C   H   C   H
    |       |       |
    CH3     CH3     CH3
       ↖    ↑    ↗
       Methyl groups
```

Olefin fibers have no polar groups. The chains are held together by crystallinity alone. The absence of polar groups makes the dyeing of the fiber a problem. Solution dyeing is expensive and not as versatile as piece dyeing and printing. An acid-dyeable olefin has been developed.

PROPERTIES

Aesthetics. Olefins are usually produced with a medium luster and smooth texture, but the luster and texture can be modified depending on the end use. Olefin has a waxy hand that some consumers find objectionable. Since olefin is seldom used in apparel, drape is not an important aspect. However, drape can be varied relative to the end use by selection of fiber modification, fabric-construction method, and finish.

Durability. Olefins may be produced with different strengths suited to the end use. Polypropylenes have a tenacity ranging from 3.5–8.0 g/d. Polyethylenes have a tenacity ranging from 1.5–7.0 g/d. Wet strength is equal to dry strength for both types. An ultra-high-strength olefin, Spectra 900 by Allied Fibers, has a tenacity of up to 30 g/d and is used in industrial end uses. Fibers produced for more-common end uses have tenacities from 4.5–6.0 g/d. Olefin fibers have very good abrasion resistance. Elongation varies with the type of olefin. For olefins normally used in apparel and home furnishings, the elongation is 10–45 percent with excellent recovery.

Comfort. Olefins have a low moisture regain, less than 0.1 percent. Hence they are not absorbent in the normal sense. Because olefins are nonpolar in nature, they are not prone to static-electricity problems. Olefins do have excellent wicking abilities, so the fibers are able to wick moisture away from the surface of the skin. For ths reason, olefin is becoming more important in some apparel end uses, such as active sportswear.

Olefin has good heat retention. However, it is olefin's ability to wick moisture that dictates its

Summary of the Performance of Olefin in Apparel Fabrics

AESTHETIC	*VARIABLE*
Luster	Medium
DURABILITY	*HIGH*
Abrasion resistance	Very good
Tenacity	High
Elongation	Variable
COMFORT	*MODERATE*
Absorbency	Low
Thermal retention	Good
APPEARANCE RETENTION	*EXCELLENT*
Resiliency	Excellent
Dimensional stability	Excellent
Elastic recovery	Excellent
RECOMMENDED CARE	*MACHINE WASH, DRY AT LOW TEMPERATURE*

use in active sportswear, socks, and underwear. It is used in active sportswear to wick perspiration away from the body and to aid in heat loss. In cold-weather wear, olefin is used to keep the skin dry by wicking moisture away from the skin's surface.

Olefin fibers are the lightest textile fibers. Polypropylene has a specific gravity of .90 to .91. Polyethylene has a specific gravity of .92 to .96. This low specific gravity provides more fiber per pound for better cover. If other problems—such as their low softening and melting temperatures, difficulty in dyeing, and unpleasant hand—could be solved, olefins would be good for warmth-without-weight fabrics for sweaters and blankets. In spite of those problems, the low specific gravity of olefin is an asset because items that are extremely lightweight can be produced. For example, olefin is used as the shell for some footwear under the tradename of Propex III by Amoco Fabrics Company and S. Starensier, Inc. It takes 1.27 pounds of nylon or 1.71 pounds of cotton to cover the same volume as 1 pound of Propex III.

Appearance Retention. Olefin has excellent resiliency and recovers quickly from wrinkling. Olefins have excellent shrinkage resistance as long as they are not heated. They also have excellent elastic recovery.

Care. Olefins have easy-care characteristics that make them suited to a number of end uses. Because they are hydrophobic, they are not affected by water-borne stains. They dry quickly after washing. Dry cleaning is seldom recommended because olefins are swollen by chlorinated hydrocarbons such as perchloroethylene. Petroleum dry-cleaning solvents are acceptable for cleaning olefins, but one needs to be sure that a petroleum solvent rather than perchloroethylene is used.

They have excellent resistance to acids, alkalis, insects, and microorganisms. Olefins are affected by sunlight, but stabilizers can be added to correct this disadvantage. Indoor/outdoor carpeting made of olefin fibers can be hosed off.

Olefins have a low melting point (325–335°F), which limits their use in apparel. Warm or cold water should be used for spot cleaning or washing. Olefin fabrics should be air dried. They are more oleophilic than nylon. Olefins should be dried and ironed at low temperatures.

Identification. Olefin melts, as do other melt-spun fibers. It burns slowly and produces a tan or off-white bead. It may burn with a blue-and-yellow flame in continued contact with the heat.

USES

Olefin is found in an ever-widening array of end uses. In apparel, it is used for underwear, socks, sweaters, and active sportswear (Figure 13–2). Thinsulate is a low-bulk, ultra-fine-microdenier fiberfill of olefin and polyester produced by 3M and used in ski jackets and other outerwear where a less-bulky silhouette is desired. In home furnishings, olefin is used in carpeting as face yarns and in tufted carpets as backing; as nonwoven, needle-punched carpets and carpet tiles; upholstery; draperies; and slipcovers. Common trade names for olefin include Herculon, Marquessa, Marvess, Patlon, Polyloom, and Vectra. In industrial end uses, olefin is found in carpet backing such as Typar, dye nets, cover stock for diapers, filter fabrics, laundry and sand bags, wall-panel fabrics such as Tyvek, envelopes (also Tyvek), banners, geotextiles, ground-control fabrics such as Mirafi and Supac, protective clothing such as Tyvek, substrate for coated fabrics, ropes and twines, and road-bed stabilizer

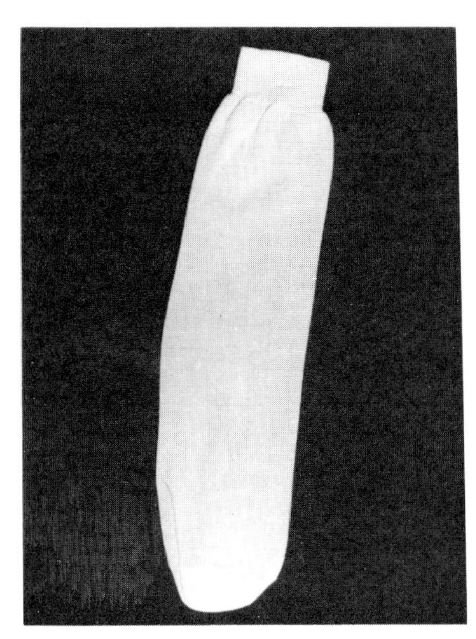

Fig. 13–2 *Sock knit of olefin.*

fabrics such as Petromat. Figure 13–3 shows a rice bag of olefin.

The following tables list modifications, trade names, and producers of olefin.

Types and Kinds of Olefin Fibers	
Heat stabilized	Acid dyeable
Light stabilized	Solution dyed
Modified cross-section	Bicomponent
Pigmented	Fibrillated

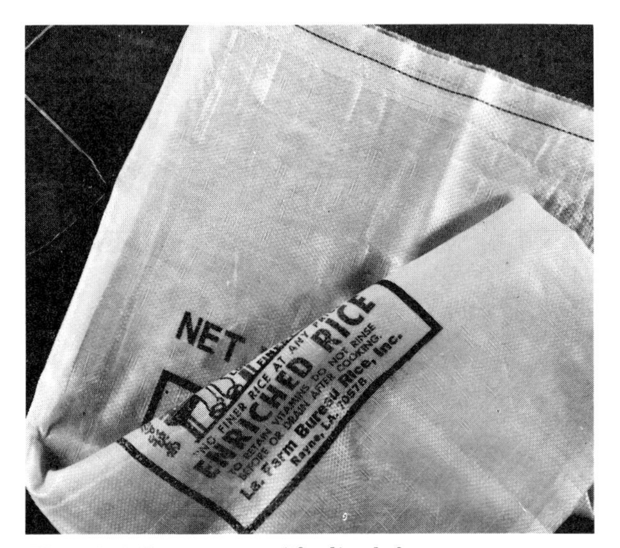

Fig. 13–3 *Bag woven with slit-olefin yarns.*

Some Trade Names and Producers	
PolyBac, Polybac FLW, Patlon, Marquessa Lana, Action Bac, Propex	Amoco Fabrics & Fibers Co.
Typar, Tyvek	du Pont
Fibrilawn, Fibrilon	Fibron Corp.
Herculon	Hercules, Inc.
Duraguard, Evolution, Evolution III	Kimberly-Clark
Marvess, Duon, Petromat, Rufon, Supac	Phillips Fibers
Polyloom	Polyloom Corp.

The following table compares the characteristics and production of nylon, polyester, and olefin.

Split-Fiber Olefins

Extrusion—forcing a liquid through a spinneret to form fine strands—is the standard method of spinning fibers. The *split-fiber method* was developed in 1965 at the Shirley Institute in England. The method is less expensive than the traditional extrusion process and can be done by small industry. However, some fiber polymers cannot be processed by the split-fiber method. Polypropylene is used extensively because of its ease of processing and economic factors.

PRODUCTION

Pellets of polypropylene are melted, extruded as a film 0.005 to 0.020 inch thick, and cooled quickly by quenching in water. The film is slit into tapes 0.1 inch wide. The slit tapes are then heat-stretched to orient the molecular chains, and the stretching is carried to a point where the film develops a tendency to fibrillate (split into fibers). Twisting or other mechanical action completes the fibrillation. Split-fiber yarns have high strength.

USES

Yarns as low as 250 denier have been made from split fibers, but these are coarse for clothing uses. Carpet backing, rope, cord, fishnets, and bagging are the major uses.

Comparison of Melt-Spun Fibers

	Nylon	*Polyester*	*Olefin*
Mill production, 1985* (Million pounds)	2,343	3,341	1,249
Breaking tenacity g/d	2.3–9.8 filament 2.9–7.2 staple	2.8–9.5 filament 2.4–7.0 staple	3.5–8.0 filament
Specific gravity	1.14	1.22 or 1.38	0.91
Moisture regain %	4.0–4.5	0.4–0.8	Less than 1
Melting point	482° or 414°F	540° or 482°F	325°–335°F
Safe ironing temperature	270–300°	325°–350°F	250°F–lowest setting
Effect of light	Poor resistance	Good resistance	Poor resistance

**Source: Chemical and Engineering News,* June 9, 1986, page 38.

Continued research and refinement of the process may result in a new generation of fibers in the future. One of the interesting outgrowths of split-film research may be the development of a new process for making nonwovens. Some of the early work with fibrillation of nylon film gave lace-like sheets that suggested possibilities in the nonwoven area.

Olefin films are slit into yarns that are used for the same textile products as split-fiber olefin.

14

Acrylic, Modacrylic, and Other Vinyl Fibers

Acrylic Fibers

Acrylonitrile, the substance from which *acrylic fibers* are made and from which the generic name is derived, was first made in Germany in 1893. It was one of the chemicals used by Carothers and his team in the fundamental research done on high polymers for the du Pont Company.

Du Pont developed an acrylic fiber in 1944 and started commercial production of this fiber in 1950. The fiber was given the trade name Orlon. Three other companies began to produce acrylics: Chemstrand Corporation (now called Monsanto Fibers) introduced Acrilan in 1952, Dow Chemical (BASF Fibers) began the production of Zefran in 1958, and American Cyanamid began the production of Creslan in 1958. These same four companies were producing acrylic fibers in 1986.

Acrylic fibers are soft, warm, lightweight, and resilient. They make easy-care fabrics. They are produced as staple fiber and used primarily in wool-like end uses.

PRODUCTION

Some acrylic fibers are dry or solvent spun and others are wet spun. In *solvent spinning,* or *dry spinning,* the polymers are dissolved in a suitable solvent, such as dimethyl formamide, extruded into warm air, and solidified by evaporation of the solvent. After spinning, the fibers are stretched hot, three to ten times their original length, and then crimped, and marketed as cut staple or tow. In *wet spinning,* the polymer is dissolved in solvent, extruded into a coagulating bath, dried, crimped, and collected as tow for use in the high-bulk process or cut into staple and baled.

PHYSICAL STRUCTURE

The cross-sectional shape of acrylic fibers varies as a result of the spinning method used to produce them (Figure 14–1). Dry spinning gives a dog-bone shape to Orlon. Wet spinning imparts a round or lima-bean shape to Acrilan, Creslan, and Zefran. Differences in cross-sectional shape affect physical and aesthetic properties and thus can be a factor in determining appropriate end use. Round and lima-bean shapes have a higher bending stiffness, which contributes to resiliency, and are appropriate for bulky sweaters and blankets. Dog-bone shape gives the softness and luster desirable for other apparel uses.

All the production of acrylic fibers in the United States is staple fiber and tow. Staple fiber is available in deniers and lengths suitable for all spinning systems. Acrylic fibers also vary in shrinkage potential. Bicomponent fibers were first produced as acrylics.

CHEMICAL COMPOSITION AND MOLECULAR ARRANGEMENT

Acrylic fibers—manufactured fibers in which the fiber-forming substance is any long-chain synthetic polymer composed of at least 85 percent by weight of acrylonitrile units $(\text{—}CH_2\text{—}\underset{\displaystyle CN}{\underset{|}{C}}H\text{—})$.—*Federal Trade Commission.*

The acrylonitrile monomer was discovered in 1893, and the polymer was first patented in 1929. The pure polymer was extremely insoluble until dimethyl formamide was discovered.

Fibers of 100 percent polyacrylonitrile have a compact, highly oriented internal structure that makes them virtually undyeable. They are an example of a *homopolymer,* a fiber composed of a single substance. Schematiclly, a homopolymer could be shown like this:

× × × × × × × × × × × × × × × **Homopolymer**

Since this structure in acrylic fibers makes dyeing so difficult, most acrylics are made as *copolymers* with up to 15 percent additives, which give a more-open structure and permit dyestuffs to be absorbed into the fiber. The additives furnish dye sites and are cationic for acid dyes and anionic for basic dyes. This makes cross dyeing possible. Copolymer fibers are composed of two substances. They could be diagrammed like this:

○×○× × ×○×○× × ×○×○× **Copolymer**

or this:

× × × × ×○× × × ×○ **Copolymer**

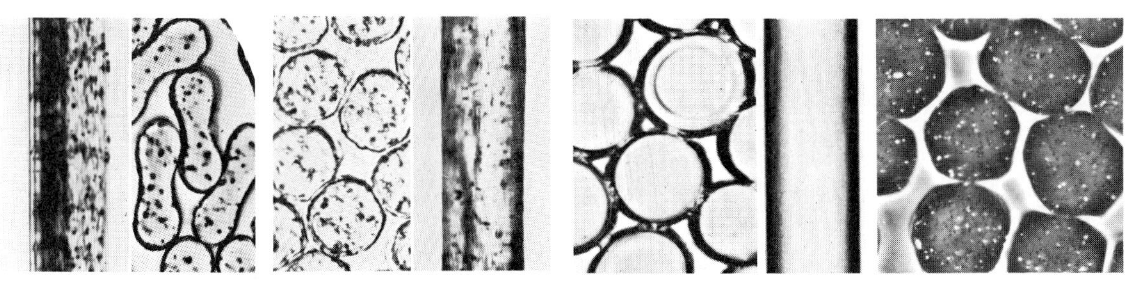

Fig. 14–1 Photomicrographs of acrylic fibers: cross-sectional and longitudinal views. (From left to right) *Orlon, Acrilan, Creslan, and Zefran. (Orlon courtesy of E. I. du Pont de Nemours & Company; Acrilan courtesy of Fibers Division of Monsanto Chemical Co., a unit of Monsanto Co.; Creslan courtesy of the Cyanamid Company; Zefran courtesy of BASF Corp. Fibers Division.)*

depending on the percentage of substances and their arrangement in relations to each other.

Zefran is a graft polymer. In graft polymerization, the additive does not become a part of the main molecular chain but is a side chain. The side branches are attached to the "backbone" chain of the molecule, which gives the molecular chains a more-open structure and less crystallinity; dye receptivity is increased.

Some fibers have molecules with chemically reactive groups; others are chemically inert. A chemically inert molecule can be made reactive by grafting it with reactive groups. It would look like this:

The copolymer acrylics are not as strong as the homopolymers or graft-polymer acrylics. Since the end uses for acrylics are mostly apparel and home furnishings, the reduced strength is not very important.

PROPERTIES

Acrylic fibers are soft, warm, lightweight, and resilient. They make easy-care fabrics. Because of their low specific gravity and high-bulk properties, the acrylics have been called the Warmth Without Weight fibers. The acrylic fibers have been very successful in end uses, such as sweaters and blankets, that were previously dominated by wool. They are superior to wool in their easy-care properties and are nonallergenic. Bulky acrylic yarns are also popular in socks, fleece fabrics, fake-fur fabrics, and hand-knitting yarns.

Aesthetic. Acrylic fibers possess favorable aesthetic properties. They look nice and have a soft, pleasant hand. The fibers are usually textured. The resulting bulky spun yarns are wool-like in texture. Indeed, acrylic fabrics imitate wool fabrics more successfully than any of the other man-made fibers.

Durability. Acrylics are not as durable as nylon, polyester, or olefin fibers, but, in apparel and household textiles, the strength of acrylics

Summary of the Performance of Acrylic in Apparel Fabrics

AESTHETIC	*WOOL-LIKE*
DURABILITY	*MODERATE*
Abrasion resistance	Moderate
Tenacity	Moderate
Elongation	Moderate–high
COMFORT	*MODERATE*
Absorbency	Low
Thermal retention	Moderate
APPEARANCE RETENTION	*MODERATE*
Resiliency	Moderate
Dimensional stability	Moderate
Elastic recovery	Moderate
RECOMMENDED CARE	*MACHINE WASH; FOLLOW CARE LABEL*

is satisfactory. Dry tenacity ranges from 2.0–3.6 g/d, which is moderate. Strength is lower than that of cotton fibers. Abrasion resistance is moderate. The elongation at break is at least 20 percent, very similar to that of wool and higher than that of many fibers. Elongation increases when the fiber is wet. The overall durability of acrylic fibers is moderate, similar to that of wool and cotton.

The first Orlon was produced in filament form with strength almost as good as nylon. Because of its exceptional resistance to weathering, it was thought that acrylics would be widely used in awnings, outdoor furniture, and curtains. (Filament-yarn acrylic is found in imported drapery fabrics.) The resistance of acrylics to dyes and the high costs of production limited its use in these end uses. Acrylic is widely used in awnings and tarpaulins now. Staple fibers with lower strength are used. The following chart shows that acrylic fibers are comparable to wool in their durability properties.

Comparison of Acrylic Fibers With Wool—Durability

Fiber Property	*Acrylic*	*Wool*
Abrasion resistance	Good	Fair
Breaking tenacity	2.0–3.6 g/d dry	1.5 g/d dry
	1.8–3.5 g/d wet	1.0 g/d wet
Elongation at break	20 percent	25 percent
Elastic recovery	92 percent	99 percent

Comfort. The fiber surface of acrylic fibers is much less regular than that of other synthetic fibers. High-magnification photographs taken by electron microscopes show mini-striations and indentations in the surface (see Fig. 14–2). In spite of the relatively low-moisture regain of 1.0–2.5 percent, acrylics are considered to be moderately comfortable because of the irregular fiber surface. Instead of absorbing moisture and becoming wet to the touch, acrylic fibers used in active sportswear tend to wick moisture from the skin side of the fabric to the outside of the fabric where it evaporates more readily. The evaporation aids in cooling the body, and many people are more comfortable in a relatively dry acrylic garment than in a wet cotton garment.

Another factor that makes acrylics comfortable is that the fibers and yarns can be made with high bulk. The resulting bulky fabrics retain body heat well so they are warm in cold temperatures. Bulky-knit sweaters are a familiar example of this.

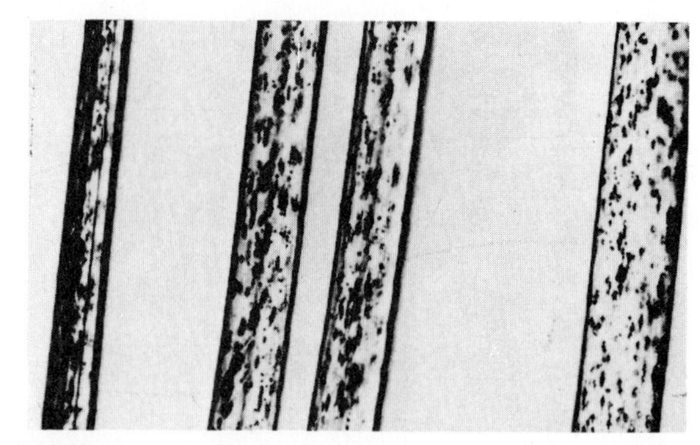

Fig. 14–2 *Acrylic that is magnified 1,000× shows a pitted and irregular surface. (Howard L. Needles,* Handbook of Textile Fibers, Dyes, and Finishes. *1981, Garland, STPM Press. Updated by Noyes Publications, Park Ridge, New Jersey. Courtesy of publisher and author.)*

However, remember that the structure of the yarn and fabric can be varied to make a warmer or cooler garment, depending on how the garment is used. In general, acrylics are more comfortable than nylon and polyester, but not as comfortable as cotton in hot, humid weather, or as wool in very cold or cold, humid weather.

High-Bulk Yarns. Acrylic fibers can be produced with a *latent shrinkage potential* and retain the bulk indefinitely at room temperature. Latent shrinkage in a fiber is achieved by heating, stretching, and then cooling while in the stretched condition. These heat-stretched fibers are called *high-shrinkage fibers.* (High-shrinkage fibers are processed on the Turbo-Stapler; see Chapter 18.) High-shrinkage fibers are combined with nonshrinkage fibers in the same yarn, which is then made into a garment. Heat treatment of the garment will cause the high-shrinkage fibers to relax or shrink, forcing the nonshrinkage fibers to bulk. This makes high-bulk sweaters and similar garments. Knitting yarns are sometimes made the same way. High-shrinkage fibers tend to migrate to the center of the yarn. Thus, if fine-denier nonshrinkage fibers are combined with coarse-denier high-shrinkage fibers, the fine-denier fibers will end up on the outer surface of the yarn. The amount of bulk can be controlled by regulating the amount of heat stretching. Zefran fibers cannot be modified for heat shrinkage and are not made into high-bulk yarns.

The high-bulk principle can be used to achieve interesting effects, such as "guard hairs" in synthetic furs and sculptured high-low effects in carpets. Carpet pile or fur-like fabrics can be made more dense by using a high-shrinkage-type fiber for the ground yarns. When the yarns shrink, the fibers are brought much closer together. Bicomponent acrylic fibers present another way of achieving bulky fabrics. Orlon Sayelle sweaters and Wintuck hand-knitting yarns are familiar examples of this (see Chapter 8).

Appearance Retention. Acrylic fibers exhibit moderate resiliency and recovery from bending. Thus they resist wrinkling during wear and care. They have moderate dimensional stability. When proper yarn and fabric structures are utilized, the dimensional stability of acrylic fabrics is good. Acrylics will shrink when exposed to boiling water, so high temperatures and steam should be avoided. The fibers have poor hot-wet properties.

Acrylic fibers cannot be heat set like nylon and polyester because acrylic does not melt, but decomposes and discolors when heated. However, some acrylics can have a crease set in a pair of slacks or pleats set in a skirt that is not affected by normal wear, laundering, or dry cleaning. With the application of heat and/or steam, the crease or pleat can be removed. This behavior is one reason why pleated skirts for winter wear are frequently made of acrylic.

Another way that acrylics differ from nylon and polyester is that stretching as well as shrinking may occur in wear or care, but these problems are infrequent. Once they occur, manufacturers generally improve the characteristics of the fiber, yarn, fabric, or finishes to be more appropriate for that end use and the performance of the garment again matches the consumer's expectations.

Pills will form on some acrylic fabrics, but, because the fiber is only moderately strong, they may drop off during subsequent wear. Acrylics tend to fibrillate, or crack, with abrasion; this may contribute to pilling.

Care. In caring for clothing or household items made of acrylic, it is especially important to follow the instructions found on care labels. There are four basic acrylic fibers with slightly differing properties due to the polymer composition and different manufacturing methods. With the additional variations available through shrinkage potentials of acrylic fibers, as well as of bicomponent fibers, there are many factors that can affect appropriate care. Many acrylic garments are imported and may have been processed in ways that are different than those U.S. manufacturers use.

The acrylics have good resistance to most chemicals except strong alkalis and chlorine bleaches. (Fibers containing nitrogen are usually susceptible to damage from alkali and chlorine.) Except for the fur-like fabrics, acrylic fabrics have good wash-and-wear characteristics. They do not wrinkle if handled properly and if directions on the label are followed.

Sweaters made of acrylic fibers do not need to be blocked to shape as is the case with wool sweaters. Some sweaters made from high-bulk yarns of bicomponent fibers need to be machine dried to regain their shape after washing. If they are blocked, dried flat, or drip dried, they may be too large or misshapen. Rewash and then tumble dry the sweater, and it should recover its original shape.

Acrylics can be dry cleaned. On some fabrics the finish is removed, which results in a harsh feel. So care labeling should be followed. Acrylics are resistant to moth damage and mildew. Acrylics have excellent resistance to sunlight.

Identification. The burning characteristics of the acrylics are similar to those of the acetates. The fibers soften, burst into flame, and burn freely, then decompose to a black, crumbly residue. There is a chemical aromatic odor that is different from the vinegar-like odor of the acetates. Differences in the flammability of the acrylics and the modacrylics is the result of the high-acrylonitrile content of the acrylics. The modacrylics, which have a much lower acrylonitrile content, are self-extinguishing.

USES

Acrylic accounted for 8 percent of the fiber produced in the United States in 1983. Approximately 631 million pounds of acrylic was produced in 1985. Acrylic is the fifth most widely used fiber in the United States. More acrylic was used in 1977 than in subsequent years. Most acrylic is used in apparel, although it is used for both home furnishings and industrial purposes as well. One source identified the uses for acrylic in 1984 as follows:

Apparel	80 percent
Home furnishings	14 percent
Industrial	6 percent

Currently the single most important use for acrylic is in fleece fabrics. One hundred twenty-three million pounds of acrylic were used in top- and bottom-weight fabrics—80 percent of those fabrics were fleece fabrics. Fleece fabrics are knits. In 1956, fleece fabrics were drab, utilitarian items of clothing used by athletes, called sweat suits. Gray was the most common color. The market for these fabrics has skyrocketed since then—largely due to the interest in physical fitness and the increased importance of leisure wear. Fleece fabrics are now a frequently used item in jogging outfits and active sportswear for all members of the family. They are available in many colors and prints. However, the demand is so great, manufacturers cannot produce enough fabric to meet the needs of apparel manufacturers.

A wide variety of fibers and fiber blends are seen in fleece fabrics. Acrylic, cotton, polyester, and blends of these fibers are found. Based on domestic fiber shipments, one company estimates that 37 percent of fleece fabrics are acrylic, 40 percent are cotton, and 23 percent are polyester. Producers of acrylic or 50 percent acrylic/50 percent cotton fleecewear state their products are better than all-cotton fleecewear because they exhibit the following:

- Better shape retention
- Improved hand, especially after several washings
- Better color
- More vibrancy, brilliance, and life

Imports of fleecewear have been increasing so rapidly they are threatening U.S.-produced fleecewear, so producers have formed an association to promote their product and encourage purchase of U.S.-made products.

The second important use for acrylic is sweaters. The main advantages acrylic has over wool are its lower cost and machine washability. As cotton and cotton-blend sweaters became more popular in the mid-1980s, acrylic had additional competition in the sweater market. A new acrylic fiber has been developed to compete directly with cotton; du Pont developed "Comfort 12," a bicomponent Orlon that is promoted as having the look and feel of cotton with year-round comfort and superior performance while being laundered.

Interestingly enough, the third important use for acrylic is in socks (see Fig. 14–3). Think of how much less yarn is used in a pair of socks than is used in making a sweater! Acrylic socks feel soft. Do you think acrylic socks are more comfortable to wear than cotton socks? Or are they both comfortable for your activities? The durability of both acrylic and cotton socks is improved by using 10–15 percent nylon as reinforcement in the heel and toe areas of the sock. Perhaps you have noticed a thin, sheer knit fabric left after most of the sock fibers have worn away in one of those areas.

Fig. 14–3 *Socks are an important use of acrylic fiber. (Courtesy of E. I. du Pont de Nemours & Company.)*

Craft yarns account for the fourth important end use of acrylic fibers. Craft yarns are often made of a heavier denier (5–6 denier). Many sweaters, vests, and afghans are knit or crocheted with these yarns. Acrylic yarns are also used for crewel embroidery.

The final important use for acrylic in apparel is for pile fabrics; thick, snuggly fun furs that are used for coats, jackets, or very warm linings. The following table gives the estimated poundages of acrylic used in 1984.

Acrylic is used in home furnishings. Uphol-

Comparison of Acrylics and Wool—Care

Fiber Property	*Acrylic*	*Wool*
Effect of alkalis	Resistant to weak	Harmed
Effect of acids	Resistant to most	Resistant to weak
Effect of solvents	Can be dry cleaned	Dry cleaning recommended
Effect of sunlight	Excellent resistance	Low resistance
Stability	Can be heat set for shape retention	Subject to felting, shrinkage
Permanence of creases	Creases can be set and removed by heat	Creases set by heat and moisture—not permanent
Effect of heat	Thermoplastic—sticks at 450–490°F	Scorches easily. Becomes brittle at high temperature
Resistance to moths and fungi	Resistant	Harmed by moths. Mildew will form on soiled, stored wool

stery fabrics are the first important use. These may be flat-woven fabrics or velvets. Usually they have a wool-like appearance. Drapery fabrics of acrylic are not yet very common, but some are available in this country and additional ones are imported from Europe. This is an appropriate and growing end use for acrylics because of their good sunlight resistance and weathering properties.

The second important home use of acrylics is in blankets. A variety of fabrics and fabric-construction methods are used. Both lightweight and winterweight blankets are available. Blankets are a very appropriate use for acrylics because the cost is lower, the bulky fabrics are lighter weight, and the care is easier than for wool blankets.

Carpets and rugs account for the third important use of acrylic fibers in homes.

Acrylics are used in a number of industrial uses for which their chemical resistance and good weathering properties make them suitable: awnings and tarpaulins, luggage, boat and other vehicle covers, outdoor furniture, tents, carbon-fiber precursors, office-room dividers, and sand bags. (Fig. 14–4.)

Uses of Acrylic in 1984

Percent	*End Use*	*Million Pounds*
80	*Apparel*	
	Top- and bottom-weight fabrics	123
	Sweaters	80
	Socks	70
	Craft yarns	55
	Pile fabrics	27
14	*Home Furnishings*	
	Upholstery and drapery fabrics	28
	Blankets	25
	Carpets and rugs	13
6	*Industrial/Other*	26

TYPES AND KINDS

Each company that produces acrylic in the United States identifies its fiber by a trade name:

Trade Name	*Company*	*Number of variants listed in* Textile Industries *"Man-made Fiber Variant Chart"*
Acrilan	Monsanto	11
Creslan	American Cyanamid	9
Orlon	du Pont	25
Zefran	BASF	8

Fiber variants that are tailored for a specific end use or differ in performance are also produced. The number of these for each company is listed in the preceding chart. These variants are identified by type number.

Fig. 14–4 Acrylics are used for luggage and outdoor furniture because of their chemical resistance and good weathering properties. (Courtesy of BASF Corp. Fibers Division.)

Du Pont identifies all of its branded acrylic fiber by the registered trade name of Orlon. In addition to that, it has 11 certification marks for acrylic fibers or yarns. These certification marks may be used on products that meet du Pont's quality standards. They are enforced by a testing program. These certification marks, what they are used on, and additional comments are in the following list:

Du Pont Certification Marks for Orlon

Ardina	yarns	wool-like
Bi-Nell	yarns	blankets
Comfort 12	yarns	cotton-like sweaters
Civona	yarns	soft feel of cashmere, sweaters, baby items
Dypac	yarns	
Jet-Spun	yarns	air-textured yarns
Lanilon	yarns	
Nomelle	yarns	cashmere-like
Orelle	yarns	
Sayelle	hand-knitting yarns	bicomponent, high bulk
Super Soft	yarns	yarns of Orlon
Ultramere	yarns	cashmere-like sweaters, ultra-soft and luxurious
Wintuk	yarns	wool-like, crisp hand, high bulk

Du Pont literature states there are now 20 different types of Orlon fibers that are available in more than 100 variants.

Modacrylic Fibers

Modacrylic fibers are modified acrylics. They are also made from acrylonitrile but have a larger proportion of other polymers added to make the copolymers.

Production of modacrylic fibers started in the United States in 1949 with Dynel produced by

Types and Kinds of Acrylic Fibers and Yarns
Homopolymer
Copolymer
Graft polymer
Bicomponent
Blends of various deniers
Blends of homopolymer and copolymer
Helical, nonreversible crimp
Reversible crimp
Surface modified
Variable cross-section-round, acorn, dog-bone
Variable dyeability-cationic, disperse, acidic, basic
Solution dyed

Union Carbide Corporation. Production of Dynel was discontinued in 1977. Verel was introduced by Eastman in 1956, but it has been discontinued. In 1971, Monsanto began production of SEF modacrylic.

Modacrylics do not support combustion, are very difficult to ignite, are self-extinguishing, and do not drip. This innate fire retardance makes them good for end uses in which compliance with the Flammable Fabrics Act is required. Modacrylics are used in children's sleepwear, contract draperies, fake furs, and wigs.

PRODUCTION

The modacrylic fibers are produced by polymerizing two components, dissolving the copolymer in a suitable solvent (acetone), pumping the solution into a column of warm air, and stretching while hot.

PHYSICAL STRUCTURE

The modacrylics are creamy white and are produced as staple or tow. They have a dog-bone or irregular cross-section (Figure 14–5). Various deniers, lengths, crimp levels, and shrinkage potentials are available to fabric producers.

CHEMICAL COMPOSITION AND MOLECULAR ARRANGEMENT

Modacrylic fibers—manufactured fibers in which the fiber-forming substance is any long-chain synthetic polymer composed of less than 85 percent but at least 35 percent by weight acrylonitrile units except when the polymer qualifies as rubber.—Federal Trade Commission.

Comparison of Modacrylic and Acrylic Fibers—Durability

Factor	*Modacrylic*	*Acrylic*
Strength	1.7–3.5 g/d	2.0–3.6 g/d
Elongation	30–60 percent	20 percent
Elastic recovery	88 percent	92 percent
Sunlight resistance	Excellent	Excellent

The chemicals, other than acrylonitrile, are vinyl chloride (CH_2CHCl), vinylidene chloride (CH_2CCl_2), or vinylidene dicyanide (CH_2CCN_2).

PROPERTIES

Modacrylics are similar to the acrylics in their properties—the major differences being fire retardance and effect of heat.

Aesthetic. Fur-like fabrics, wigs, hairpieces, and fleece-type pile fabrics are important end uses for modacrylic fibers. Some are produced with different amounts of crimp and shrinkage potential. By mixing different fiber types it is possible to obtain fibers of different pile heights; long, polished fibers (guard hairs); and soft, highly crimped undercoat fibers much like real fur (Figure 14–6). Fabrics can be sheared, embossed, and printed to resemble fur.

Durability. Modacrylics have adequate durability for their end uses. A comparison of durability factors of modacrylics and acrylics is listed.

Comfort. Modacrylics are poor conductors of heat. Fabrics are soft, warm, and resilient. Fab-

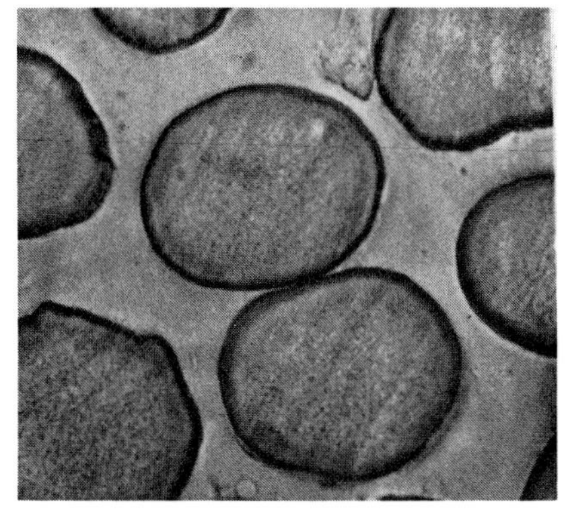

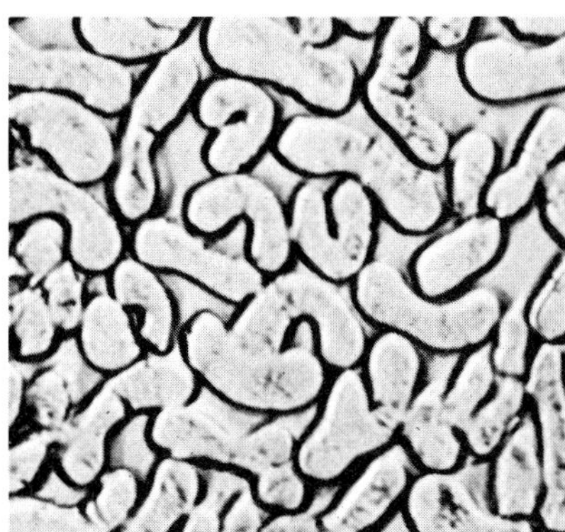

Fig. 14–5 Photomicrographs of modacrylic fibers. (Courtesy of American Association of Textile Chemists and Colorists.)

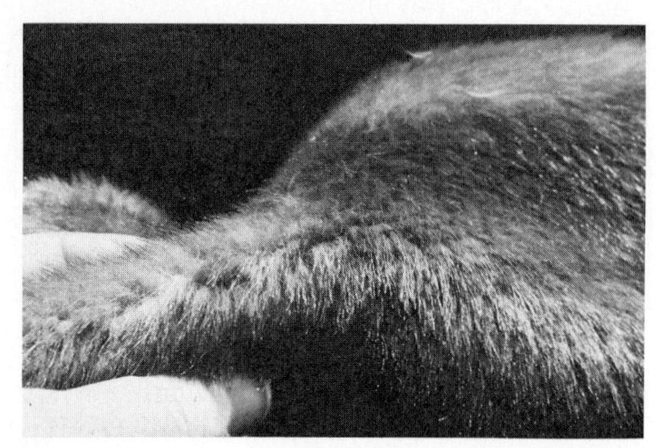

Fig. 14–6 *Fur-like fabric of modacrylic. Notice the sleek guard hairs and the soft, fine undercoat.*

Fig. 14–7 *Label giving directions for cleaning fur-like fabric.*

rics have a tendency to pill. Their absorbency is low, varying from 2–4 percent moisture regain.

Appearance Retention. Modacrylic fibers exhibit moderate resiliency. In typical end uses, they do not wrinkle. They have moderate dimensional stability and high elastic recovery.

Care. Modacrylics are resistant to acids, weak alkalis, and most organic solvents. Most modacrylics dissolve in boiling acetone. They are resistant to mildew and moths. They have very good resistance to sunlight and very good flame resistance.

Modacrylics can be washed or dry cleaned, but special care must be taken. Excessive rubbing may cause fabrics to pill. Fibers are heat sensitive; they shrink at 250°F and stiffen at temperatures over 300°F. If machine washed, warm water should be used. Tumble dry at low setting. Most fabrics need little ironing; lowest setting should be used if ironed. The fur-like fabrics vary in methods used for cleaning. Some are dry-cleanable; some require special care in dry cleaning (no steam, no tumble, tumble cold); some should be cleaned by furrier method. Look for instructions on labels (Figure 14–7).

Other Vinyl Fibers

In 1958 and in later amendments, the Textile Fiber Products Identification Act gave generic names to the various vinyl fibers, based on their chemical composition.

SARAN

Saran—a manufactured fiber in which the fiber-forming substance is any long-chain synthetic polymer composed of at least 80 percent by weight of vinylidene chloride units. (CH_2CCl_2)—Federal Trade Commission.

Saran is a vinylidene-chloride/vinyl-chloride copolymer developed in 1940 by the Dow Chemical Company. The raw material is melt spun and stretched to orient the molecules. Both filament and staple forms are produced. Much of the filament fiber is produced as a monofilament for seat covers, furniture webbing, screenings, luggage, shoes, and handbags. Monofilaments are also used in doll's hair and wigs. The staple form is made straight, curled, or crimped. The curled form is unique in that the curl is inherent and closely resembles the curl of natural wool.

Saran production is low because of competition from olefin, which has similar properties at a lower cost. Saran is used as an agricultural protective fabric to shade delicate plants such as tobacco and ginseng. Saran is used in rugs, draperies, and upholsery (see Figure 14–8). In addition to its use as a fiber, saran has wide use in the plastics field.

Saran has good weathering properties, chemical resistance, and stretch resistance. It is an unusually tough, durable fiber. Saran has a tenacity of 1.4–2.4 g/d with no change when wet. Saran has an elongation of 15–30 percent with excellent recovery. Saran also has good resiliency. It is an off-white fiber with a slight yellowish tint.

The fiber as seen under the microscope is perfectly round and smooth. It has a moisture re-

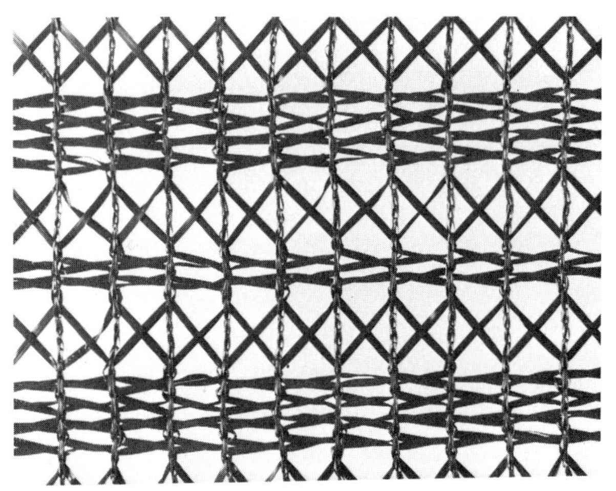

Fig. 14–8 *Saran casement cloth designed by Jack Lenore Larson.*

gain of less than 0.1 percent. Saran absorbs little or no moisture, so it dries rapidly. It is difficult to dye, and, for this reason, solution dyeing is used. Saran also has no static charge. It is heavy, with a specific gravity of 1.7. Saran does not support combustion. When exposed to flame, it will soften, char, and decompose. It softens at 115°C (240°F). It has excellent size and shape retention and is resistant to acids, alkalies, and organic solvents. Exposure to sunlight causes light-colored objects to darken, but no strength loss occurs. Saran is also immune to biological attack.

VINYON

Vinyon—a manufactured fiber in which the fiber-forming substance is any long-chain synthetic polymer composed of at least 85 percent by weight of vinyl chloride units. (—CH_2CHCl—)—Federal Trade Commission.

A patent for making fiber from a copolymer of vinyl chloride (86 percent) and vinyl acetate (14 percent) was obtained by the Union Carbide Corporation in 1937. The raw material is dissolved in acetone and dry spun. The name *vinyon* was adopted as a trade name and later released for generic usage. Commercial production of vinyon was begun in 1939.

Vinyon has an irregular-, round-, dog-bone- or dumbbell-shaped cross-section. The fiber is white and somewhat transluscent. Vinyon is very sensitive to heat. The fibers soften at 150–170°F, shrink at 175°F, and do not withstand boiling water or normal pressing and ironing temperatures. They are unaffected by moisture, chemically stable, resistant to moths and biological attack, poor conductors of electricity, and do not burn. These properties make vinyon especially good for bonding agent for rugs, papers, and nonwoven fabrics. The fibers have a tenacity of 0.7–1.0 g/d, which indicates that they are not stretched after spinning. These fibers, which are amorphous, have a warm, pleasant hand. Elongation ranges from 12–125 percent. Specific gravity ranges from 1.33–1.43. Moisture regain is 0.1 percent.

Vinyon is also used for wigs, flame-retardant Christmas trees, filter pads, fishing lines and nets, and protective clothing. Some trade names of vinyon are Leavil, Teviron, and Viclon.

One modification of vinyon is a bicomponent-bigeneric fiber called Cordelan. It is a matrix-fibril fiber of 50 percent vinal/50 percent vinyon. It is produced in Japan and used for flame-retardant children's sleepwear, airplane blankets, and home furnishings.

VINAL

Vinal—a manufactured fiber in which the fiber-forming substance is any long-chain synthetic polymer composed of at least 50 percent by weight of vinyl alcohol units (—CH_2—$CHOH$—) and in which the total of the vinyl alcohol units and any one or more of the various acetal units is at least 85 percent by weight of the fiber.—Federal Trade Commission.

No vinal fibers are produced in the United States. Modified vinal fibers are imported for use in children's sleepwear because of their inherent flame-retardant properties. Vinal is made in Japan and Germany under the trade names of Kuralon, Mewlon, Solvron, Vilon, Vinol, and Vinylal.

When the fibers are extruded, they are water soluble. The fibers must be treated with formaldehyde in order to form crosslinks, which make the fiber nonwater soluble. The fiber has a smooth, slightly grainy appearance. The cross-section is most often U-shaped. Vinal has a tenacity of 3.5–6.5 g/d and an elongation of 15–30 percent, and is 25 percent weaker when wet. The specific gravity of vinal is 1.26. It has a moisture regain of 5.0 percent. It does not support combustion, but softens at 200°C (390°F) and melts at 220°C (425°F). It has good chemical resistance and is unaffected by alkalies and common solvents. Concentrated acids will harm the fiber. Vinal has excellent resistance to biological attack. Mass pigmentation is about the only way to color the fiber.

In other countries, vinal is used in protective apparel. Major uses in the United States are industrial and include fishing nets, filter fabrics, tarpaulins, and brush bristles. In water-soluble forms, the fiber is used as a ground fabric to create laces and other sheer fabrics. Once the fabric has been produced, the vinal ground is dissolved and the sheer fabric remains.

One modification of vinal is a bicomponent-bigeneric fiber called Cordelan, which is a matrix-fibril fiber of 50 percent vinal/50 percent vinyon. See the preceding discussion of vinyon for examples of end uses for Cordelan.

NYTRIL—PRODUCTION DISCONTINUED

Nytril—a manufactured fiber containing at least 85 percent of a long-chain polymer of vinylidene dinitrile ($—CH_2C(CN)_2—$) *where the vinylidene dinitrile content is no less than every other unit in the polymer chain.—Federal Trade Commission.*

Nytril is no longer produced anywhere in the world. Other fibers with similar properties and lower costs have replaced nytril in all end uses.

TETRAFLUOROETHYLENE

Tetrafluoroethylene is not defined by the Textile Fiber Products Identification Act. Tetrafluoroethylene is common as a coating for cooking pans under the trade name Teflon. It is also used as a coating in plastic forms.

Tetrafluoroethylene is polymerized under pressure and heat in the presence of a catalyst. Emulsion spinning can be used. In *emulsion spinning,* polymerization and extrusion occur simultaneously. Tetrafluoroethylene has the following repeat unit ($—CF_2—CF_2—$). It has a tenacity of 1.6 g/d, with low elongation and good pliability. The fiber is heavy, with a specific gravity of 2.3. The fiber can withstand temperatures up to 260°C (500°F) without damage. It is resistant to chemicals, sunlight, weathering, and aging. The fibers are chemically inert. Tetrafluoroethylene does produce electrical charges. It is tan in color, but can be bleached white with sulfuric acid. Gore-Tex is a trade name for fabrics that have a thin microporous film of tetrafluoroethylene applied to a fabric for use in outerwear. Tetrafluoroethylene is used industrially in filter fabrics (to reduce smokestack emissions), packing fabrics, gaskets, industrial felts, covers for presses in commercial laundries, electric tape, and part of some protective fabrics.

CELIOX

Celiox is a heat-stabilized cyclic and cross-linked polyacrylonitrile. The fiber has exceptional heat resistance and does not ignite or melt. It maintains its full strength of 1.5 g/d after prolonged exposure to temperatures of more than 200°C. Celiox has a density of 1.4 and a moisture regain of 10 percent. The fiber has an elongation of 10 percent. Because of these properties and its comfortable hand, Celiox is used in protective clothing and as a substitute for asbestos in industrial products.

15

Spandex and Other Elastomeric Fibers

According to the Society of Testing and Materials (ASTM), an *elastomer* is a natural or synthetic polymer that, at room temperature, can be stretched repeatedly to at least twice its original length and that, after removal of the tensile load, will immediately and forcibly return to approximately its original length. Fibers that will be discussed in this chapter are spandex, rubber, and anidex. Of these, anidex is no longer produced in the United States.

Kinds of Stretch

Every apparel covering of the body needs some stretch or elasticity. Skin is very elastic and will stretch when the body turns, twists, and bends. The amount of stretch in stretch fabric should be adequate for this elongation.

There are two kinds of stretch: power stretch and comfort stretch. *Power stretch* is important in end uses where holding power and elasticity are needed. Elastic fibers that have a high retractive force must be used to attain this kind of stretch. Some end uses are foundation garments, surgical-support garments, swim suits, garters, belts, and suspenders. Power-stretch garments have about 200 percent extensibility and support muscles and body organs, reduce apparent body size, and firm and shape body flesh.

Comfort stretch is important in outerwear where only elasticity is desired. Comfort-stretch garments look no different than garments made from nonstretch fabrics. They have 10–15 percent extensibility and provide comfort as well as fit and neatness retention. Comfort-stretch fabrics are usually lighter in weight than power-stretch fabrics.

Rubber

Rubber—manufactured fiber in which the fiber-forming substance is comprised of natural or synthetic rubber, including:

1. *A manufactured fiber in which the fiber-forming substance is a hydrocarbon such as natural rubber, polyisoprene, polybutadiene, copolymers of dienes and hydrocarbons, or amorphous (noncrystalline) polyolefins.*

2. *A manufactured fiber in which the fiber-forming substance is a copolymer of acrylonitrile and a diene (such as butadiene) composed of not more than 50 percent but at least 10 percent by weight of acrylonitrile units* $(\text{—}CH_2\text{—}\underset{\displaystyle CN}{\underset{|}{C}}H\text{—})$*. The term* lastrile *may be used as a generic description for fibers falling in this category.*

3. *A manufactured fiber in which the fiber-forming substance is a polychloroprene or a copolymer or chloroprene in which at least 35 percent by weight of the fiber-forming substance is composed of chloroprene units* $(\text{—}CH_2\text{—}\underset{\displaystyle Cl}{\underset{|}{C}}{=}CH\text{—}CH_2)\text{.—}$

Federal Trade Commission.

Natural *rubber* is the oldest elastomer and the least expensive. It is obtained by coagulation of the latex from the rubber tree *Hevea brasiliensis*. In 1905, sheets of rubber were cut into strips that made the yarns used in foundation garments and the like. Before this time, whale bone and lacings had been used for corsets. During and shortly after World War II, synthetic rubbers were developed. These synthetic rubbers are crosslinked diene polymers, copolymers containing dienes or amorphous polyolefins. Both synthetic and natural rubbers must be vulcanized or crosslinked with sulfur in order to develop elastomeric properties. Natural and synthetic rubbers are large in cross-section. The shape of the cross-section will be round if extruded or rectangular if cut.

Rubbers have excellent elongation characteristics of 700–900 percent with excellent recovery. Rubber has low tenacity ranging from .5–1.0 g/d. The low-tenacity limits the use of rubber in light-weight garments. The finest rubber yarns must be three times as large as spandex yarns to be comparable in strength. Because of rubber's low-dye acceptance, hand, and appearance, yarns of monofilament or multifilament fibers are almost always covered by a yarn of another fiber content or by other yarns used to produce the fabric. In the latter case, the fabric is made so that the rubber yarn is almost completely covered by the interlacing or interlooping pattern of the ground yarns in the fabric.

Rubber has been replaced in many uses by spandex. Rubber continues to be used in narrow elastic fabrics. It is more common to find synthetic rubber in these elastic fabrics than it is to find natural rubber.

Although antioxidants are incorporated in the spinning solution, rubber still does not have good resistance to oxidizing agents and is damaged by aging, sunlight, oil, and perspiration. Rubber's resistance to alkali is generally good, but it is damaged by heat, chlorine, and solvents, so it should be washed with care and should not be dry cleaned.

Spandex

After many years of research, du Pont introduced the first man-made elastic fiber, Lycra, in 1958.

There was much interest in *spandex* fibers; they were superior to rubber in strength and durability. By 1965, eight companies were producing spandex. It was assumed that spandex would be widely used in all apparel to make fabrics more comfortable. At this time, durable press was introduced. Both durable-press and spandex fabrics required special or different cutting, sewing, and pressing techniques, and the ready-to-wear industry could only cope with one completely new development at a time. Efforts were concentrated on durable press. At this time also, knitted fabrics were being made in greater quantity and often with textured stretch yarns (see Chapter 18). Knits have comfort stretch because of their fabric construction. At present, spandex is produced by du Pont under the trade name of Lycra and by Globe Manufacturing Company under the trade names of Glospan and Cleerspan.

PRODUCTION

Spandex fibers are made by reacting preformed polyester or polyether molecules with di-isocyanate and then polymerizing into long molecular chains. Filaments are obtained by wet or solvent spinning. Like all man-made fibers, the spinning solution may contain delustering agents, dye receptors, whiteners, and lubricants.

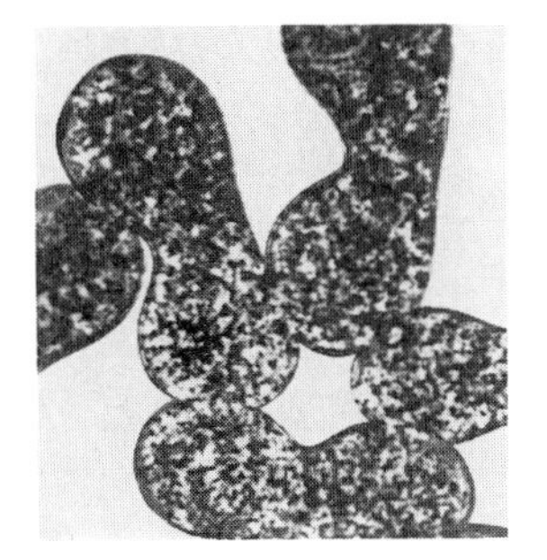

Fig. 15–1 *Lycra spandex fiber: cross-section* (left)*; lengthwise* (right). *(Courtesy of E. I. du Pont de Nemours & Company.)*

PHYSICAL STRUCTURE

Spandex is produced as monofilament or multifilament yarns in a variety of deniers. Monofilaments are round in cross-section whereas multifilaments are coalesced or partly fused together at intervals (Figures 15–1 and 15–2). A pin inserted in the yarn cannot be pulled through the entire length but will be stopped by the joinings. The advantage of multifilament yarns is, when sewing, the needle will go between the fine filaments; thus there is no danger of breaking the filaments. The advantage of the monofilament yarns is, when sewing with a ballpoint needle, the needle pushes the monofilament aside; thus there is no danger of rupture. Spandex fibers are delustered and usually white.

Deniers range from 20–4,300. Twenty-denier spandex is used in lightweight support hosiery where a large amount of stretch is desirable. Much coarser yarns, 1,500 to 2,240 denier, stretch less and are used for support in hosiery tops, swim wear, and foundation garments.

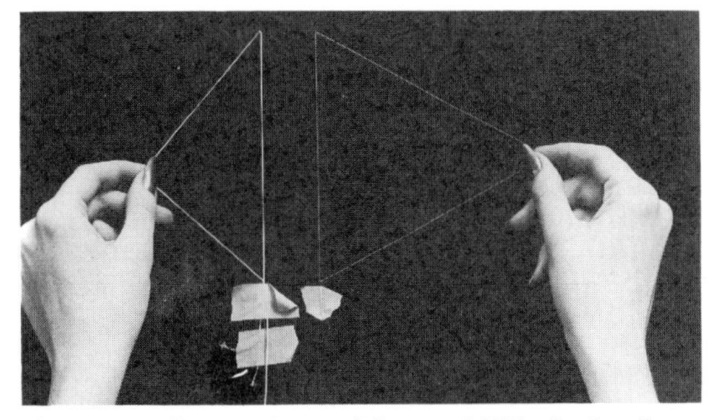

Fig. 15–2 *Comparison of heavy 1,500 denier Lycra spandex fibers* (left) *and fine 20 denier yarn* (right). *(Courtesy of E. I. du Pont de Nemours & Company.)*

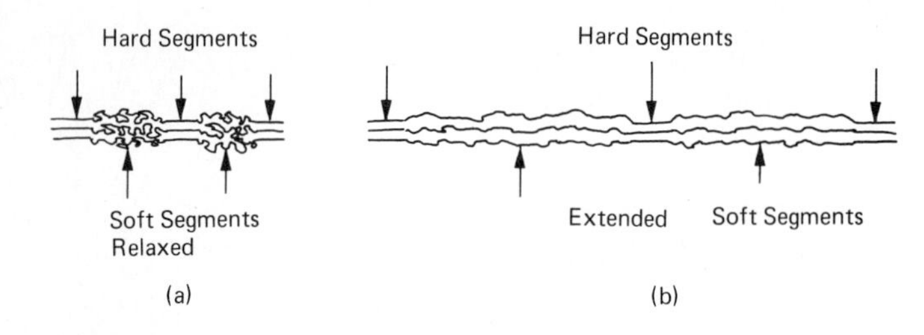

Fig. 15–3 Spandex molecular chains: (a) relaxed; (b) extended.

CHEMICAL COMPOSITION AND MOLECULAR ARRANGEMENT

Spandex—a manufactured fiber in which the fiber-forming substance is a long-chain synthetic polymer consisting of at least 85 percent of a segmented polyurethane.—Federal Trade Commission.

Spandex is a generic name, but it is not derived from the chemical nature of the fiber as are most of the man-made fibers (rayon and nylon are the exceptions); it was coined by shifting the syllables of the word expand.

Spandex is made up of rigid and flexible segments in the polymer chain; the soft segments provide the stretch and the rigid segments hold the chain together. When force is applied, the folded, or coiled, segments straighten out; when force is removed, they go back to their original positions (Figure 15–3). Varying proportions of hard and soft segments are used to control the amount of stretch.

PROPERTIES

Aesthetics. Spandex is seldom used alone in fabrics. Other yarns or fibers are used to achieve the desired hand and appearance. Even in power-stretch fabrics for foundation garments and surgical hose, where beauty is not of major importance, nylon or other yarns are used. The characteristics of spandex that contribute to beauty in fabrics are dyeability of the fiber and good strength, making it possible to have fashion in color and prints and sheer garments. Rubber does not take dye, which previously limited the use of color in foundation garments. Covered yarns and core-spun yarns were necessary not only to absorb dye but also to protect the rubber from body oils and secretions and to ensure the degree of stretch.

Spandex needs no cover yarns since it will take dye. Eliminating the cover yarn, or core-spun yarn, reduces the cost and results in lighter weight garments. This is important not only for beauty but also for comfort.

Summary of the Performance of Spandex and Rubber in Apparel Fabrics

	Spandex	*Rubber*
AESTHETIC	*ADEQUATE*	*POOR*
DURABILITY	*ADEQUATE*	*POOR*
Abrasion resistance	Low	Low
Tenacity	Low	Low
Elongation	Excellent	Excellent
COMFORT	*ADEQUATE*	*POOR*
APPEARANCE RETENTION	*GOOD*	*GOOD*
Resiliency	Good	Good
Dimensional stability	Good	Good
Elastic recovery	Excellent	Excellent
RECOMMENDED CARE	*MACHINE WASH OR DRY CLEAN*	*WASH WITH CARE*

Durability Factors

Fiber Property	*Spandex*	*Rubber*	*Nylon*
Breaking tenacity g/d	0.6–0.9	0.34	3.0–9.5
Breaking elongation	400–700 percent	500–600 percent	23 percent
Flex life	Excellent	Fair	Excellent
Recovery from stretch	99 percent	97 percent	100 percent

Durability. Spandex is more durable than rubber because it is not deteriorated by aging. Durability factors are compared in the following chart. Nylon is included in the chart because it has more stretch than other man-made filaments and illustrates the difference between a hard fiber and an elastomeric fiber.

Spandex is resistant to body oils, perspiration, and cosmetics, which cause degradation of rubber. It also has good shelf life; that is, it does not deteriorate with age.

Comfort. Spandex fibers have a moisture regain of 0.75–1.3 percent, making them uncomfortable for skin-contact apparel. Power-stretch garments are uncomfortable to most people. Lighter-weight foundation garments of spandex have the same holding power as heavy garments of rubber. Spandex has a specific gravity of 1.2–1.25, which is greater than that of rubber. However, because of the greater tenacity of spandex, this higher specific gravity does not create unduly heavy garments.

Care. Spandex is resistant to dilute acids and to alkalis. It has good resistance to cosmetic oils and lotions. Most spandex fibers are resistant to bleaches. They have good resistance to dry-cleaning solvents. Spandex is thermoplastic, with a melting point of 446–518°F.

USES

Spandex is used in foundation garments, active sportswear—such as swim suits and leotards—hosiery, and narrow fabrics. It also has medical uses, such as surgical and support hose, bandages, and surgical wraps. It is also used in fitted sheets and slipcovers.

See the following tables for fiber modifications and a summary of end uses related to stretch properties.

Types and Kinds of Spandex
White (delustered)
Transparent—clear luster
20–210 denier—support hosiery
140–560 denier (core spun)—men's hosiery
70–2240 denier—laces, foundation garments, swim wear, narrow fabrics, hosiery tops
Bicomponent

Spandex

Major End Uses	*Important Properties*
Athletic apparel	Power stretch, washability
Foundation garments (Power net, tricot)	Power stretch, washability, lightweight
Bathing suits	Power stretch, resistance to salt and chlorine-treated water, dyeability
Golf jackets	Comfort stretch
Ski pants	Comfort stretch
Support and surgical hose	Power stretch, lightweight
Elastic webbing	Power stretch

Other Elastomers

Anidex—a manufactured fiber in which the fiber-forming substance is any long-chain synthetic polymer composed of at least 50 percent by weight of one of more esters of a monohydric alcohol and acrylic acid. ($CH_2{=}CH{-}COOH$)

Anim 8, an anidex fiber, was produced by Rohm and Haas Company between 1970 and 1975. *Lastrile* was a generic name established by the Federal Trade Commission for an elastomeric fiber; Rohm and Haas experimented with lastrile but did not produce it commercially.

Lastrile fibers are made from copolymers of acrylonitrile or polychloroprene. The properties of lastrile fibers are similar to those of the other rubber fibers. At present, no lastrile fibers are produced in the United States.

16

Special-Use Fibers

In previous chapters, many fibers have been discussed. In this chapter, fibers with specialized end uses (but only those that have been defined in the Textile Fiber Products Identification Act) will be discussed.

Aramid

Aramid—a manufactured fiber in which the fiber-forming substance is a long-chain synthetic polyamide in which at least 85 percent of the amide linkages $\left(\begin{array}{c}-\underset{\underset{O}{\|}}{C}-NH-\end{array}\right)$ *are attached directly to two aromatic rings.—Federal Trade Commission.*

Nylon is a polyamide fiber; *aramid* is an aromatic polyamide fiber. When researchers at the du Pont Company were working on nylon variants, they produced a fiber that had exceptional heat and flame resistance. Du Pont introduced this fiber in 1963 under the trade name Nomex nylon. Another variant of nylon was introduced by du Pont in 1973 as Kevlar. This fiber had exceptional strength in addition to fire resistance. In response to a petition by du Pont for a new generic-name classification for these fibers that were uniquely different from nylon, the Federal Trade Commission (FTC) established the generic classification of aramid in 1974.

Aramid can be wet or dry spun. Aramid fibers have high tenacity and high resistance to stretch, to most chemicals, and to high temperatures. The fiber can be produced as a high-tenacity fiber. The following table compares normal-tenacity aramids with high-tenacity aramids. These fibers maintain their shape and form at high temperatures. Aramid fibers have excellent impact and abrasion resistance. Aramid fibers are usually round- or dog-bone-shaped, depending on spinning methods (Figure 16–1).

Hollow aramid fibers are used to produce fresh water from sea water through reverse osmosis. The thin, dense skin of the fiber allows only water to pass through. Aramids are difficult to dye. Aramids have poor resistance to acids. Common trade names for aramid fibers are Nomex, Kevlar, Conex, Fenilon, Arenka, and Kermel. Nomex and Kevlar are trade names owned by du Pont, and they are the most common trade names found in the United States.

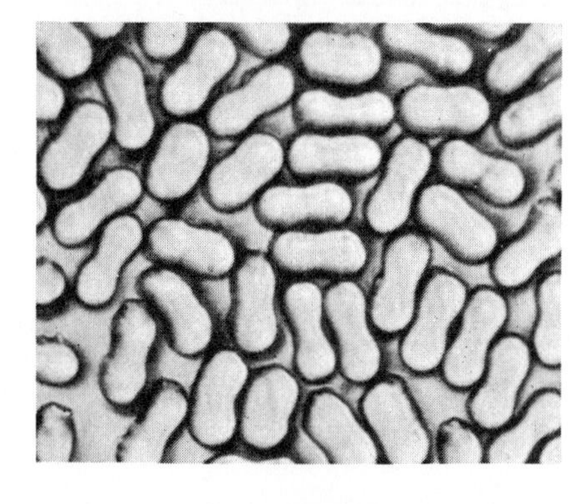

Fig. 16–1 *Photomicrographs of aramid. (Courtesy of American Association of Textile Chemists and Colorists.)*

Kevlar fibers are found as Kevlar, Kevlar 29, and Kevlar 49. Kevlar is used primarily in reinforcements of radial tires and other mechanical rubber goods. Kevlar 29 is found in protective apparel, cables, and cordage, and as a replacement for asbestos, such as in brake linings and gaskets. A seven-layer, body-armor undervest of Kevlar 29 weighing 2.5 pounds can deflect a knife slash and stop a .38-caliber bullet fired from 10 feet (Figure 16–2). Kevlar 49 has the highest tenacity of the aramids and is found as a plastic-reinforcement fiber for boat hulls and aerospace uses.

Nomex is used where resistance to heat and combustion with low-smoke generation are required. Protective clothing, such as firefighters' apparel and race-car drivers' suits, and flame-retardant furnishings for aircraft are made of Nomex. Hot-gas filtration systems and electrical

Properties of Aramid

Property	*Normal Tenacity*	*High Tenacity*
Breaking tenacity	4.3–5.1 g/d–filament 3.7–5.3 g/d–staple	21.5 g/d
Specific gravity	1.38	1.44
Moisture regain	4.5 percent	3.5–7.0 percent
Effect of heat	Carbonizes above 800°F Very resistant to flame Does not melt	
Resistance to acids	Better than nylon	
Resistance to alkalis	Good	
Resistance to organic solvents	Good	
Resistance to sunlight	Poor	
Oleophilic	Yes, unless special finishes are used	
Static buildup	Yes, unless special finishes are used	

insulation are constructed of Nomex. This heat-resistant fiber is also found in covers for laundry presses and ironing boards.

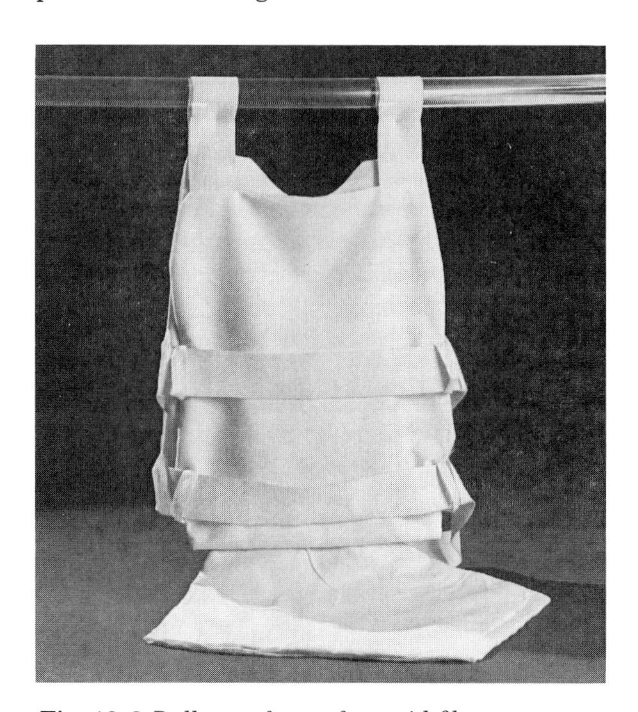

Fig. 16–2 *Bulletproof vest of aramid fibers.*

Glass

Glass—a manufactured fiber in which the fiber-forming substance is glass.—Federal Trade Commission.

Glass is an *incombustible* textile fiber; it cannot burn. This makes it especially suitable for end uses where the danger of fire is a problem—such as in draperies for motels, nursing homes, public buildings, and homes. Glass fibers have also been used in bedspreads and tablecloths and in interlinings for coats and mittens. The possibility of severe skin irritation for tiny broken fibers has limited the use of glass fibers in wearing apparel.

The process of drawing out glass into hair-like strands dates back to ancient times. It is thought that Phoenician fishermen noticed small pools of molten material among the coals of the fires they built on the sands of the Aegean beaches and while poking at the strange substances, they drew out a long strand—the first glass fiber.

The raw materials for glass are sand, silica, and limestone, combined with additives of feldspar and boric acid. These materials are melted in large electric furnaces (2,400°F). For filament yarns, each furnace has holes in the base of the melting chamber. Fine streams of glass flow through the holes and are carried through a hole in the floor to a winder in the room below. The winder revolves faster than the glass comes from the furnace, thus stretching the fibers and reducing them in size before they harden. The round rod-like filaments are shown in Figure 16–3. When staple yarn is spun, the glass flows out in thin streams from holes in the base of the furnace, and jets of high-pressure air or steam yank the glass into fibers 8–10 inches long. These fibers are collected on a revolving drum

Fig. 16–3 Photomicrograph of fiberglass. (Courtesy of the Owens-Corning Fiberglas Corporation.)

and made into a thin web, which is then formed into a sliver, or soft, untwisted yarn.

Beta Fiberglas was introduced by the Owens-Corning Fiberglas Corporation in 1964. It has one-sixth the denier of common glass fibers. The extremely fine filaments are resistant to breaking and thus more resistant to abrasion. Beta Fiberglas has about half the strength of regular glass fiber, but its tenacity of 8.2 is still greater than most fibers.

CORONIZING

Coronizing is a process for heat setting, dyeing, and finishing glass fiber in one continuous operation. Since glass is low in flexibility, the yarns resist bending around one another in the woven fabric. Heat setting at a temperature of 1100°F softens the yarns so that they will bend and assume yarn crimp. Coronized fabrics have greater wrinkle resistance and softer draping qualities.

After heat setting, the glass fabric is treated with a lubricating oil; then color and a water-repellent finish are added. For this treatment, the Hycar-Quilon process is used. Hycar is an acrylic latex resin, which, with the colored pigment, is padded on the fabric and then cured at a temperature of 320°F. This is followed by a treatment with Quilon, a water-repellent substance and the fabric is again cured. The resin used in the color treatment increases the flexibility of the fiber but is damaged by chlorinated dry-cleaning solutions, so dry cleaning should be done with Stoddard solvent.

Screen printing as well as roller printing can be done by the Hycar-Quilon process, since the color paste dries fast enough to allow one screen to follow another rapidly. The Hycar-Quilon process gives good resistance to rubbing off (crocking), which is one of the disadvantages of other coloring methods.

Glass has a tenacity of 6–10 g/d dry and 5–8 g/d wet. Glass has a low elongation of only 3–4 percent but has excellent elasticity in this narrow range. Glass fibers are brittle and break when bent. Thus they exhibit poor flex resistance to abrasion. These fibers are very heavy, with a specific gravity of 2.5. The fibers have no absorbency and are resistant to most chemicals. Trade names include Fiberglas, Beta glass, Chemglass, J-M fiberglass, PPG fiberglass, and Vitron.

Glass fiber is used in the decorator field for curtains and draperies. Here the fiber performs best if bending and abrasion can be kept at a minimum. Curtains or draperies should not be used at windows that will be kept open, allowing the wind to whip the curtains. The bottom of the curtains or draperies should not be allowed to touch the floor or windowsill.

The weight of the fabrics may mean that special rods are necessary, especially if large areas are draped.

Glass fiber has wide industrial use where noise abatement, fire protection, temperature control (insulation), and air purification are needed. Glass is common as a reinforcement fiber in molded plastics in boat and airplane parts. Insulation in buildings and transportation vehicles, such as boats and railway cars, are made of glass. In addition, glass is found in ironing-board covers and space suits. Flame-resistant-glass mattress covers are produced for hotels, dormitories, and hospitals. A silica-glass quilt called AFRSI (Advanced Flexible Reuseable Surface Insulation) was used on several space shuttles. Glass is used in geotextiles. Filters, fire blankets, and heat- and electrical-resistant tapes and braids are other industrial products made of glass. Owens-Corning is researching glass yarns suitable for apparel.

Additional end uses are the optical fibers that are very fine fibers of pure glass. Very rapid laser beams, rather than electricity, activate the fibers. Glass optical fibers are free of electrical interference. Optical fibers are found in communication and medical equipment.

CARE

Hand washing is preferred to machine washing, which causes excessive breaking of the fibers. Figure 16–4 shows the remnants of a fiberglass laundry bag that was machine washed. A residue of tiny glass fibers in the washing machine will contaminate the next load and cause severe skin irritation for those who wear that clothing. Even with hand washing, severe skin irritation can occur. The Federal Trade Commission proposed in 1967 that the label should disclose the possibility of severe skin irritation.

Curtains should not require frequent washing because glass fibers resist soil, and spots and stains can be wiped off with a damp cloth. No ironing is necessary. Curtains can be smoothed and put on the rod to dry. Oils used in finishing have caused graying in white curtains. Oil holds the dirt persistently and also oxidizes with age. Washing has not proved to be a very satisfactory way to whiten the material, and dry cleaning is not recommended unless Stoddard solvent is used.

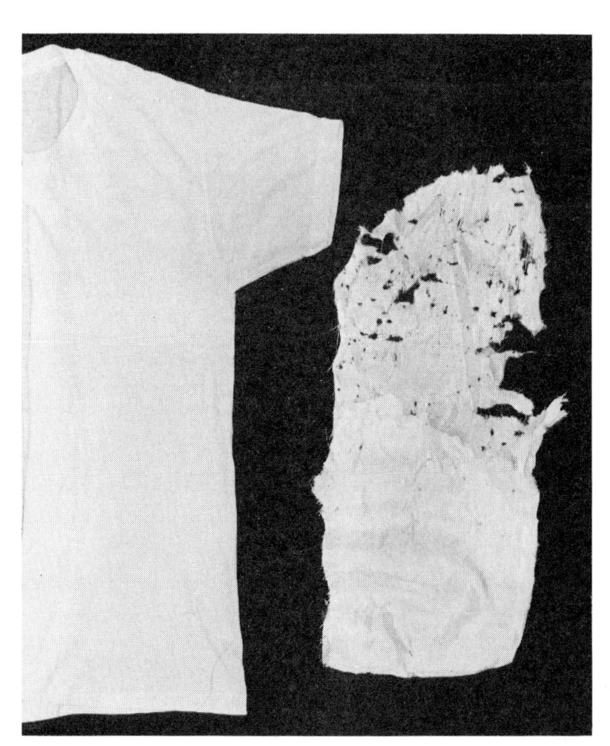

Fig. 16–4 *Glass-fiber laundry bag after being washed with regular family wash.*

Glass-Fiber Properties Important in Draperies	
Flexibility	Breaks easily
Specific gravity	Heavy
Absorbency, percent of moisture regain	None
Effect of sunlight	None
Effect of acid and alkali	None
Effect of heat	Flameproof

Metal and Metallic Fibers

Metallic—a manufactured fiber composed of metal, plastic-coated metal, metal-coated plastic, or a core completely covered by metal.—Federal Trade Commission.

Gold and silver have been used since ancient times as yarns for fabric decoration. More recently, aluminum yarns, aluminized plastic yarns, and aluminized nylon yarns have taken the place of gold and silver. Metallic filaments can be coated with transparent films to minimize tarnishing. A common film is polyester with a trade name of Lurex. These fibers are often found as a decorative touch in apparel and home furnishings.

Stainless-steel fibers were developed in 1960, and other metal fibers have also been made into fibers and yarns. Stainless steel has had the most extensive development.

The use of stainless steel as a textile fiber was an outgrowth of research for fibers to meet aerospace requirements. Superfine-stainless-steel filaments (3–15 micrometers) are made by the following process. A bundle of fine wires (0.002 inch) is sheathed with dissimilar alloys and drawn to its final diameter, pickled in nitric acid to remove the sheath, and the yarn of several filaments is then ready for sizing, warping, and weaving. The early fibers were costly, $25 per pound. Several problems had to be solved, one of which involved twist. Each filament tended to act like a tiny coil spring, so the yarns required special treatment to deaden the twist.

Stainless-steel fibers are produced as both filament and staple. They can be woven or knitted and can be used as either the wrap or core of core yarns. The staple fiber can be blended with other textile fibers to reduce static permanently. Only 1–3 percent of the stainless-steel fiber is needed. The limitation on the use of stainless steel in clothing is its inability to be dyed, although some producers claim that such a small amount will not affect the color of white fabrics. Stainless steel has been used in carpets to reduce static and has been used experimentally in men's suits. It is also suitable for this purpose for use in upholstery, blankets, and work clothing, but until the cost of stainless-steel fibers is reduced, its use in such articles will not be practical. Stainless-steel fibers are used for industrial purposes, such as tire cord, wiring, and missile nose cones, and in corrective heart surgery (Figure 16–5).

Metal fibers are blended with other fibers to produce static-free clothing worn in operating rooms and computer-production facilities. Metals do not have many of the properties usually attributed to textile fibers. They are much heavier than the organic materials that compose most fibers—specific gravity of metal fibers is 7.88 g/cc as compared to 1.14 for nylon. They cannot be folded and unfolded without leaving permanent crease lines, have very little or no drape characteristics, and do not have the hand associated with textiles. Reduction in the denier of the fiber improves its properties, but the finer fibers are more expensive.

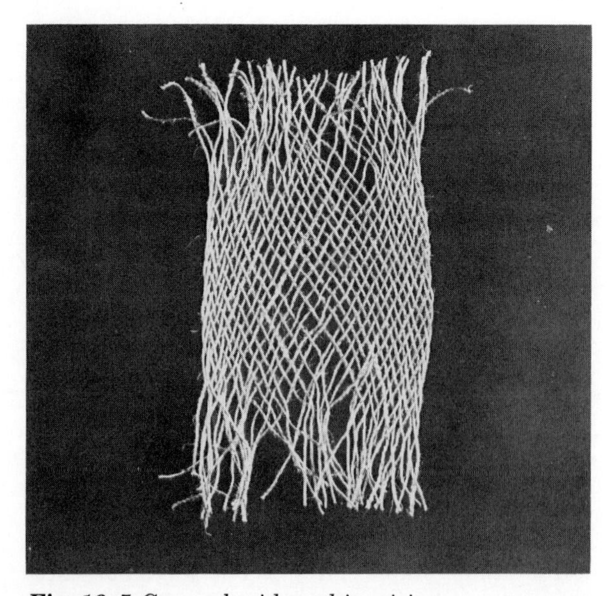

Fig. 16–5 Copper braid used in wiring.

Novoloid

Novoloid—a manufactured fiber in which the fiber-forming substance contains at least 35 percent by weight of crosslinked novolac (a crosslinked phenolformaldehyde polymer).—Federal Trade Commission.

Kynol, a type of novoloid fiber, was introduced by the Carborundum Company (now American Kynol) in 1969 with commercial production beginning in 1972.

Novoloid shows outstanding flame resistance to a blaze of 2500°C from an oxyacetylene torch, and it can meet any flammability requirements that might be legislated. The yarns do not melt, burn, or fuse but turn to carbonized yarns that maintain their construction.

Novoloid has an elasticity of 35 percent. It has good resistance to sunlight. Novoloid is inert to

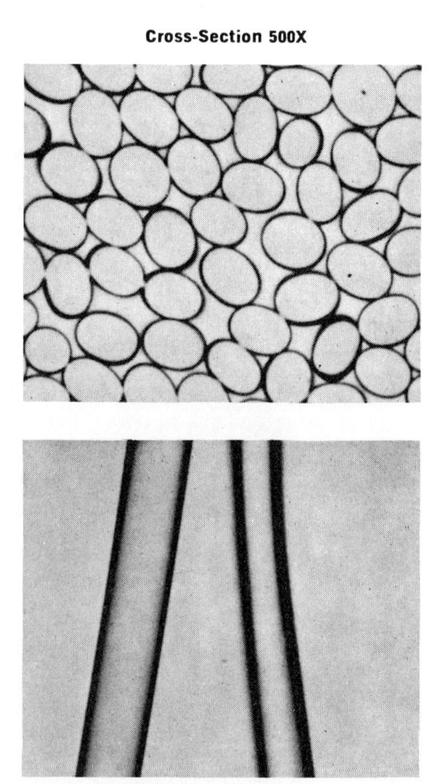

Fig. 16–6 Photomicrographs of novoloid. (Courtesy of American Association of Textile Chemists and Colorists.)

Properties

Cross-section	Round
Tenacity (grams/denier)	1.5–2.5
Specific gravity (grams/cc)	1.25
Moisture regain (percent)	5.5
Chemical resistance	Excellent
Uses	
Fireproof clothing and fabrics Chemical filters Blankets and draperies Institutional bedding, carpets, and household furnishings	

acids and organic solvents but susceptible to alkaline substances.

Novoloid fiber is gold in color, so its dyeing possibilities are somewhat limited—darker shades present no problem. It can be bleached to a good white. It was first produced as staple and later as filament. (See Figures 16–6 and 16–7.)

Fig. 16–7 *Heat and flame resistant apparel of Kynol™ novoloid. (Courtesy of American Kynol, Inc.)*

PBI

PBI is one of the most recent fibers to be defined. In 1986, the Textile Fiber Products Identification Act was amended with this definition of PBI: a manufactured fiber in which the fiber-forming substance is a long-chain aromatic polymer having reoccurring imidazole groups as an integral part of the polymer chain. PBI is produced by the Celanese Corporation. The trade name Arazole may be used. PBI is a condensation polymer that is dry spun. The specific gravity of PBI is 1.39; if the fiber has been stabilized, the specific gravity is 1.43. PBI has a tenacity of 3.1–4.2 g/d and a breaking elongation of 30 percent. PBI has a high moisture regain of 15 percent, but it is difficult to dye. Hence the most common coloration method currently used is mass pigmentation. PBI does not burn or melt and has very low shrinkage when exposed to flame. Even when charred, PBI fabrics remain supple and intact. Because of its heat resistance, it is ideal for use in heat-resistant apparel for firefighters, astronauts, fuel handlers, race-car drivers, welders, foundry workers, and hospital workers (Figure 16–8). The fiber is also found in upholstery, window-treatment fabrics, and carpets for aircraft, hospitals, and submarines. PBI is being evaluated as a flue-gas filter in coal-fired boilers and as a reverse-osmosis membrane material.

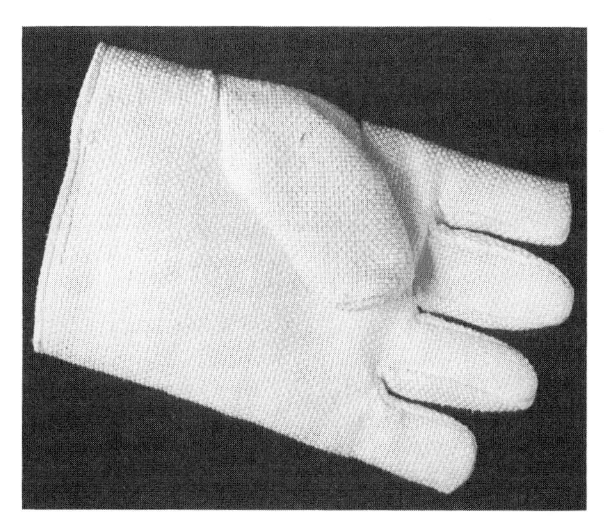

Fig. 16–8 *Heat-resistant glove of PBI.*

Sulfar

Sulfar is another newly defined fiber. In 1986, the FTC defined *sulfar* as a manufactured fiber in which the fiber-forming substance is a long-chain synthetic polysulfide in which at least 85 percent of the sulfide (—S—) linkages are attached directly to two (2) aromatic rings. Sulfar is produced by Phillips Fibers Corporation by melt spinning. The fiber is gold in color. The trade name Ryton PPS may be used. Sulfar has a tenacity of 3.0–3.5 g/d and a breaking elongation of 25–35 percent. It has excellent elasticity. Moisture regain is low (0.6 percent), and specific gravity is 1.37. Sulfar is highly resistant to acids and alkalis and not soluble in any known solvent below 200°C (392°F). Sulfar is used in filtration fabrics, paper-making felts, electrolysis membranes, high-performance membranes, rubber reinforcement, electrical insulation, and fire-fighting suits, and other protective clothing (Figure 16–9).

Fig. 16–9 *Fire fighting outfits made from sulfar fiber. (Courtesy of Thomas D. Lowes, 491 Delaware Ave., Buffalo, N.Y. 14202; Tel. 716-883-2650.)*

17

Yarn Classification

A *yarn* is a continuous strand of textile fibers, filaments, or material in a form suitable for knitting, weaving, or otherwise intertwining to form a textile fabric (ASTM 1984, vol. 07.01, p. 76.)

A yarn is classified and identified by the similarity or irregularity along its length, the composition of staple or filament fibers, the number of parts in the yarn, the direction and amount of twist, and the yarn size. These factors contribute to the performance of the fabric made from the yarn.

A wide variety of yarns is available to the fabric manufacturer. The selection of yarn is based on end use. Yarns play a very important part in determining the hand and performance of the fabric. For example, yarns with very high twist are used to create the texture in true crepe fabrics, and yarns with very low twist are used in fabrics that are to be napped, such as flannel fabrics and blankets. Yarn may enhance good fiber performance or compensate for poor fiber performance. The effectiveness of a finish may depend on the proper choice of yarn. Most yarns can be easily recognized and identified.

SPUN AND FILAMENT YARNS

The names of the most common yarns are based on the length of the fibers within them, and the appearance and parallelism of those fibers. When a typical yarn is unraveled from a fabric and held up to a light, it will appear uniformly smooth, or uniformly wavy, or it will appear to be fuzzy with protruding fiber ends. It can be untwisted until it falls apart into fibers. These fibers will be either short, typically ½–2½ inches, or as long as the piece of fabric from which the yarn was pulled.

A *spun yarn* is composed of short staple fibers that are twisted together, resulting in a fuzzy yarn with protruding fiber ends. Spun yarns of cotton are the most common in apparel, followed by yarns of polyester/cotton blend.

A *filament yarn* is composed of long fibers that are grouped together or slightly twisted together. Filament yarns may be smooth—with straight, almost parallel fibers—or be uniformly wavy, in which case they are called *textured-bulk-filament yarns* or just *textured-bulk yarns.* Filament yarns used in apparel are most frequently polyester. Filament, textured-bulk filament, and spun yarns are discussed in Chapters 18 and 19. The chart "Comparison of Spun, Smooth Filament and, Textured-Bulk Yarns" summarizes the properties that result from the use of these yarns in fabrics.

Simple Yarns

A *simple yarn* is alike in all its parts. It can be described as a spun or filament yarn based on the length of fibers present. A simple yarn also can be described by the number of parts it has, by the direction and amount of twist in the yarn and by the size of the yarn.

The following chart outlines the relation of yarns within this category to each other.

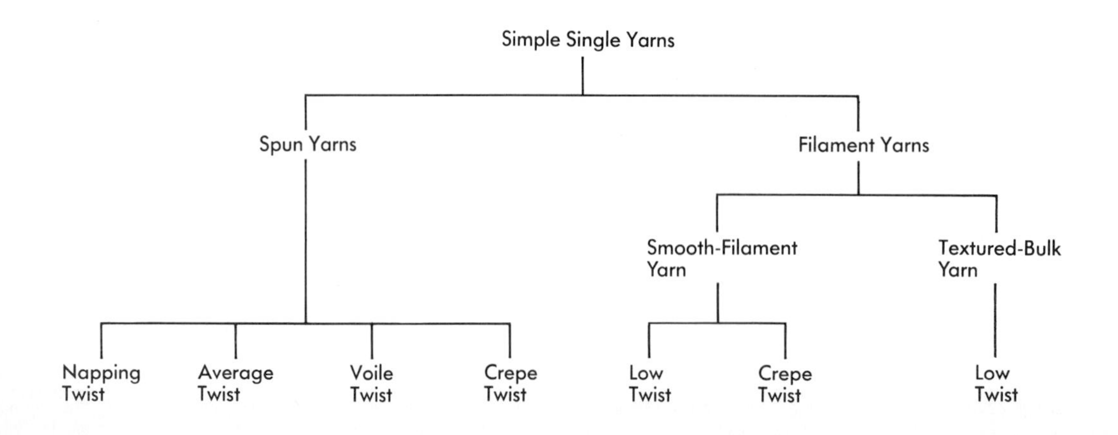

Number of Parts

Simple yarns are classified as single, ply, or cord yarns. A *single yarn* is the product of the first twisting operation that is performed by the spinning machine (Figure 17–1a).

Spun, filament, and textured yarns are each examples of simple, single yarns. Since these three types of yarns are most commonly found in apparel and home-furnishing fabrics, they usually are referred to as spun, filament, or textured-bulk yarns. It is understood that they are simple yarns, alike in all parts; and that they are single yarns, consisting of one strand of fibers.

A *ply yarn* is made by a second twisting operation, which combines two or more singles (Figure 17–1b). Each part of the yarn is called a *ply*. The twist is inserted by a machine called a *twister*. Most ply yarns are twisted in the opposite direction to the twist of the singles from which they are made; thus the first few revolutions tend to untwist the singles and straighten the fibers somewhat from their spiral position and the yarn becomes softer.

Plying tends to increase the diameter, strength, uniformity, and quality of the yarn. Ply yarns are commonly used in the warp direction of woven fabrics to increase the strength of the fabric. Two-ply yarns are found in the best-quality men's broadcloth shirts. Ply yarns are frequently seen in women's sweaters, both hand and machine knit. Two-ply and three-ply yarns are found in sewing thread and string used to tie packages. If simple, ply yarns are used only and are in the filling direction, they are used for some fabric effect other than strength.

A *cord* is made by a third twisting operation, which twists ply yarns together (Figure 17–1c). Some types of sewing thread and some ropes belong to this group. Cord yarns are seldom used in apparel fabrics, but are used in industrial weight fabrics such as duck and canvas.

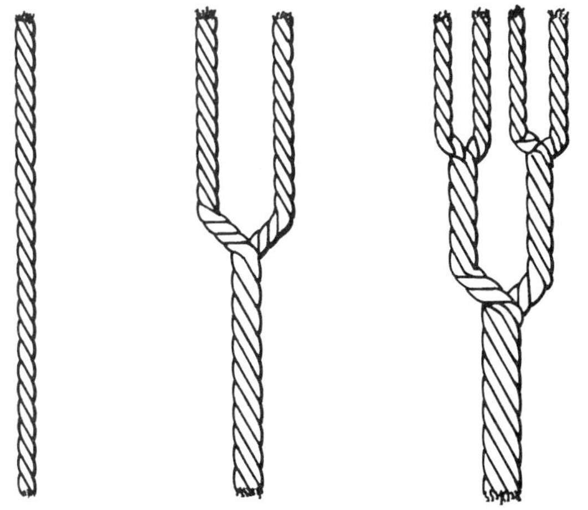

(a) A Single Yarn (b) Two-ply Yarn (c) A Cord Yarn

Fig. 17–1 *Parts of a yarn: (a) single yarn; (b) two-ply yarn; (c) cord yarn.*

Yarn Twist

Twist is the spiral arrangement of the fibers around the axis of the yarn. Twist is produced by revolving one end of a fiber strand while the other end is held stationary. Twist binds the fibers together and gives the spun yarn strength. It is a way to vary the appearance of fabrics.

The number of twists is referred to as *turns per inch.* They have a direct bearing on the cost of the yarn because higher twist yields lower productivity.

DIRECTION

The direction of twist is described as S-twist and Z-twist. A yarn has S-twist if, when held in a vertical position, the spirals conform to the direction of slope of the central portion of the letter "S." It is called Z-twist if the direction of spirals conforms to the slope of the central portion of the letter "Z." Z-twist is the standard twist used for weaving yarns (Figure 17–2).

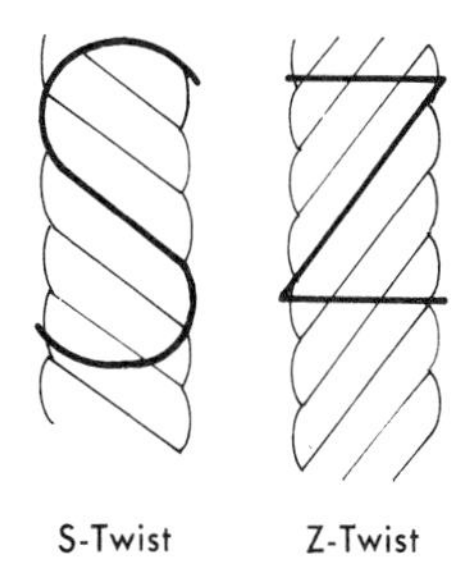

Fig. 17–2 *S- and Z-twist yarns.*

Comparison of Spun, Smooth Filament, and Textured-Bulk Yarns

Spun Yarns	*Smooth-Filament Yarns*	*Textured-Bulk Yarns*
I. Fabrics are cotton-like or wool-like.	I. Fabrics are silk-like.	I. Fabrics have the strength of filament yarns and the appearance of spun yarns.
II. Strength of fibers is not completely utilized.	II. Strength of fiber is completely utilized.	II. Strength may or may not be completely utilized.
III. Short fibers twisted into continuous strand, has protruding ends.	III. Long continuous, smooth, closely packed strand.	III. Long continuous, irregular, porous, flexible strand.
1. Dull, fuzzy look.	1. Smooth, lustrous.	1. Fuzzy, dull.
2. Lint.	2. Do not lint.	2. Do not lint.
3. Subject to pilling.	3. Do not pill readily.	3. Pilling depends on fabric construction.
4. Soil readily.	4. Shed soil.	4. Soil more easily than plain filament.
5. Warm (not slippery).	5. Cool, slick.	5. Warmer than plain filament.
6. Loft and bulk depend on size and twist.	6. Little loft or bulk.	6. Loft, bulk, and/or stretch.
7. Do not snag readily.	7. Snagging depends on fabric construction.	7. Snag easily.
8. Stretch depends on amount of twist.	8. Stretch depends on amount of twist.	8. Stretch depends on method of processing.
9. More cover (more opaque).	9. Less cover (less opaque).	9. More cover (more opaque).
IV. Are absorbent.	IV. Absorbency depends on fiber content.	IV. More absorbent than plain-filament yarns of same fiber content.
1. Good for skin contact (most absorbent).	1. Thermoplastics are low in absorbency.	1. Most manufacturing processes require thermoplastic fibers.
2. Less static buildup.	2. Static buildup high in thermoplastics.	2. Static buildup.
V. Size often expressed in yarn number.	V. Size in denier.	V. Size in denier.
VI. Various amounts of twist used.	VI. Usually very low or very high twist.	VI. Usually low twist.
VII. Most complex manufacturing process.	VII. Least complicated manufacturing process.	VII. Manufacturing process is more complex than plain filament.

AMOUNT

The amount of twist varies with (1) the length of the fibers, (2) the size of the yarn, and (3) the intended use. Increasing the amount of twist up to the point of perfect fiber-to-fiber cohesion will increase the strength of the yarns. Too much twist places the fibers at right angles to the axis of the yarn and causes a shearing action between fibers, and the yarn will lose strength (Figure 17–3).

Combed yarns with long fibers do not require as much twist as carded yarns with short fibers because they establish more points of contact per fiber and give a stronger yarn for the same amount of twist. Fine yarns require more twist than coarse yarns. Knitting yarns have less twist than the filling yarns used in weaving. The following chart and discussion give examples of uses for yarns with different amounts of twist.

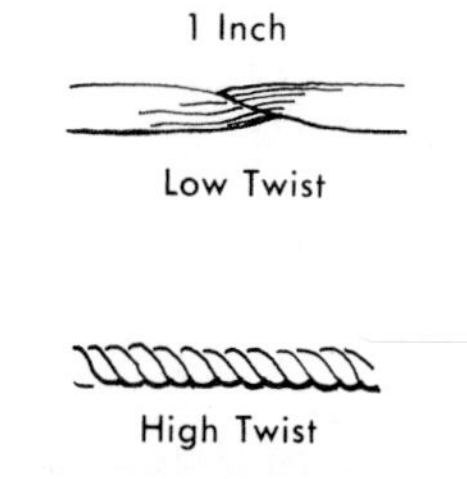

Fig. 17–3 *Low and high turns per inch.*

Amount of Twist

Amount	*Example*
Low twist	Filament yarns: 2–3 tpi*
Napping twist	Blanket warps: 12 tpi Filling: 6–8 tpi
Average twist	Percale warps: 25 tpi Filling: 20 tpi Nylon hosiery: 25–30 tpi
Voile twist	Hard-twist singles: 35–40 tpi are plied with 16–18 tpi
Crepe twist	Singles: 40–80 or more tpi are plied with 2–5 tpi

*Turns per inch.

Low twist in spun yarns results in lofty yarns. It is used in filling yarns of fabrics that are to be napped. The low twist permits the napping machine to tease out the ends of the staple fibers and create the soft, fuzzy surface. (See "Napping," Chapter 33).

Average twist is frequently used for yarns made of staple fibers and is very seldom used for filament yarns. The amount of twist that gives warp yarns maximum strength is referred to as *standard warp twist.* Warp yarns need more twist than filling yarns because warp yarns are under high tension on the loom and they must resist wear caused by the action of the loom. The lower twist of the filling yarns makes them softer and less apt to kink.

Hard twist (voile twist) yarns have 30–40 turns per inch. The hardness of the yarn results when twist brings the fibers closer together and makes the yarn more compact. This effect is more pronounced when a twist-on-twist ply yarn is used. *Twist-on-twist* means that the direction of twist in the singles is the same as that of plying twist (Figure 17–4). This results in a buildup of the total amount of twist in the yarn (see "Voile," page 202).

Crepe yarns are made of either staple or filament fiber. They are made with a high number of turns per inch (40–80) inserted in the yarn. This makes the yarn so lively and kinky that it must be twist-set before it can be woven or knitted. *Twist-setting* is a finishing process in which

Fig. 17–4 *Twist-on-twist two-ply yarn.*

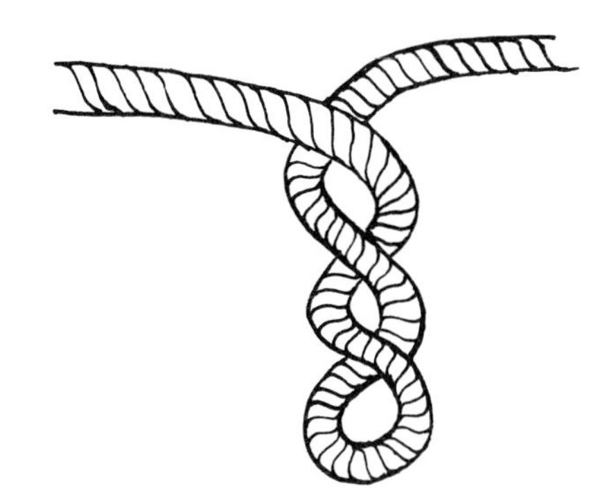

Fig. 17–5 *Kink in crepe yarn.*

the yarns are moistened and then dried in a straightened condition. After weaving, the cloth is moistened and the yarns become lively and kinky once more, thus producing the crinkle characteristic of true crepe fabrics. All the common natural fibers and rayon can be used in crepe-twist yarns because they can be twist-set in water. The thermoplastic fibers are twisted and set by heat. They do not have as much stretch as other crepe yarns, and fabrics made from them do not shrink. Increasing the amount of crepe-yarn twist and alternating the direction of twist will increase the amount of crinkle in a crepe fabric. For example, 6S filling yarns in a band followed by 6Z filling yarns in a band will give a more prominent crinkle than 2S filling yarns in a band followed by 2Z filling yarns in a band.

To identify crepe yarns, ravel adjacent sides to obtain a fringe on each of the two edges. Test the yarns that are removed by pulling on the yarn and then letting one end go. The yarn will "kink up" as shown in Figure 17–5.

Do not confuse kink with yarn crimp. Examine the fringe of the fabric. If yarns other than crepe yarns are used in the fabric, they will probably be of very low twist. The majority of crepe fabrics have crepe yarns in the crosswise direction, although some are in the lengthwise direction, and some have crepe yarns in both directions. Crepe fabrics are discussed in Chapter 26.

Yarn Size

YARN NUMBER

Spun-yarn size is referred to as *number* and is expressed in terms of length per unit of weight.

It differs according to the kind of fiber. The cotton system is given here. Many weaving yarns and sewing thread are numbered by the cotton system. It is an indirect system; the finer the yarn, the larger the number. The count is based on the number of hanks (1 hank is 840 yards) in 1 pound of yarn (see the following table). Some examples that show how the size of the weaving yarn affects the weight of the fabric are given in the second half of the table.

Cotton sewing threads provide an example of yarn number. The most commonly used mercerized thread is number 50. Years ago, when a wider selection of cotton threads was readily available for home sewing, experts would select number 60 thread for finer fabrics; number 40 thread for heavier fabrics like denim; and number 8, 16, or 20 thread for making buttonholes or for sewing on buttons. The woolen and worsted systems are similar to the cotton system, except that hanks are of different lengths.

Cotton System

Number or Count of Spun Yarn	*Hanks*	*Weight (pounds)*
No. 1	1 (840 yards)	1
No. 2	2 (1,680 yards)	1
No. 3	3 (2,520 yards)	1
Examples of Fabric Weight	*Yarn Size*	
	Warp	*Filling*
Sheer lawn	70s*	100s
Dress-weight percale	30s	40s
Suiting-weight sailcloth	13s	20s

*The "s" after the number means that the yarn is single.

DENIER

Filament yarn size is dependent partly on the size of the holes in the spinneret and partly on the rate at which the solution is pumped through the spinneret and the rate at which it is withdrawn. The size of filament yarns is based on the size of the fibers in the yarn and the number of those fibers grouped into the yarn. The size of both filament fibers and filament yarns is expressed in terms of weight per unit of length—denier (pronounced "den'yer"). In this system, the unit of length remains constant. The numbering system is direct because the finer the yarn, the smaller is the number.

Filament Yarn Size

1 denier	9,000 meters weigh 1 gram
2 denier	9,000 meters weigh 2 grams
3 denier	9,000 meters weigh 3 grams

Filament yarns are made in a specific denier for certain end uses. For example:

Yarn Denier	*Use*
20	Sheer hosiery
40–70	Tricot lingerie, blouses, shirts, support hosiery
140–520	Outerwear
520–840	Upholstery
1,040	Carpets, some knitting yarns

These are yarn denier sizes. Each yarn is made up of many filament fibers. When polyester double knits were so popular in the early 1970s, the typical filament yarn was 150/32—a 150 denier yarn consisting of 32 filaments. One can calculate the denier of each individual filament to be 4.5 (150 ÷ 32.) After the yarn was textured, the denier increased to 170. As demand for double-knit fabrics with more drape and less body became greater, one change that was made in the yarns was to increase the number of filaments per yarn and to decrease the denier of each filament. Thus 150/68 or 150/96 yarns were used. The resulting fabrics had a softer hand and a better drape. Even finer, lighter-weight knits were made with yarns of 100 denier or smaller.

TEX SYSTEM

The International Organization for Standardization has adopted the Tex system, which determines yarn count or number in the same way for all fiber yarns and uses metric units (weight in grams of 1 thousand meters of yarn.) This system has not yet been adopted in the United States.

Tex System	
1 Tex	1,000 meters weigh 1 gram
2 Tex	1,000 meters weigh 2 grams
3 Tex	1,000 meters weigh 3 grams

Fancy Yarns

Fancy yarns may be defined as yarns that have unlike parts and that are irregular at regular intervals. The regular intervals may or may not be obvious at first to the observer.

Fancy yarns may be single, plied, or cord yarns. They may be spun, filament, or textured yarns—or any combination of yarn types. They are called fancy yarns, or novelty yarns, because of their appearance; they lend an interesting or novel effect to fabrics made with them.

Fancy yarns are named by their appearance. The yarn can be classified according to the number of parts and named for the effect that dominates the fabric. Fancy yarns have been more common in drapery and upholstery fabrics than in apparel fabrics in the recent past. Fancy yarns are used by hand weavers and hand knitters to create interest in otherwise plain fabrics. They are seen in wool fabrics, especially suitings and coatings.

Fancy yarns are made on twisters with special attachments for giving different tensions and rates of delivery to the different plies (parts) and thus allow loose, curled, twisted, or looped areas in the yarn. Slubs and flakes of color are introduced into the yarn by special attachments. Knots or slubs, are made at regular cycles of the machine operation (Figure 17–6).

The following are some facts about novelty yarns:

1. Novelty yarns are usually plied yarns, but they are not used to add strength to the fabric.

2. If novelty yarns are used in one direction only, they are usually in the filling direction. They are more economical in that direction (there is less waste) and are subject to less strain and are easier to vary for design purposes.

3. Novelty yarns add interest to plain weave fabrics at lower cost than if effects were obtained from variations in weave. Novelty-yarn effects are permanent.

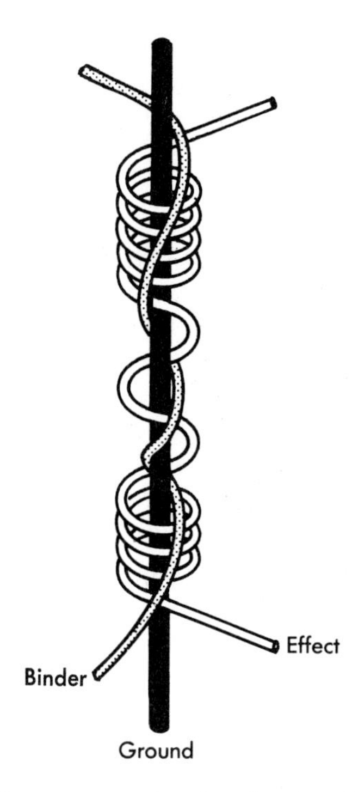

Fig. 17–6 *Fancy yarn, showing the three basic parts.*

4. Novelty yarns that are loose and bulky give crease resistance to a fabric, but they make the fabric spongy and hard to sew.

5. The durability of novelty-yarn fabrics is dependent on the size of the novelty effect, how well the novelty effect is held in the yarn, and on the firmness of the weave of the fabric. Generally speaking, the smaller the novelty effect, the more durable the fabric is, since the yarns are less affected by abrasion and do not tend to catch and snag so readily.

The following chart lists the most common fancy yarns and classifies them according to whether they are typically single or ply yarns.

Tweed yarns are an example of a single, spun,

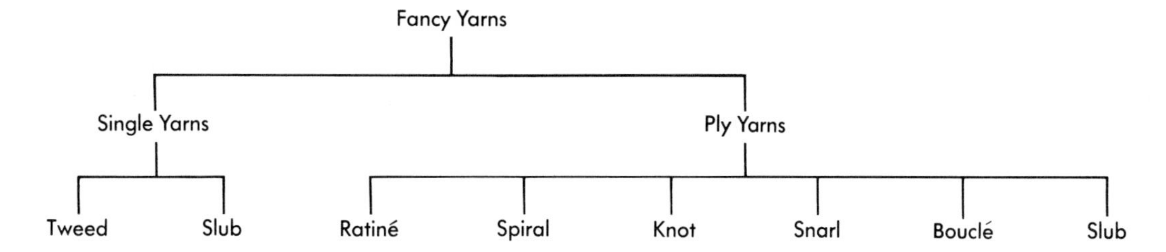

fancy yarn. Flecks of color from short fibers are twisted into the yarn for special interest. The cohesiveness of wool fibers explains why tweeds are often wool.

A *slub yarn* is another example of a single, spun, novelty yarn. This thick-and-thin yarn is made by varying the amount of twist in the yarn at regular intervals. The thicker part of the yarn is twisted less; the thinner part of the yarn is twisted more. Slub yarns can be found in shantung, drapery, and upholstery fabrics as well as in hand-knitting yarns and sweaters.

Other fancy yarns have two plies. One ply may be of one color, the other of a different color. The plies may be of differing thicknesses. A two-ply fancy yarn may have one spun ply combined with a filament ply.

Frequently a fancy yarn will have three basic parts:

1. The ground, or foundation, or core
2. The effect, or fancy
3. The binder

As a fancy yarn is examined, the *binder* is the first ply that can be unwound from the yarn. Its purpose is to hold the effect ply in place. The *effect ply* is mainly responsible for the appearance of the yarn, as well as the name given to the yarn. The *ground ply* forms the foundation of the yarn. Figure 17–6 shows a three-ply novelty yarn. In the diagram, each ply appears to be a simple, single, monofilament (single-fiber) yarn. Other novelty yarns may be three ply, but the ground may be a simple, single, filament yarn; the effect could be a two-ply spun yarn; and the binder could be a simple, single, spun yarn. Each part of the novelty yarn needs to be examined and classified. Endless varieties are possible!

The following are typical novelty yarns:

1. *Ratiné* is a typical novelty yarn. The effect ply is twisted in a somewhat spiral arrangement around the ground ply; but at intervals a longer loop is thrown out, kinks back on itself, and is held in place by the binder (Figure 17–7a).

2. The *spiral* or *corkscrew yarn* is made by twisting together two plies that differ in size, type, or twist. These two parts may be delivered to the twister at different rates of speed (Figure 17–7b).

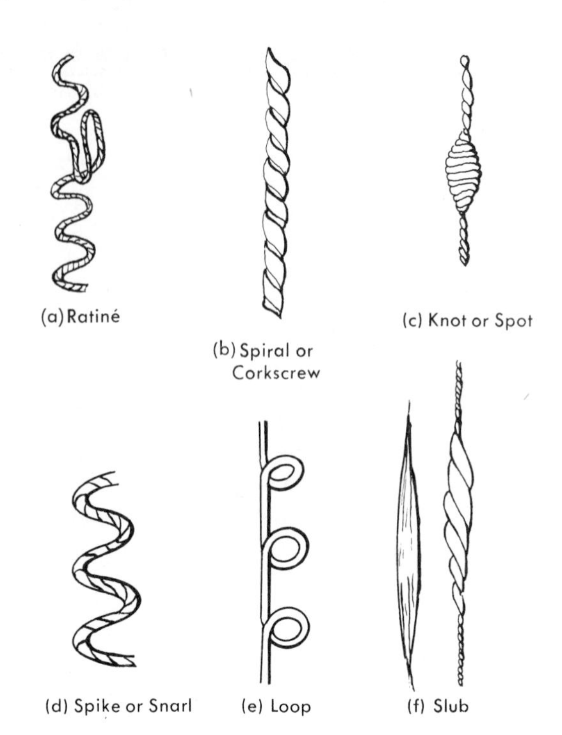

Fig. 17–7 *Effect ply of several kinds of fancy yarns.*

3. The *knot, spot, nub,* or *knop yarn* is made by twisting the effect ply many times in the same place (Figure 17–7c). Two effect plies of different colors may be used and the knots arranged so the colored spots are alternated along the length of the yarn. A binder is added during the twisting operation.

4. In the *spike* or *snarl,* the effect ply forms alternating unclosed loops along both sides of the yarn (Figure 17–7d).

5. The *loop, curl,* or *bouclé yarn* has closed loops at regular intervals along the yarn (Figure 17–7e). These yarns are used in woven or knit fabrics to create a looped pile that resembles caracul lambskin and is called *astrakhan cloth.* They are used to give textured effects to other coatings and dress fabrics. Mohair makes the best loop. Rayon and acetate are good also.

6. *Slub effects* are achieved in two ways (Figure 17–7f). True slubs are made by varying the amount of the twist at regular intervals. Intermittently spun flake or slub effects are made by incorporating soft, thick, elongated tufts of fiber into the yarn at regular intervals. A core or binder is needed in the latter.

7. *Metallic* yarns have been used for thousands of years. The older yarns were made of pure

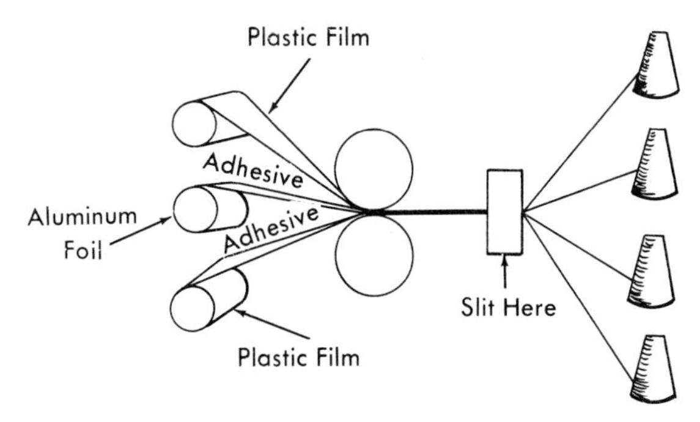

Fig. 17–8 *Laminating layers to produce a metal yarn.*

metal (lamé) and were heavy, brittle, expensive, and had the disadvantage of tarnishing.

Two processes are now used to make metallic yarns. The *laminating process* seals a layer of aluminum between two layers of acetate or polyester (mylar) film, which is then cut into strips for yarn (Figure 17–8). The film may be colorless, so the aluminum foil shows through, or the film and/or the adhesive may be colored before the laminating process. The *metalizing process* vaporizes the aluminum under high pressure and deposits it on the polyester film.

Fabric containing a large amount of metal can be embossed. Ironing is a problem when metallic film yarns are used with cotton, because a temperature high enough to take the wrinkles out of cotton will melt the plastic. The best way to remove wrinkles is to set the iron on its end and draw the edge of the fabric across the sole of the iron.

The table "Performance of Yarns in Fabrics" summarizes information about spun, smooth-filament, textured-bulk, and fancy yarns relative to aesthetics, durability, comfort, and care. The table should facilitate review of the major yarns and help to compare their performance.

Composite Yarns

Composite yarns are regular in appearance along their length. Composite yarns have both staple-fiber and filament-fiber components; they have unlike parts. Before these parts are combined to form the yarn, they may be staple fibers, staple fibers in a roving, filament fibers, or a molten-polymer stream. Composite yarns include covered yarns, core-spun yarns, filament-wrapped spun yarns, filament-wrapped filament yarns, and molten-polymer yarns.

The relationship of the basic composite yarns is shown in the following chart.

Covered yarns have a central yarn that is completely covered by fiber or another yarn. These yarns were developed to make rubber more comfortable in foundation garments and surgical hose. These stretch-covered yarns consist of a central core of rubber or spandex covered with yarns. There are two kinds: single covered and double covered. Covered yarns are wrapped with yarn. The yarn may be filament or spun and may be of any suitable fiber content. Single-covered yarns have a single yarn wrapped around them. Unless the torque is controlled, the yarns may be unbalanced. They are lighter, more resilient, and more economical than double-covered yarns and can be used in satin, batiste, broadcloth, and suiting as well as for lightweight foundation garments. Most ordinary elastic yarns are double-covered to give them balance and better coverage. Fabrics made with these yarns are heavier. A double-covered yarn is shown in Figure 17–9. Covered yarns are subject to "grin-through," which happens as the fabric gets older and the elastic core gets weaker

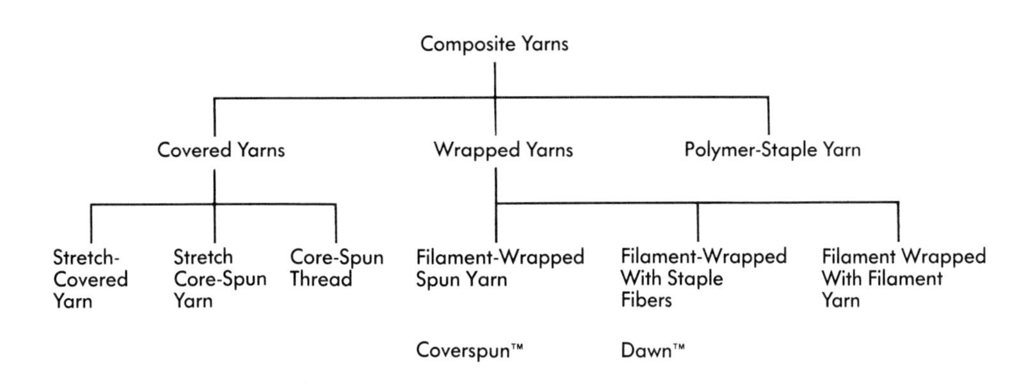

Performance of Yarns in Fabrics

Yarn Type	*Aesthetics*	*Durability*	*Comfort*	*Care*
Spun yarns	Fabrics are cotton-like or wool-like in appearance Fabrics lint and pill	Weaker than filament yarns of same fiber Ply yarns stronger than simple yarns Yarns are more cohesive so fabrics tend to resist raveling and running	Warmer More absorbent because of larger surface area	Yarns do not snag readily Soil readily
Smooth-filament yarns	Fabrics are smooth and lustrous Fabrics do not lint or pill readily	Stronger than spun yarns of same fiber Fabrics ravel and run readily	Cooler Least absorbent but more likely to wick moisture	Yarns may snag Resist soiling
Bulk yarns	Fabrics are less lustrous; more similar to those made of spun yarns Fabrics do not lint, but may pill	Stronger than spun yarns of same fiber Yarns are more cohesive so fabrics ravel and run less than those made with filament yarns, but more than those made with spun yarns	Bulkier and warmer than smooth-filament yarns More absorbent than filament yarns Stretch more than other yarns	Yarns likely to snag Soil more readily than filament yarns
Fancy yarns	Interesting texture Greater novelty effects show wear sooner than smaller novelty effects Fabrics lint and pill	Weaker than filament yarns Most resist raveling Less abrasion-resistant	Warmer More absorbent if part is spun	Yarns likely to snag Soil readily
Composite	Varies, yarns may be large, may have spun or filament appearance	Related to process	May have stretch, larger than many other yarns	Large yarns may snag

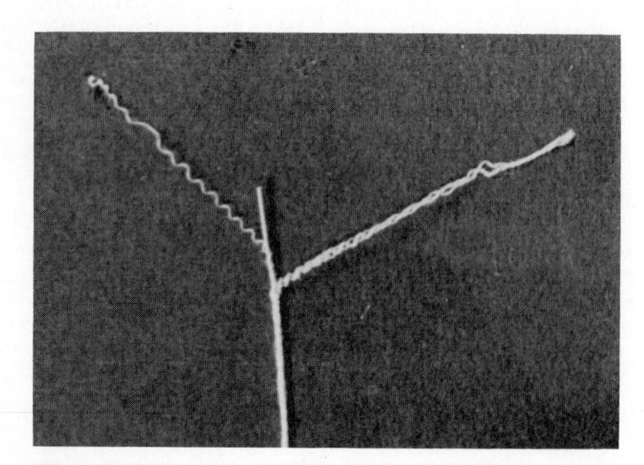

Fig. 17–9 Double-wrapped elastic yarn.

or when the covering sheath is disturbed and lets the core show through.

An alternate way of making a stretch yarn is a stretch core-spun yarn. *Core spinning* is a technique of spinning a sheath of staple fibers (roving) around a core. When working with elastomeric cores, the core is stretched while the sheath is spun around it. Any kind of fiber or blend of fibers can be used. The core is completely hidden and does not change the fabric surface. The sheath gives aesthetic properties to the yarn, and the core gives just enough stretch for comfort. Core-spun yarns can be used to give woven fabrics—such as batiste, broadcloth, and

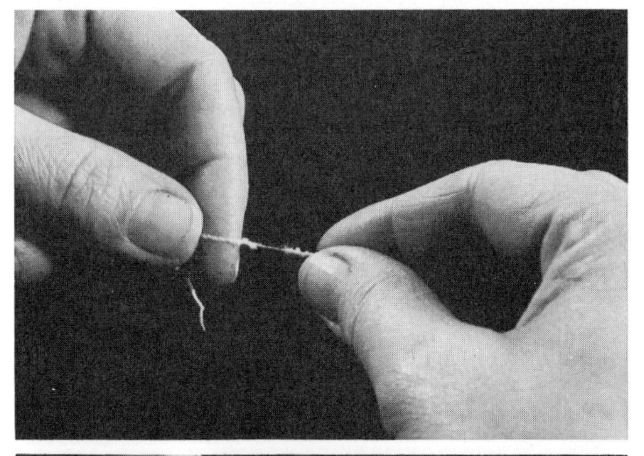

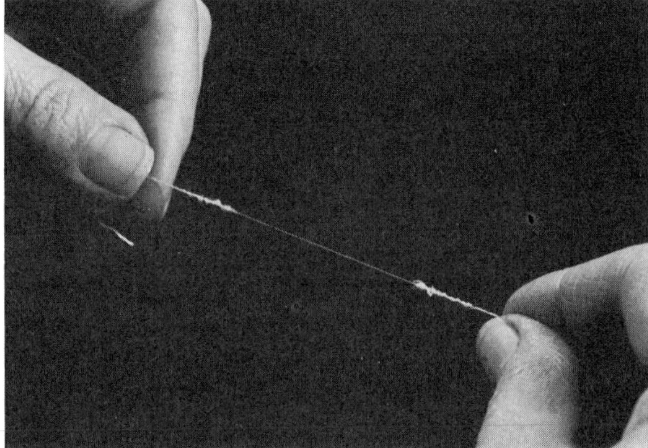

Fig. 17–10 *Core-spun spandex yarn: relaxed yarn with sheath removed to show core* (top); *extended yarn* (bottom).

suiting—an elasticity more like that of the knits.

In the core-spinning process, Figure 17–10, a spandex filament is stretched before it is fed to a modified conventional-spinning frame where it is combined with a roving that is wrapped around it. The tension on the spandex causes it to keep to the center of the bundle. The core must be completely covered to prevent "grin-through," which makes the core visible.

Another core-spun yarn was developed to meet the need for a strong sewing thread that had good resistance to damage by resin finishes. Cotton thread was weakened by resin finishes so it was not resistant to abrasion, and it also lacked "give" and would break when the seams stretched. Thread shinkage caused unsightly puckering of seams and zipper areas—especially those on the straight grain of the fabric.

Polyester/cotton core-spun thread has a high-strength filament polyester core around which is spun a sheath of high-quality cotton. This thread combines the good characteristics of polyester and cotton fibers. The cotton outer cover gives the thread excellent sewability, and the polyester core provides high strength and resistance to abrasion. Polyester/cotton thread also provides the slight "give" that is necessary in knits.

In *stretch-covered yarns,* stretch core-spun yarns and core-spun sewing thread, the outer component completely covers the central component in order to meet the performance required by these yarns. In other yarns, the outer component only partially covers the central component and performs satisfactorily.

Coverspun is a trade name for a new process of producing core-spun yarns. In this process, a core of staple fibers (often a twistless yarn) is wrapped with filament fibers that serve as a binder. These yarns can be produced economically to have good evenness, strength, appearance, and finishing properties. This yarn could be described as a *filament-wrapped spun yarn.* Other yarns of this type are being researched.

Another type of yarn is a *filament-yarn wrapped with staple fibers.* Fasciated yarns are described as a bundle of filaments wrapped with staple fibers. They are produced by running the filaments and stretch-broken tow through a device that spins the staple around the bundle of filaments. Production of these yarns is very fast. The purpose of these yarns is to give better texture and hand to fabrics. The yarns are combinations of coarse filaments for strength and fine broken filaments for softness (Figure 17–11).

Trevira Dawn is a polyester yarn composed of high-strength and low-strength polyester fibers. The yarn is crimped and twisted causing the low-strength component to break. Two-thirds of

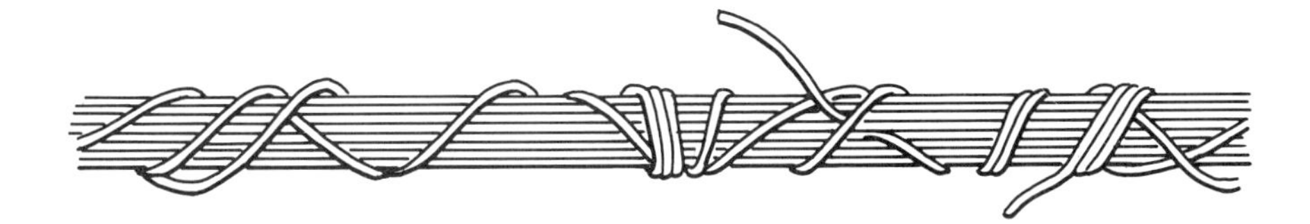

Fig. 17–11 *Fasciated yarn.*

the fibers are broken. The other third maintains the filament form and becomes the core of the yarn. Still another variation is a *filament-wrapped filament yarn* (Figure 17–12). Finally, a yarn can be produced by pressing staple fibers of any length or generic class into a molten polymer stream. As the polymer solidifies, the fibers that are partially embedded become firmly attached and form a sheath of staple fiber. The resultant yarn is about two-thirds staple fiber and one-third coagulated polymer. The polymer, which is extruded as man-made fibers are, is a less-expensive product than other melt-spun filaments.

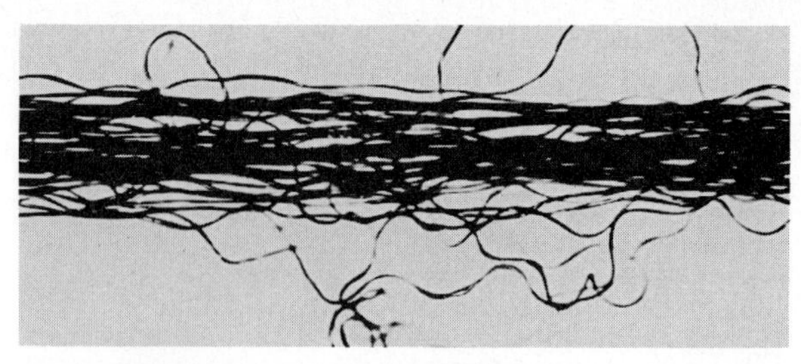

Fig. 17–12 *Trevira® polyester Dawn yarn. (Courtesy of Hercules, Inc.)*

18

Filament Yarns: Smooth and Bulky

Filament yarns are primarily man-made because silk is the only natural filament and acounts for less than 1 percent of fiber and yarn production. Man-made filament yarns are made by chemical spinning, the process in which a polymer solution is extruded through a spinneret, solidified in fiber form, and then the individual filaments are immediately brought together with or without a slight twist (Figure 18–1). The bringing together of the filaments and/or the addition of twist creates the filament yarn. The spinning machine winds the yarn on a bobbin. The yarn is then rewound on spools or cones and is a finished product unless some additional treatment is required, such as crimping, twisting, texturing, or finishing.

Throwing was originally a process for twisting silk filaments but evolved into the twisting of man-made fibers and then into texturing (page 165). Throwing provides the weaver or knitter with the kind of yarn needed for a particular product; high twist for crepe or ply yarns for men's suit fabrics are examples. The throwster performs a service for the industry and, until the advent of textured yarns, did not use a trade name for his products. Some throwsters now have trade names that are used on garments made from their yarns. The consumer sometimes incorrectly assumes that these names represent new fibers, but the trade name simply indicates the yarn-texturing process. The yarns are made of nylon, polyester, acetate, and other fibers. Recently there has been a trend for the fiber producer to texture yarns as a final step in the fiber-spinning process. A texturing device is installed as part of the fiber-spinning machine.

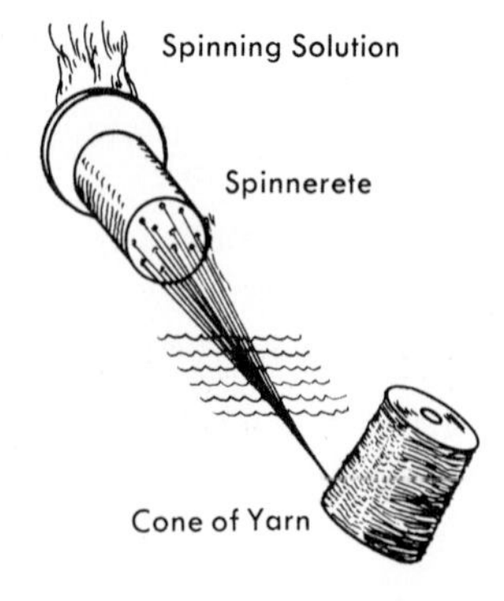

Fig. 18–1 *Chemical spinning of filament yarn.*

Smooth-Filament Yarn

In the development of a new fiber, *smooth-filament yarn* production usually precedes production of tow for staple. Filament yarns are more expensive in price per pound; however, the cost of making tow into staple and then spinning it into yarn by the mechanical spinning process usually makes the final cost about the same. The number of holes in the spinneret determines the number of filaments in the yarn.

Regular-filament, or conventional-filament, yarns are smooth and silk-like as they come from the spinneret. Their smooth nature gives them more luster than spun yarns, but the luster varies with the amount of delustering agent used in the fiber-spinning solution and the amount of twist in the yarn. Maximum luster is obtained by the use of bright filaments, which are laid together with little or no twist. Crepe yarns, of very high twist, were developed as a means of reducing the luster of the filaments. Filament yarns are generally used with either high twist or low twist.

Filament yarns have no protruding ends, so they do not shed lint; they resist pilling; and fabrics made from them tend to shed soil. Filaments of round cross-section pack well into compact yarns that give little bulk, loft, or cover to fabric. Compactness is a disadvantage in some end uses, where bulk and absorbency are necessary for comfort. Nonround or lobal filaments create more open space for air and moisture permeability and give greater cover.

The strength of a filament yarn depends on the strength of the individual fibers and on the number of filaments in the yarn. Filament fiber strength is usually greater than that of staple fibers. For example:

Tensile Strength

Polyester Filament	*Polyester Staple*
5–8 grams/denier	3–5.5 grams/denier

The strength of each filament is fully utilized. In order to break the yarn, all the filaments must be broken. Therefore it is possible to make hosiery and very sheer fabrics of fine filaments that have good strength. Filament yarns reach their maximum strength at about 3–6 turns per inch; then strength remains constant or decreases.

Fine-filament yarns are soft and supple. However, they are not as resistant to abrasion as coarse filaments, so for durability it may be desirable to have fewer, but coarser filaments in the yarn.

Monofilament Yarns

Monofilament yarns are rarely used except in industrial uses. These yarns usually are coarse-filament fibers. End uses include fruit and vegetable bags, nets, and other woven or knitted fabrics where low cost and high durability are the most important characteristic desired.

Tape Yarns

Tape yarns are produced from extruded polymer films. These yarns can be produced by slitting the film into the desired width or by fibrillating the film so that it splits parallel to the long direction of the film. Slit-film-tape yarns are much more regular than fibrillated film-tape yarns. However, fibrillated yarns are less expensive and quicker to produce. These tape yarns are found most often in industrial textiles such as carpet backing and bagging.

Bulk Yarns

A *bulk yarn* is defined by the American Society for Testing and Materials (ASTM) as a yarn that has been prepared to have greater covering power, or apparent volume than that of a conventional yarn of equal linear density and of the same basic material with normal twist. Often these bulk yarns will be referred to as *bulk-continuous-filament yarns.* A common shorthand notation used when referring to these yarns is BCF. BCF yarns include any continuous-filament yarn whose smooth, straight fibers have been displaced from their closely packed, parallel position by the introduction of some form of crimp, curl, loop, or coil (Figure 18–2).

The texturing processes discussed in this chapter are mechanical-texturing methods based on the use of thermoplastic fibers and heat. The most common fibers used to produce bulk yarns are polyester, olefin, and nylon. Turn to page 75 for a discussion of chemical methods of achieving texture by means of bicomponent fibers.

Bulk yarns have characteristics that are quite different from smooth-filament yarns. Bulking gives slippery filaments the aesthetic properties of spun yarns by altering the surface characteristics and creating space between the fibers. This gives the fabric more breathability and more permeability to moisture, and the fabric is more absorbent, more comfortable and has less static buildup. Bulk, cover, and elasticity, or stretch, are increased. Excellent ease of care (dependent on fiber content) is an advantage for the consumer, and for the manufacturer there is tremendous flexibility. Bulk yarns can have spun-like qualities without the pilling and shedding that occurs with spun yarns. The first bulk nylon, Helanca, was produced in the United

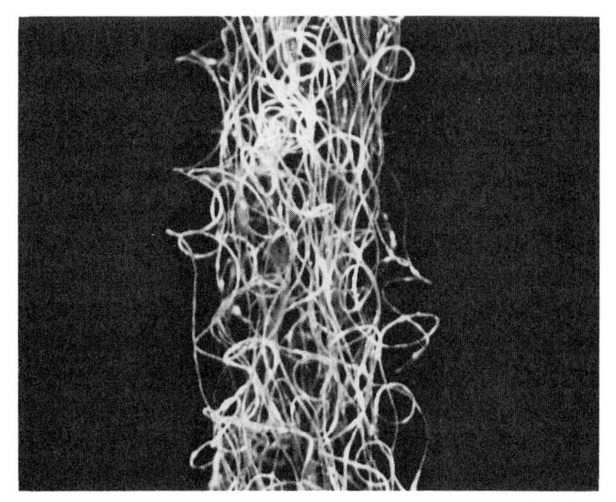

Fig. 18–2 *Typical bulk yarn. (Courtesy of the Fibers Division of Monsanto Chemical Co., a unit of Monsanto Co.)*

States in 1952 and used in stretch hosiery for men. The stretch eliminated the need for garters.

Three classes of yarns are bulk made:

1. Bulky yarns
2. Stretch yarns
3. Textured yarns

BULKY YARNS

Bulky yarns are defined by ASTM as yarns formed from inherently bulky fibers, such as man-made fibers that are hollow along part or all of their length, or yarns formed from fibers that cannot be closely packed because of their cross-sectional shape, fiber alignment, stiffness, resilience, or natural crimp.

Bulky texturing processes can be used with any kind of filament fiber or spun yarn. The yarns have less stretch than either stretch or textured yarns. Bulky texturing is done by:

1. Gear crimping.
2. Stuffer-box process.
3. Air-jet process.
4. Draw texturing.
5. Friction texturing.

Gear Crimp. The filament yarn passes between the teeth of two heated gears that mesh and thus give the yarn the shape of the gear teeth—saw-tooth crimp (Figure 18–3). The bends are angular in contrast to the rounded waves and bends of natural crimp. This method has been used in the past to increase the crimp of carpet wools. When it is used with thermoplastic fibers, the crimp is permanent.

The Stevetex crimping process crimps multiple-filament yarns arranged in warp formation rather than one yarn at a time. Crimp is uniform along the whole warp length and the yarns are free of torque. The number of crimps can be controlled—increased or decreased. Some end uses are tricot for lingerie, blouses, pajamas, and dresses.

Stuffer Box. The stuffer box produces a saw-tooth crimp of considerable bulk. Straight-filament yarns are literally stuffed into one end of a heated box (see Figure 18–4) and are then

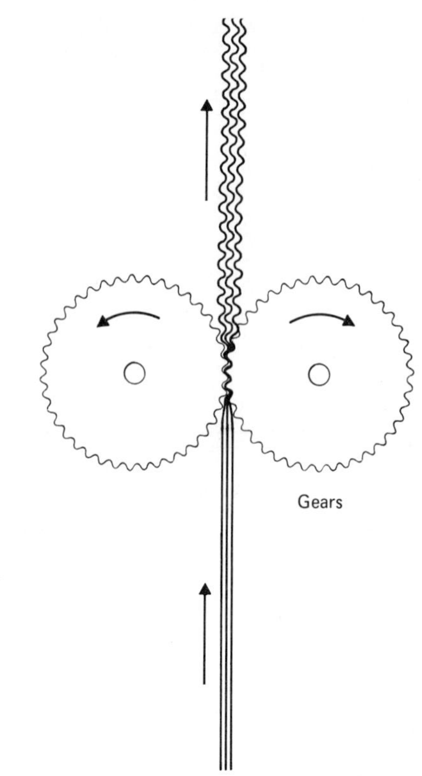

Fig. 18–3 *Filament yarn passes through heated gears to receive crimped shape.*

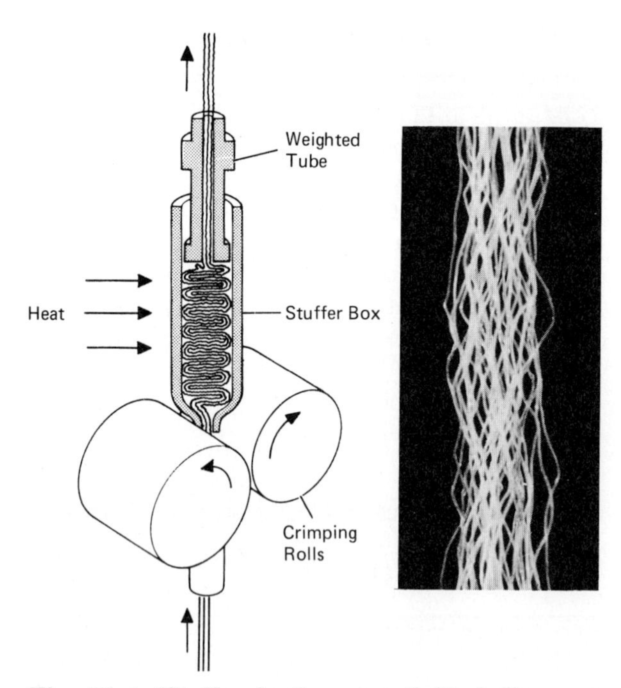

Fig. 18–4 *"Stuffing-box" process* (left)*; bulky yarn used in apparel* (right). *(Courtesy of the Fibers Division of Monsanto Chemical Co., a unit of Monsanto Co.)*

withdrawn at the other end in crimped form. Bulked single yarns are usually plied to hold the filaments together and minimize snagging. The apparent volume increase is 200–300 percent. The yarns have some elasticity but not enough to be classified as stretch yarns. The stuffer box is a fast method and one of the least costly. It is the most widely used bulking process.

End products of yarns crimped by this process are carpets, upholstery, sweaters, and knitted dresses.

Air Jet. The air-jet process produces loop-type displacement of the fibers. Conventional filament is fed over an air jet (Figure 18–5) at a faster rate than it is drawn off. The blast of air forces some of the filaments into very tiny loops; the velocity of the air affects the size of the loops. This is not a high-speed process and is relatively costly, but it is commonly used.

Volume increase in the yarn is between 50–150 percent. A fabric made from these yarns may seem to have more texture or luster change than it has change in bulk. The process is very versatile. Any kind of fiber can be used and styling possibilities are unlimited—silk-like, worsted, heather, slub, and blends with spun or textured.

Air-jet yarns will maintain their size and bulk under tension, because the straight sections of the fiber bear the strain and allow the loops to remain relatively unaffected. The yarns have little or no stretch. Heat setting is not necessary, so the process can be used for non-heat-settable yarns. The yarn under magnification looks like a novelty yarn. Some end uses for these yarns are shirts, blouses, women's wear, shoelaces, and household furnishings.

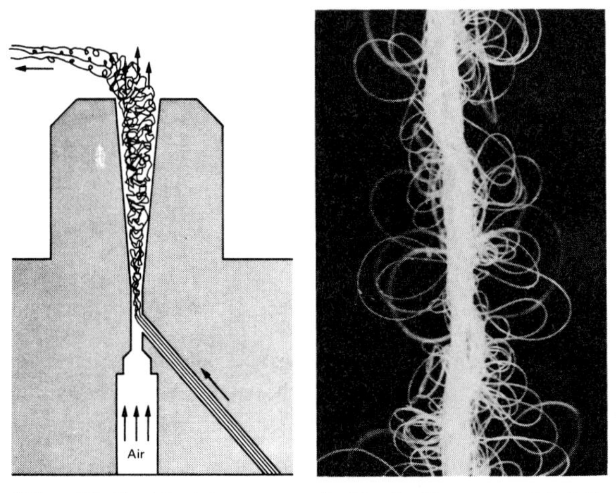

Fig. 18–5 *Air-jet process* (left)*; bulky yarn* (right). *(Courtesy of the Fibers Division of Monsanto Chemical Co., a unit of Monsanto Co.)*

Draw-Texturing. Prior to 1970, all textured polyester yarn was produced by throwsters by this process. In 1970, fiber producers developed a process by which the drawing and texturing could be done in one process. In the traditional melt-spinning process, filaments from the spinneret are gathered together into a yarn and wound on spools.

Polyester and nylon yarns are then stretched to orient the molecules and develop fiber strength and are finally heat set to stabilize them. These are the flat yarns used in the traditional bulk process.

In the draw-texturing process, the unoriented filaments or partially oriented filaments are fed directly through the double-heater false-twist spinner. This eliminates the separate stretching process.

Draw texturing is a much faster and cheaper way of making textured bulk yarns and the finished yarns are as good as, or better than, those done by conventional means. Throwsters are changing from the conventional process and are modifying their machines and buying POY yarns from fiber producers.

Friction-Texturing. Friction texturing is a process in which POY yarns (partially oriented yarns) are fed into machines that put twist into the yarns by friction surfaces rather than false twisters or spindles.

STRETCH YARNS

Stretch yarns are defined by ASTM as a generic term for thermoplastic filament or spun yarns having a high degree of potential elastic stretch and rapid recovery, and characterized by a high degree of yarn curl. These yarns are usually formed by an appropriate combination of deforming, heat setting, and developing treatments to attain elastic properties.

Stretch yarns are characterized by high elongation—300–500 percent, rapid recovery, and moderate bulk per unit of weight. The original or "classical" method of producing a stretch yarn was developed by the Heberlein Company of

Switzerland. It was done in three separate steps, which involved twisting the yarn 60–130 turns per inch, heat setting in an autoclave, and untwisting the yarn. The process made an excellent stretch yarn, but it was costly and has been largely replaced by the false-twist method.

Stretch yarns are made primarily of nylon fiber and have been used extensively in men's and women's hosiery, pantyhose, leotards, swimwear, ski pants, football pants and jerseys. Stretch yarns make it possible to manufacture fewer sizes, as one-size items will fit wearers of different sizes.

There are two types of stretch yarns: torque and nontorque. *Torque yarns* are defined by ASTM as stretch yarns that, when permitted to hang freely, rotate in the direction of the unrelieved torque resulting from previous deformation. Torque yarns are produced by twisting, setting, and untwisting the yarns. *Nontorque yarns* are defined by ASTM as stretch yarns that have no tendency to rotate when permitted to hang freely. Nontorque yarns include those where the deformation is mechanically produced and heat set resulting in a potential stretch. It also includes plied yarns whose single yarn torque properties have been balanced by plying. Stretch yarns should not be confused with yarns made with elastomeric fibers.

Stretch yarns are made by:

1. False-twist.
2. Edge-crimped.
3. Knit-de-knit.
4. Draw texturing.
5. Friction texturing.

False-Twist. The false-twist spindle is an ideal device for making stretch yarns. (It is also used for textured yarns.) The spindle whirls at 600,000 revolutions per minute and generates sound comparable to a jet engine. The effect of this sound on health and hearing has been a matter of concern in the industry. The process is continuous; the yarn is twisted, heat set, and untwisted as it travels through the spindle (Figure 18–6). As the false-twist spindle turns, there will be an S-twist in the yarn on one side and a Z-twist in the yarn on the other side (Figure 18–7). When the yarn untwists, the filaments are essentially in the form of a helical coil distorted by the untwisting (Figure 18–6). If the yarn is pulled at each end, the coils will straighten out—thus the stretch.

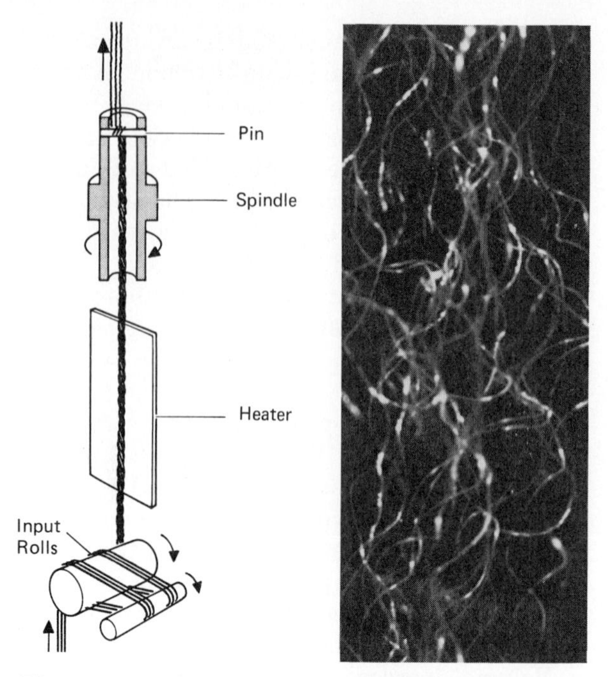

Fig. 18–6 *"False-twist" process* (left); *yarn* (right). *(Courtesy of the Fibers Division of Monsanto Chemical Co., a unit of Monsanto Co.)*

A wide range of stretch-yarn properties can be achieved by differences in the amount of false twist and differences in the degree of tension on the feed roll. The spindle can be twisted to the right or the left and can twist alternately to the right or left by reversing at controlled time intervals. The yarn can be used as a single yarn or can be plied by combining singles of right and left twist. They are usually made of nylon fiber. False-twist is the most widely used method of making stretch yarns.

The Duo-Twist process by the Turbo Machine Company is similar to the false-twist method

Fig. 18–7 *False-twist spindle.*

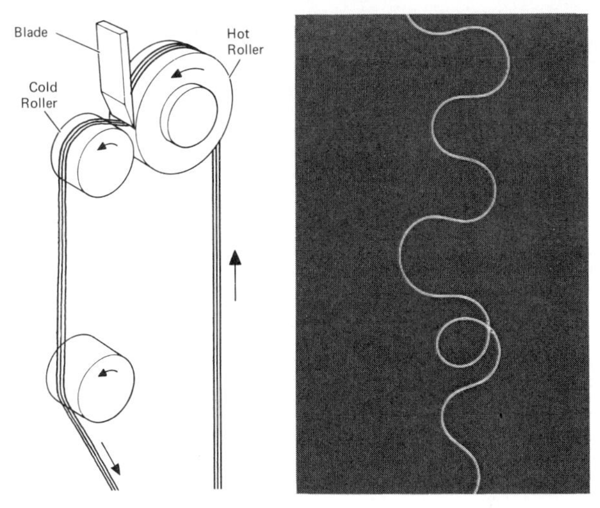

Fig. 18–8 Edge-crimp yarn process (left); *fiber* (right). *(Photograph courtesy of Milliken Research Corporation.)*

but uses no spindle and yarns have lower twist and torque. Two yarns are twisted together, heat set and untwisted, and then wound on individual spools.

Edge-Crimp. Curl-type stretch yarns are made by drawing heated filaments over a knife-like edge (Figure 18–8), which flattens the filaments on one side and causes the yarn to curl, with an effect like that obtained by pulling a Christmas ribbon over scissors to curl it. The filament cross-section changes from round to flattened on one side (Figure 18–9). The flattened side is shortened, so differential shrinkage of the sides causes a nontorque curl. The effect is similar to that of bicomponent stretch.

The process is low cost and speedy. It can be used on monofilaments as well as multifilaments. The primary end use is hosiery.

Knit-de-Knit. *Knit-de-knit* was one of the older methods but was not used much until 1965, when it became very popular. A small-diameter tube like a seamless stocking is knit at a rapid speed (Figure 18–10), (Figure 18–11). It is then

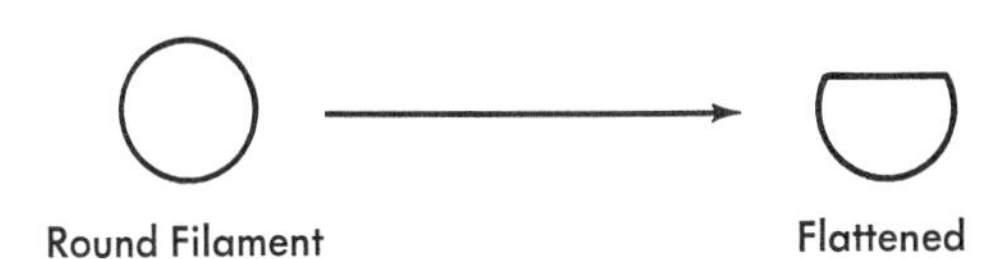

Fig. 18–9 Edge-crimping flattens one side of the filament.

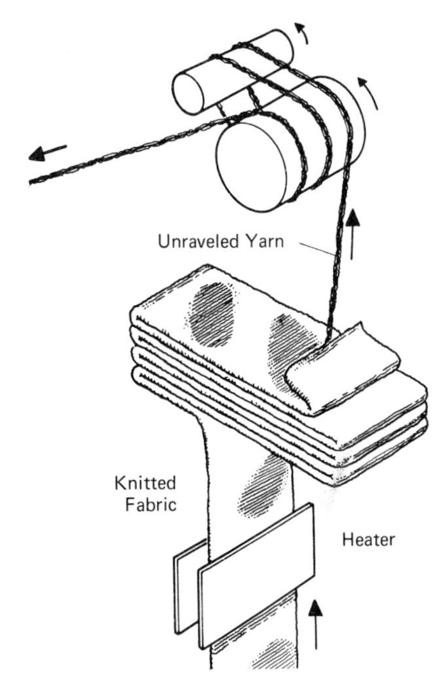

Fig. 18–10 Knit-de-knit fabric is heat set then unraveled.

heat set, unraveled, and wound on cones (Figure 18–12). Crimp size and frequency can be varied by difference in stitch size and tension. The knitting stitch used to make the garment must be of different gauge than that of the knit-de-knit tube or pinholes will form where the crimp gauge and knit gauge match. Knit-de-knit pro-

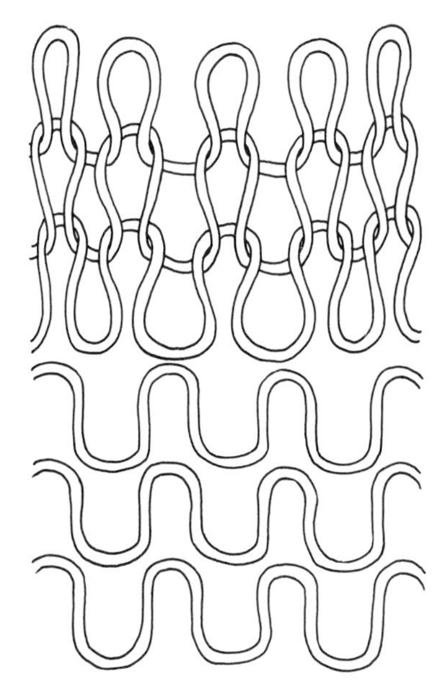

Fig. 18–11 Knit-de-knit crimp.

Fig. 18–12 The knit-de-knit machine knits tubes for crinkle-type yarns. (Courtesy of Scott & Williams, Inc.)

duces a bouclé or crepe effect and is used in outerwear, swimwear, and upholstery.

TEXTURED YARNS

Textured yarns are defined by ASTM as filament or spun yarns that have been given notably greater apparent volume than a conventional yarn of similar fiber (filament) count and linear density. These yarns have much lower elastic stretch than stretch yarns, but greater stretch than bulky yarns. They are sufficiently stable to present no unusual problems in subsequent processing or in use by the ultimate consumer. The apparent increase in volume may be achieved through physical, chemical, or heat treatments or a combination of these. In some countries, the term *bulked yarn* may be used. The following are the types of textured yarns:

1. Loopy yarns
2. High-bulk yarns
3. Crimped yarns
4. Flat-drawn textured yarns
5. Draw texturing and friction texturing yarns

Loopy Yarns. *Loopy yarns* are defined by ASTM as yarns essentially free from stretch that are characterized by a relatively large number of randomly spaced and randomly sized loops along the fibers or filaments. The majority of these yarns are produced by air-jet texturing. In air-jet texturing, the yarn is fed through an air jet at greater speed than the final yarn take-up. These textured yarns also may have a tightly twisted center core.

High-Bulk Yarns. *High-bulk yarns* are defined by ASTM as yarns essentially free from stretch in which a fraction of the fibers (in any cross-section) has been forced to assume a relatively high random crimp by shrinkage of the remaining fibers that, in general, have very low crimp. Since these fibers are cut or broken to staple lengths, these yarns are discussed in the spun-yarn chapter.

Crimped Yarns. *Crimped yarns* are of two types. The first type is defined by ASTM as thermoplastic textured yarns having relatively low elastic stretch (usually under 20 percent) and frequently characterized by high sawtooth crimp or curl. Methods used to produce these yarns are identified below.

The second type of crimped yarns is defined by ASTM as nonthermoplastic textured yarns with irregular crimp and relatively high elastic stretch but low power of contraction. The crimp in these yarns is produced by release of internal strains following immersion of the fabric in water, by heat treatment, or by chemical treatment. See Chapter 8, page 74, for a more-detailed discussion of the process.

Textured yarns, primarily polyester, comprise more than half of all yarns used in apparel. They are used in both knit and woven fabrics. They have true wash-and wear characteristics and give excellent hand and drape to fabrics.

Textured yarns are stabilized stretch yarns—they have bulk and some comfort stretch. Fabrics made from these yarns maintain their original size and shape during wear and care.

Textured yarns are made from (1) flat-drawn yarn (thermoplastic filaments drawn after spinning), (2) undrawn or unoriented filament yarns, and (3) partially drawn or oriented (POY) yarn. Texturing and stabilizing are done by (1) a double-heater false-twist machine or (2) by a double-heater friction-twist machine.

Flat-Drawn Textured Yarns. These yarns are made by a machine that is essentially the same as that shown in Figure 18–6 with a second heating zone. The false-twisted yarn is stretched slightly and stabilized at a temperature higher than that used for texturing.

Draw Texturing and Friction Texturing. Draw texturing and friction texturing were discussed in the section on bulky yarns.

The following table summarizes the three major types of bulk-filament yarns.

Bulk Yarns

	Bulky Yarns	*Stretch Yarns*	*Textured Yarns*
Nature	Inherently bulky	High degree of yarn curl	High degree of bulk
Fiber type	May be hollow or crimped fibers	Any thermoplastic fiber	Any fiber that can be treated with moisture, heat, or chemical to develop crimp
Stretch	Least stretch	300–500 percent stretch	Moderate amount of stretch
Characteristics	Sawtooth, loops in individual fibers	Torque and nontorque	Loopy, high bulk, crimped
Processes	Gear crimping, stuffer box, air jet, draw texturing, friction texturing	False twist, edge crimp, knit de knit, draw texturing, friction texturing	Air jet, flat-drawn textured, draw texturing, friction texturing

19

Spun Yarns and Blends

Spun yarns are a continuous strand of staple fibers held together by some mechanism. Often the mechanism is a mechanical twist that takes advantage of the fiber's irregularities and natural cohesiveness to bind the fibers together into one yarn. The process of producing yarns from staple fibers by twisting is an old one. The initial discovery of twisting fibers together has been lost in history. In recent years, ways of producing spun yarns without twist have been developed.

In this chapter, production methods and characteristics of spun yarns will be discussed. In addition, fiber blends will be discussed, because fiber blends are most often found in spun yarns. Spun yarns are suited to clothing fabrics in which absorbency, bulk, warmth, or cotton-like or wool-like textures are desired.

Spun yarns are characterized by protruding fiber ends. The fiber ends hold the yarn away from close contact with the skin; thus a fabric made of spun yarn is more comfortable on a hot, humid day than a fabric of smooth-filament yarns that traps body moisture next to the skin and does not allow perspiration to evaporate.

Many of the insulating characteristics of a fabric are due to the structure of the yarns used to produce that fabric. There is more space between fibers in a spun yarn than in a filament yarn. A spun yarn with low twist has more space than a spun yarn with high twist. Hence a spun yarn with low twist will be better at insulating than a highly twisted yarn. For that reason, most fabrics for cold-weather wear have lower twist yarns. (If wind resistance is desired fabrics with high-twist yarns and a high count are more desirable because air permeability is reduced.)

Carded yarns, made of short-staple fibers, have more protruding fiber ends than *combed yarns,* which are made of long-staple fibers. Protruding ends contribute to a dull, fuzzy appearance; to the shedding of lint; and to the formation of pills on the surface of the fabric. Fuzzy ends can be removed from the yarn or from the fabric by singeing (Chapter 32).

The strength of the individual staple fiber is less important as a factor in yarn strength than it is in filament yarns. Instead spun-yarn strength is dependent on the cohesive or clinging power of the fibers and on the points of contact resulting from pressure of twist or other binding mechanism used to produce the spun yarn. The greater the number of points of contact, the greater is the resistance to fiber slippage within the yarn. Fibers with crimp or convolutions have a greater number of points of contact. The friction of one fiber against another gives resistance to lengthwise fiber slippage. A fiber with a rough surface—wool scales, for example—creates more friction than a smooth fiber.

The mechanical spinning of staple fibers into yarns is one of the oldest manufacturing arts and has been described as an invention as significant as that of the wheel. The first spun yarns were made from flax, wool, and cotton, all of which are staple fibers. The basic principles of spinning are the same now as they were when yarns were first made.

The earliest primitive spinning consisted of drawing out the fibers, which were held on a stick called a distaff, twisting them by the rotation of a spindle, which could be spun like a top, and then winding up the spun yarn (Figure 19–1). The spinning wheel was invented by the spinners of India and was introduced into Europe in the 14th century. The factory system began in the 18th century when spinning was done by a class distinct from the weavers. In 1764, an Englishman named James Hargreaves invented the first spinning jenny—a machine that could turn more than one spinning wheel at a time. Other inventions for improving the spinning process followed and led to the Industrial Revolution, when power machines took over hand processes and made mass production possible.

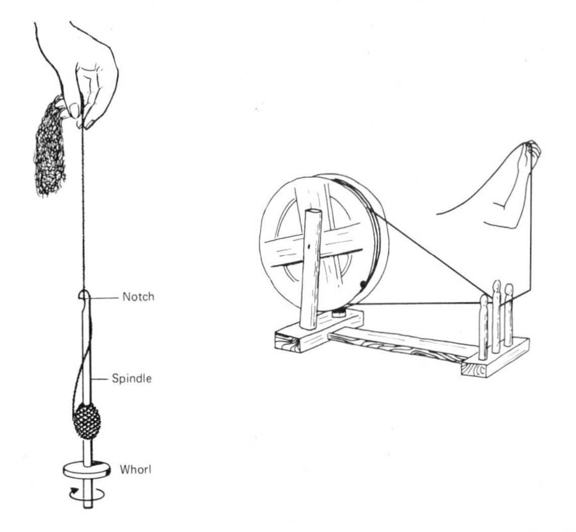

Fig. 19–1 *Hand spinning* (left)*; early spinning wheel* (right).

Machines were developed for each separate step in the spinning process.

Spinning is currently in a process of evolution. Progress in conventional ring spinning has been in the area of reduction in the number of steps involved in the combination of individual steps—continuous spinning. Several steps have been automated. Other spinning methods have excited a great deal of interest as being faster, simpler, and more economical than ring spinning. Spun-yarn processes are shown in the following chart.

Spun Yarn Processes

From Staple Fiber	*From Filament Tow*
Conventional Ring	Tow-to-top
Direct	Tow-to-yarn
Open-end	High-bulk yarns
Friction	
Twistless	
Self-twist	

Spinning Staple Fibers

CONVENTIONAL, OR RING, SPINNING

Conventional, or ring, spinning has traditionally consisted of a series of operations performed by individual machines and involved a great deal of hand labor. Although continuous spinning and some automation have come into use, spinning is still a long and expensive process. The different operations are designed to (1) clean and parallel staple fibers, (2) draw them out into a fine strand, and (3) twist them to keep them together and give them strength. Spinning may be done by any one of five conventional systems (cotton, woolen, French, Bradford, and American) that are adapted to the characteristics of the fiber—length, cohesiveness, diameter, elasticity, and surface contour. Because the cotton system is representative of the rest, it is discussed here in detail. The following chart summarizes the steps in producing a spun yarn.

Fig. 19–2 *Karousel opener-picker machine. (Courtesy of the Reiter Company, Inc.)*

The Cotton System

Operation	*Purpose*
Opening	Loosens, blends, cleans, forms *lap.*
Carding	Cleans, straightens, forms *carded sliver.*
Drawing	Parallels, blends, forms *drawn sliver.*
Combing	Parallels, removes short fibers, forms *combed sliver* (used for long-staple cotton only).
Roving	Reduces size, inserts slight twist, forms *roving.*
Spinning	Reduces size, twists, winds the finished *yarn* on a bobbin.
Winding	Rewinds yarn from bobbins to spools or cones.

Opening. Opening loosens, cleans, and blends the fibers. The fibers have been compressed very tightly in the bale and may have been stored in this state for a year or more. Machine-picked cotton contains a much higher percentage of trash and dirt than does hand-picked cotton; consequently, the work of cleaning has become more complicated. Part of it is done at the gin. Cotton varies from bale to bale, so the fibers from several bales are blended together to give yarns of more-uniform quality. Two types of opening units are used. One is a chute-feed system and the other (Figure 19–2) is a merry-go-round-like unit. In both systems the bales travel over bale pluckers (Figure 19–3) that pull small tufts of fiber from the underside of the bales and drop the tufts on a screen or lattice. High-velocity air removes dirt and trash. The loosened cleaned fibers are fed to the carding machine in sheet form.

There are several possibilities for blending different generic fiber types in the opening operation. In one method, several bales of fiber are laid around the picker and an armful from each bale is fed alternately into the machine. Another method is called *sandwich blending.* The desired amounts of each fiber are weighed out and a layer of each is spread over the preceding layer to build up a sandwich composed of many layers. Vertical sections are then taken through the sandwich and fed into the picker. *Feeder blending* is an automatic process in which each type of fiber is fed to a mixing apron from individual hoppers (Figure 19–4).

Carding. Carding partially straightens the fibers and forms them into a thin web, which is brought together as a soft rope of fibers called a *carded sliver* (Figure 19–5). The carding machine consists of cylinders covered with heavy fabric embedded with specially bent wires or with granular cards that are covered with a rough surface similar to rough sandpaper.

Drawing. Drawing increases the parallelism of the fibers and combines several carded slivers into one *drawn sliver.* This is a blending operation that contributes to greater yarn uniformity. Drawing is done by sets of rollers, each set running successively faster than the preceding set (Figure 19–6). As slivers are combined, their size is reduced and a small amount of twist is added.

The drawing process may be repeated more than once. It is often at this stage that fibers of different generic types are blended because of

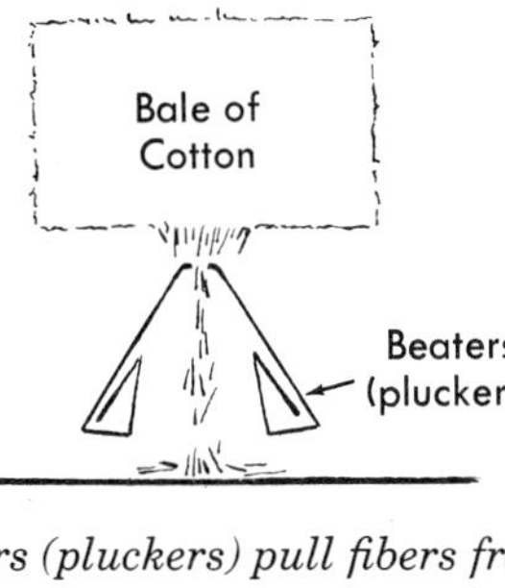

Bale of Cotton

Beaters (pluckers)

Fig. 19–3 *Beaters (pluckers) pull fibers from the bale.*

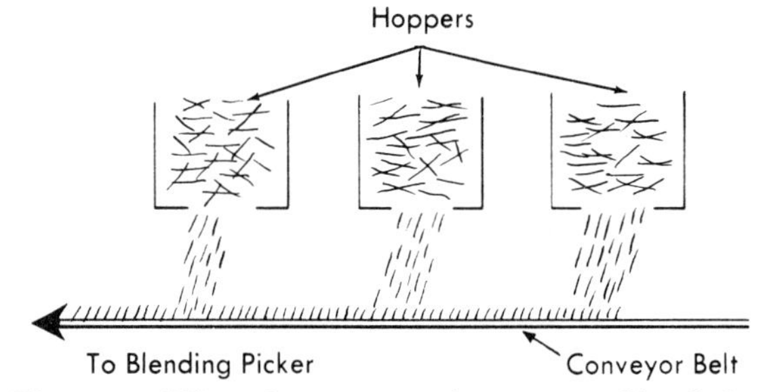

Fig. 19–4 *Fibers from various hoppers are blended on a conveyor belt.*

Fig. 19–5 Carding. (Courtesy of Coats & Clark, Inc.)

the differences in physical properties. Blending during the drawing process also eliminates mixed wastes.

Combing. If long-staple fibers are to be spun, carding and drawing will be followed by combing. The fundamental purpose of combing is to parallel fibers and to remove any short fibers from the long staple so that the combed fibers will be more uniform in length. Fibers emerge from the combing machine as *combed sliver*. The combing operation and the long-staple fiber are costly and as much as one-fourth of the fiber is combed out as waste. Combing is done to achieve a yarn that is superior to a carded yarn in smoothness, fineness, evenness, and strength. When working with cotton or cotton blends, the term *combed yarn* is used. When working with wool or wool blends, the terms *worsted yarn* is used.

Roving. Roving reduces the drawn sliver, increases the parallelism of the fibers, and inserts a small amount of twist (Figure 19–7). The product is called a *roving*. It is a softly twisted strand of fibers about the size of a pencil. Successive roving operations that gradually reduce the size of the strand may be used.

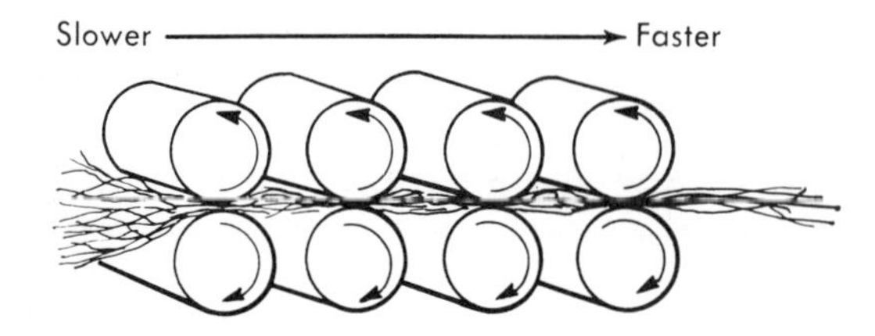

Fig. 19–6 Drawing rolls.

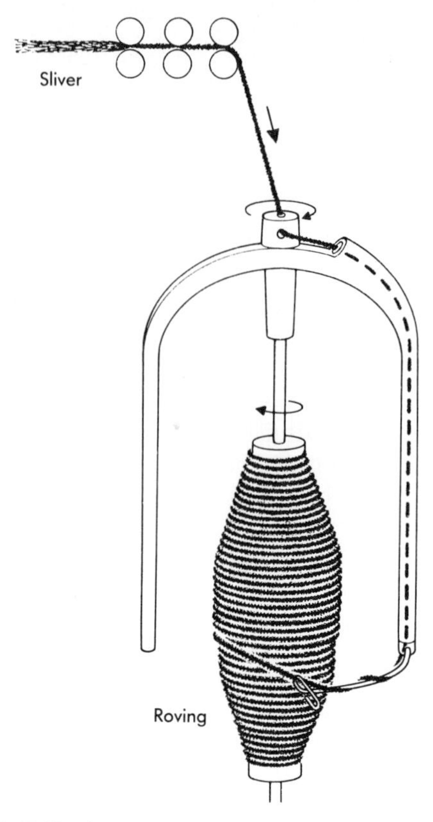

Fig. 19–7 Roving.

Spinning. Spinning adds the twist that makes the yarn—a single-spun yarn. Ring spinning draws, twists, and winds in one continuous operation. The traveler (Figure 19–8) carries the yarn as it slides around the ring, thus inserting the twist. Because ring spinning is a slow textile process in terms of productivity per unit produced—it has been limited in traveler speed, package size, and adaptability to automation—much interest has been shown in *open-end* spinning (page 178), which in many ways resembles whorlless primitive spinning.

Figure 19–9 shows a ring-spinning frame—a multiple-spinning machine that holds a number of individual units.

Blending can also occur during roving or spinning because several fiber strands are combined in these processes. When blending is done during roving or spinning, it is usually done to achieve a blending of color. Mule spinning is an intermittent action that may be used in the

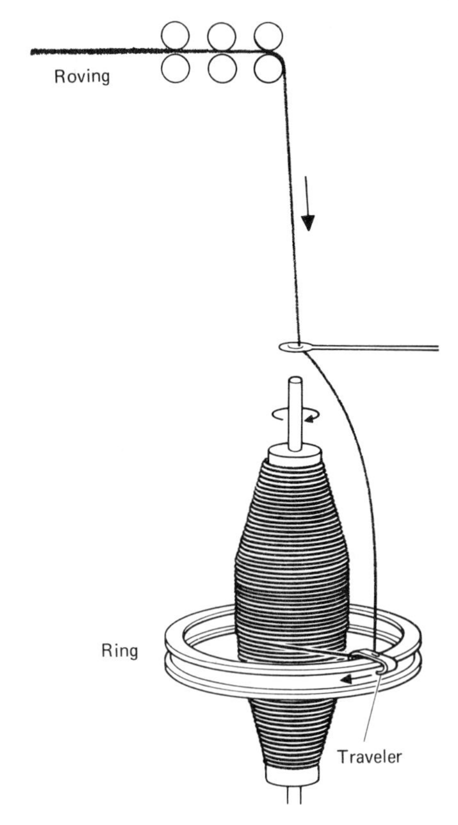

Fig. 19–8 Ring spinning.

woolen spinning system. The yarn is drawn out and twisted, then the twisting stops while the twisted portion of yarn is wound on the bobbin. This intermittent action is the basic fault of mule spinning.

Fig. 19–9 A spinning frame holds multiple ring spinners.

Some alternatives to mule spinning are stand spinning and ring-frame spinning. At present, ring-frame spinning is most commonly used for woolen yarns. It is similar to ring spinning of cotton yarns.

COMPARISON OF CARDED AND COMBED YARNS

The length and parallelism of fibers in spun yarns is a major factor in the kind of fabric, the cost of the yarns and fabrics, and the terminology used to designate these characteristics.

Yarns made from carded sliver are called *carded* yarns. Carded sliver of short wool fibers is made into *woolen* yarns and the fabrics are called *woolen* fabrics. (*Note:* The term *woolen* has a specific meaning and should not be used as a synonym for wool, the name of the fiber. For example, if one says "This is a woolen dress," it means that the dress fabric is made from the shorter wool fibers.)

Yarns made from combed sliver are called *combed* yarns except in the case of wool, in which case combed sliver is referred to as *top* and the yarns made from top are called *worsted* yarns. The short fibers that are combed out are called *noils* and are a source of fiber supply for carded yarns. Fine combed cotton yarns are made from the fibers that measure more than 1⅛ inches. In many fabrics combed cotton is being replaced by long-staple high-performance rayon or polyester fibers.

Carded and combed yarns are compared in the chart and in Figure 19–10.

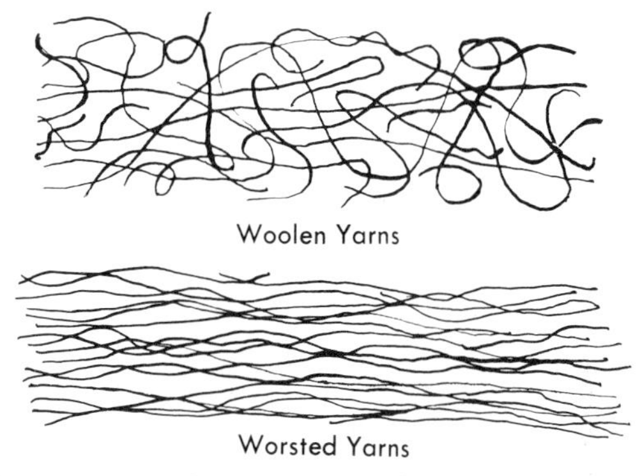

Fig. 19–10 Woolen and worsted yarns: short staple-wool fibers in carded or woolen yarn (top); *long parallel wool fibers in combed or worsted yarn* (bottom).

Carded and Combed Yarn Comparison

	Carded	*Combed*
Fibers used:	Short staple	Long staple
Yarns:	Medium to low twist	Medium to high twist
	More protruding ends	Fewer protruding ends
	Bulkier, softer, fuzzier	Parallel fibers; finer count
		Longer wearing, stronger
Fabrics:	May become baggy in areas of stress	Smoother surface, lighter weight
	Fabrics may be soft to firm	Do not sag
	Blankets always carded	Take and hold press
	Wide range of uses	Fabrics range from sheers to suitings

New Developments in Spun-Yarn Production

Much research has been done to develop a spinning system for staple fibers that will shorten or simplify yarn spinning by eliminating or bypassing some of the steps in the conventional ring-spinning system. The goal of most of the current developmental work is to use fiber directly from the card machine, thus eliminating drawing, roving, ring spinning, and rewinding. Some of these goals have been achieved.

DIRECT SPINNING

The Mackie direct spinner (1960) eliminates the roving but still uses the ring-spinning device for inserting the twist. Notice in Figure 19–11 that the sliver is fed directly to the spinning frame. This machine is used to make heavier yarn for pile fabrics and carpets. Do not confuse this machine with the direct spinner for processing filament tow (page 180).

OPEN-END SPINNING

Open-end spinning eliminates the roving and twisting by the ring. Knots are eliminated, larger packages of yarn are formed, less operator supervision is needed, and higher production speeds (about four times that of ring spinning) are achieved.

In the rotor-spinning process, fibers from a sliver are fed through rollers or over a spiked roller that breaks up the sliver so that individual fibers are fed by an air stream and deposited on the inner surface of a rotating device driven at high speed. As the fibers are drawn off, twist is inserted by the rotation of the rotor making a yarn (Figure 19–12).

The air-vortex-spinning process is similar to the rotor process except that the yarn is formed by moving air rather than the rotor. These yarns are even and regular. They are stronger than rotor yarns, but tend to pill more readily.

FRICTION SPINNING

Friction spinning makes yarns from a modification of open-end spinning that combines the

Fig. 19–11 *Mackie direct spinner turns sliver into finished yarn—eliminates roving. (Courtesy of James Mackie & Sons, Ltd.)*

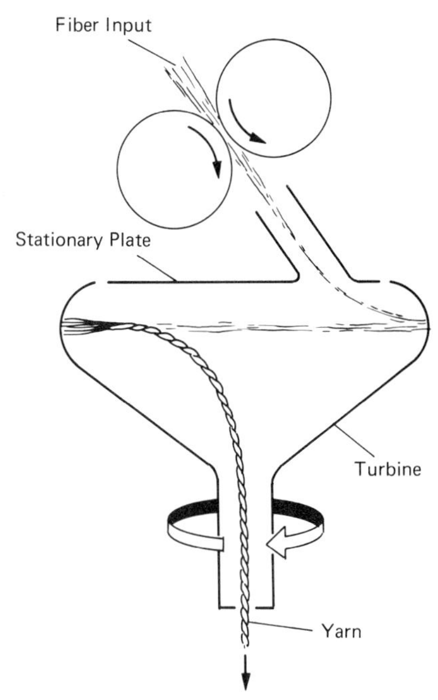

Fig. 19–12 *Open-end, or break, spinning.*

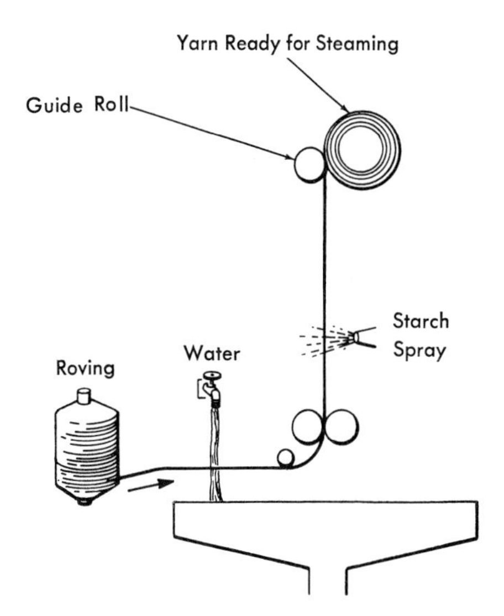

Fig. 19–13 *Twistless spinning.*

rotor and air techniques. The sliver is separated into fibers that are spread onto carding or combing rolls. The carded or combed sliver is delivered by air to two cylinders that rotate in the same direction. These two cylinders pull the fibers into a yarn. The feed angle into the cylinders controls fiber alignment. If the angle is small, the fibers will be more parallel. Friction-spun yarns are more even, freer of lint and other debris, and loftier, but they are weaker when compared to conventional yarns.

TWISTLESS SPINNING

Twistless spinning eliminates the twisting process. A roving is wetted, drawn out, sprayed with sizing or adhesive, and wound on a package. The package is steamed to bond the fibers together (Figure 19–13).

The yarns are flat and ribbon-like in shape and are quite stiff because of the sizing. They lack strength as individual yarns but gain strength in the fabric from the pressure between the warp and filling. The absence of twist gives the yarns a softness and good luster after the sizing is removed. The notable feature of the twistless yarns is their opaqueness. The yarns are open to dye and have very good durability but are not suitable to very open weaves.

SELF-TWIST SPINNING

Self-twist spinning is a process in which two strands of roving are carried between two rollers that move forward to draw out the roving and sideways to put in twist. The yarns have areas of S-twist and areas of Z-twist. When the two twisted yarns are brought together, they intermesh and entangle, and, when pressure is released, the yarns try to untwist and this causes them to ply over each other (Figure 19–14). This process can be performed with two strands of staple fibers, one strand of staple fibers and one strand of filament fibers, or two strands of filament fibers.

Spinning Filament Tow into Spun Yarns

Any kind of man-made fiber in filament-tow form can be processed by the direct-spinning process. Filament tow can be made into spun yarns, without disrupting the continuity of the strand, by either the tow-to-top system or by the direct-spun tow-to-yarn system.

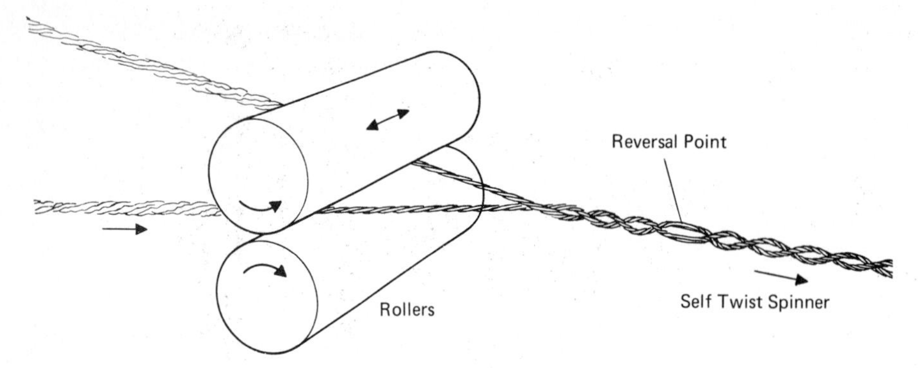

Fig. 19–14 Self-twist spinning.

TOW-TO-TOP (SLIVER) SYSTEM

The *tow-to-top system* bypasses the opening, picking, and carding steps of conventional spinning. In this system, the filament tow is reduced to staple and formed into sliver (or top) by either diagonal cutting on a Pacific Converter or break stretching on a Perlock machine. The sliver from either machine is made into regular-spun yarn by conventional spinning.

The diagonal cutting stapler changes tow into staple of equal or variable lengths and forms it into a crimped sliver (Figure 19–15). Yarn spinning is completed later on the conventional spinning system.

The break-stretch stapler operates on the principle that, when tow is stretched, the fibers will break at their weakest points (random breakage) without disrupting the continuity of the strand. The resultant staple will be of various lengths.

TOW-TO-YARN SYSTEM

Tow-to-yarn spinning is done by a machine called a direct spinner. Light tow (4,400 denier) is fed into the machine through leveling rolls, passes between two nip rolls, and then passes across a conveyor belt to a second pair of nip rolls, which travel at a faster rate of speed and create tension that causes the fibers to break at their weakest points. The strand is then drawn out to yarn size, twisted, and wound on a bobbin (Figure 19–16).

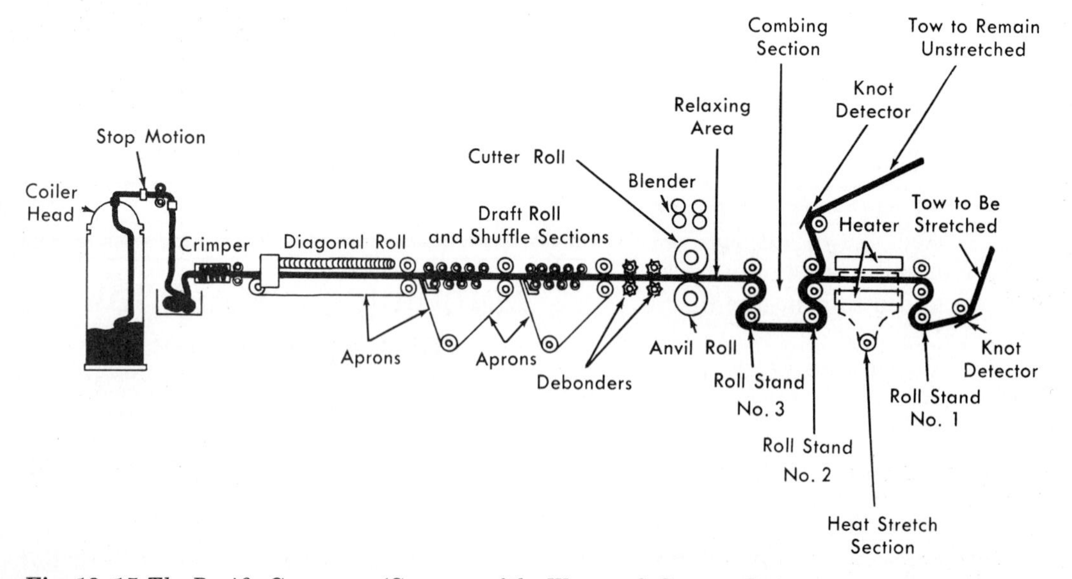

Fig. 19–15 The Pacific Converter. (Courtesy of the Warner & Swasey Co.)

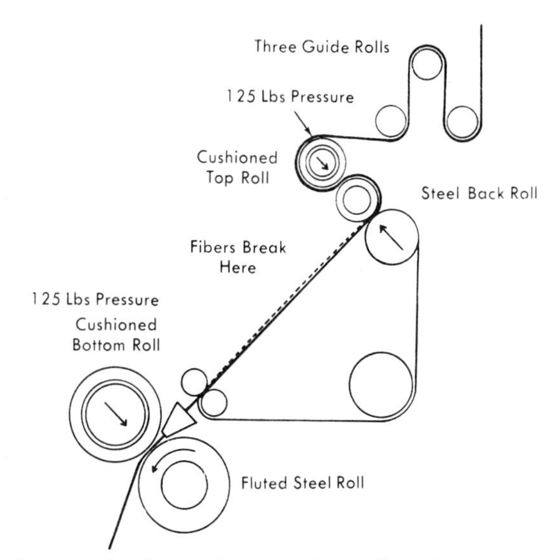

Fig. 19–16 *Direct spinning of yarn from filament tow.*

High-Bulk Yarns

High-bulk yarns are defined by ASTM as yarns essentially free from stretch in which a fraction of the fibers (in any cross-section) has been forced to assume a relatively high random crimp by shrinkage of the remaining fibers that, in general, have very low crimp. The shrinkage is produced by heating or steaming a yarn containing a proportion of thermally unstable fibers so that the latter shrink, producing crimp in the other fibers. This process was explained in more detail under fiber modifications of acrylic fibers.

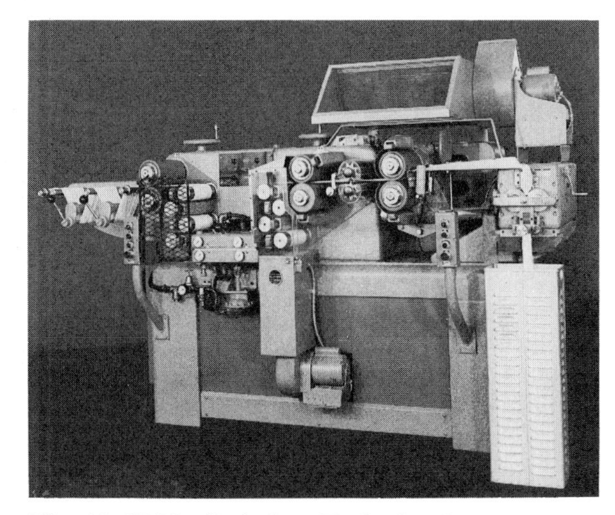

Fig. 19–17 *The Perlock or Turbo-Stapler.*

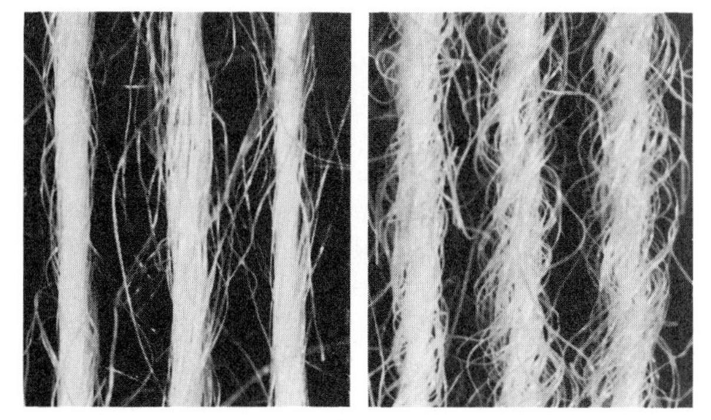

Fig. 19–18 *High-bulk yarn before* (left) *and after* (right) *steaming.*

On the Perlock or Turbo-Stapler (Figure 19–17), the tow is heat stretched and passed through the breaker zone. The fibers are crimped, a portion of the top is relaxed by steam, and the two portions are brought together and spun into yarns. Both of these methods make yarns that are composed of high- and low-shrinkage fibers. The yarns containing high-shrinkage and low-shrinkage fibers are first made into a garment such as a sweater. The garment is then immersed in boiling water and the high-shrinkage component of the yarn contracts—shrinking as much as 20 percent—and draws in toward the center causing the low-shrinkage fibers to buckle and create loft and bulk (Figure 19–18). The heat treatment will stabilize the garment and no further shrinkage will occur in use.

Fiber Blends

A *blend* is an intimate mixture of fibers of different composition, length, diameter, or color spun together into a yarn. A *mixture* is a fabric that has yarn of one fiber content in the warp and yarn of a different fiber content in the filling. A *combination* yarn has two unlike fiber strands twisted together as a ply. Blends, mixtures, and combinations give properties to fabrics that are different from those obtained with one fiber only. The following discussion relates to blends, although most of the facts are true for mixtures and combinations as well.

Blending is done for several reasons:

1. To produce fabrics with a better combination of performance characteristics in the product.

This is perhaps the most important reason for blending. In end uses where durability is very important, nylon or polyester blended with cotton or wool provide strength and resistance to abrasion, while the wool or cotton look is maintained. A classic example is in durable-press garments, where 100 percent cotton fabrics are not as durable as polyester/cotton blends.

2. To improve spinning, weaving, and finishing efficiency for uniformity of product, as with self-blends of natural fibers to improve uniformity.

3. To obtain better texture, hand, or fabric appearance. A small amount of a specialty wool may be used to give a buttery or slick hand to wool fabrics, or a small amount of rayon may give luster and softness to a cotton fabric. Fibers with different shrinkage properties are blended to produce bulky and lofty fabrics or fur-like fabrics with guard hairs.

4. For economic reasons. Expensive fibers can be extended by blending them with more plentiful fibers. This use is sometimes unfair to the consumer, especially when the expensive fiber is used in small amounts but advertised in large print; for example, CASHMERE and wool.

5. To obtain cross-dyed effects or create new color effects such as heather, when fibers with unlike dye affinity are blended together and then piece dyed.

Blending is a complicated and expensive process, but it makes it possible to build in a combination of properties that are permanent. Not only are blends used for better serviceability of fabrics but they are also used for improved appearance and hand.

In the following chart, some fiber properties are rated. Notice that each fiber is deficient in one or more important property. Try different fiber combinations to see how a blend of two fibers might give different performance than either fiber used alone.

BLEND LEVELS

For a specific end use, a blend of fibers that complement each other will give more satisfactory all-round performance than a 100 percent fiber fabric.

M. J. Caplan, in his article "Fiber Translation in Blends" (*Modern Textiles Magazine,* **40**:39 July 1959), used the following example to show that a blend will yield a fabric with intermediate values. He took two fibers, *A* and *B*, each of which could be used to make a similar fabric, measured five performance properties of each of these 100 percent fabrics, and then predicted the performance of a blended fabric 50/50 *A* and *B* by averaging the values of each fabric in the blend.

Notice that the predicted value for the blend is lower than the high value of one fabric, and is greater than the low value of the other fabric. By blending them, a fabric with intermediate values is obtained. Unfortunately, the real values do not come out in the same proportion as the respective percentage in a blend.

Much research has been done by the fiber manufacturers to determine just how much of each fiber is necessary in the various fiber constructions. It is very difficult to generalize about

Fiber Properties

Properties	*Cotton*	*Rayon*	*Wool*	*Acetate*	*Nylon*	*Polyester*	*Acrylic*	*Modacrylic*	*Olefin*
Bulk and loft	−	−	+ + +		−	−	+ + +	+ + +	
Wrinkle recovery	−	−	+ + +	+ +	+ +	+ + +	+ +	+ +	+ +
Press (wet) retention	−	−	−	+	+ +	+ + +			
Absorbency	+ + +	+ + +	+ + +	+	−	−	−	−	−
Static resistance	+ + +	+ + +	+ +	+	+	−	+	+	+ +
Resistance to pilling	+ + +	+ + +	+	+ + +	+				+ +
Strength	+ +	+	+	+	+ + +	+ + +	+	+	+ + +
Abrasion resistance	+	−	+ +	−	+ + +	+ + +	+	+	+ + +
Stability	+ +	−	−	+ + +	+ + +	+ + +	+ + +	+ + +	+ + +
Resistance to heat	+ + +	+ + +	+ +	+	+	+	+ +	−	−

+ + +, excellent; + +, good; +, fair; −, deficient.

Property	Known Values A	Known Values B	Predicted Values 50/50 A and B
1	12	4	8
2	9	12	10.5
3	15	2	8.5
4	7	9	8
5	12	8	10

percentages, because the percentage varies with the kind of fiber, the fiber construction, and the expected performance. For example, a very small amount of nylon (15 percent) improves the strength of wool, but 60 percent nylon is needed to improve the strength of rayon. For stability, 50 percent Orlon blended with wool in a woven fabric is satisfactory, but 75 percent Orlon is necessary in knitted fabrics.

Fiber producers have controlled blend levels fairly well by setting standards for apparel or fabrics identified with their trademark. For example, the du Pont Company recommends a blend level of 65 percent Dacron polyester/35 percent cotton in light- or medium-weight fabrics, whereas 50/50 Dacron/cotton is satisfactory for suiting-weight fabrics. This assures satisfactory performance of the fabric and maintains a good fiber "image" for Dacron. The fabric manufacturer profits from large-scale promotion carried on by the fiber producer.

By using specially designed fiber variants, it is possible to obtain desired performance and appearance in fabrics. For example, fading, shrinking, and softening in time are desirable characteristics to the people who buy denim jeans. (These characteristics are undesirable in most apparel). Du Pont has developed a Dacron polyester variant that, when blended with cotton, will fade, shrink uniformly, and become softer.

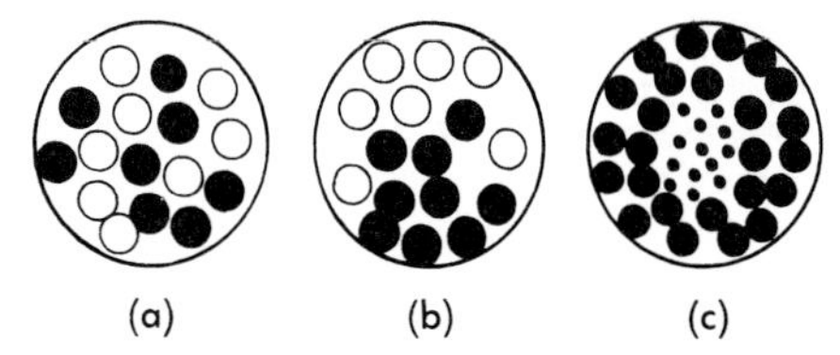

Fig. 19–19 *Cross-section of yarn showing the location of the fiber in the blend.*

BLENDING METHODS

Blending can be done at any stage prior to the spinning operation. Blending can be done during opening-picking, drawing, or roving. One of the disadvantages of direct spinning is that blending cannot be done before the sliver is formed.

The earlier the fibers are blended in processing, the better is the blend. Figure 19–19 shows a cross-section of yarn (a), in which the fibers were blended in opening, and yarn (b), a yarn in which the fibers were blended at the roving stage.

Variations occur from spot to spot in the yarn and also from inside to outside. Long, fine fibers tend to move to the center of a yarn, whereas coarse, short fibers migrate to the periphery of yarn (c). The older methods of blending involve much hand labor.

It has become common to produce fabrics that are mixtures with bulk-filament yarns in one area and spun yarns in another area. These fabrics may have filament yarns in only one direction and spun in the other or these fabrics may have filament yarns in bands in the warp or filling to create a design in the fabric.

BLENDED FILAMENT YARNS

A *blended filament yarn* is one in which unlike filaments of different deniers or generic types are blended together. This is usually done to improve performance and appearance of fabrics.

20

Introduction to Fabric Construction

Fabric-Construction Processes

A *fabric* is a pliable, plane-like structure that can be made into garments and household textiles, and for industrial uses in which some shaping or flexibility is needed. Fabrics are made from solutions, fibers, yarns, and combinations of these elements with a previously made fabric or cloth. Fabrics are usually available to the consumer by the yard or meter.

The fabric-forming process determines the appearance and texture, the performance during use and care, and the cost of a fabric. The process often determines the name of the fabric; for example, felt, lace, double knit, jersey. The cost of fabrics in relation to the construction process depends on the number of steps involved and the speed of the process; the fewer the steps and the faster the process the cheaper is the fabric.

This chapter provides brief descriptions of the fabric-construction processes and major characteristics of typical fabrics. Each method will be discussed more completely in the following chapters.

FABRICS MADE FROM SOLUTIONS

Films

- Solution is extruded through narrow slits into warm air or cast onto a revolving drum. Molding powders may be pressed between hot rolls (see Figure 20–1).
- Waterproof. Low cost. Resistant to soil. Nonfibrous.
- May lack permeability.
- May lack strength unless supported by a fabric back.
- Low drapability.
- Finished to look like every other fabric as well as having its own characteristic appearance.
- Used for shoes, shower curtains, upholstery fabrics, and plastic bags.

Foam

- Made by incorporating air into an elastic-like substance. Rubber and polyurethane are most commonly used (see Figure 30–1).

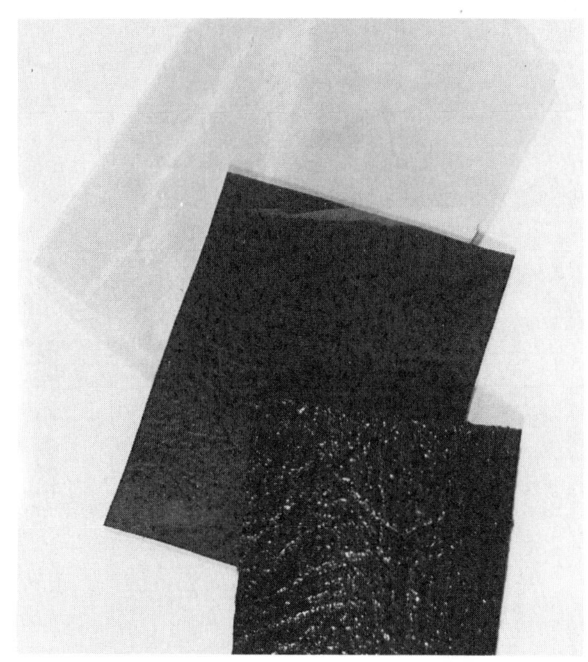

Fig. 20–1 *Films.*

- Lofty, springy, bulky material. Too weak to be used without backing or covering. Usually combined with fabric to give warmth without weight in jackets and all-weather coats.
- Used in pillows, chair cushions, mattresses, and carpet padding.

FABRICS MADE DIRECTLY FROM FIBERS

Felt

- Wool fibers are carded (and combed), laid down in a thick batt, sprayed with water, and run through hot agitating plates, which cause the fibers to become entangled and matted together (see Figure 20–2).
- No grain. Does not fray or ravel.
- Absorbs sound. Lacks pliability, strength, and stretch recovery.
- Used in crafts and industrial matting.

Fiberwebs

- Produced by bonding and/or interlocking fibers by mechanical, chemical, thermal, or solvent means, or combinations of these processes (see Figure 20–3).

Fig. 20–2 *Felt (magnified).*

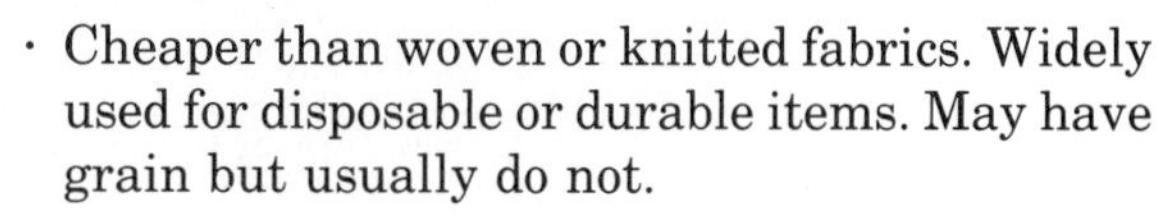

- Cheaper than woven or knitted fabrics. Widely used for disposable or durable items. May have grain but usually do not.
- Used for interfacings and industrial purposes.

FABRICS MADE FROM YARNS

Braid

- Yarns are interlaced diagonally and lengthwise. Fabrics are narrow (see Figure 20–4).
- Stretchy.
- Easy shaping.
- Made circular for shoelaces.
- Used for trim and industrial purposes.

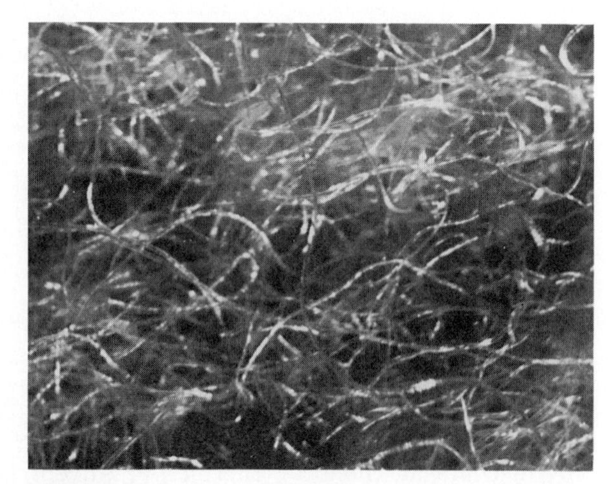

Fig. 20–3 *Fiberweb (magnified).*

Fig. 20–4 *Braid.*

Knit

- One or more yarns are formed into a series of interlocking loops (see Figure 20–5).
- Faster technique than weaving. Requires more yarn per unit of cover.
- Stretchy, elastic fabrics.
- Porous. Resilient.
- Used for apparel, household textiles, and industrial purposes.

Lace

- Yarns are knotted, interlaced, interlooped, or twisted to form open-work fabrics—usually with some figures (see Figure 20–6).
- Open porous structure.

Fig. 20–5 *Knit.*

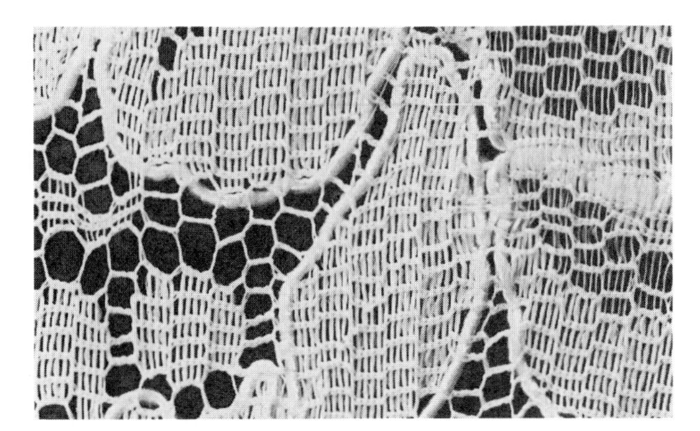

Fig. 20–6 *Lace.*

- Decorative edgings or entire fabric.
- Used for apparel and decorative aspects in apparel and furnishings.

Woven

- Two or more sets of yarns are interlaced at right angles to each other (see Figure 20–7).
- Most widely used construction technique.
- Many different interlacing patterns give interest to fabric.
- Fabrics can be raveled from adjacent sides.
- Fabrics have grain.
- Relatively rigid fabrics—do not have much stretch in warp or filling.
- Used in apparel, furnishings, and industrial products.

MULTIPLEX FABRICS

Coated

- Application of semiliquid material to a fabric substrate. Rubber, polyvinyl chloride, and polyurethane are usual coating materials (see Figure 20–8).

Fig. 20–7 *Woven fabric.*

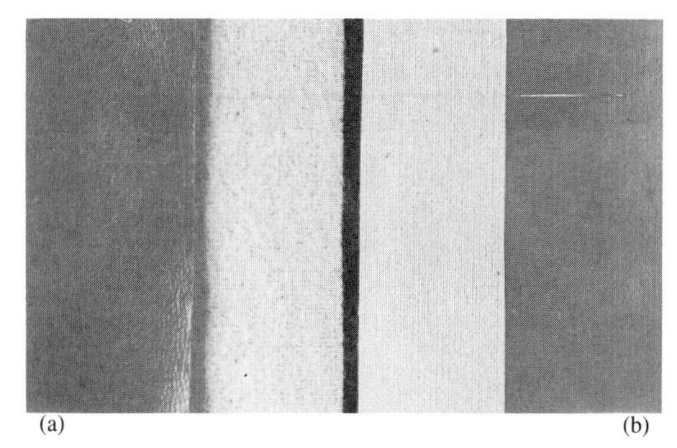

Fig. 20–8 *Coated fabrics: (a) knitted base fabric; (b) woven base fabric. The outer portions show the technical face of the fabric. The inner portions show the technical back or base of the fabric.*

- Stronger and more stable than unsupported films.
- Used for upholstery, handbags, and leather-like apparel.

Flocked

- Fibers are forced into a fabric substrate and held by an adhesive or electronic bonding to make a pile figure or overall pile on fabric.
- Used in apparel, upholstery, and automotive fabrics.

Foam and Fiber

- Fibers and polyurethane solution are mixed together, cast on a drum, or forced through a slit to make fabric. Napped on both sides. Fibers used are polyester, nylon, and rayon (see Figure 20–9).

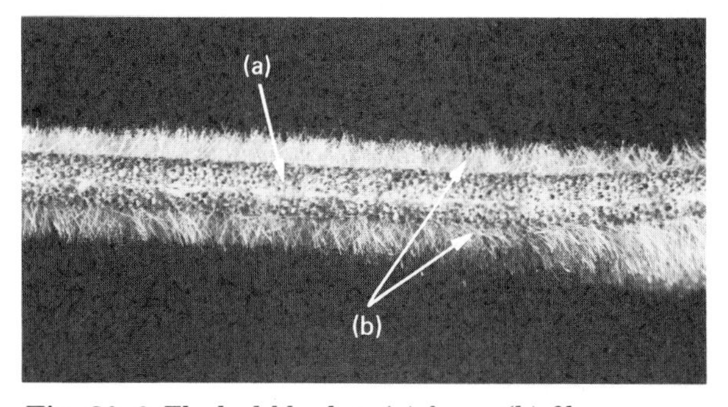

Fig. 20–9 *Flocked blanket: (a) foam; (b) fiber.*

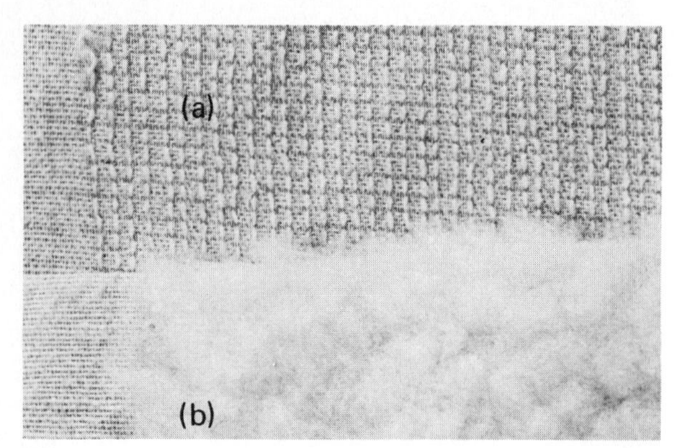

Fig. 20–10 Tufted fabric: (a) reverse side; (b) right side, with fiber surface formed by cut pile.

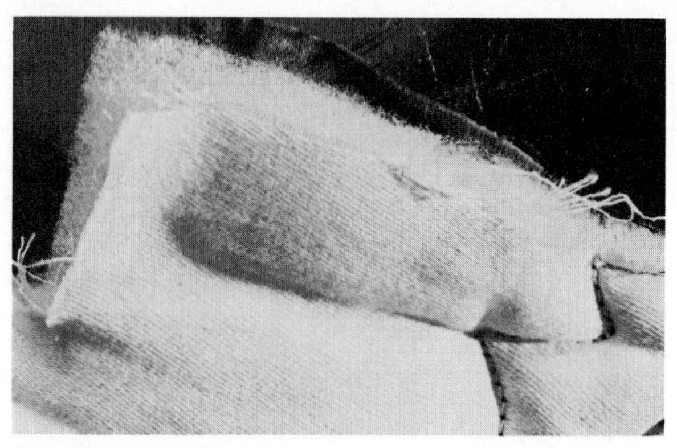
Fig. 20–12 Quilted fabric; reverse side in picture.

- Looks and feels like suede.
- Machine washable and dry cleanable.
- Uniform in thickness and quality and sold by the yard or meter.
- Used in apparel and furnishings.

Tufted

- Yarns carried by needles are forced through a fabric substrate and formed into cut or uncut loops (see Figure 20–10).
- Cheaper than woven or knitted pile fabrics.
- Used in carpets and rugs, upholstery, coat linings, and bedspreads.

Bonded or Laminated

- Two or more fabrics are made to adhere together by an adhesive or flame-foam process (see Figure 20–11).
- Lower cost than double woven or double knit.

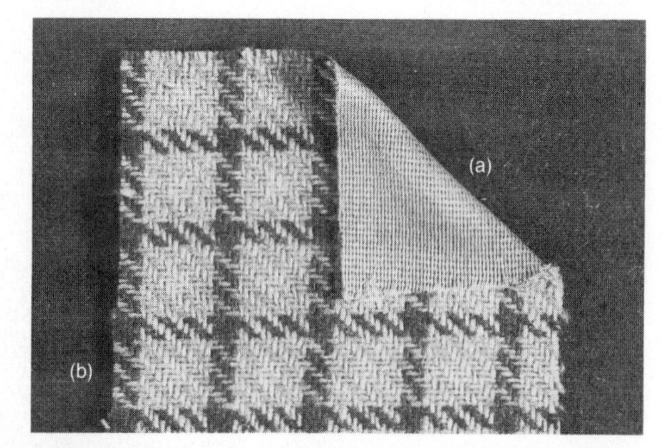

Fig. 20–11 Bonded fabric: (a) reverse side—tricot; (b) right side—woven plaid.

- Warmth without weight, when foam is one layer.
- Makes possible the use of lightweight fabrics for outerwear.
- Fabrics have body.
- Do not hold sharp creases.
- May be laminated off grain, and may come apart (delaminate).
- Used in apparel, shoes, and industrial products.

Quilted

- One or two fabrics and wadding, batting, or foam are stitched together by machine or hand or welded by sonic vibrations (see Figure 20–12).
- Bulky, warm, and decorative. Stitches may break.
- Used in ski jackets, robes, comforters, quilts, and upholstery.

21

Weaving and the Loom

With the exception of triaxial fabrics, all woven fabrics are made with two or more sets of yarns interlaced at right angles to each other. The yarns running in the lengthwise direction are called *warp* yarns or *ends,* and the yarns running crosswise are called *filling, weft,* or *picks.* The right-angle position of the yarns gives the cloth more firmness and rigidity than yarn arrangements in knits, braids, or laces. Because of this structure, yarns can be raveled from adjacent sides. Woven fabrics vary in interlacing pattern, count (number of yarns per inch), and balance (ratio of warp to filling yarns).

Woven fabrics are widely used, and weaving is one of the oldest methods of making cloth. Names were given to fabrics based on the end use, the town in which the fabric was woven, or the person who originated or was noted for that fabric. For example: Hopsacking (cloth used in bags for collecting hops), tobacco cloth (used to give shade to tobacco plants), cheese cloth (to wrap cheeses) and ticking (cloth used to cover mattresses, which were called "ticks") are named for their original end use; bedford cord (New Bedford, Massachusetts), calico (Calcutta, India), chambray (Cambrai, France), and shantung (Shantung, China) are named for the town in which they were first woven; batiste (Jean Baptiste a linen weaver) and jacquard (Joseph Jacquard) are named for people. Fabrics that look like the woven basic fabric but are constructed differently often carry the name of the woven fabric. Denim, for example, is a yarn-dyed woven twill fabric, but denim-looking fabrics are also made by knitting.

The Loom

Weaving is done on a machine called a *loom.* All the weaves that are known today have been made by the primitive weaver. The loom has changed in many ways, but the basic principles and operations remain the same. Warp yarns are held between two beams, and filling yarns are inserted and pushed back to make the cloth.

In primitive looms, the warp yarns were kept upright or horizontal (Figure 21–1). In backstrap looms, which are still used for hand weaving in many countries, the warp yarns are kept taut by attaching one beam to a tree or post and the other beam to a strap that fits around the weaver's hips as the weaver squats or sits (Figure 21–1). Filling yarns were inserted over and under the warp by the fingers and later by a shuttle batted through raised warp yarns. To separate the warp yarns and make hand weaving faster, alternate warp yarns were attached to wooden bars that could be raised, bringing alternate warp yarns up. A comb, very much like a hair comb, was used to beat up the filling yarns. The wooden bar mechanism developed into heddles and harnesses that were attached to foot pedals so that the weaver could separate the warp yarns by foot, leaving the hands free for inserting the filling yarns.

During the Industrial Revolution, mass production high-speed looms were developed. The basic modern loom consists of two beams, a warp

Fig. 21–1 *Primitive looms: upright loom* (left)*; horizontal loom* (center)*; backstrap loom* (right). *(Courtesy of Prodesco, Inc.)*

beam and a cloth beam, holding the warp yarns between them (Figure 21–2). The warp is raised and lowered by a harness-heddle arrangement. A *harness* is a frame to hold the heddles. A *heddle* (headle) is a wire with a hole in its center through which the yarn goes. There are as many heddles as there are warp yarns in the cloth, and the heddles are held in two or more harnesses. As can be seen in the diagram of a two-harness loom, as one harness is raised, the yarns form a *shed* through which the filling can be inserted. A *shuttle* carries the filling yarn through the shed. A *reed,* or *batten,* beats the filling yarn into the cloth to make the fabric firm. A reed is a set of wires in a fame, and the spaces between the wires are called *dents.*

Weaving consists of the following steps:

1. *Shedding:* raising one or more harnesses to separate the warp yarns and form a shed.
2. *Picking:* passing the shuttle through the shed to insert the filling.
3. *Beating up:* pushing the filling yarn into place in the cloth with the reed.
4. *Take-up:* winding finished cloth on the cloth beam.

Additional harnesses are used to make more-intricate designs. The number of harnesses that a loom can operate efficiently is limited. When the repeat of the woven pattern requires more than six harnesses, a *dobby attachment* is added to the loom to control the raising and lowering of the warp yarns. More-complicated designs are made on a jacquard loom.

PREPARATION OF YARNS FOR WEAVING

Winding. After staple and filament yarns are spun, they are wound on packages. The first step in preparing yarn for weaving or knitting is to repackage the yarn so it can be used on the particular equipment that will be utilized in making the fabric. This step is called *winding.* As yarns are rewound, spun yarns can be given more twist or combined with other singles to make ply yarns.

Creeling. Spools of yarn are placed on a large frame called a *creel* (Figure 21–3). The purpose of the creel is to hold the spools of yarn as they are wound onto a warp beam. The yarn from the warp beam will be run through a "slash" bath, dried, and rewound on the warp beam before it is put on the loom for weaving. The "slash" solution is a sizing to protect the yarns from mechanical action in weaving. After weaving, the cloth is removed from the cloth beam, washed to remove the slashing, finished to specification and wound on bolts or tubes for sale to cutters or consumers.

LOOM DEVELOPMENTS

Loom developments over the years have centered on (1) devices to separate the warp to make intricate designs; (2) computers and electronic monitoring systems to increase speed and quality; (3) speedier methods of inserting the filling, such as multiple sheds; and (4) different interlacing of yarns.

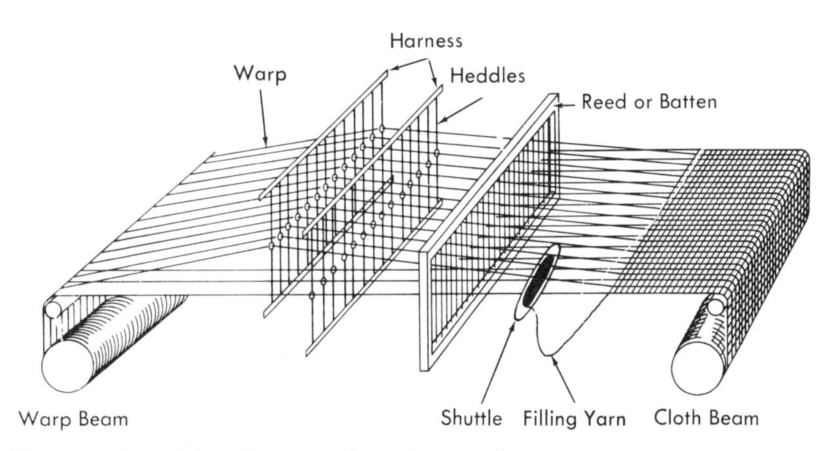

Fig. 21–2 *Simplified drawing of two-harness loom.*

Fig. 21–3 Creeling. Spools of yarns are placed on a large creel and wound onto a warp beam.

Warp Shedding. *Warp-shedding devices* have included the dobby, doup, lappet and leno attachments, and the jacquard loom. These have become so sophisticated that pictures can be woven in cloth (see Chapter 25).

Computer Systems. *Computer* and *electronic devices* play an important part in developing design tables for setting up "maximum weavability" properties, such as tightness and compactness in wind-repellent fabrics or tickings. The computer plays a part in textile designing. Designs can be programmed by means of punched tapes that control the operation of individual warp yarns.

Shuttleless Looms. *Shuttleless looms* were developed as a way to replace the shuttle. In the simple loom, a "flying" shuttle is "batted" back and forth through the warp shed by picker sticks at both sides of the machine. The speed with which the shuttle is sent back and forth is limited—usually about 200 picks per minute. Manufacturers have long sought a way to replace the shuttle and increase the speed of weaving.

Looms using shuttles are often called *power looms* or *conventional looms.* They are still the backbone of the weaving industry, comprising 80–85 percent of all looms in place. Many different kinds of looms are used, as will become evident as specific fabrics are discussed. Perhaps the greatest advantage conventional looms have over shuttleless looms is that the individual loom can generally be used for a wider variety of yarns and weaves. The newer looms increase the choices available as mills seek to find the loom best suited for the type of fabric being woven, the length of run expected to be made, and the speed of fashion change expected for the fabrics.

Four different types of shuttleless looms have been developed—water-jet, air-jet, rapier, and projectile looms. They give higher weaving speeds and reduced noise levels—a factor of great importance to the worker. In all these looms, the filling yarns are measured and cut, thus leaving a fringe along the side. This fringe may be fused to make a selvage if the yarns are thermoplastic, or the ends may be turned back into the cloth. These selvages are not always as usable as conventional selvages because of a tendency toward puckering that requires slitting.

At the end of 1983, there were 214,693 looms in place in the United States. Of these, a total of 19 percent were shuttleless. Water-jet looms accounted for 3 percent of all looms in place; air-jet looms for 2 percent; rapier looms for 4 percent; and projectile, or missile, looms for 9 percent.

Water-Jet Loom. The water-jet loom uses a high-pressure jet of water to carry the filling yarn across the warp (Figure 21–4). It works on the principle of continuous feed and minimum tension of the filling yarns, so it can weave fabrics without barré or streaks. The filling yarn comes from a stationary package at the side of the loom, goes to a measuring drum that controls the length of each filling, and continues through a guide to the water nozzle, where a jet of water carries it across through the warp shed. After the filling is beaten back, it is cut off. If the fibers are thermoplastic, a hot wire is used to cut the yarn, fusing the ends so they serve as a selvage.

The water is removed from the loom by a suction device. Water from the jet will dissolve regular warp sizings so one of the problems has been that of developing water-resistant sizings that can be removed easily in cloth-finishing processes. The fabric is wet when it comes from the loom and must be dried—an added expense.

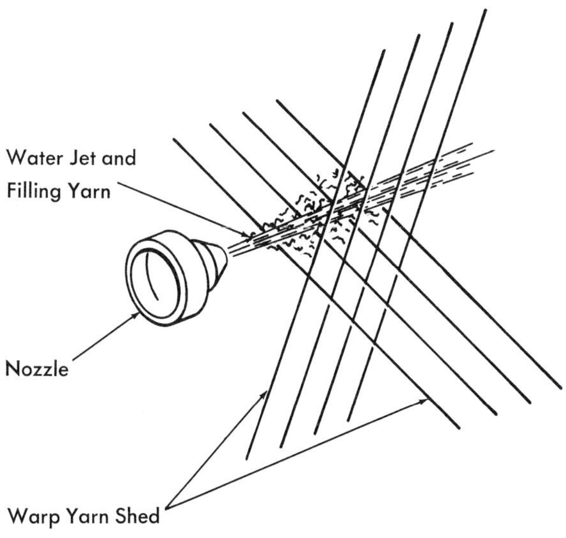

Fig. 21–4 *Water jet carries filling yarn through warp shed.*

The water-resistant sizings may reduce the wettability of the yarns and help solve this problem. The water-jet loom has proved ideal for fashion fabrics, using both smooth and bulk filaments, and for men's slacks and shirtings. It will use practically every natural or synthetic fiber in several different deniers. The loom is more compact, less noisy, and takes up less room than the conventional loom. It can operate at 400–600 picks per minute—two or three times faster than the conventional loom. Maintenance is relatively easy.

Air-Jet Loom. The air-jet loom, or pneumatic method, was developed in Sweden by a textile engineer who got the idea while sailing. He noticed the short regular puffs that came from the exhaust of a diesel motor. His first loom used a bicycle pump to furnish the compressed air. The filling is premeasured and guided through a nozzle, where a blast of air sends it across. The loom can operate at 320 picks per minute and is suitable for spun yarns. Good warp preparation is required. There are limitations in the types of filling yarns it can handle. Air-jet looms can now weave fabrics up to 400 centimeters (157 inches) in width.

Rapier Loom. The rapier loom weaves primarily spun yarns at up to 510 picks per minute. The double-rapier loom has two metal arms, about the size of a small penknife, called carriers or "dummy shuttles," one on the right side and the other on the left side of the loom. A measuring mechanism on the right side of the loom measures and cuts the correct length of filling yarn to be drawn into the shed by the carriers. The two carriers enter the warp shed at the same time and meet in the center. The left-side carrier takes the yarn from the right-side carrier and pulls it across to the left side of the loom. After each insertion, the filling threads are cut near the edge and the protruding ends tucked back into the cloth to reinforce the edge. Figure 21–5 shows the carrier arms. This loom has found wide acceptance for use with basic cotton and woolen/worsted fabrics and is considered to be more flexible than the air-jet loom.

Projectile Loom. The projectile loom uses a projectile with grippers to carry the yarn through the shed. The yarn is carried across the full width of the shed by one projectile. The yarn may be inserted from one or both sides. This loom is also called the missile, or gripper, loom.

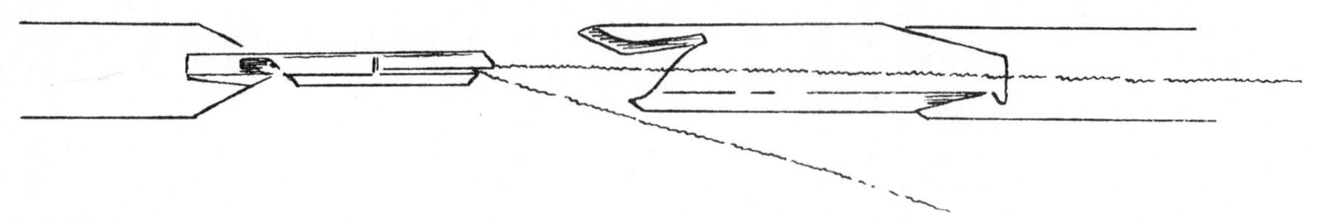

Fig. 21–5 Carrier arms of a rapier shuttleless loom.

Multiple-Shed Loom. Each of the looms discussed so far has formed one shed at a time, then one filling yarn has been placed in the fabric. This process is about as fast as is practical.

In *multiple-shed weaving,* more than one shed is formed at a time. In one loom, for example, as a yarn carrier enters one portion of the warp, a shed is formed; as the carrier leaves that area, the shed changes. This action may occur simultaneously across the width of the warp several times. In another type of multiple-shed loom, several sheds are formed along the length of the warp yarns; the sheds would open at the same time, one filling yarn would be inserted into each shed, and then the sheds would change. These are the newest types of looms.

As many as 16–20 filling carriers insert the precut filling that is stored inside them in a continuous process instead of the intermittent process of single-shed weaving. Beating up and shedding arrangements are different and the process is said to put less stress on warp and filling yarns and to decrease the noise level and energy consumption. In this continuous-weaving process, the picks per minute (ppm) is doubled.

Circular Loom. Because most looms are flat machines and weave flat fabrics, few people realize that circular looms have been developed that make tubular fabric. The *circular loom* in Figure 21–6 weaves sacks of split-film polypropylene. Pillowcases are also tubular woven.

Fig. 21–6 *Circular loom used to weave bagging. (Courtesy of* Textile Industries.*)*

Fig. 21–7 Triaxial-weave pattern. (Courtesy of Barbara Colman Co.)

Triaxial Loom. A new weaving system, Doweave, which weaves three sets of yarns at 60-degree angles to each other, has been developed (Figure 21–7).

The resulting fabric is *triaxial.* The outstanding property of these fabrics is their stability in all directions—horizontally, vertically, and on the bias. Biaxial-woven fabrics, in which the sets of yarns are woven at right angles to each other, do not have stability on the bias.

In triaxial weaving, all of the yarns are usually alike in size and twist. Two thirds of the yarns are warp and one third are filling. Fabrics can be produced more quickly than other weaves because there are fewer picks per inch and speed of weaving is based on the number of picks per minute.

The end uses of triaxial fabrics include balloons, air structures, sailcloth, diaphragms, truck covers, and outerwear apparel. Some projected end uses of this fabric are upholstery, filter cloth, shoe uppers, canvas shoes, foundation garments, slacks, swim suits, and athletic uniforms.

Characteristics of Woven Fabrics

All yarns in woven fabrics interlace at right angles to one another (Figure 21–8). An *interlacing*

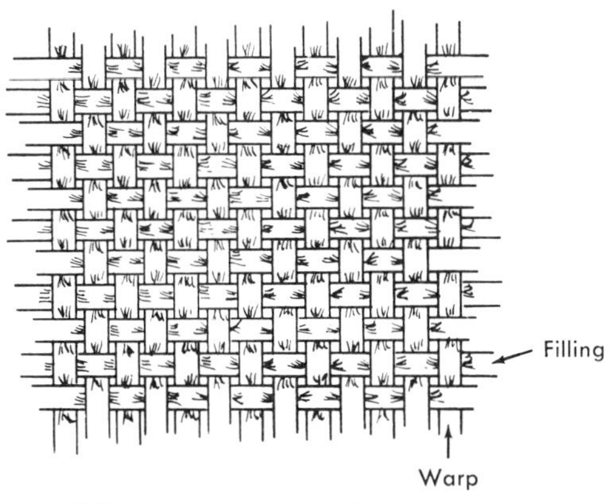

Fig. 21–8 Yarn arrangement in weaving.

is the point at which a yarn changes its position from the surface of the cloth to the underside and vice versa. When a yarn crosses over more than one yarn at a time, *floats* are formed and the fabric has fewer interlacings.

YARNS

Warp and *filling yarns* have different characteristics, and thc fabric performs differently in the warp and filling directions. The warp must resist the high tensions of the loom and the abrasion of the shuttle as it flies back and forth, so the warp yarns are stronger, of better quality and have higher twist. Filling yarns are more apt to be decorative or special-function yarns, such as high-twist crepe yarns or low-twist napping yarns.

The following are ways to differentiate between warp and filling:

1. The selvage, or "fringe" from shuttleless looms, always runs in the lengthwise (warp) direction of all fabrics.

2. Most fabrics stretch less in the warp direction.

3. The warp yarns lie straighter in the fabric because of loom tension. They show less crimp.

4. Decorative or special-function yarns are usually the filling.

5. Specific characteristics may indicate the warp and filling directions. For example: poplin always has a filling rib, satin always has warp floats, and flat crepe has high-twist yarns in the filling and low-twist yarns in the warp.

6. Warp yarns tend to be smaller, with higher twist.

GRAIN

Grain is a term used to indicate the warp and filling positions in a woven fabric. Figure 21–9 shows why a bias edge will ravel more than any of the other grain positions.

The grain positions are as follows:

- Lengthwise grain is a position along any warp yarn.
- Crosswise grain is a position along any filling yarn.
- True bias (Figure 21–9) is the diagonal of a square.
- Bias is any position on the cloth between true bias and either lengthwise or crosswise grain (Figure 21–9).

Off-Grain. *Off-grain fabrics* have been a troublesome problem for the industry and the consumer. At the factory, off-grain causes reruns and lower fabric quality. For the consumer, it means that products will not drape properly or hang evenly and that printed designs will not be straight. If the fabric has a durable-press finish or is made of heat-sensitive fibers, the home sewer will not be able to straighten the fabric, so it is on-grain. Wool fabrics can be straightened, but only with much extra work. The fabric must be dampened and allowed to stand until it relaxes and then pressed carefully so that the yarns are at right angles. Figure 21–10 shows a plaid design that has been printed slightly off grain—the printed lines do not follow the yarns or a torn edge.

There are two kinds of off-grain. *Skew* results when one side of the fabric travels ahead of the other. The fabric in Figure 21–10 is skewed.

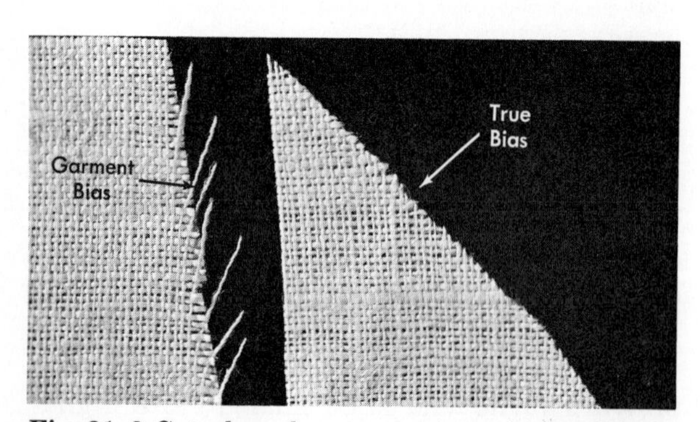

Fig. 21–9 *Cut edges show two grain positions of cloth.*

Fig. 21–10 *"Off-grain" fabric.*

Bow occurs when the center of the fabric lags behind but the two sides keep even as the fabric travels through the tenter frame. The consumer should always examine a fabric or product to see if it is off-grain.

Cloth Straighteners. It is now possible for the manufacturer to control the straightness of the fabric to within 0.25 inch. This is done by cloth straighteners—sensitive detectors attached to the tenter frame that gently ride on the cloth and "feel" the bowed or skewed threads (Figure 21–11).

FABRIC COUNT

Fabric count, or *count,* is the number of warp and filling yarns per square inch of gray goods (fabric as it comes from the loom). This may be

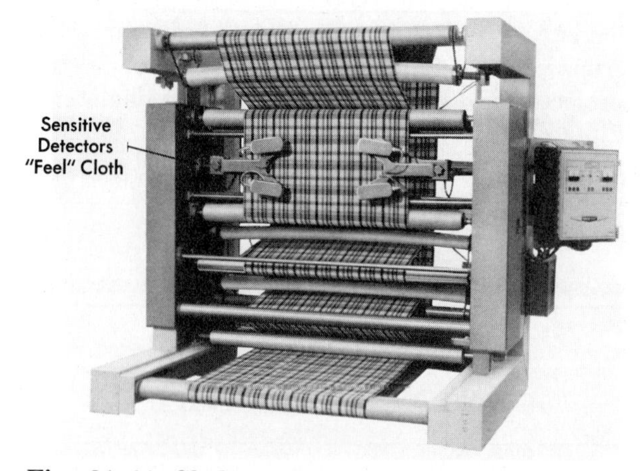

Fig. 21–11 *Cloth straightener. (Courtesy of Automated Energy Systems, Inc.)*

changed by shrinkage during dyeing and finishing. Count is written with the warp number first, for example 80 × 76; or it may be written as the total of the two, as 156. (Count should not be confused with yarn number, which is a measure of yarn size. See the discussion on page 155).

Fabric count is an indication of the quality of the fabric—the higher the count, the better the quality for any one fabric—and can be used in judging raveling, shrinkage, and durability. Higher count also means less potential shrinkage and less raveling of seam edges. Catalogs sometimes give count because the customer must judge the quality from printed information rather than from the fabric itself.

A standard method of determining count was developed by the American Society for Testing and Materials. The count is made with a fabric counter (Figure 21–12). It is possible to use a "hand" method, by which the area is measured by a ruler and counted by sight or yarns raveled and counted.

Print cloth has, in the past, had a standard count of 80 × 80 and was called 80-square fabrics. In February 1960, *Women's Wear Daily* announced that, as a result of cost factors, a 78 × 78 fabric would be the basic print cloth used. (Print cloth is a gray goods fabric as it comes from the loom before any finishing has been done.)

BALANCE

Balance is the ratio of the warp yarns to filling yarns in a fabric. A well-balanced fabric has approximately one warp yarn for every filling yarn, or a ratio of 1:1. Examples of typical unbalanced fabrics are cotton broadcloth, with a count of 144 × 76 and a ratio of about 2:1, and nylon satin, with a count of 210 × 80 and a ratio of about 3:1.

Balance is helpful in recognizing and naming fabrics and in distinguishing the warp direction of a fabric. Balance is not always related to quality. Balance plus count is helpful in predicting slippage. If the count is low, there will be more slippage in unbalanced fabrics than in balanced fabrics.

SELVAGES

A *selvage* is the self-edge of a fabric formed by the filling yarn when it turns to go back across the fabric. The conventional loom makes the same kind of selvage on both sides of the fabric, but shuttleless looms have different selvages because the filling yarn is cut and the selvage looks like a fringe. In some fabrics, stronger yarns, a basket-weave arrangement, or a leno-weave selvage are used.

Fig. 21–12 *Fabric counter.*

Plain selvages are similar to the rest of the fabric. They do not shrink and can be used for seam edges in garment construction. *Tape selvages* are made of larger and/or plied yarns to give strength. They are wider than the plain selvage and may be a basket weave for flatness. An example is the selvage on bed sheets. *Split selvages* are used when narrow items such as towels are made by weaving two or more side by side and cutting them apart after weaving. The cut edges are finished by a machine chain stitch or a hem. *Fused selvages* are the heat-sealed edges of ribbon or tricot yard goods made from wide fabric and cut into narrower widths. On some shuttleless looms, the filling yarns are cut at the edge of the fabric leaving a fringe (Figure 21–13).

FABRIC WIDTH

The loom determines the width of the fabric. Hand-woven fabrics are usually 27–36 inches wide. Before the 1950s, machine-woven cottons were traditionally 36 inches wide. However, wider fabrics are more economical to weave and the garment cutter can lay out patterns to better advantage. The new looms weave cotton 45 or 60 inches wide. Wool fabrics are 54–60 inches

Basic Weaves

Name	*Interlacing Pattern*	*General Characteristics*	*Typical Fabrics*	*Chapter Reference*
Plain $\frac{1}{1}$	Each warp interlaces with each filling.	Most interlacings per square inch. Balanced or unbalanced. Wrinkles most. Ravels most. Less absorbent.	Batiste Voile Percale Gingham Broadcloth Crash Cretonne Printcloth	22
Basket $\frac{2}{1}$ $\frac{2}{2}$ $\frac{4}{4}$	Two or more yarns in either warp or filling or both woven as one in a plain weave	Looks balanced. Fewer interlacings than plain weave. Flat looking. Wrinkles less. Ravels more.	Oxford Monk's cloth Duck Sailcloth	22
Twill $\frac{2}{1}$ $\frac{2}{2}$ $\frac{3}{1}$	Warp and filling yarns float over two or more yarns from the opposite direction in a regular progression to the right or left.	Diagonal lines. Fewer interlacings than plain weave. Wrinkles less. Ravels more. More pliable than plain weave. Can have higher count.	Serge Surah Denim Gabardine Herringbone Flannel	23
Satin $\frac{4}{1}$ $\frac{1}{4}$	Warp and filling yarns float over four or more yarns from the opposite direction in a progression of two to the right or left.	Flat surface. Most are lustrous. Can have high count. Fewer interlacings. Long floats—subject to slippage and snagging.	Satin Sateen	23
Momie or Crepe	An irregular interlacing of yarns. Floats of unequal lengths in no discernable pattern.	Rough-looking surface. Crepe looking.	Granite cloth Moss crepe Sand crepe	26
Dobby	Special loom attachment allows up to 32 different interlacings.	Small figures. Cord-type fabrics.	Shirting madras Huck toweling Waffle cloth Piqué	25
Jacquard	Each warp yarn controlled individually. An infinite number of interlacings is possible.	Large figures.	Damask Brocade Tapestry	25
Pile	Extra warp or filling yarns are woven in to give a cut or an uncut three-dimensional fabric.	Plush or looped surface. Warm. Wrinkles less. Pile may flatten.	Velvet Velveteen Corduroy Fur-like fabrics Wilton rugs Terrycloth	24

Basic Weaves (continued)

Name	*Interlacing Pattern*	*General Characteristics*	*Typical Fabrics*	*Chapter Reference*
Slack-tension	A type of pile weave. Some warp yarns can be released from tension to form raised areas in the cloth or a pile surface.	Crinkle stripes or pile surface. Absorbent. Nonwrinkling.	Seersucker Terrycloth Friezé	24, 25
Leno	A doup attachment on the loom causes one of two warp yarns to be carried over the other on alternate passings of the filling yarns.	Mesh-like fabric. Lower count fabrics that are resistant to slippage.	Marquisette Curtain fabrics	27
Swivel	An attachment to the loom. Small shuttles carrying extra filling yarns weave in small dots.	Dots on both sides of fabric as filling floats.	Dotted swiss	25

wide and silk-type fabrics are 40–45 inches wide. Single and double knits are usually about 60 inches wide.

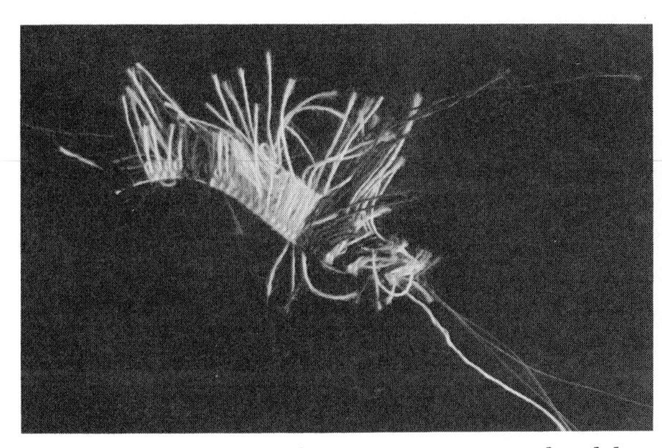

Fig. 21–13 Fringe selvage woven on a shuttleless loom.

PROPERTIES

Fabric properties resulting from weaving variables are summarized in the following chart.

The type of weave or interlacing pattern influences fabric properties as well as fabric appearance. The Basic Weaves chart is an introduction to the various weaves and also a summary that may be used as a reference. The next several chapters deal with woven fabrics.

Properties of Woven Fabrics

Fabric Type	*Properties*
High count	Firmness, strength, cover, body, compactness, stability, wind repellency, water repellency, fire retardancy, reduced raveling of seams.
Low count	Flexibility, permeability, pliability, better drape, higher shrinkage potential, more seam raveling.
Balanced	Less seam slippage, warp and filling wear evenly resulting in holes.
Unbalanced (usually more warp)	Seam slippage in low count; warp yarns wear out first, leaving strings (common in upholstery fabrics). In plain weave, crosswise ribs give interesting surface.
Floats	Luster, smoothness, flexibility, resiliency, tendency to ravel and snag, seam slippage in low count.

22

Plain-Weave Fabrics

Plain Weave

Plain weave is the simplest of the three basic weaves that can be made on a simple loom without the use of any attachment. It is formed by yarns at right angles passing alternately over and under each other. Each warp yarn interlaces with each filling yarn to form the maximum number of interlacings (Figure 22–1). Plain weave requires only a two-harness loom and is the least expensive weave to produce. It is described as a $\frac{1}{1}$ weave: one harness up and one harness down when the weaving shed is formed.

Figure 22–1 shows several ways of diagramming a plain weave. The top drawing is a cross-sectional view of a fabric cut parallel to a filling yarn. The cut ends of the warp yarns appear as black circles. The filling yarn goes over the first warp yarn and under the second warp yarn. In a plain-weave fabric, this same pattern is repeated until the filling yarn has interlaced with all the warp yarns across the width of the loom. The second filling yarn goes under the first warp yarn and over the second. This pattern also is repeated across the width of the loom. Notice how these two filling yarns have interlaced with the warp yarns to have the maximum number of interlacings. In a plain weave, all odd-numbered filling yarns will have the same interlacing pattern as the first filling yarn, and all even-numbered filling yarns will have the same interlacing pattern as the second filling yarn.

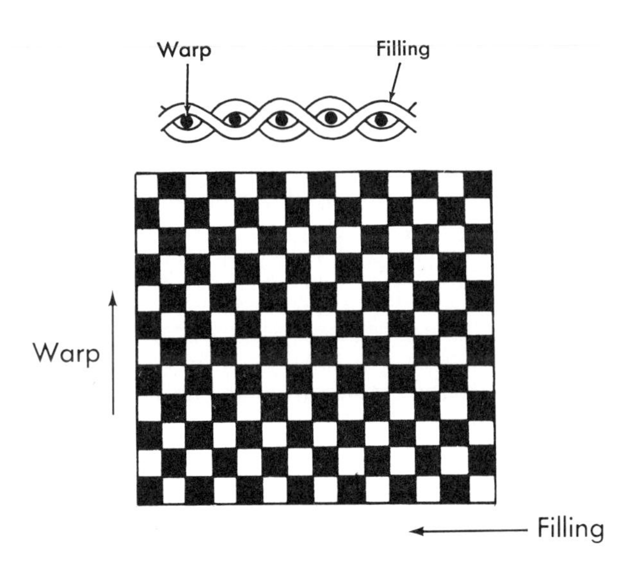

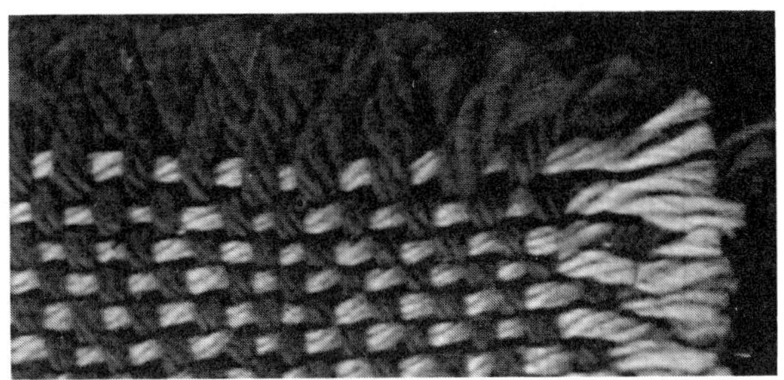

Fig. 22–1 Three ways to show the yarn-interlacing pattern of plain weave: cross-section (top)*; checkerboard* (center)*; photograph* (bottom).

The photograph of the fabric at the bottom of Figure 22–1 shows the same interlacing pattern as in the cross-section. In the photograph, the yarns are opaque and only the yarn on the surface is visible. Hence, a pattern of dark warp yarns and light filling yarns develops. The checkerboard pattern in the center of Figure 22–1 is a simple representation of the plain-weave fabric in the photograph. In the checkerboard pattern, each square represents one yarn on the surface of the fabric. A dark square represents a warp yarn and a light square represents a filling yarn. The top row in the checkerboard is light, then dark, then the pattern repeats across the fabric. The checkerboard represents a filling yarn on the surface, then a warp yarn on the surface, and so on. The second row is just the opposite and represents the interlacing pattern of the second filling yarn. All woven fabrics are diagrammed using this technique. These diagrams are an easy way to represent the interlacing patterns and to help students identify the weave in a fabric.

Plain-weave fabrics have no right or wrong side unless they are printed or given a surface finish. Their plain uninteresting surface serves as a good background for printed designs, embossing, and puckered and glazed finishes. Because there are many interlacings per square inch, plain-weave fabrics tend to wrinkle more, ravel less, and be less absorbent than other weaves. Interesting effects can be achieved by the use of different fiber contents, novelty or textured yarns, yarns of different sizes, high- or low-twist yarns, filament or staple yarns, and different finishes.

The simplest form of plain weave is one in which warp and filling yarns are the same size and the same distance apart so they show equally on the surface—balanced plain weave.

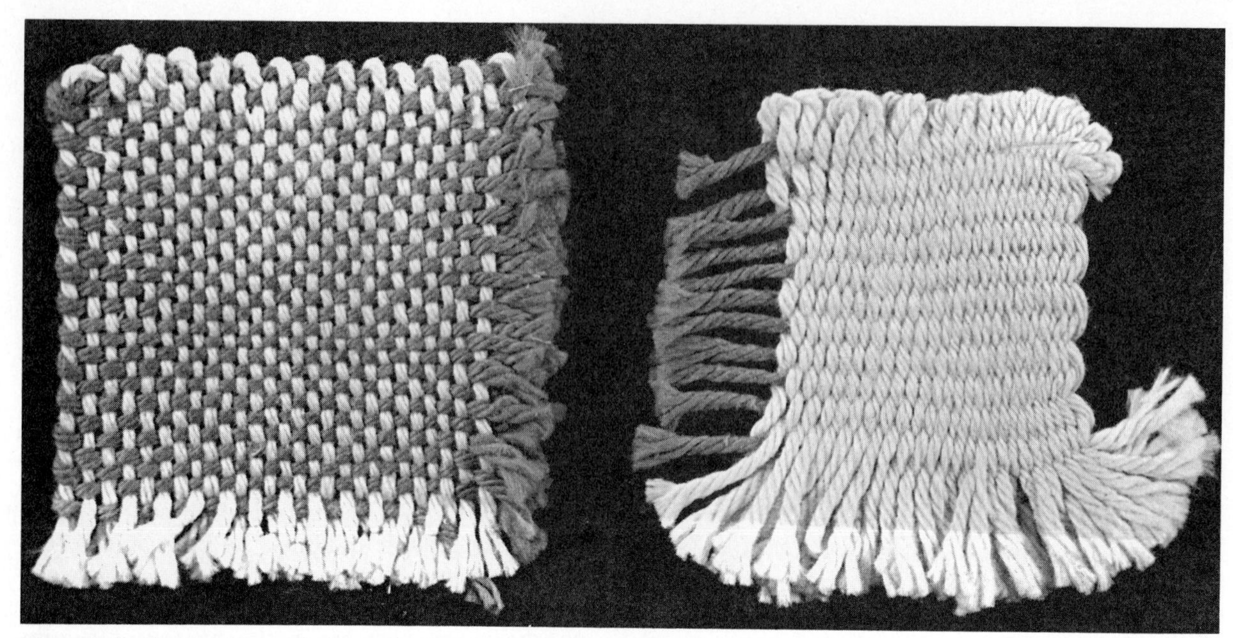

Fig. 22–2 Comparison of balanced plain weave fabric (left) *and unbalanced plain weave fabric* (right).

Other forms have warp yarns so numerous as to cover the filling that are obvious only in the form of ridges called *ribs*—unbalanced plain weave, and a variation that has two or more yarns interlaced as one—basket weave.

- Balanced plain weave.
- Unbalanced plain weave (Figure 22–2).
- Variation: basket weave.

BALANCED PLAIN WEAVE

Balanced plain-weave fabrics have a wider range of end uses than fabrics of any other weave and are therefore the largest group of woven fabrics. They can be made in any weight, from very sheer to very heavy.

One convenient way of grouping fabrics to study and compare them is by fabric weight. Plain-weave fabrics will be discussed in five groups: lightweight sheer, lightweight opaque, low-count sheer, medium weight, and heavyweight.

Lightweight Sheer Fabrics. *Lightweight sheer fabrics* are very thin, light, and transparent or semitransparent. High-count sheers are characterized by transparency as a result of the fineness of yarns.

Filament-yarn sheers are often designated by the fiber content; for example, polyester sheer or nylon sheer.

Ninon is a filament sheer that is widely used for curtains. It is usually 100 percent polyester because of that fiber's good resistance to sunlight and excellent resiliency and easy washability. Frequently, ninon appears to have paired warp yarns. Although ninon is a plain weave, two yarns are a little closer to each other and spaced a little farther away from the yarns on either side. Ninon has medium body and hangs well.

Georgette and *chiffon* are made with filament-crepe yarns, the latter being smoother and more lustrous. Both are very lightweight and drape well. Many yards of chiffon can be made into a very graceful, nicely draping formal gown. Georgette is more-commonly seen in feminine blouses and dresses. Both fabrics were originally made of silk but now often are made from man-made filament yarns.

Voile is a sheer made with special high-twist or twist-on-twist spun yarns (see page 155). Voile was originally a cotton or wool fabric, but it is now found on the market in other fiber contents.

Organdy is the sheerest cotton cloth made. Its sheerness and crispness are the result of an acid finish (see Chapter 33). Because of its stiffness and fiber content, it wrinkles badly.

High-Count Sheers

Fabric	Typical Count	Yarn Size or Number — Warp	Yarn Size or Number — Filling
Lawn	88 × 80	70s*	100s*
Organdy	Similar to lawn	Similar to lawn	
Batiste	Similar to lawn	Similar to lawn	

*The "s" after the number means that the yarn is a single yarn. If a ply yarn is used in the fabric, it is indicated by writing the number as 38/2 or 44/2, and so forth.

Organza is the filament-yarn counterpart to organdy. It has a lot of body. Wrinkling depends on fiber content and finishes, but is not as severe a problem as it is for organdy.

Sheer fabrics are used for glass curtains (which give privacy but let in light); for summer-weight shirts, blouses, and dresses; and for baby dresses. Short wash cycles and low ironing temperatures are required. Sheer fabrics dry faster, so they may need to be redampened during ironing. Small hems and enclosed seams are used to enhance the daintiness of the fabric and to prevent pulling out during use and care.

Lightweight Opaque Fabrics. *Lightweight opaque fabrics* are very thin and light, but are not as transparent as sheer fabrics. The distinction between the two groups of fabrics is not always pronounced; the fabrics are grouped because, more often than not, they fall within these descriptions of opacity or transparency.

Organdy (a sheer fabric), lawn, and batiste are finished from the same gray goods (lawn gray goods). They differ from one another in the way they are finished. The better qualities are made of combed cotton or cotton/polyester yarns. Organza has already been described as a sheer fabric. *Lawn* is a fabric that is often printed. Lawn is usually all cotton or cotton/polyester.

Batiste is the softest of the lightweight, opaque fabrics. Cotton batiste is highly mercerized and often available in whites, pastels, or as other solid colors. Cotton batiste is used for fine hand smocking for apparel. Batiste fabrics are also made of wool, polyester, and polyester/cotton fiber. *Tissue ginghams* and *chambray* are similar in weight to lawn but are yarn-dyed.

Handkerchief linen is similar in count and luster to batiste but is made from flax or slub yarns so it is linen-like in appearance. It is used for handkerchiefs, blouses, and summer dresses.

China silk is also similar to batiste, except that it is made from fine-filament yarns. It is a soft fabric that was originally made of silk and used for women's suit linings and matching blouses. *Habutai* is slightly heavier than China silk.

Crepe de Chine is traditionally a filling crepe-silk fabric that drapes beautifully. It has a dry and very pleasant hand and medium luster. The fabric is more-commonly found now as a filament polyester for blouses and fine linings. Coupe de Ville is a trade name.

Challis tends to be heavier than the fabrics discussed so far, and, depending on fiber content and fashion, it may be a medium-weight fabric. A classic challis fabric is wool in a paisley print. It is soft and drapes very well. Challis usually is printed and slightly napped and it frequently is made from rayon.

Low-Count Sheer Fabrics. *Low-count sheer fabrics* are characterized by open spaces between the yarns, making them transparent. They are made of carded yarns of size 28s and 30s in the warp and 39s and 42s in the filling. Count ranges from 10 × 12 to 48 × 44. They are neither strong nor durable, are seldom printed, and differ in the way they are finished. They are functional fabrics that may be used for decorative and industrial fabrics, or shaping fabrics in apparel.

Cheesecloth is a very loosely woven plain-weave fabric with only slightly twisted yarns. It is used as a wrap for cheese or cured meats such as bacon during processing and as a cleaning cloth. It makes a good pressing cloth for home sewing. It is also used to shade young tobacco plants and for surgical bandages.

Crinoline is similar to cheesecloth but it is much stiffer and used in book binding and millinery (hat making), and also for stiff petticoats or underskirts for bridal and formal gowns.

Buckram is also a very heavily sized fabric. It may be a low-count, heavy-yarn fabric bonded to a finer, higher-count fabric. It is commonly seen in headings of pinch-pleated draperies. It is also used in styling hats, in lampshades, and in book binding, although it is being replaced in these uses by fiber webs.

Bunting is a cheesecloth-like fabric in bright colors that is used for flags and banners.

Medium-Weight Fabrics. *Medium-weight fabrics* comprise the largest group of woven fabrics because they have many more uses than either lightweight or heavyweight fabrics. These fabrics have medium-sized yarns, a medium count, and carded or combed yarns, and they may be finished in different ways or woven from dyed yarns. They are also called top-weight fabrics because they are frequently used for blouses, shirts, and many home-furnishing items—such as window-treatment fabrics, household linens, and some upholstery fabrics.

The carded-yarn fabrics in this group are converted from a gray-goods cloth called *print cloth.* Look at the photos in Figure 33–1 to see the way print cloth looks and how different it can look after a variety of finishes have been applied. Chapter 33 is a useful supplement in understanding some of the differences in fabrics discussed in this chapter. Both plissé and embossed fabrics are discussed there; they are both converted from print cloth.

Most print cloth is made into *percale,* a smooth, slightly crisp, printed or plain-colored fabric. It is called *calico* if it has a small, quaint, printed design; *chintz* if it has a printed design; and *cretonne* if it has a large-scale floral design. When a solid-color fabric is given a glazed calendar finish, it is called *polished cotton.* When chintz is glazed, it is called *glazed chintz.* Glazed chintz is made in solid colors as well as prints. The name chintz comes from the Hindu word meaning spotted. These fabrics are often made with blends of cotton and polyester or high-wet-modulus rayon. They are used for shirts, dresses, blouses, pajamas, and aprons. They are also used in home furnishings for matching curtains and bedspreads, upholstery, slipcovers, draperies, and wall coverings.

Any plain-woven, balanced fabric ranging in weight from lawn to heavy bed sheeting may be called *muslin.* This is also a specific name for medium-weight fabric that is unbleached or white.

Gauze, or *Indian gauze,* is also a variable-weight fabric. It is usually white or piece dyed. It is characterized by a crinkled look that looks like the fabric was crushed together lengthwise. This effect is achieved by finishing, or occasionally by the use of crepe yarns.

Napped fabrics may be either medium or heavy weight. Flannel and outing flannel may be either plain weave or twill weave. *Flannelette*

Converted (Finished) Fabrics

Fabrics	*Range in Count*	*Yarn Size*	
		Warp	*Filling*
Percale, carded (muslin, plissé, calico, chintz, and so forth)	80 × 80 to 44 × 48	30s	42s
Combed cotton	96 × 80	40s	50s

is consistently a plain-weave fabric that is lightly napped on one side. It is cotton or cotton blend, unless it is intended for children's sleepwear—then it is usually a flame-retardant polyester. *Outing flannel* is heavier and stiffer than flannelette; it may be napped on one or both sides. It is used for shirts, dresses, lightweight jackets and jacket linings. Some outing flannels are made with a twill weave.

Ginghams are yarn-dyed fabrics with checks and plaids, or they may appear to be solid in color (Figure 22–3). *Chambrays* appear to be solid color but have white filling and colored-warp yarns, or they may have darker yarns in the filling (iridescent chambray), or they may have stripes. Some chambray is unbalanced with high warp count, which produces a filling rib similar to that of broadcloth and poplin.

Ginghams and chambrays are usually made of cotton or cotton blends and usually given a durable-press treatment. When they are made of fibers other than cotton, the fiber content is

Fig. 22–3 *Gingham fabrics: yarn-dyed checks or plaids.*

included in the name; for example, silk gingham. In filament rayon, these fabrics are given a crisp finish and called *taffetas*. In wool, similar fabrics are called wool checks, plaids, and shepherd's checks. *Madras,* or *Indian madras,* is frequently all cotton, and often of a lower count than gingham. It is a yarn-dyed plaid fabric.

The construction of gingham and chambray is a more-costly process than the making of converted goods because the loom must be rethreaded for each new design, and threading a loom for yarn-dyed fabrics requires more skill than threading it with undyed yarns. It is necessary also for the manufacturer to carry a larger inventory, which requires more storage space.

Stripes, plaids, and checks present problems that are not present in solid-colored fabrics. Crosswise yarns must be parallel to the floor in draperies, lined up with the edges of furniture in upholstery, and properly balanced in apparel. Ginghams may have a design with an up and down, a right and left, or both. These are called *unbalanced plaids*. Compare the unbalanced plaid in the picture in Figure 22–2 with the balanced plaid. More time is needed to cut out an item in plaid than in plain material, and more attention must be given to the choice of design. Plaids in inexpensive garments seldom match except at the center front and back seams, places where failure to match is more noticeable.

Imitations of gingham are made with printed designs. There is, however, a technical face and a technical back to the print, whereas true gingham is the same on both sides. Lengthwise printed stripes are usually on-grain, but the crosswise stripes are frequently off-grain (see page 347). These printed fabrics are not a gingham, but may be a percale if the fabric was converted from a print cloth.

Pongee is a filament-yarn, medium-weight fabric. It has a fine warp of regular yarns with filling yarns that are irregular in size. It was originally silk with slub-filling yarns, but is now made of a variety of fibers. *Honan* is similar to pongee, but it is characterized by slub yarns in both the warp and the filling.

Suiting-Weight Fabrics. *Suiting-weight,* or *bottom-weight, fabrics* are heavy enough to tailor well. Filling yarns are usually larger than the warp yarns because of slightly lower twist.

Yarn-Dyed Fabrics

Fabrics	*Range in Count*	*Yarn Size*	
		Warp	*Filling*
Gingham, cotton Carded	64 × 76 to 48 × 44	Same as percale	
Combed	88 × 84 to 84 × 76		

Chambray and gingham madras are similar to gingham.

Because of their weight, they are more durable and more resistant to wrinkling than sheer or medium-weight fabrics, but they tend to ravel more because of their low count.

Cotton suiting is converted from a gray goods called *coarse narrow sheeting*. Kettlecloth is a manufacturer's trade name. Cotton suiting is plain in color or printed.

Weaver's cloth is a more-general name used for cotton or cotton/polyester suiting.

Homespun is a name used for home-furnishing fabrics with slightly irregular yarns, a lower count, and a hand-woven look.

Crash is made with yarns that have thick-and-thin areas, giving it an uneven nubby look. It shows wrinkles less than a plain surface does.

Butcher rayon is a crash-like fabric of 100 percent rayon or rayon/polyester. In heavier weights, it looks like linen suiting.

Burlap has a much lower count than crash, and is used in home furnishings rather than apparel. It has characteristic coarse thick-and-thin yarns and is usually made of jute.

Osnaburg is a variable-weight fabric most often found in suiting weight. Like muslin, it may be unbleached or bleached. It is a lower-quality fabric than muslin generally, with a lower count and bits of leaf and bark from the cotton plant, giving it a characteristic spotted appearance. It is a utility fabric that occasionally becomes fashionable. It is seen as a drapery-lining fabric.

Flannel is a plain-weave suiting fabric that is napped. Made in woolen yarns, it is used for women's suits, slacks, skirts, and jackets. It may have a plain or twill weave.

Tweed is made of any fiber or mixture of fibers and is always characterized by nubs of different

Suiting-Weight Fabrics

Fabric	Typical Count	Typical Yarn Size
Cotton suiting	48 × 48 to 66 × 76	13s to 20s

colors. The name comes from the Tweed River in Scotland. Harris tweed is hand woven in the Outer Hebrides Islands, and Donegal tweed is handwoven in Donegal County, Ireland.

Tropical worsted suitings are made from long-fiber worsted yarns in the lightest weight, wool-like fabric made for men's suits. These suitings are intended for use in tropical countries or for summer use in temperate climates. They are frequently made of fiber blends.

UNBALANCED PLAIN WEAVE

Increasing the number of warp yarns in a plain woven fabric until the count is about twice that of the filling yarns creates a crosswise ridge called a filling rib, as well as a warp surface in which the warp yarns completely cover the filling yarns. Small ridges are formed when the warp and filling yarns are the same size, and larger ridges are formed where the filling yarns are larger than the warp. Yarn sizes are given in the fabric chart (see page 207).

If the yarns are of different colors, the only color showing on the surface will be that of the warp yarns. Figure 22–2 shows the high warp count, the warp surface, and a difference in color in a ribbed fabric.

Ribbed fabrics such as broadcloth look very much like percale. If the following technique of analysis is used, the difference between the two fabrics will become evident.

1. Use a 2-inch square of broadcloth and of percale.

2. *Ravel* adjacent sides of each fabric to make a ¼-inch fringe.

3. Observe the difference in the number of yarns in each fringe. Broadcloth will have a very thick fringe of warp yarns (144 × 76); whereas, in the percale fabric (78 × 78), the fringe of warp yarns will be about the same as the fringe of filling yarns.

Slippage is a problem in ribbed fabrics made with filament yarns, especially those of lower quality and lower count, as shown in Figure 22–4. This occurs at points of wear and tension, such as at seams and buttonholes. If the yarns are of different colors, as they are in iridescent taffeta with black warp and bright red filling, a bright red streak would show along a seam where slippage had occurred, and the main portion of the garment would remain black.

Wear occurs on the surface of the ribs. The warp yarns wear out first and splits occur in the fabric. The filling yarns, which are covered by the warp, are protected from wear.

Ribbed fabrics with fine ribs are softer and more drapable than comparable balanced fabrics—broadcloth is softer than percale. Those with large ribs have more body and less drapability and are good for garments where a bouffant look is desired. A few sheer rib fabrics are used in glass curtains.

Medium-Weight Ribbed Fabrics. Medium weight is the largest group of ribbed fabrics. *Broadcloth* has the finest rib of any of the staple-fiber fabrics because the warp and filling yarns are the same in size. The better qualities are made of long-staple combed cotton, plied yarns, and are usually mercerized for luster. They have a very silky appearance. The term "Pima Broad-

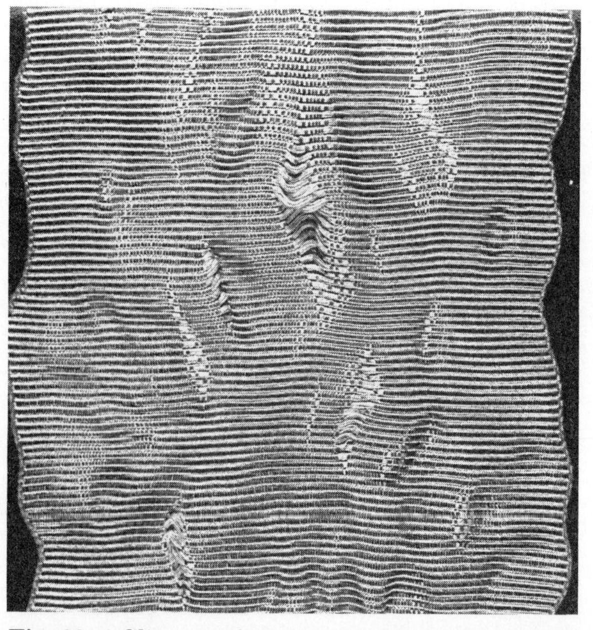

Fig. 22–4 *Slippage of yarns in a ribbed fabric.*

Medium-Weight Unbalanced Ribbed Fabrics

Fabric	Count*	Yarn Size	
		Warp	*Filling*
Staple Fiber			
Combed broadcloth	144 × 76	100/2	100/2
Carded broadcloth	100 × 60	40s	40s
Filament Fiber			
Rayon taffeta	60 × 15	10/2	3s
Acetate taffeta	140 × 64	75 denier	150 denier
Faille	200 × 64	75 denier	200 denier

*Counts may be high for polyester/cotton and durable-press fabrics.

cloth" on a label refers to the use of long-staple cotton fiber. Combed broadcloth will cost from two to four times as much as carded broadcloth. *Slub broadcloth* is made with a yarn that contains slubs at regular intervals. *Silk broadcloth* has filament warp and staple filling.

Taffeta is a fine-rib, filament-yarn fabric with crispness and body. *Moiré taffeta* has a watermarked, embossed design that is durable on acetate taffeta but temporary on rayon taffeta, unless it is resin treated (Figure 22–5).

Shantung has an irregular rib surface produced by long, irregular areas in the yarn. It may be made in medium or suiting weight and of various kinds of fiber.

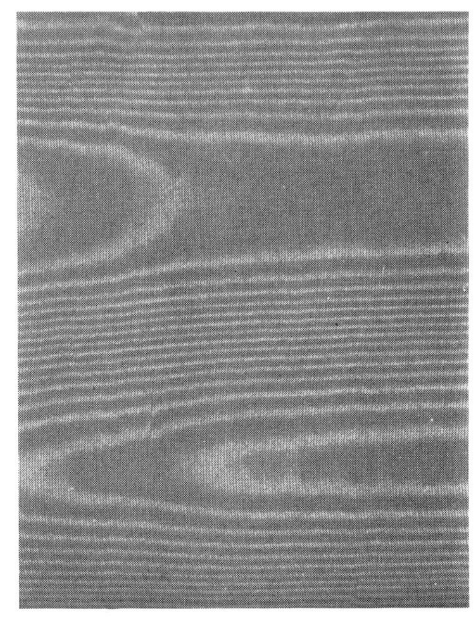

Fig. 22–5 *Moiré taffeta.*

Suiting-Weight Ribbed Fabrics. *Poplin* is similar to broadcloth, but the ribs are heavier because of larger filling yarns, and it is usually suiting weight. Polyester/cotton blends are widely used.

Faille (pronounced file) is made of filament-warp and staple-filling yarns. The filament yarns are usually acetate, rayon, polyester, or nylon.

Rep is a heavy, coarse fabric with a pronounced rib effect. *Bengaline* is similar to faille and often made with rayon warp and cotton filling. It is sometimes woven with two warps at a time to emphasize the rib. *Ottoman* has large-and-small ribs that are adjacent to each other, created by using filling yarns of different sizes or using different numbers of filling yarns in adjacent ribs. *Grosgrain* (pronounced grow'-grane) has a rounder rib than faille. Grosgrain ribbon may shrink as much as 2–4 inches per yard depending on its fiber content. It is often used at the button closure of sweaters and causes an unattractive pucker when it shrinks.

Bedford cord is seen occasionally in apparel fabrics, but more commonly in home-furnishing fabrics like summer-weight bedspreads. It has spun warp yarns that are larger than the filling yarns. Cords, located at intervals across the fabric are formed by extra filling yarn floating across the back giving a raised effect. Stuffer yarns are sometimes used to give a more pronounced cord. The lengthwise cords may be the same size, or alternately larger and smaller.

Suiting-Weight Unbalanced Ribbed Fabrics

Fabric	Count	Yarn Size	
		Warp	*Filling*
Rep	88 × 31	30/2	5s
Bengaline	92 × 40	150 denier	15s spun
Shantung	140 × 44	150 denier	30/2

PLAIN-WEAVE VARIATION: BASKET WEAVE

Basket weaves are made with two or more warps used as one, and with two or more fillings placed in the same shed. The most common basket weaves are 2 × 2 and 4 × 4, but variations of the basket weave include 2 × 1 and 2 × 3. A full basket would have the basket feature used in both warp and filling. A half basket would have the basket feature in only warp or filling. These fabrics have greater flexibility and more wrinkle resistance because there are few interlacings per square inch. The fabrics have a flatter appearance than a comparable plain-weave fabric would have. Long floats will snag easily. Figure 22–6 shows a 2 × 2 basket weave.

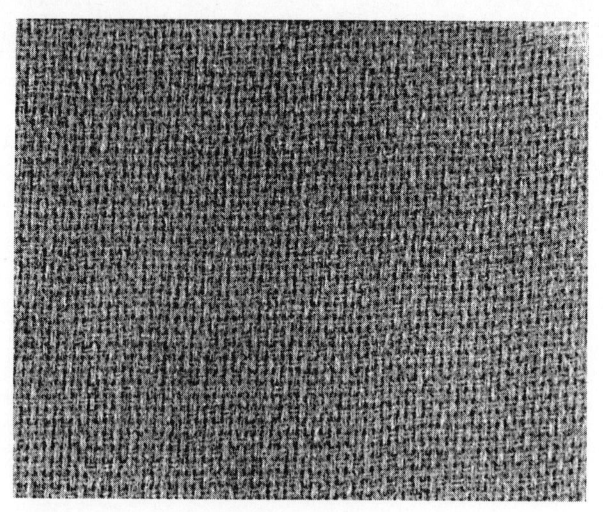

Fig. 22–6 *A 2 × 2 basket weave.*

Dimity is a sheer unbalanced fabric used for apparel and for window treatments. It has heavy-warp cords at intervals across the fabric. The cords may be formed by heavy yarns, or by grouping yarns together in that area. Dimity is often white or printed.

Oxford is usually a 2 × 1 or 3 × 2 basket weave. It is most common as a 2 × 1 half basket. It may have a yarn-dyed warp and white filling and be called an *oxford chambray.* Oxford looks-like a balanced fabric because the warp yarns are finer and have higher twist than the filling. Because of soft yarns and loose weave, yarn slippage is likely to occur at the seams and within the fabric itself. Loose-weave fabrics will snag and pill. Filling yarns have a little higher breaking strength than the warp. Oxford fabrics are soft, porous, and lustrous. Oxford is usually made of polyester/cotton today. Oxfords are medium-weight fabrics.

Most basket weaves are suiting-weight fabrics. The ones most commonly used in apparel include sailcloth, duck, or canvas. Sailcloth is lighter weight than duck or canvas. Sailcloth is usually a blend of polyester and cotton and is used in slacks, skirts, and summer-weight suits. Duck is usually 100 percent cotton. Home-furnishing and industrial variations of these fabrics come in many different weights and range from soft to stiff in drape. They are usually 2 × 1 or 3 × 2 basket weaves that are used for slipcovers, boat covers, shoe fabrics, and house and store awnings.

Hopsacking is an open basket-weave fabric made of cotton, linen, or wool. It is primarily used for coats and suits. It gets its name from the sacks used to gather hops.

Monk's cloth, friar's cloth, druid's cloth, and mission cloth are some of the oldest full-basket-weave fabrics. They are usually brownish white or oatmeal color. These fabrics are usually 2 × 2, 3 × 3, 4 × 4, or 6 × 6. They are used primarily in home furnishings.

23

Twill- and Satin-Weave Fabrics

Twill- and satin-weave fabrics are more complex and require a greater number of harnesses on the loom than plain-weave fabrics.

Twill Weave

Twill weave is one in which each warp or filling yarn floats across two or more filling or warp yarns with a progression of interlacings by one to the right or left, forming a distinct diagonal line, or *wale*. A *float* is the portion of a yarn that crosses over two or more yarns from the opposite direction. Twill weaves vary in the number of harnesses used. The simplest twill requires three harnesses. The more-complex twills may have as many as 18 filling yarns inserted before repeating a shed and are woven on a loom with a dobby attachment. Twill weave is the second basic weave that can be made on the simple loom.

Twill weave is often designated by a fraction—such as $\frac{2}{1}$—in which the numerator indicates the number of harnesses that are raised and the denominator indicates the number of harnesses that are lowered when a filling yarn is inserted. The fraction $\frac{2}{1}$ would be read as "two up, one down." The number of harnesses needed to produce a simple twill can be determined by adding the numerator and the denominator. A $\frac{2}{1}$ twill is shown in Figure 23–1. The floats on the surface are warp yarns, making it a warp surface; it is classified as a warp-faced twill.

Fig. 23–1 *A $\frac{2}{1}$ twill weave.*

CHARACTERISTICS

Twill fabrics have a technical face and a technical back. If there are warp floats on the technical face, there will be filling floats on the technical back. If the twill wale goes up to the right on one side, it will go up to the left on the other side. Twill fabrics have no up and down as they are woven. Check this fact by turning the fabric end to end and then examining the direction of the twill wale.

Sheer fabrics are seldom made with a twill weave. Printed designs are seldom used, except in silk and lightweight twills, because a twill surface has interesting texture and design. Soil shows less on the uneven surface of twills than it does on smooth surfaces such as plain weaves.

Fewer interlacings permit the yarns to move more freely and give the fabric more softness, pliability, and wrinkle recovery than a comparable plain-weave fabric. When there are fewer interlacings, yarns can be packed closer together to produce a higher-count fabric (Figure 23–2). If a plain-weave fabric and a twill-weave fabric had the same kind and number of yarns, the plain-weave fabric would be stronger because it has more interlacings. It is, however, possible to crowd more yarns into the same space in twill, in which case the twill fabric would be stronger.

The prominence of a twill wale may be increased by the use of long floats, combed yarns, plied yarns, hard-twist yarns, twist yarns opposite to the direction of the twill line, and high counts. Fabrics with prominent wales, such as gabardine, may become shiny because of flattening caused by pressure and wear. If the ridges have been flattened by pressure, steaming will raise them to remove the shine. Pure white vinegar (5 percent) or sandpaper may be used to remove shine caused by pressure or wear. Dip a piece of terrycloth in the vinegar, wring it out, and rub hard and fast in both directions of the cloth in the shiny area. As the cloth dries, the vinegar odor will disappear.

The direction of the twill wale usually goes from lower left to upper right in wool and wool-

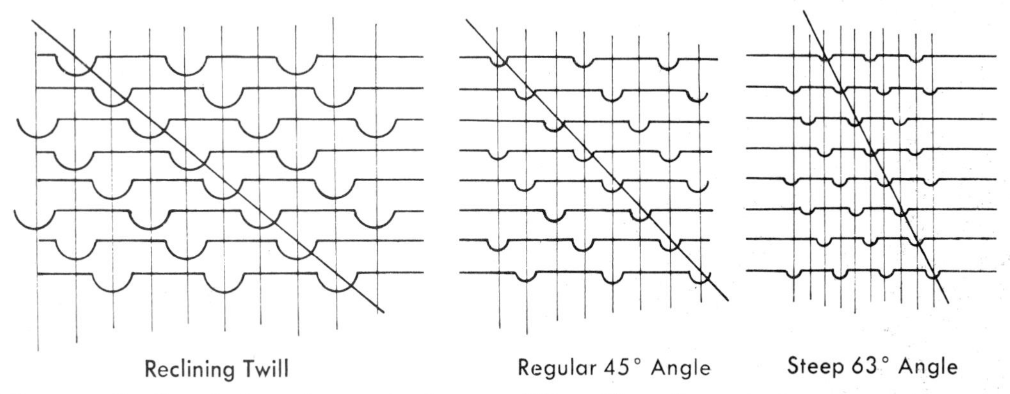

Fig. 23–2 *Twill angle depends on ratio of warp to filling. These diagrams show left-handed twills.*

like fabrics—right-hand twills—and from lower right to upper left in cotton or cotton-like fabrics—left-hand twills. This fact is important only in deciding which is the right and wrong side of a twill fabric. In some fabrics that have a very prominent wale or are made with white and colored yarns, the two lapels of a coat or suit will not look the same (Figure 23–3). This cannot be avoided, and, if it is disturbing, a garment of a different design or a different fabric should be chosen.

The degree of angle of the wale depends on the balance of the cloth. The twill line may be steep, regular, or reclining. The greater the difference between the number of warp and filling yarns, the steeper the twill line. Steep-twill fabrics have a high warp count and therefore are stronger in the warp direction. The importance of the angle is that it serves as a guide in determining the strength of a fabric. Figure 23–2 is a diagram that shows how the twill line changes in steepness when the number of warp yarns changes and the filling yarns remain the same in number.

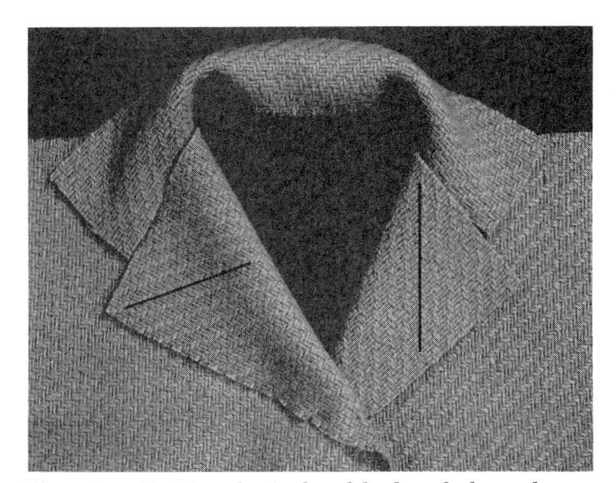

Fig. 23–3 *Twill wales in lapel look unbalanced.*

Filling-faced twills are not discussed in this text because they are seldom used. They are usually reclining twills and are less durable than the others.

EVEN-SIDED TWILLS

Even-sided twills have the same amount of warp and filling yarn exposed on both sides of the fabric. They are sometimes called *reversible twills* because they look alike on both sides, although the direction of the twill line differs. Better-quality filling yarns must be used in these fabrics than in the warp-faced twills because both sets of yarn are exposed to wear. They are $\frac{2}{2}$ twills and have the best balance of all the twill weaves (see the following chart).

Surah is a printed filament-twill fabric of $\frac{2}{2}$ construction that is used in silk-like dresses, linings, ties, and scarves. It is soft, smooth, and lightweight. Sometimes it is piece dyed instead of printed.

Viyella® is a medium-weight twill-weave fabric made of an intimate blend of 55 percent wool and 45 percent cotton. It is manufactured by William Hollins & Co., Ltd., of London, and the fabric name is a trademark of that company. Viyella® has the appearance of a very fine flannel and is most frequently seen in small prints or plaids. It was developed to be a washable fabric and is used for children's wear, sleepwear, shirts, and dresses. Because of its cost in the United States after importing, Viyella® is seen

Even-Sided Twills

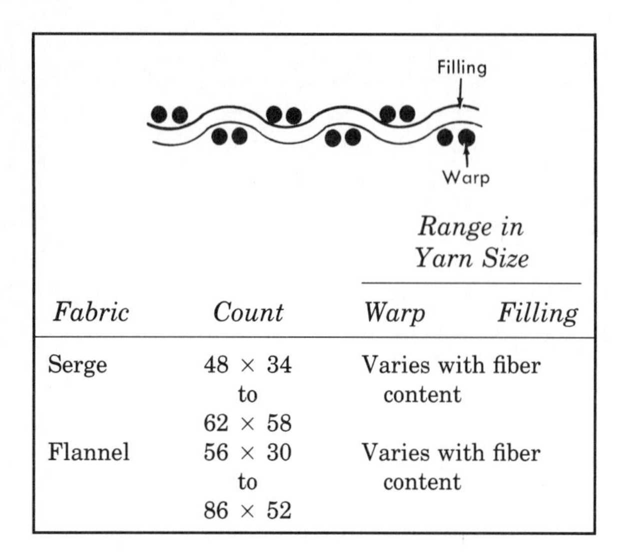

Fabric	Count	Range in Yarn Size: Warp	Range in Yarn Size: Filling
Serge	48 × 34 to 62 × 58	Varies with fiber content	
Flannel	56 × 30 to 86 × 52	Varies with fiber content	

in better clothing for men and women and the label may state it should be dry cleaned.

Serge is a $\frac{2}{2}$ twill with a rather subdued wale and a clear finish. Cotton serge of fine yarn and high count is often given a water-repellent finish and used for jackets, snowsuits, and raincoats. Heavy-yarn cotton serge is used for work pants. Wool serge gets shiny from abrasion and repeated pressing but is not subject to flattening of the wale.

Twill flannel is a $\frac{2}{2}$ or $\frac{2}{1}$ twill. The filling yarns are low-twist, larger yarns, specially made for napping. They may be either woolen or worsted. Worsted flannels have less nap, will take and hold a sharp crease better, and are less apt to show wear or get baggy, as compared with woolen flannels.

Sharkskin is a $\frac{2}{2}$ twill with a sleek appearance. It has a small-step pattern because yarns in both the warp and the filling alternate one white yarn with one colored yarn. Sharkskin is used primarily for men's slacks and suits.

Herringbone fabrics have the twill line reversed at regular intervals to give a design that resembles the backbone of a fish (Figure 23–4). Usually two different color yarns are used to accentuate the pattern.

Another woven-in pattern is called *houndstooth*. This $\frac{2}{2}$ twill fabric is basically a check, but is unique in appearance because it is pointed—rather like an eight-point star. A photo of this fabric is shown in Figure 33–13.

Fig. 23–4 *Herringbone.*

WARP-FACED TWILLS

Warp-faced twills have a predominance of warp yarns on the right side of the cloth. Since warp yarns are made with higher twist, these fabrics are stronger and more resistant to abrasion.

Warp-Faced Twills

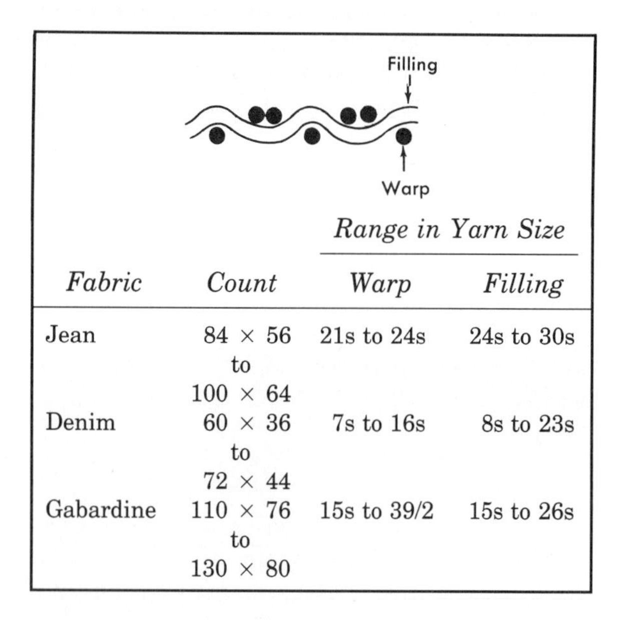

Fabric	Count	Range in Yarn Size: Warp	Range in Yarn Size: Filling
Jean	84 × 56 to 100 × 64	21s to 24s	24s to 30s
Denim	60 × 36 to 72 × 44	7s to 16s	8s to 23s
Gabardine	110 × 76 to 130 × 80	15s to 39/2	15s to 26s

Lining twill is a medium-weight $\frac{2}{1}$ fabric made from filament yarns and usually piece dyed or printed in a small pattern. It is similar to surah in appearance and use.

Denim was traditionally a yarn-dyed cotton twill made in two weights: for sportswear and for overalls. Its use in jeans has increased tremendously, and the nature of denim has also changed. It is often napped, printed, and made with stretch yarns.

Jean is a piece dyed or printed medium-weight twill used for children's playclothes, draperies, slipcovers, and work shirts. Jean is not heavy enough for work pants.

Drill is a strong, medium- to heavy-weight twill fabric. It is usually a $\frac{2}{1}$ twill that is piece dyed. It is also constructed in a $\frac{3}{1}$ twill. It is usually seen in work clothing and industrial fabrics.

Covert is a twill fabric with a mottled appearance resulting from two colors of fibers being used in the yarns or from two colors of plies twisted together to form the yarns. It is usually a $\frac{2}{1}$ twill in suiting weight.

Chino is a hard-wearing steep-twill fabric of cotton or a blend of cotton/polyester. It has a slight sheen and is frequently made from combed yarns. Usually two-ply yarns are used in both the warp and the filling directions. It is frequently vat-dyed khaki and is typically a summer-weight military-uniform fabric.

Gabardine is a warp-faced steep twill with a very prominent, distinct wale. It has a 63° angle or greater and always has many more warp than filling. Cotton gabardine is made with 11, 13, or 15 harnesses. Rayon and wool gabardine are sometimes made with a three-harness arrangement in which the warp yarns are crowded close together, giving a steep twill.

Calvary twill also has a pronounced steep-twill line. Its unique characteristic is that it has a double-twill line; that is, two diagonal lines that are very close together separated by a little space from the next pair of diagonal lines.

Satin Weave

Satin weave is one in which each warp yarn floats over four filling yarns ($\frac{4}{1}$) and interlaces with the fifth filling yarn, with a progression of interlacings by two to the right or the left (Figure 23–5). Or each filling yarn floats

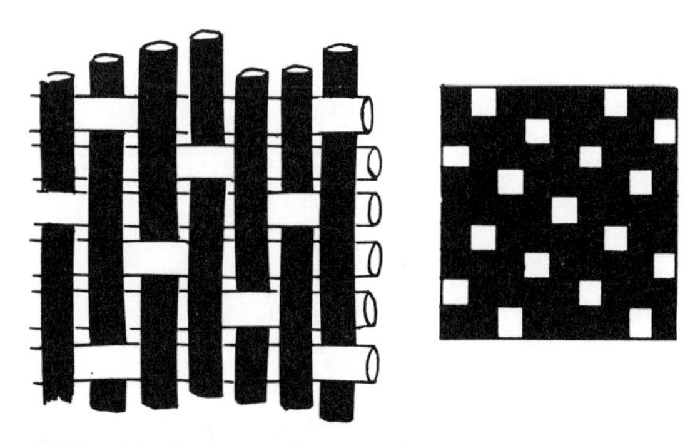

Fig. 23–5 *Warp-faced satin weave:* $\frac{4}{1}$ *yarn arrangement.*

over four warps and interlaces with the fifth warp ($\frac{1}{4}$) with a progression of interlacings by two to the right or left (Figure 23–6). In certain fabrics, each yarn floats across seven yarns and interlaces with the eighth yarn. Satin weave is the third basic weave that can be made on the simple loom, and the basic fabrics made with this weave are *satin* and *sateen.*

Satin-weave fabrics are characterized by luster because of the long floats that cover the surface. Note in the checkerboard designs that (1) there are few interlacings—thus the yarns can be packed close together to produce a very-high-count fabric; and (2) no two interlacings are adjacent to one another, so no twill effect results from the progression of interlacings unless the count is low.

When warp yarns cover the surface, the fabric is a warp-faced fabric—satin—and the warp count is high. When filling floats cover the surface, the fabric is a filling-faced fabric—sateen—and the filling count is high. These fabrics are therefore unbalanced, but the high count compensates for the lack of balance.

All these fabrics have a right and wrong side. A high count gives them strength, durability,

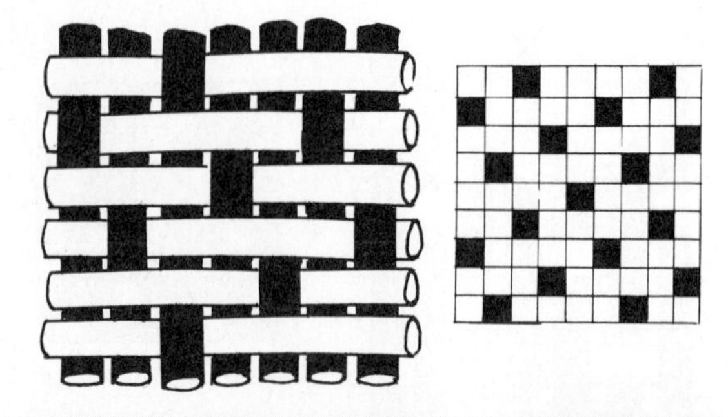

Fig. 23–6 Filling-faced satin weave: $\frac{1}{4}$ yarn arrangement.

body, firmness, and wind repellency. Fewer interlacings give pliability and resistance to wrinkling but may permit yarn slippage and raveling.

SATIN FABRICS

Satin fabrics are usually made of bright filament yarns with very low twist. Warp floats almost completely cover the surface. Because of the bright fibers, low twist, and long floats, satin is one of the most lustrous fabrics made. It is made in many weights (see the following table) for use in dresses, linings, lingerie, draperies, and upholstery. It seldom has printed designs. It is especially good for linings because the high count makes it very durable and the smooth surface makes the garment easy to slip on and off. Satin makes a more-pliable lining than taffeta and thus does not split as readily at the hem edges of coats and suits. Quality is particularly important in linings. The higher the count, the better the quality. Low-count satins will pull at the seams and rough up in wear, and the floats will shift in position to make bubbly areas and wrinkled effects on the surface of the cloth. Satin upholstery should be applied so that one "sits with" the floats.

Typical Satin Fabrics

		Kind of Yarn	
Fabric	*Count*	*Warp*	*Filling*
Satin	200 × 65	100-denier acetate	100-denier acetate
Slipper satin	300 × 74	75-denier acetate	300-denier acetate
Crepe-backed satin	128 × 68	100-denier acetate	100-denier rayon crepe

In crepe-back satin, crepe yarns are used in the filling and the low-twist warp floats give the smooth satiny surface to the fabric. The crepe yarns give softness and drapability.

Care of satin fabrics should be directed toward maintenance of luster and the prevention of distortion of the floats. Wash or dry clean as indicated by the fiber content, and press on the wrong side or iron with the direction of the floats.

SATEEN

Sateen is a lustrous fabric made of spun yarns. In order to achieve luster with staple fibers, low twist must be used in the yarns forming the float surface. These yarns are the filling yarns because, if warp yarns were made with twist low enough to produce luster, the yarns would not be strong enough to resist the tensions of weaving. A resin finish is used also on the woven cloth to enhance the luster and make it durable.

Filling sateen is a smooth, lustrous cotton fabric used for draperies, drapery linings, and dress fabrics. It is often made with carded yarns with a high filling count. Yarns are similar in size to those used in print cloth, but the filling yarns have a low twist and are larger in size than the warp yarns. (This factor can be used to help identify the warp and filling direction of the fabric.)

Luster is obtained by the Schreiner finish (see Chapter 33). *Schreinering* is a mechanical finish in which fine lines, visible only under a hand lens, are embossed on the surface. Unless a resin finish is applied at the same time, the finish is

Typical Sateen Fabrics

Fabric	*Count*	*Kind of Yarn*	
		Warp	*Filling*
Filling sateen	60 × 104 carded	32s	38s
	84 × 136 carded	40s	50s
	96 × 108 combed	40s	60s
Warp sateen	84 × 64 carded	12s	11s
	160 × 96 carded	52s	44s

only temporary. Combed sateens are usually mercerized as well as Schreinered. Carded sateens are Schreinered only, because short fibers are used; mercerization would not produce enough luster to justify the cost.

Warp sateens are cotton fabrics made with warp floats in $\frac{4}{1}$ interlacing pattern. They have a rounded wale effect that makes them resemble a twill fabric. They are stronger and heavier than filling sateens because of the high-warp count. They are less lustrous than filling sateen and used where durability is more important than luster. Large amounts of warp sateens are used in slacks, skirts, pillow and bed tickings, draperies, and upholstery fabrics. Warp sateens are the most likely type of satin fabrics to be printed.

24

Pile Fabrics

Pile fabrics are three-dimensional fabrics that have fibers from a sliver or yarns forming a dense cover of the ground fabric. Pile fabrics can be both functional and beautiful.

- A high pile is used to give warmth as either the shell or the lining of coats, jackets, gloves, and boots.
- High-count fabrics give durability and beauty in carpets, upholstery, and bedspreads.
- Low-twist yarns give absorbency in towels and washcloths.
- Other uses for pile fabrics are stuffed toys, wigs, paint rollers, buffing and polishing cloths, and decubicare pads for bed-ridden patients.

Interesting effects can be achieved by combinations of the following:

- Cut and uncut pile (Figure 24–1).
- Pile of various heights.
- High- and low-twist yarns.
- Areas of pile on a flat surface.
- Curling and crushing or forcing pile into a position other than upright.

In pile fabrics, the pile receives the surface abrasion and the base weave receives the stress. It is necessary to have a durable base structure in order to have a satisfactory pile fabric. A tight ground or base weave increases the resistance of a looped pile to snagging and of a cut pile to shedding and pulling out. A dense pile will stand erect, resist crushing, and give better cover. Care must be taken in cleaning and pressing to keep the pile erect. Cut pile usually looks better if dry cleaned, but some pile fabrics—such as pinwale corduroy—can be washed if the laundry procedures are suited to the fiber content. All pile fabrics are softer and less wrinkled if tumble dried or line dried on a breezy day. Pressing should be done with a minimum of pressure—or none at all. Flattening of the pile causes the fabric to appear lighter in color. For fabrics such as velvet and velveteen, if pressing is done at home, a "needle" board should be used. Steaming and pressing lightly on a terry towel is a less-desirable alternative when a needle board is not available.

Many pile fabrics are pressed during finishing so that the pile slants in one direction, giving an up and down (Figure 24–2). Garments cut with the pile directed up wear better, give a richer color, and prevent garments from "working up" under a coat or jacket. However, it is more important that the pile be directed the same way in all pieces of a garment. Otherwise, light will be reflected differently and the garment will appear to be made of two colors. Direction of the pile can be determined by running the hand over the fabric.

The Comparison of Pile Fabrics chart gives the methods of making pile fabrics.

Loop Pile

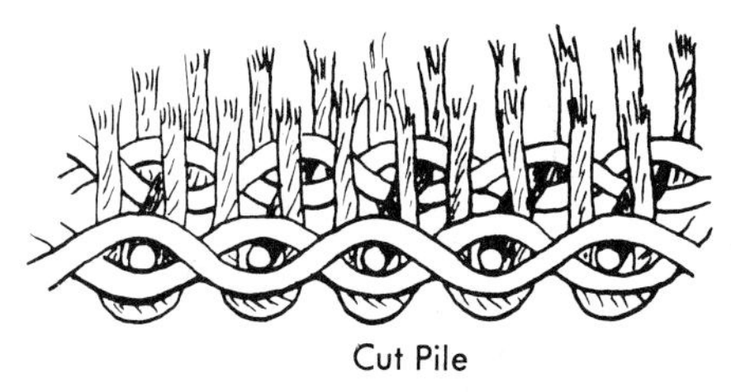

Cut Pile

Fig. 24–1 *Cut pile and loop pile: woven fabric.*

Woven-Pile Fabrics

Woven-pile fabrics are three-dimensional structures made by weaving an extra set of warp or filling yarns into the ground yarns to make loops or cut ends on the surface (Figure 24–1). The

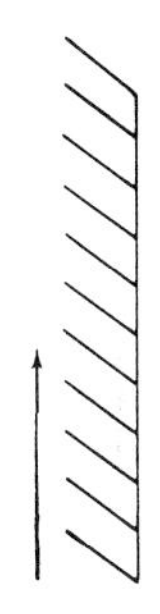

Fig. 24–2 *Pile should be directed up.*

Comparison of Pile Fabrics

Method	Types and Kinds	Fabrics—End Uses	Identification
Weaving	1. Filling floats cut and brushed-up	1. Velveteen, corduroy	Filling pile around warp
	2. Made as double cloth and cut apart	2. Velvet, velour	Warp pile around filling
	3. Over wires	3. Friezé, Wilton and velvet carpets	
	4. Slack tension	4. Terrycloth, friezé	
Knitting	Filling knit: laid-in yarn Sliver knit Warp knit: laid-in yarn or pile loops	Velour, terry, fake fur	Stretchy—Rows of knit stitches on wrong side More stable
Tufting	Yarns punched into substrate	Rugs and carpets Robes, bedspreads Upholstery Fake furs	Rows of stitches (like machine stitches) on wrong side
Flocking	Fibers anchored to substrate	Blankets, jackets Designs on outerwear	Fiber surface rather stiff
Chenille yarns	Pile-type yarns made by weaving Chenille yarns woven or knitted into fabric	Upholstery Outerwear fabric	Ravel adjacent yarns and examine novelty yarn

pile is usually ½ inch or less in height. Woven pile is less pliable than knitted or tufted pile and sometimes when the fabric is folded, the rows of tufts permit the back to show, or "grin through."

FILLING-PILE FABRICS

The pile in *filling-pile fabrics* is made by cutting floats on the surface after weaving (Figure 24–3). Two sets of filling yarns and one set of warp are used. The extra filling yarns float across the ground yarns in weaving. In *corduroy,* the floats are arranged in lengthwise rows; in *velveteen,* they are scattered over the base fabric.

Cutting is done by a special machine consisting of guides that lift the individual floating yarns from the ground fabric and revolving knives that cut the floats (Figure 24–4). A gray-goods corduroy with some of the floats cut is shown in Figure 24–5. When wide-wale corduroy is cut, the guides and knives can be set to cut all the floats in one operation. For pin-wale corduroy and velveteen, alternate rows are cut and the cloth must be run through the machine twice.

After cutting, the surface is brushed crosswise and lengthwise to open and raise the pile and intermesh the cut yarns from separate floats. It is then singed and waxed. The final pressing lays the pile at a slight angle in one

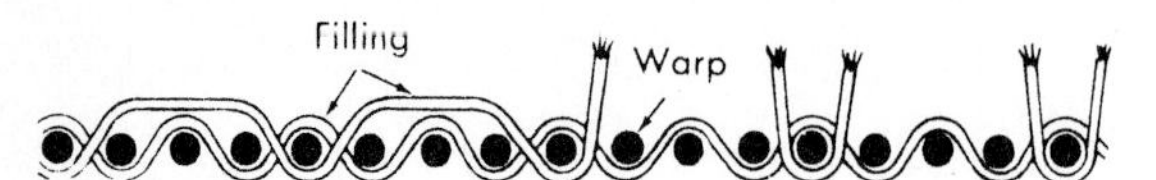

Fig. 24–3 Filling pile. Cross-section of weave in corduroy.

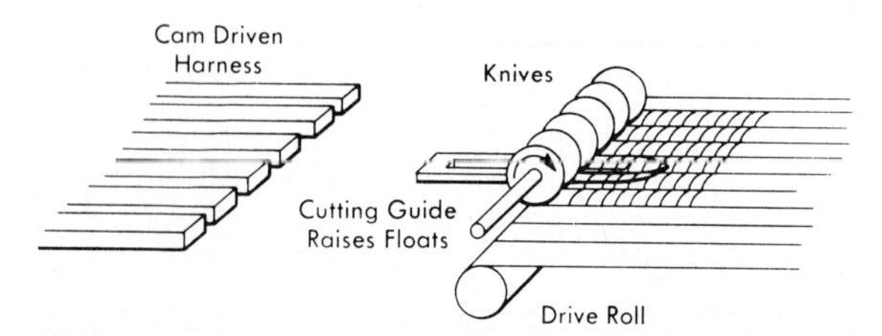

Fig. 24–4 Knives for cutting floats to make corduroy.

Fig. 24–5 Corduroy gray goods showing some of the floats cut.

direction, thus giving the up and down to the pile. The back of both velveteen and corduroy is given a slight nap. In no-wale corduroy, the evidence of wales is nearly eliminated by napping and shearing.

Both velveteen and corduroy are made with long-staple, combed, mercerized cotton for the pile. In good-quality fabrics, long-staple cotton is used for the ground as well. Polyester/cotton corduroy became available in 1976 with polyester in the ground warp yarns for strength. The ground may be made with plain- or twill-weave interlacing patterns. With the twill pattern, it is possible to have a higher count and therefore a denser pile. Corduroy can be recognized by lengthwise wales. It is warm, durable, washable, and, if tumble-dried, needs no ironing. Velveteen has more body and less drapability than velvet. The pile is not over ⅛ inch high. Both corduroy and velveteen are commonly seen as solid-color and printed fabrics.

WARP-PILE FABRICS

Warp-pile fabrics are made with two sets of warp yarns and one set of filling. The extra set of warp makes the pile. Several methods are used.

Double-Cloth Method. Two fabrics are woven, one above the other, with the extra set of yarns interlacing with both fabrics. There are two sheds, one above the other, and two shuttles are thrown with each pick. The fabrics are cut apart while still on the loom by a traveling knife that passes back and forth across the loom. With the *double-cloth method* of weaving, the depth of the pile is determined by the space between the two fabrics (Figure 24–6).

Velvet was originally made of silk and was a compact, heavy fabric. Today velvet is made of rayon, nylon, or silk filaments with a pile 1⁄16 inch high or shorter. Velvet is not wound on bolts, as are other fabrics, but is attached to hooks at the top and bottom of a special bolt so that there are no folds or creases in the fabric.

Velvet and velveteen, the hard-to-tell-apart fabrics, can often be distinguished by fiber content, since velvet is usually made with filaments and velveteen with staple. To tell warp directions in these fabrics, ravel adjacent sides. In velvet, the tufts will be interlaced with a filling yarn (Figure 24–7). Another way to tell warp direction is to bend the fabric. In velveteen, the pile "breaks" into lengthwise rows because the filling tufts are around the warp yarns. In velvet, the pile breaks in crosswise rows because the warp tufts are around the ground-filling yarns. This technique works best with medium- to poor-quality fabrics. To distinguish cotton

Kinds of Corduroy

	Wales per Inch	*Ounces per Yard*	*Characteristics*
Feather wale	18–19	5 ±	Shallow pile, flexible
Pin wale	14–16	7 ±	Shallow pile, flexible
Midwale	11	10 ±	Men's and women's outerwear
Wide wale	3–9	12 ±	Coats, most durable corduroy made

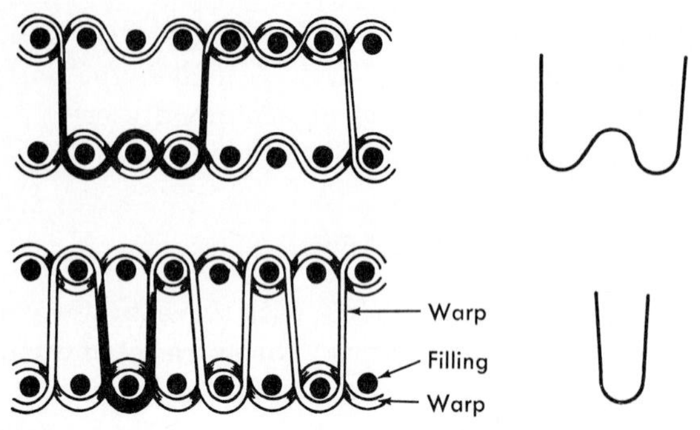

Fig. 24–6 Warp pile: double-cloth method. W-interlacing (above); *V-interlacing* (below).

velvet from velveteen, pull the fabric to determine the filling direction that will have more stretch and then ravel adjacent sides.

Crushed velvet is made by mechanically twisting the wet cloth. The surface yarns are randomly flattened in different directions. The fabric has a more-casual look than regular velvet.

Panné velvet is an elegant fabric that has had the pile pressed flat, by heavy pressure, in one direction to give high luster. It is definitely not an easy-care fabric. If the pile is disturbed or brushed in the other direction, the smooth, lustrous look is destroyed.

Velour is a warp-pile cotton fabric used primarily for upholstery and draperies. It has a deeper pile than velveteen and is heavier in weight. (Velour can also be made by knitting.)

Plush may be cotton, wool, or rayon. It has a deeper pile than velour or velvet usually longer than 1/4 inch.

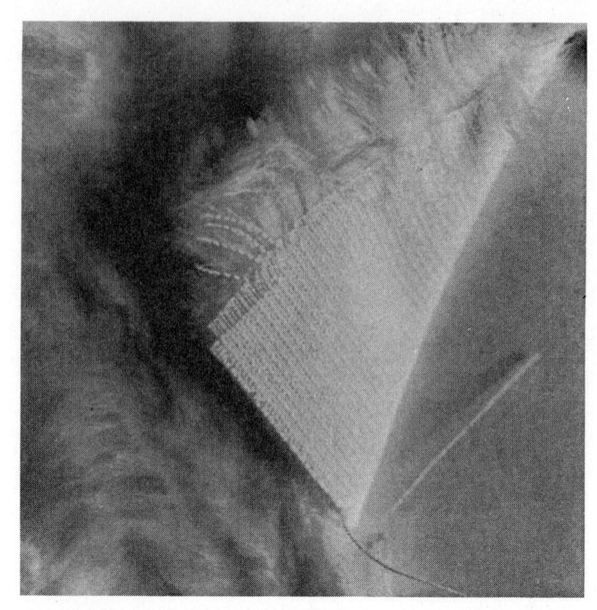

Fig. 24–8 Woven fur-like fabric.

Fur-like fabrics may be finished by curling, shearing, sculpturing, or printing to resemble different kinds of real fur (see Figure 24–8). (Most fur-like fabrics are made by knitting; see Figure 28–14.)

Over-Wire Method. In the *over-wire method,* a single cloth is woven with wires placed across the width of the loom over the ground warp and under the pile warp. Each wire has a knife edge

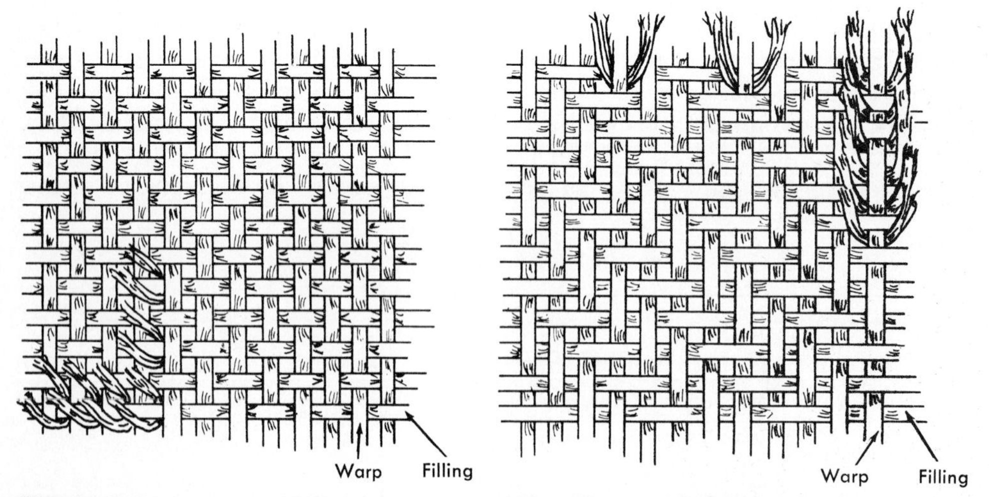

Fig. 24–7 Comparison of filling pile and warp pile. Velvet: warp-pile yarn is around ground filling (left); *velveteen: filling-pile yarn is around ground warp* (right).

Fig. 24–9 Friezé is woven over wires.

which cuts all the yarns looped over it as it is withdrawn. Uncut pile can be made over wires without knives or over waste picks of filling yarns. The wires are removed before the cloth is off the loom, and the waste picks are removed after the fabric is off the loom. Friezé and mohair-pile plush are made in this way. Most woven carpets are made over wires.

Friezé, an uncut pile fabric, is an upholstery fabric usually made of mohair, nylon, or cotton with a cotton back. Durability of friezé depends on the closeness of the weave (Figure 24–9).

Velvet can also be made by the over-wire method. Complex patterns using different-color yarns and loops combined with cut pile result in a wide variety of fabrics.

Slack-Tension Pile Method. The pile in terrycloth is formed by a special weaving arrangement in which three picks are put through and beaten up with one motion of the reed. After the second pick in a set is inserted, there is a let-off motion that causes the yarns on the warp-pile beam to slacken, while the yarns on the ground-pile beam are held at tension. The third pick is inserted, and the reed moves forward all the way and all three picks are beaten up firmly into the fell of the cloth (Figure 24–10). These picks move along the ground warp and push the pile-warp yarns into loops. The loops can be on one side only or on both sides. The height of the loops is determined by the distance the first two picks are left back from the fell of the cloth.

Terrycloth is highly absorbent cotton or cotton/polyester fabric used for bath towels, beach robes, and sportswear. Each loop acts as a tiny sponge. When the loops are sheared and the surface is brushed to loosen and intermesh the fibers of adjacent yarns, the surface becomes more

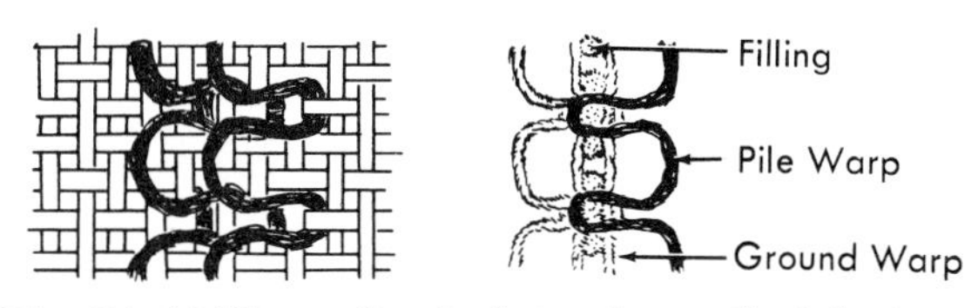

Fig. 24–10 Warp pile: slack-tension method for terrycloth.

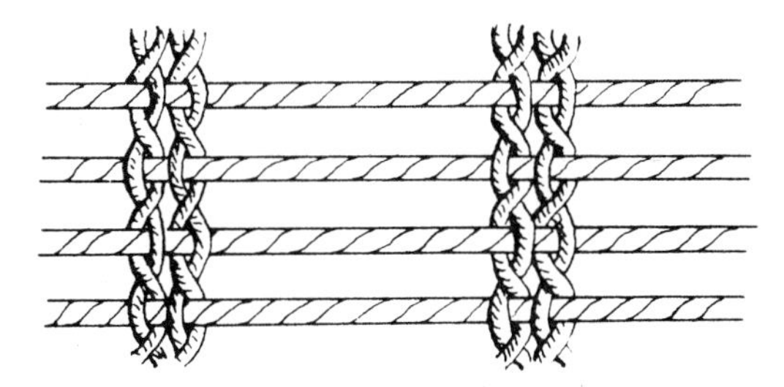

Fig. 24–11 Fabric from which chenille yarn is cut.

compact and less porous and therefore less absorbent than the loop-pile terry. Cotton/polyester terry towels have blended ground yarns and cotton pile, the pile yarns for absorbency and the polyester ground yarns for strength and durability—especially in selvages.

There is no up and down in terrycloth unless the cloth is printed. Some friezés are made by the terrycloth method. *Shagbark,* which has spaced rows of loops, is also made this way.

Chenille-Yarn Pile Fabrics

Chenille yarn is made by cutting a specially woven ladder-like fabric into warpwise strips (Figures 24–11 and 24–12). The cut ends of the softly twisted yarns loosen and form a pile-like fringe. This fringed yarn may be woven to make a fabric with pile on one side or on both sides.

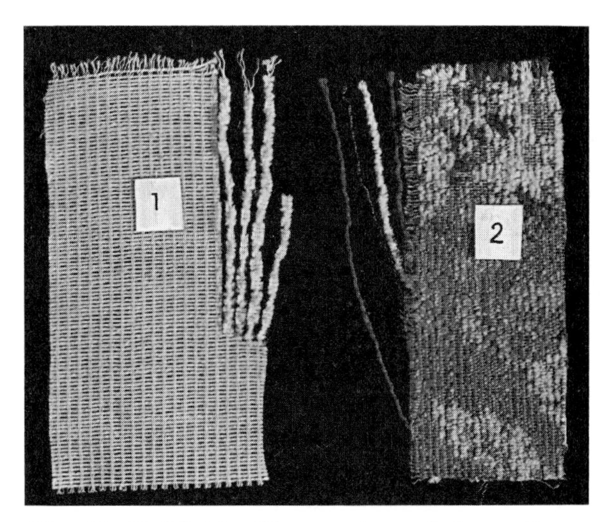

Fig. 24–12 (1) Chenille yarn is made by cutting a specially woven fabric into strips. (2) Fabric made from chenille yarn.

If the pile is on one side only, the yarn must be folded before it is woven. The yarn is sometimes referred to as a "caterpillar" yarn. Chenille-type yarns can be made by flocking. Other chenille-type yarns are novelty yarns made by twisting. As the effect yarn is wound around a core yarn and secured by a binder yarn, it is cut at the same time. Chenille yarns are mainly used in home furnishings and apparel.

Tufted-Pile Fabrics

Tufting is a process of making pile fabrics by punching extra yarns into fabric. The ground fabric ranges from thin sheeting to heavy burlap, and may be woven, knitted, or a fiber web. The pile yarns can be of any fiber content. Textured-filament nylon yarns gave great impetus to the tufting industry. Tufting developed in the southeastern United States as a hand craft. It is said that the early settlers used candle wicks and carefully worked them into bedspreads to create interesting textures and designs, and the making of candlewick bedspreads grew into a cottage industry. Hooked rugs were also made by hand in the same way. In the 1930s, machinery was developed to convert the hand technique to mass production (see Figure 24–13). Cotton rugs, bedspreads, and robes were produced in many patterns and colors at low cost.

Tufting is done by a series of needles (see Figures 24–14 and 24–15), each carrying a yarn from a series of spools held in a creel. The backing fabric is held in a horizontal position, the needles all come down at once and go through the fabric at a predetermined distance, much as a sewing-machine needle goes through cloth. Under each needle is a hook that moves forward to hold the loop as the needle is retracted. For cut-loop pile, a knife is attached to the hook and it moves forward as the needles are retracted to cut the loop. The fabric moves forward at a predetermined rate, and the needles move downward again to form another row of tufts.

The yarns must be opened and the fibers teased from the yarn. The tufts are held in place by the "blooming" (untwisting) of the yarn, by

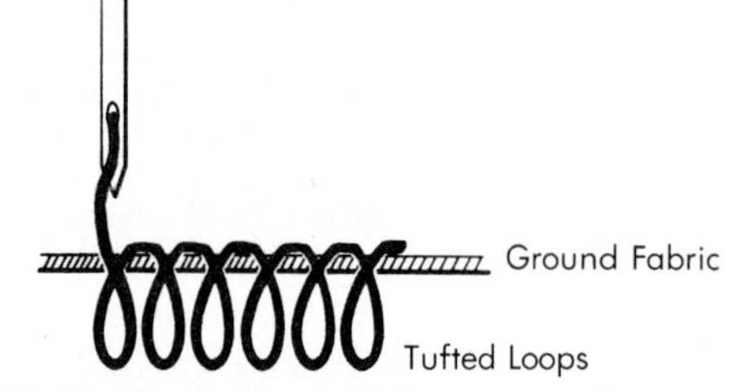

Fig. 24–13 *How tufting stitches are made.*

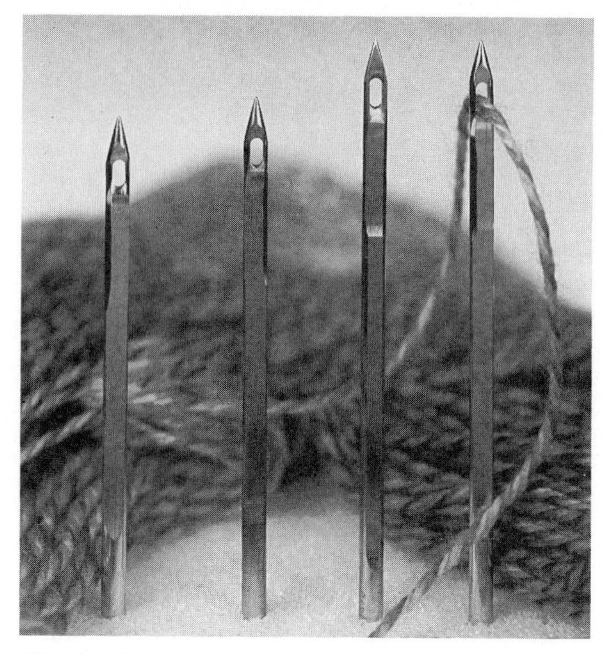

Fig. 24–14 *Tufting needles and yarn.*

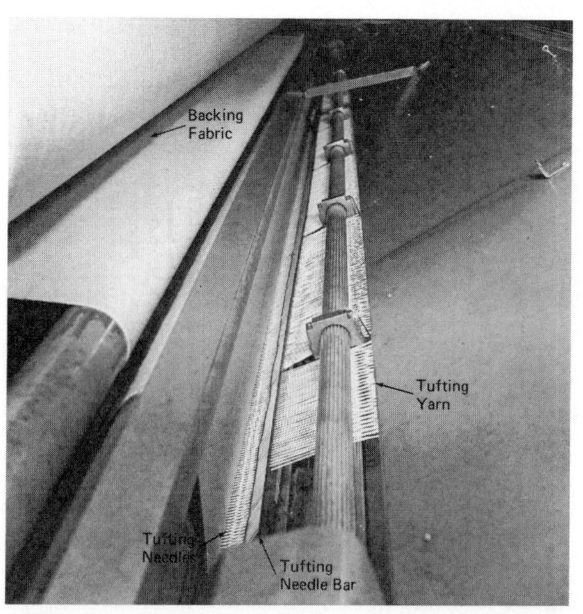

Fig. 24–15 *Needle area of tufting machine.*

shrinkage of the ground fabric in finishing, and frequently by using a coating on the back of the ground fabric.

Tufting is a less-costly method of making pile fabrics because it is an extremely fast process and involves less labor and time to create new designs. Tufted apparel fabrics are made on 5/64-gauge machines. This gauge is the distance in inches between the tufting needles. Normal tufting specifications on this gauge call for 10–11 stitches per inch and a pile height of 1/8 inch. Tufted fabrics with a woven base are usually 1/2 inch or less in pile height. Fur-like tufted fabrics are used for shells or linings of coats and jackets (Figure 24–16). There are few other tufted-apparel fabrics at present.

Tufted bed-size blankets can be made in two minutes. In 1960, Barwich Mills, Inc., started pilot-plant operations to make tufted blankets. This end use combines pile construction with napped finish. It has the advantage over traditional blanket fabric of maintaining a strong, firm ground fabric because the fibers are teased from the pile yarns to create the nap. Also, the thickness of the blanket is determined by the height of the pile rather than by the thickness of the yarn. Tufted blankets have not been successful in the United States but are being produced in Europe.

Tufted upholstery fabric is made in both cut and uncut pile. It is widely available now. The back is coated to hold the yarns in place.

Carpeting of room-width size was first made by tufting in 1950; in 1984, 95 percent of broadloom carpeting was made by tufting. A tufting machine can produce approximately 645 square yards of carpeting per hour compared to an Axminster loom, which can weave about 14 square yards per hour. Variations in texture can be made by loops of different heights. Cut and uncut tufts can be combined. Tweed textures are made by the use of different-colored plies in the tufting yarns. New techniques of dyeing have been developed to produce colored patterns or figures in which the color penetrates the tufts completely. In carpeting, a latex coating is put on the back to help hold the tufts in place (Figure 24–17).

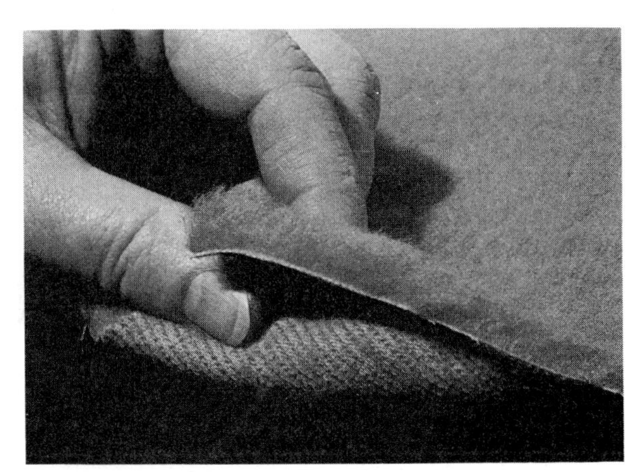

Fig. 24–16 *Cut-pile tufted fabric. Notice machine-like stitches on wrong side.*

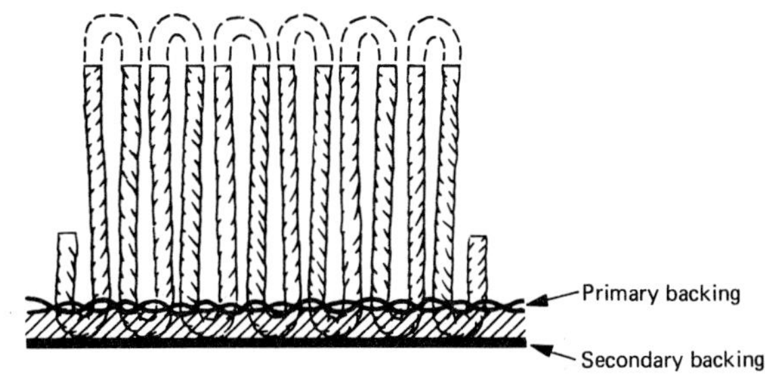

Fig. 24–17 *Tufted carpet. Tufts are punched through primary backing. Secondary backing is bonded to primary backing to lock-in pile.*

Knitted-Pile Fabrics

Knitted-pile fabrics will be discussed in Chapter 28, "Weft-Knit Fabrics" and Chapter 29, "Warp-Knit Fabrics."

Flocked-Pile Fabrics

Flock are very short fibers attached to the surface of a fabric by an adhesive to make a pile-like design or fabric. Flocking was used as a technique for wall decoration in the 14th century. Short silk fibers were applied to freshly painted walls by a bellows. Flock can be applied to many base materials—cloth, foam, wood, metal, and concrete—or it can be applied to an adhesive film that can then be laminated to a base fabric.

Cotton and rayon flocks have been used for dress and curtain fabrics since 1920. Interest in flocking was intensified in the 1960s with the development of new, improved adhesives that will withstand repeated washings or dry cleanings. The new emulsions are not stiff and thick as were the early adhesives; they have good flexibility, durability, drape, and hand. They are also colorless and free of undesirable odor.

Rayon fibers are inexpensive and easy to cut, and are thus used in large quantities. Rayon can be made fire retardant. Nylon has excellent abrasion resistance and durability. Acrylics, polyesters, and olefins can also be used. Fibers for flocking must be straight, not crimped. As the fiber length is increased, the denier also must be increased so that the fiber will stand up straight in the fabric. Fibers that are cut square at the ends will anchor more firmly in the adhesive (Figure 24–18). Adhesives were for many years the "bottleneck" to satisfactory flocking. Most of the problems have been overcome. Over half of the adhesives used are aqueous-based acrylic resins.

Flocked products are used in all areas of textiles. The largest use is in household items, the second largest is children's clothing, and the third largest women's clothes. The potential for flocked fabrics is great.

An interesting flocked product is the Vellux blanket, a revolutionary product made by West-Point Pepperell and copied by other manufacturers. The blanket is made of soft nylon fibers electronically bonded to two layers of polyurethane foam that are permanently sealed to nylon scrim (Figure 24–19). The blanket is light in weight and very warm. It can withstand over 50 washings and dryings with no ill effect if a gentle-agitation cycle is used. The price is competitive with other types of blankets.

The following are some of the major end uses of flocking:

- Velvet upholstery fabrics
- Draperies, bedspreads, blankets

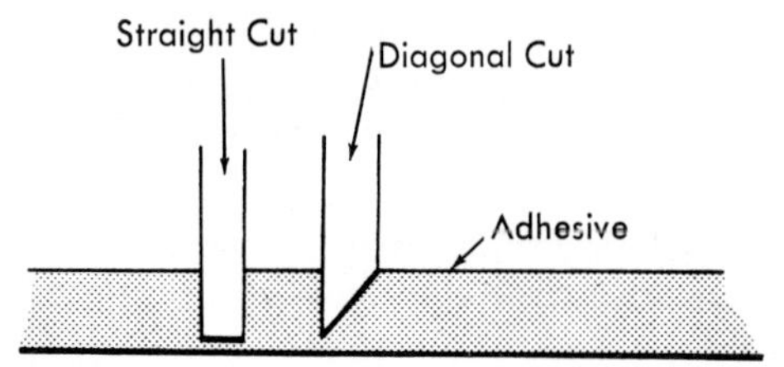

Fig. 24–18 Flock with square-cut ends anchors more firmly.

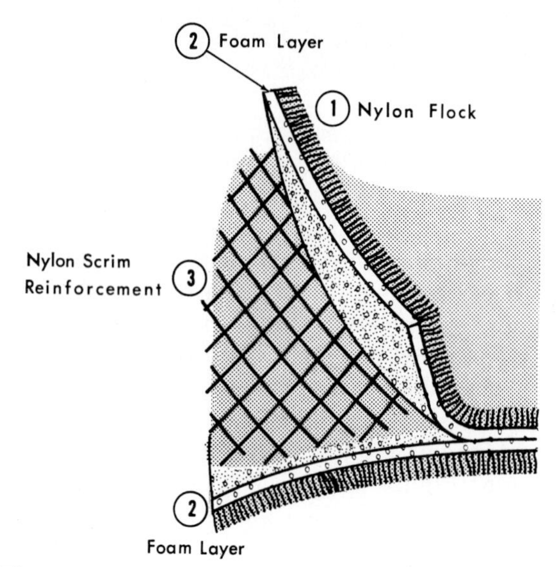

Fig. 24–19 Foam-flocked construction. Pile surface made of fibers.

- Flocked designs on dress fabrics
- Flocked carpets
- Flocked wall coverings for aesthetics and noise reduction
- Automotive fabrics
- Toys, books, shoes, hats
- Industrial uses such as conveyor belts and air filters

The two basic methods of applying the flock fibers are mechanical and electrostatic. In both processes, the flock is placed in an erect position, and, after flocking, the fabric is sent to an oven to dry the adhesive. A comparison of the two methods is given in the table, Flocking Process. Overall flocking, or area flocking, can be done by either method. A rotating screen is used to deposit the flock.

COMPARISONS OF LOOK-ALIKE FABRICS

Pile fabrics, no matter how they are made, are called by woven-fabric names. A recognition of the construction process may help to evaluate a fabric, at the point of sale, in terms of expected performance, realistic cost, and care required. See the table, Look-Alike Pile Fabrics.

In all woven and tufted fabrics, yarns make the pile surface and they must be untwisted by a finishing process to obtain a fiber surface. Both

Look-Alike Pile Fabrics

Fur-like Fabrics—Used for Shells, Linings, Decubicare Pads, and Accent Rugs			
	Woven-Warp Pile	*Sliver Knit*	*Tufted*
Fibers used	Cotton ground. Wool, acrylic, rayon, polyester pile	Cotton, olefin, or modacrylic ground. Acrylic, modacrylic pile	Cheesecloth or soft, filled sheeting substrate. Acrylic, modacrylic pile
Cost	Most expensive	Variable	Least expensive
Characteristics	Pile firmly held in place. Tendency to "grin-through" in low-count fabrics	Most widely used Dense underfibers and guard hairs possible, like real fur. Denser surface possible	Mostly used for sheepskin-type goods. Blooming of yarns and shrinkage of ground holds tufts in place

Carpets					
	Woven-Warp Pile	*Filling Knit Raschel Knit*	*Chenille Yarns*	*Tufted*	*Flocked*
Types and kinds	Wilton Axminster Velvet	Laid-in yarn	Usually woven to order	Most widely used. All kinds of textures. Tufts held in by double back	Not very durable. Limited pile height.
Cost	Most expensive	Raschel knits expensive	Expensive	Inexpensive	Inexpensive

Velour			
	Woven-Warp Pile	*Filling Knit—Jersey*	*Warp Knit*
Fiber content	Cotton	Cotton, polyester	Nylon, acetate
Characteristics	Heavy fabric, durable	Medium weight, soft, drapable	Medium to heavy weight
End uses	Upholstery, draperies	Robes, shirts, jogging suits	Robes, nightwear

Velvet			
	Woven-Warp Pile	*Tufted*	*Flocked*
Fibers	Rayon, nylon, cotton	Nylon	Nylon
Characteristics	Filament—dressy, pile flattens Cotton or nylon—durable Rich looking—heavy weight fabric	Pile not held in as firmly as woven, less expensive	Least expensive
End uses	Filament—apparel. Cotton, nylon—upholstery	Upholstery	Upholstery, draperies, bedspreads

Terrycloth		
	Slack-Tension Weave	*Filling Knit—Jersey*
Characteristics	Usually cotton Holds it shape	All fibers. Soft, stretchy. Cotton does not hold its shape well. Very pliable
End uses	Towels, washcloths, robes	Baby towels and washcloths Baby sleepers, adult sportswear, socks

Flocking Process

Mechanical Flocking	*Electrostatic Flocking*

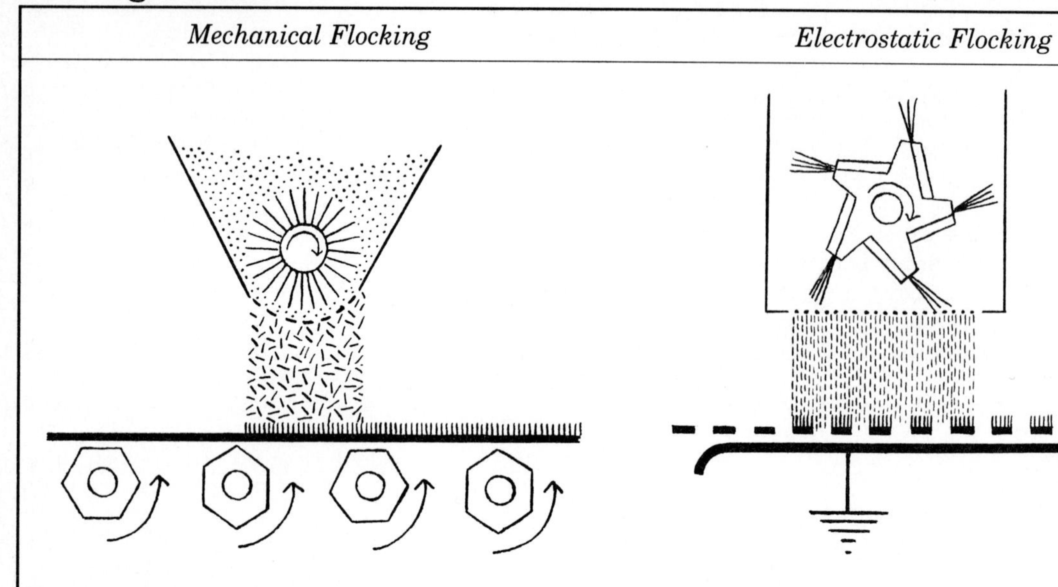

Mechanical Flocking

1. Short fibers are sifted onto the adhesive-coated fabric. Vibration of beater bars causes fibers that do not fall flat against the adhesive to stand erect, and, once erect, the fibers penetrate fully into the adhesive. The erect fibers help the free fibers to align themselves and to work down to the adhesive.
2. Most units consist of 6–20 beater bars and one or more sifting hoppers and run as high as 10 or more yards per minute.
3. Simpler in design, usually less expensive, and most widely used in the United States.

Electrostatic Flocking

1. Flock passes through an electrostatic field that orients the fibers. In coating irregular surfaces, the lines of force are always perpendicular to the substrate, so this method is best for three-dimensional surfaces.
2. Most units operate at speeds of 3–5 yards per minutes.
3. Can "up-flock" as well, so both sides of a fabric can be flocked.
4. Requires generators, proper insulation, gives better end-on-end fiber orientation, and higher densities are possible.

Fig. 24–20 *(Courtesy of the Fibers Division of Monsanto Chemical Co., a unit of Monsanto Co.)*

of these techniques also make fabrics with uncut loops. Knitted pile fabrics have the pile surface made from yarns or fibers. Flocked fabrics have the pile surface made directly from fibers.

Napped fabrics (see Chapter 33) often are confused with pile fabrics.

25

Structural-Design Fabrics

Structural designs are formed as the fabric is produced. This chapter will discuss fabrics with designs or figures other than stripes, checks, and plaids. The techniques used to obtain figured fabrics vary in complexity and influence the cost and serviceability of the fabric. Structural designs are permanent designs, whereas applied designs are less durable. Applied designs are discussed in Chapter 33 but referred to in this chapter for comparison with structural-design fabrics. Recognition of the technique used in manufacturing the fabric should be helpful in selecting a serviceable fabric for the intended end use. Understanding the techniques used may help in naming a fabric correctly; although, in today's market, fabric names are not very precise. A comparison of structural and applied designs is made in the following chart to clarify the differences between the two types of designs.

Comparison of Structural and Applied Designs

Structural Designs	*Applied Designs*
Usually more expensive because decisions must be made farther in advance of market and process is more time consuming	Usually less expensive
Permanent design	Permanent, durable, or temporary
Woven figures are always on-grain (circular-knit jacquards may be skewed)	Figures may be off-grain
Kinds and Types	
Woven—jacquard, dobby, extra yarns, swivel dots, lappet designs, piqué, double cloth	Printed Flocked Embroidered Burned out
Knitted—jacquard single knits, jacquard dobule knits	Embossed Plissé
Lace	
Typical Fabrics	
Huck, damask, brocade, tapestry, shirting madras, piqué, dotted swiss, matelassé	Flock dotted swiss, embroidered linen, burned out

Structural Designs

WOVEN FIGURES

Woven figures are made by changing the interlacing pattern in the design from that of the background. The interlacing pattern is controlled by the warp yarns in the harnesses. In a three-harness loom, there are three possible arrangements of the warp yarns; in a four-harness loom, there are as many as 12 different arrangements. As the number of harnesses increases, the number of possible different interlacings also increases. But there is a limit to the number of harnesses that can be used efficiently. Consider the number of interlacings needed to make a figure: If the figure is ¼ inch in length, there may need to be 20 different interlacings in an 80-square cloth; if the figure is ½ inch long on a nylon-satin background (320 × 140), there may need to be 70 different interlacings. To make woven figures by mass production, therefore, special looms or special attachments to the loom are necessary.

Dobby Weaves. Small-figured designs, which require fewer than 25 different yarn arrangements to complete one repeat of the design, are made on a loom with a dobby attachment—usually referred to as a *dobby loom* (Figure 25–1).

The weave pattern is controlled by a plastic tape with punched holes (Figure 25–2). These tapes somewhat resemble the rolls for a player piano. The holes control the raising and lowering of the warp yarns. Many designs made on the dobby loom are small geometric figures.

Bird's-eye has a small diamond-shaped filling-float design with a dot in the center that resembles the eye of a bird. This design was originally used in costly white silk fabric for ecclesiastical vestments. At one time, it was widely used for kitchen and hand towels and diapers. *Huck,* or *huck-a-back* has a pebbly surface made by filling floats. It is used primarily in roller, face, and medical-office towels.

Shirting madras has small, satin-float designs on a ribbed or plain ground.

Waffle cloth is made with a dobby attachment and has a three-dimensional honeycomb appearance.

Fig. 25–1 Loom with a dobby attachment (upper left).

Extra-Yarn Weaves. When yarns of various colors or types different from the background are wanted for a figure, extra yarns are woven into the fabric. The figure portion has warp or filling floats. When not used in the figure, the extra yarns float across the back of the fabric and are usually cut away during finishing. In hand-woven fabrics, the warp yarns can be manipulated by hand and the extra yarns can be laid in where wanted by using small shuttles. But, in power looms, an automatic attachment must be used.

Fig. 25–2 Pattern roll that controls warp shedding on a dobby loom. (Courtesy of the Crompton & Knowles Corp.)

Extra-warp yarns are wound on a separate beam and threaded into separate heddles. The extra yarns interlace with the regular filling yarns to form a design and float above the fabric until needed for the repeat. The floats are then clipped close to the design or clipped long enough to give an eyelash effect. Figure 25–3 shows a fabric before and after clipping.

Extra-filling yarns are inserted in several ways. *Clipped-dot designs* are made with low-twist filling yarns inserted by separate shuttles. The shedding is done so that the extra yarns interlace with some warp yarns and float across

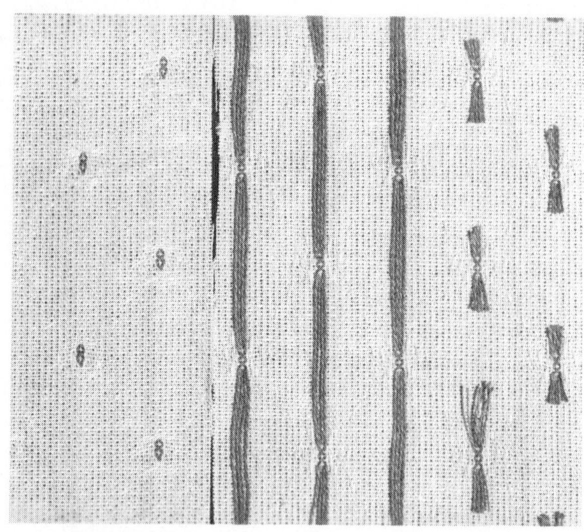

Fig. 25–3 Fabric made with extra warp yarns: face side of fabric (left)*; back side, before and after clipping* (right).

the back of other warp yarns. Clipped spots are woven on a box loom that has a wire along the edge to hold the extra yarns so that they need not be woven in the selvage. Figure 25–4 shows a clipped fabric, dotted swiss, before and after clipping.

Many of the fabrics that have small-dot designs are called dotted swiss. The dots may be structural designs: clipped-dot designs, as described above, or swivel-dot designs. *Swivel-dot designs* are made on a loom that has an attachment holding tiny shuttles. The fabric is woven so the shuttles and extra yarns are above the

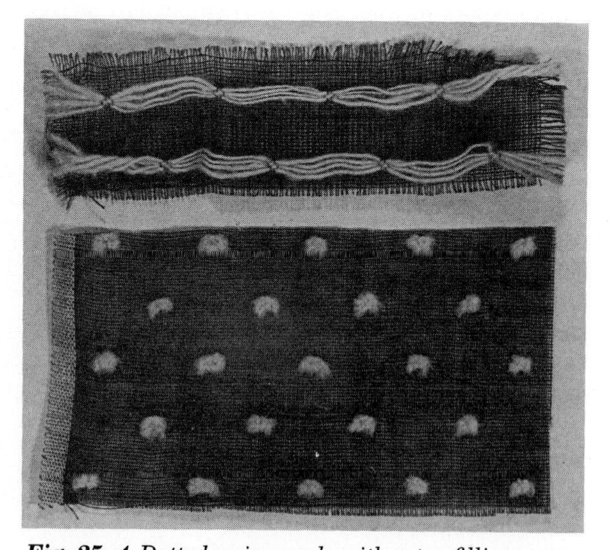

Fig. 25–4 Dotted swiss made with extra filling yarns: before (above) *and after clipping* (below).

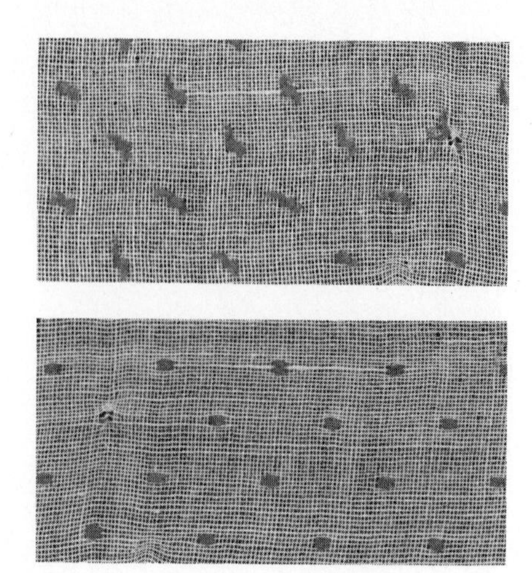

Fig. 25–5 Dotted swiss showing both sides of a swivel-dot fabric.

ground fabric. Each shuttle carrying the extra yarn goes four times around the warp yarns in the ground fabric and then the yarn is carried along the surface to the next spot. The yarn is sheared off between the spots (Figure 25–5). Swivel-dot fabrics are rarely seen in the United States, except as imported designer fabrics.

Swivel-dot fabrics, clipped-dot fabrics, and the larger clipped-spot fabrics may use either side of the fabric as the "right" side. Fashion may dictate which side is correct—or personal preference may determine which side to use.

Dotted swiss may also be an applied design: the dots may be printed or, for a three-dimensional effect, flocked, or made with expanded foam. (See Chapter 33 for more details.) A comparison of several different types of dotted swiss is an interesting exercise in determining the serviceability of fabrics.

Piqué Weaves. The word *piqué* comes from the French word meaning quilted; the raised effect in these fabrics is similar to that in quilts.

Piqué weave produces a fabric with ridges, called wales or cords, that are held up by floats on the back. The wales vary in width. *Wide-wale piqué* (0.25 inch) is woven with 20 or more warp yarns in the face of the wale and then two warps in between. *Pin-wale piqué* (0.05 inch) is a six-warp wale with two consecutive filling yarns floating across the back of the odd-numbered wales and then woven in the face of the even-numbered wales. The next two consecutive picks alternate with the first two by floating across

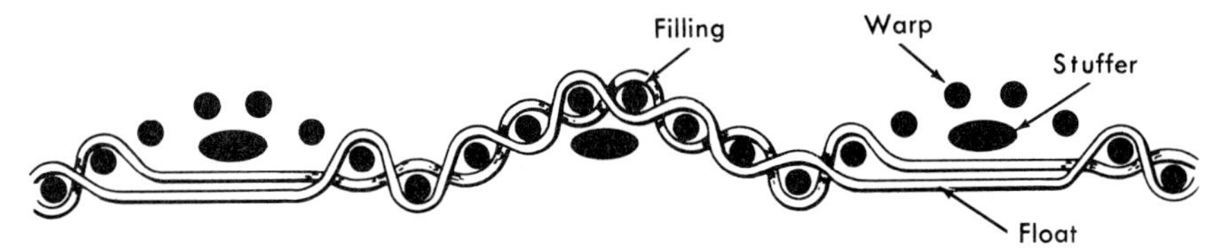

Fig. 25–6 *Six-warp pinwale piqué.*

the back of the even-numbered wales. Figure 25–6 shows a six-warp pin-wale piqué.

Stuffer yarns are laid under the ridges in the better-quality piqué fabrics to emphasize the roundness or quilted effect, and their presence or absence is one way of determining quality. The stuffer yarns are not woven in the main part of the fabric and may be easily removed when analyzing a swatch of fabric. Piqué fabrics, depending on the complexity of the design, are woven on either a dobby or jacquard loom.

Cords or wales usually run in the lengthwise direction, with the exception of bird's-eye and bull's-eye piqués, in which the cords run crosswise. Cord fabrics have a definite technical face and technical back. The fabric tears more easily in the lengthwise direction. If there are stuffer yarns, it is especially difficult to tear the fabric crosswise. In wear, the floats on the wrong side usually wear out first. Figure 25–7 shows the right and wrong side of a piqué fabric.

Piqué fabrics are more resistant to wrinkling, have more body than flat fabrics, and, for these reasons, have less need to be given a durable press finish. Piqué fabrics should be ironed on the wrong side because the beauty of the fabric is in the roundness of the cord and pressing on the right side will flatten it.

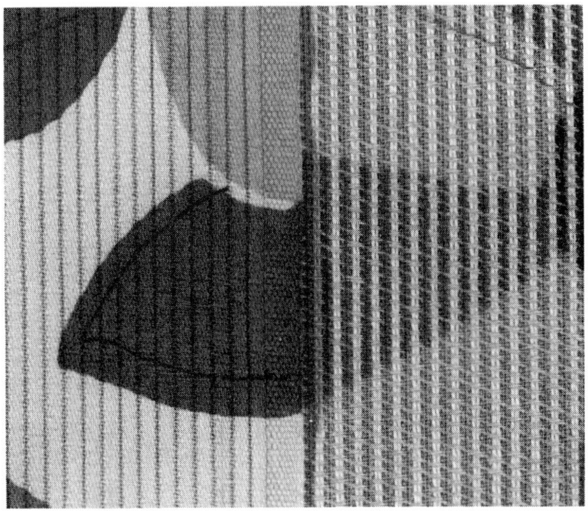

Fig. 25–7 *Piqué: right side of fabric* (left)*; reverse side of fabric* (right)*. Note stuffer yarns running vertically.*

Fabrics in this group are called piqué, with the exception of a wide-wale fabric called bedford cord. *Bedford cord* is a heavy fabric with warp cords. It is used for slacks, trousers, uniforms, bedspreads, and upholstery. It is made with carded-cotton yarns, woolen or worsted yarns, rayon or acetate, or combinations. The wales are wide and stuffer yarns are usually present.

Piqué is lighter in weight than bedford cord and has a narrower wale. The better-quality fabrics are made with long-staple, combed, mercerized yarns and have one stuffer yarn. The carded yarn piqués are made without the stuffer and are sometimes printed.

Bird's-eye piqué has the tiny design formed by the wavy arrangements of the cords and by the use of stuffer yarns. *Bull's-eye piqué* is made like bird's-eye but has a much larger design. Both these fabrics have crosswise rather than lengthwise cords. They are used for apparel and home-furnishing fabrics.

Jacquard Weaves. Large-figured designs, which require more than 25 different arrangements of the warp yarns to complete one repeat design, are woven on the jacquard loom (Figure 25–8). Each warp is controlled independently by punched cards that are laced together in a continuous strip. As the cards move over the loom, all the warp yarns are raised by rods attached to them. When the rods hit the cards, some will go through the holes and thus raise the warp yarns; others will remain down. In this manner, the shed is formed for the passage of the filling yarn. Figure 25–9 shows a picture woven with fine silk yarns on a jacquard loom. Notice that there is no repeat of the pattern from top to bottom or from side to side. The repeat would be another picture.

Fabrics made on a jacquard loom include damask, brocade, and tapestry. *Damask* has

Fig. 25–8 Jacquard loom for weaving large-figured fabrics. (Courtesy of the Crompton & Knowles Corp.)

satin floats on a satin background, with the floats in the design opposite those in the background. It can be made from any fiber and in many different weights for apparel and home furnishings. Damask is the flattest-looking of the jacquard fabrics and is often finished to maintain that flat look.

Quality and durability are dependent on high count. Low-count damask is not durable because the long floats rough up, snag, and shift during use. *Brocade* has satin or twill floats on a plain, ribbed, twill, or satin background (Figure 25–10). Brocade differs from damask in that the floats in the design are more varied in length and are often of several colors.

Brocatelle fabrics are similar to brocade fabrics, except that they have a raised pattern. This jacquard-woven fabric frequently is made with filament yarns, using a warp-faced pattern and filling-faced ground. Coarse cotton stuffer filling yarns may be used to help maintain the three-dimensional appearance of the fabric when intended for upholstery.

Originally, *tapestry* was an intricate picture that was hand woven with discontinuous filling yarns. It was usually a wall hanging and took years to weave. The jacquard tapestry is mass produced for upholstery, handbags, and the like. It is a complicated structure consisting of two or more sets of warp and two or more sets of filling interlaced so that the face warp is never woven into the back and the back filling does not show on the face. Upholstery tapestry is durable if warp and filling yarns are comparable. Very often, however, fine yarns are combined with coarse yarns, and when the fine yarns are abraded away, the coarse yarns are left loose and are unsightly.

Wilton rugs are figured pile fabrics made on a jacquard loom. These rugs, once considered imitations of Oriental rugs, are so expensive to weave that the tufting industry has found a way to create similar figures by printing techniques.

Fig. 25–9 Jacquard-woven picture.

Fig. 25–10 Brocade.

Double Cloth

Double-cloth fabrics have a different appearance on the two sides due to the method of producing the fabric. Double-cloth fabrics tend to be heavier and have more body than single cloths. A single cloth, such as percale, is made from one set of warp yarns and one set of filling yarns. Another way of stating this is to say that percale is made from two sets of yarns. Double cloth is a woven fabric made from three or more sets of yarns.

The following are three types of woven double cloth:

1. Double cloth—coat fabrics: melton and kersey.

2. Double weave—apparel and upholstery fabrics: matelassé.

3. Double-faced—blanket cloth, double-satin ribbon, Sunbac®, and silence cloth

DOUBLE CLOTH

Double cloth is made with five sets of yarns: two fabrics woven one above the other on the same loom with the fifth yarn (warp) interlacing with both cloths (Figure 25–11). True double cloth can be separated by pulling out the yarns holding the two cloths together. It can be used in reversible garments such as capes and skirts.

Double cloth is expensive to make because it requires special looms and the production rate is slower than for single fabrics. Double cloth is more pliable than the same weight single fabric because finer yarns can be used. Designers of high-fashion women's wear use true double cloth for coats and capes. The two specific fabrics that may be either true double cloth or single cloth are melton and kersey. Both of these heavy-weight-wool coating fabrics are twill-weave fabrics that have been heavily fulled and felted. Because of these processes, it is difficult to see and identify the weave. In addition, the weave is obscured because the fabrics often are napped, then closely sheared. Melton tends to have a smoother surface than kersey. Kersey is usually heavier than melton and has a shorter, more-lustrous nap. Both fabrics are used in winter coats, overcoats, riding habits, and military uniforms.

Fig. 25–11 *Double cloth made with five sets of yarns.*

DOUBLE WEAVE

Double weave is made with four sets of yarns creating two separate layers of fabric that periodically reverse position from top to bottom, thus interlocking the two layers of fabric. Between the interlocking points, the two layers are completely separate, creating pockets in the fabric (Figure 25–12).

Double-weave fabrics can also be called *pocket fabrics, pocket cloth,* or *pocket weave.* They are most commonly seen in high-quality upholstery

Fig. 25–12 *Double weave: right side of cloth* (bottom); *reverse side of cloth* (top).

fabrics. Instead of using specific fabric names, interior designers group these fabrics together. The main advantage of the fabrics is the designs that can be achieved. An additional advantage is their heavier weight. They are usually closely woven, durable fabrics.

Double-weave fabrics in apparel are less common, but they are seen in designer fashions. An example is matelassé which has a three-dimensional texture. *Matelassé* fabrics are made on dobby or jacquard looms with either crepe yarns or very-coarse cotton yarns. When the fabrics are finished, the crepe or cotton yarns shrink, causing the fabric to have a puckered appearance. Heavy cotton yarns are used as stuffer yarns beneath the fabric face to emphasize the three-dimensional appearance of the fabric. Heavier-weight matelassé fabrics also are found in upholstery fabrics.

DOUBLE-FACED

Double-faced fabrics are made with three sets of yarns: two warp and one filling, or two sets of filling and one set of warp. Blankets, satin ribbons, interlinings, and silence cloth can be made by this process (Figure 25–13).

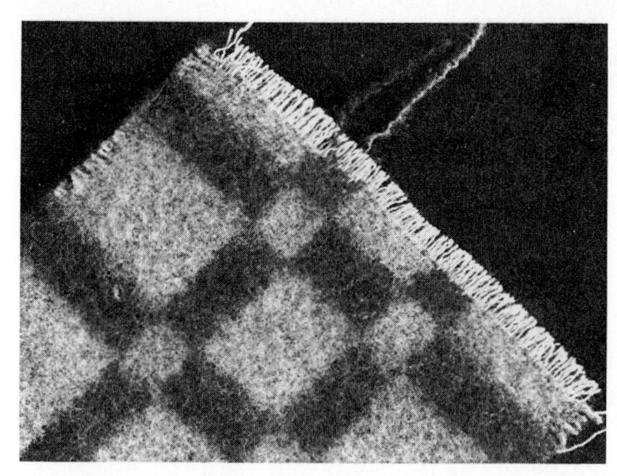

Fig. 25–13 *Double-faced blanket. One set of warp yarns and two sets of filling yarns.*

Blankets with one color on one side of the blanket and another color on the other side of the blanket are usually double-faced blankets. One set of warp yarns is used, with two sets of different-colored filling yarns. Sometimes designs are made by interchanging the colors from the one side to the other. Double-faced blankets are usually expensive woven-wool blankets.

Satin ribbons, which have a lustrous satin face on both sides of the ribbon, are used in designer lingerie and evening wear. These ribbons have two sets of warp yarns that form the surface on both sides of the ribbon—they are interlaced with one set of filling yarns.

Sunbak® is the trade name of a double-faced interlining fabric used to add warmth to winter jackets and coats. The face of the fabric is a filament-yarn satin weave that takes the place of a lining fabric and slides easily over other clothing. The back of the fabric uses a third set of low-twist yarns that are heavily napped for warmth. Thus the fabric functions as a combination lining and interlining fabric. Sunbak® may be found in fabric stores in areas of the country that have cold winters.

Silence cloth is a heavy cotton fabric that has been napped on both sides. Available in white, it is used under fine table cloths to silence the noises of china and silverware while dining. Although difficult to find it is still available.

26

Crepe Fabrics

Crepe is a French word meaning *crinkle.* A crepe crinkle is obtained in several ways, so this chapter deals with a *family* of crepe fabrics. Crepe fabrics can be classified according to the way the crinkle is obtained.

1. Yarn.
   a. High-twist natural-fiber yarn.
   b. High-twist man-made fiber yarn.
   c. High-twist thermoplastic-fiber yarn.
   d. Low-twist bulky yarn.
2. Weave.
   a. Crepe or momie weave.
   b. Slack-tension weave.
3. Finish.
   a. Embossing calendar.
   b. Printing with sodium hydroxide.
   c. Printing with phenol.

High-Twist Yarn Crepes

The first crepe fabrics were made with *high-twist natural-fiber yarns.* These crepe fabrics are made on a loom with a box attachment that can insert alternating groups of S- and Z-twist yarns to enhance the amount of crinkle. The high-twist crepe yarns are made of rayon, cotton, flax, wool, and silk fibers, because the liveliness of the high twist can be set by wetting and drying before weaving.

The warp yarns of a crepe fabric may be low-twist yarns. Low twist in the warp enhances the crinkle achieved by the crepe yarns in the filling.

Gray-goods crepe fabric is smooth as it comes from the loom. It is woven wide and then shrunk to develop the crinkle. Immersion in water causes the crepe-twist yarns to regain their liveliness and contract or shrink. For example, the fabric is 47 inches wide on the loom, contracts to 30–32 inches in boil-off, and is finished at 39 inches. This explains why a crepe fabric will shrink when it gets wet and why garment size is so much more easily controlled by dry cleaning than by washing. High-twist crepe fabrics are described by the position of the crepe yarn, as filling crepes, warp crepes, balanced crepes, and double cloth.

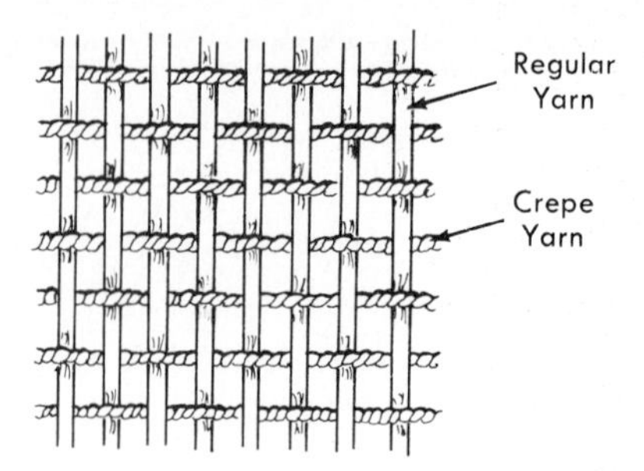

Fig. 26–1 *Filling crepe.*

FILLING-CREPE FABRICS

Filling-crepe fabrics have high-twist crepe yarns in the filling direction and low-twist yarns in the warp direction (Figure 26–1).

Filling-crepe fabrics have a dull crepe-like surface. One fabric is made with a very-low-twist acetate filament in the warp and a high-twist rayon filament in the filling. The rayon-crepe yarns alternate with S- and Z-twist or with 2S- and 2Z-twist. A high warp count and low filling count give a crosswise rib effect. Low count in the filling gives the crepe yarns room to contract, so the amount of crinkle will be greater. Figure 26–2 shows a filling crepe. Analysis of a filling-crepe fabric will show that it is easy to distinguish between the warp and filling yarns because the crepe yarns will kink up. Wool crepes and silk crepes, such as silk crepe de Chine, are used in designer clothes. The hand and drape of these fabrics are outstanding.

When handling cellulosic and protein high-twist crepe-yarn fabrics, the manufacturers

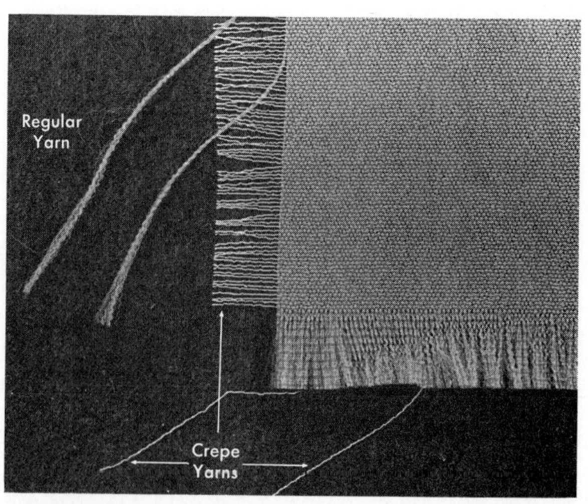

Fig. 26–2 *Filling crepe. Notice crimp of regular yarn caused by pressure of crepe yarn when fabric was pressed.*

should keep the fabric in a relaxed position. The fabric has much more give than typical woven fabrics. Pressing should be done with as little pressure and as little steam as possible to avoid distortion. It is best to dry clean wool and silk crepes to avoid excessive shrinkage in laundering or pressing.

High-twist thermoplastic yarns are also used to make crepe fabrics. They may be used in either the filling or the warp direction, or woven into a balanced-crepe fabric. Polyester- and nylon-crepe fabrics are much less likely to shrink when wet, and thus may be laundered. They are heat set for stability and so act differently from the other high-twist crepe fabrics. They perform very well.

WARP-CREPE FABRICS

Warp-crepe fabrics are made with crepe yarns in the warp and regular yarns in the filling direction. Very few warp crepes are on the market because of the difficulty in weaving with warp yarns that are made from high-twist crepe yarns. Warp-crepe fabrics also shrink more in the warp direction, so it is difficult to keep an even hemline in washable fabrics.

BALANCED-CREPE FABRICS

Balanced-crepe fabrics have crepe yarns in both directions and are usually balanced in count. They are often made in sheers and the crepiness of the yarns in both directions helps prevent yarn slippage. Figure 26–3 shows a balanced crepe. *Chiffon* is a very sheer crepe made with alternate S- and Z-twist yarns in a plain weave. *Georgette* is a sheer crepe originally made with two S- and two Z-twist yarns alternating. It is duller and heavier than chiffon.

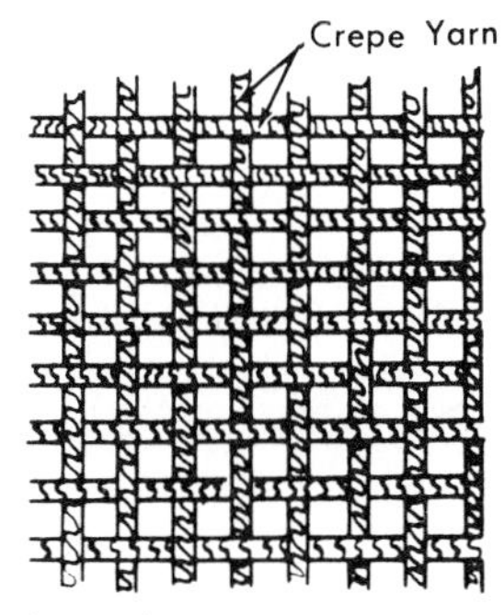

Fig. 26–3 *Balanced crepe.*

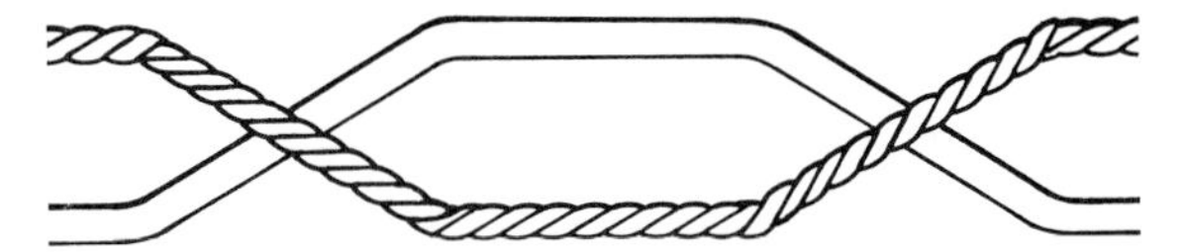

Fig. 26–4 *Interlacing of yarns between fabric surfaces of matelassé.*

DOUBLE-CLOTH FABRIC MADE WITH CREPE YARNS

Matelassé crepe is a double-cloth construction with either three or four sets of yarns. Two of the sets are always the regular warp and filling yarns and the others are crepe yarns. They are woven together so that the two sets crisscross, as shown in Figure 26–4. The other sets of yarns may be crepe yarns or they may just be coarser yarns that shrink more during finishing to create puffy areas in the regular-yarn part of the fabric. Matelassé may be a rayon/acetate combination fabric. For a better understanding of double-cloth fabrics, see that section in Chapter 25.

Low-Twist Textured-Yarn Crepe

Filament yarns textured by the false-twist process are woven as the filling yarns in a plain-weave fabric with standard filament yarns in the warp. These textured filament yarns are *low twist.* The crinkled appearance forms during the wet finishing when the textured yarn shrinks.

The finished fabric has a high level of crinkle, good hand, and exceptional performance for the consumer. It packs well and never needs ironing. The fabric is relatively stiff or crisp when compared with high-twist-yarn crepes—it does not drape well.

Crepe Weaves

Two kinds of weaves are used: the momie weave and slack-tension weave. The crinkle is permanent.

MOMIE WEAVE

Momie is the name given to a class of weaves that present no twilled or other distinct weave effect but give the cloth the appearance of being sprinkled with small spots or seeds. The appearance resembles crepe made from yarns of high twist. Fabrics are made on a loom with a dobby attachment. Some are variations of satin weave, with filling yarns forming the irregular floats. Some are even-sided and some have a decided warp effect. Momie weave is also called *granite,* or *crepe weave.* Fibers that do not lend themselves to high-twist-yarn crepe fabrics are often used in making crepe-weave fabrics. Wool and cotton fibers are also used frequently because the woven-crepe fabric is easier to care for than the high-twist-yarn crepes. For a comparison of characteristics, refer to the table on page 239. An irregular interlacing pattern of crepe weave is shown in Figure 26–5.

Sand crepe is a common momie-weave fabric. It has a repeat pattern of 16 warp and 16 filling and requires 16 harnesses. No float is greater than two yarns in length. It is woven of either spun or filament yarns. A silk-like acetate-sand crepe is widely used.

Granite cloth is made with the momie weave, based on the satin weave, and is an even-sided fabric with no long floats and no twilled effect. It is used for furnishings and apparel.

Moss crepe is a combination of high-twist crepe yarns and crepe weave. The yarns are plied yarns with one ply made of a crepe-twist single yarn. Regular yarns may be alternated with the plied yarns or they may be used in one direction while the plied yarns are used in the other direction. This fabric should be treated as a high-twist crepe fabric. Moss crepe is used in dresses and blouses.

Polyester-crepe fabrics are frequently used in blouses and dresses. They are usually crepe-weave fabrics made with textured yarns.

SLACK-TENSION WEAVE

In slack-tension weaving, two warp beams are used. The yarns on one beam are held at regular tension and those on the other beam are held at slack tension. As the reed beats the filling yarn into place, the slack yarns crinkle or buckle to form the puckered stripe and the regular-tensioned yarns form the flat stripe. (Loop-pile fabrics, such as terrycloth, are made by a similar weave; see page 221). *Seersucker* is the fabric made by slack-tension weave (Figure 26–6).

The yarns are wound onto the two warp beams in groups of 10–16 yarns for a narrow stripe. The crinkle stripe may have slightly larger yarns to enhance the crinkle, and this stripe may also have a 2 × 1 basket weave. The stripes are always in the warp direction. Seersucker is produced by a limited number of manufacturers. It is a low-profit, high-cost item to produce because of slow weaving speed. Most seersuckers are made in 45 or 60 inch widths in plain colors, stripes, plaids, and checks. Cotton, polyester, and acetate fibers are used singly or in blends. Seersucker is used in large amounts in the men's-wear trade for summer suiting and for women's and children's dresses and sportswear. Seersucker is frequently printed for children's wear.

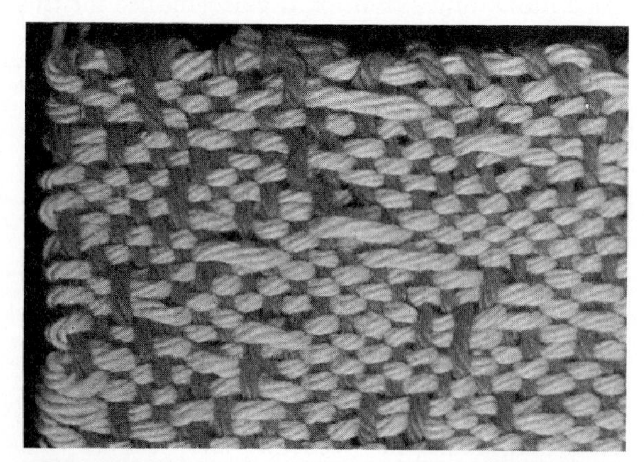

Fig. 26–5 *Crepe weave with irregular interlacings.*

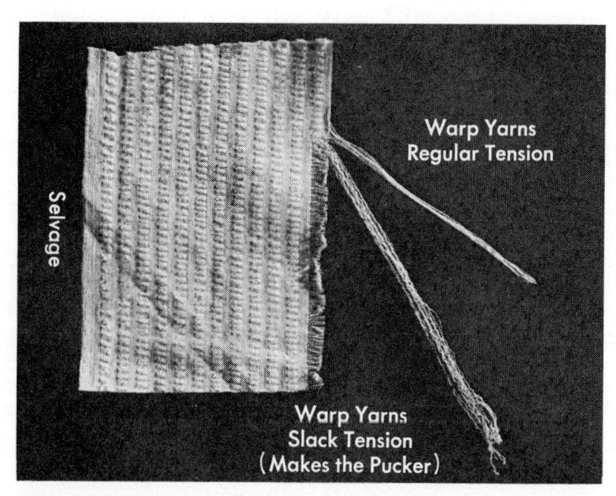

Fig. 26–6 *Seersucker, showing the difference in length of slack- and regular-tension yarns.*

Crepe Finishes

Crepe finishes achieve the crinkled look after the fabric is woven. *Plissé* fabric is converted from either lawn or print-cloth gray goods by printing sodium hydroxide (caustic soda) on 100 percent cotton cloth in the form of stripes or designs. The chemical causes the fabric to shrink in the treated areas. As the treated stripe shrinks, it causes the untreated stripe to pucker. Shrinkage causes a slight difference in count between the two stripes. The treated or flat stripe increases in count as it shrinks. The upper portion of the cloth in Figure 33–9 shows how the cloth looks before finishing, and the lower portion shows the crinkle produced by the caustic soda treatment. The roller failed to print the chemical in the unpuckered area.

The crinkle stripes can be narrow, as shown in Figure 33–9, or wide. In piece-dyed, wide-crinkle-stripe fabrics, the flat treated area may be a deeper color than the crinkled area. The crinkle is permanent, but can be flattened somewhat by steam and pressure such as may occur during ironing.

Embossed fabric is made by pressing a three-dimensional design onto the surface of the cloth. Cotton cloth must be given a resin finish to make the design durable. Cotton/polyester blends are used most commonly. The design is durable; however, with steam and pressure, the design can be pressed flat. Thermoplastic fibers can be heat set to make the design permanent.

Both plissé and embossed fabrics will retain their original appearance best if they are tumble dried and *not* pressed. This is one reason they are enjoyed in summer sports clothes and summer sleepwear.

Seersucker, plissé, and embossed fabrics can be very similar in appearance, and are frequently found in the same price range. Compare the serviceability of these fabrics for a specific end use. Will one fabric outperform the others?

The following chart compares crepe fabrics.

Comparison of Crepe Fabrics

High-Twist Natural or Man-Made Yarns (40–80 tpi)	*High-Twist Thermoplastic Yarns*	*Bulky Yarn*	*Weave*	*Finish*
Permanent crinkle. Will flatten during use. Moisture will restore it	Permanent crinkle. Retains appearance during wear and care	Permanent crinkle. does not sit out or need ironing	Crinkle does not flatten in use	Crinkle may sit out or be less prominent after washing
High potential shrinkage	Low potential shrinkage	Low potential shrinkage	Lower potential shrinkage	Lower potential shrinkage
Good drapability	Good drapability	Less drapable	Less drapable	Less drapable
Stretches	Moderate stretch	Low stretch	Low stretch	Low stretch
Resilient, recovers from wrinkles	Resilient	Does not wrinkle	Wrinkles do not show because of rough surface	Wrinkles do not show because of rough surface
Dry cleaning preferable	Easy care	Easy care	Washable unless fiber content requires dry cleaning	Washable unless fiber content requires dry cleaning
Typical fabrics Wool crepe Crepe de Chine Matelassé Chiffon Georgette Silk crepe	Typical fabrics Chiffon Georgette	Typical fabric Silk-like synthetics	Typical fabrics Sand crepe Granite cloth Seersucker	Typical fabrics Plissé Embossed

Lace, Leno Weave, and Narrow Fabrics

Lace

Lace is the second basic method of making fabric from yarns. Weaving has interlaced yarns, knitting has interlooped yarns, and lace has intermeshed yarns—yarns twisted around each other. Lace is an open-work fabric with complex patterns or figures that is hand made or machine made on special lace machines or on Raschel knitting machines.

It can be difficult to tell whether a fine-lace fabric was made by a Leavers machine or on a loom or by a Raschel knitting machine without the aid of a microscope. However, with many laces, it is a fairly simple matter to determine their origin. Some lace-like fabrics are made by printing or flocking (Figure 27–1).

Lace was very important in men's and women's fashion between the 16th and 19th centuries, and all countries in Europe developed lace industries. The names given to lace often originated from the town in which the laces were made. For example, the best quality needlepoint lace was made in Venice in the 16th century—hence the name Venetian lace. Alençon and Valenciennes are also laces made in these French towns.

HAND-MADE LACE

Hand-made lace has always been as it is today—a prestige textile. Some people use old lace on garments or as decorative wall hangings. With the interest in crafts today, many of the old lace-making techniques are being used to make less delicate lace. For example, macramé, crochet, tatting, and hairpin laces are made as wall hangings, belts, bags, shawls, afgans, bedspreads, and tablecloths.

Lace is classified according to the way it is made and the way it appears. Hand-made laces are needlepoint, bobbin, crochet, and darned.

Needlepoint Lace. *Needlepoint lace* is made by drawing a pattern on paper, laying down threads over the pattern, and, with a needle and thread, making buttonhole or blanket stitches over the threads. The network of fine threads making the ground is called *reseau* or *brides*. The solid part of the pattern, which may be made with buttonhole stitches or interlaced as a woven area, is called *toile*. Needlepoint laces are Alençon, which has a hexagonal mesh, Rosepoint, and Venetian Point, which have an irregular mesh. Needlepoint laces have birds, flowers, and vases as the design (Figure 27–2).

Bobbin Lace. *Bobbin lace* is made on a pillow. The pattern is drawn on paper and pins are inserted at various points. The threads, which are on many bobbins, are twisted and plaited around the pins to form the mesh and the design (Figure 27–3). Cluny is a coarse, strong lace; Duchesse has a fine net ground with raised patterns; Maltese has the Maltese cross in the pattern; Mechlin has a small hexagonal mesh and very fine threads; Torchon is a rugged lace with very simple patterns; Valenciennes has a diamond-shaped mesh; Chantilly has a double ground with filling of flowers, baskets, or vases.

Crocheted Lace. *Crocheted lace* is a combinatin of loops, but it differs from knitting in that the loops are thrown off and finished, whereas in knitting the entire series of loops is held on needles while new loops are made. Crocheted laces are Irish lace and Syrian lace. Crocheting is done with a crochet hook.

Darned Lace. *Darned lace* has a chain stitch outlining the design on a mesh background. The needle carries the thread around the yarns in the mesh. The mesh is square in Filet and rectangular in Antique.

MACHINE-MADE LACE

In England in 1802, Robert Brown perfected a machine that made nets on which lace motifs could be worked by hand. In 1808, John Heathcoat made the first true lace machine by developing brass bobbins to make bobbinet. In 1813, John Leavers developed a machine that made patterns and background simultaneously. A card system, based on the jacquard loom, made it possible to produce intricate designs with the Leavers machine (Figure 27–4).

Leavers. The *Leavers machine,* a machine of tremendous size and weight, consists of warp yarns placed in the machine and oscillating bobbins that are set in frames called *carriages*. The carriages move back and forth with the bobbins swinging around the warp to form a pattern. These little brass bobbins, holding 60–300 yards

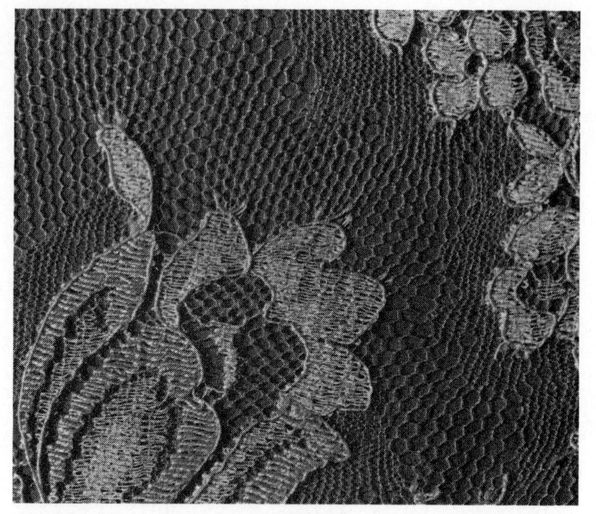

(a) Machine-made lace. Alençon.

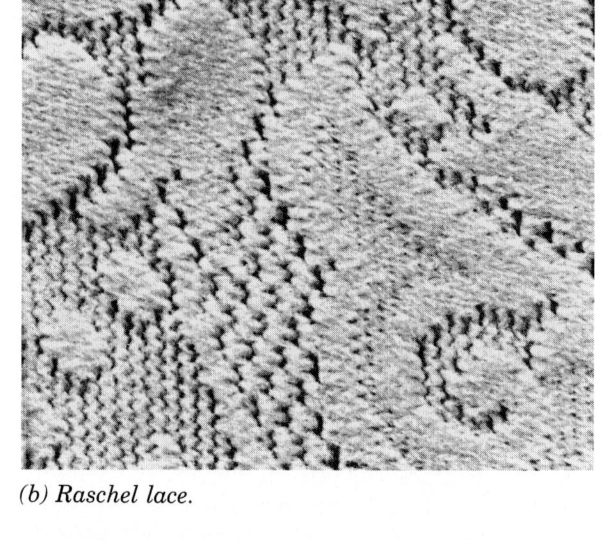

(b) Raschel lace.

(c) Filling knitted lace-like fabric.

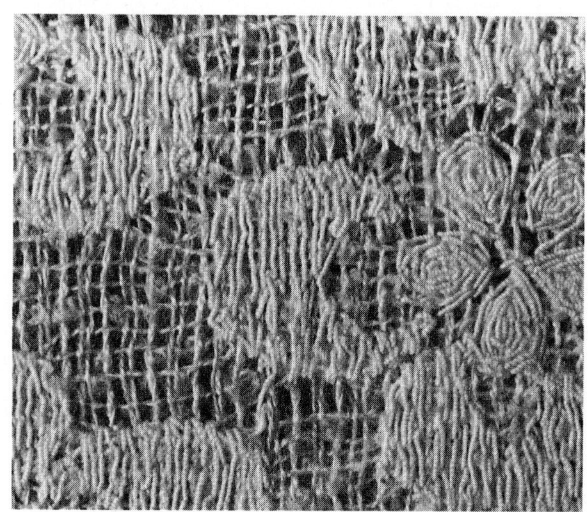

(d) Woven lace-like fabric.

(e) Imitation lace.

Fig. 27–1 *Lace and lace-like fabrics: (a) re-embroidered lace made on Leavers lace machine; (b) lace fabric made on Raschel-warp knitting machine; (c) lace-like fabric made on purl knitting machine; (d) woven lace-like fabric; (e) imitation lace. Cotton percale printed with a lace-like design.*

Fig. 27–2 *Needlepoint lace: hand made* (left)*; machine made* (right).

of yarn, are thin enough to swing between adjacent warp yarns and twist themselves around one warp before moving to another yarn (Figures 27–5 and 27–6). The Leavers machine has approximately 20 brass bobbins for each inch width of the machine. A machine 200 inches wide would have 4,000 brass bobbins side by side.

Laces made on the Leavers machine are fairly expensive, depending on the quality of yarns used and the intricacy of the design. On some of the dress fabrics, a yarn or cord outlines the design. These are called *Cordonnet,* or *re-embroidered, lace* (Figure 27–1a).

Raschel. *Raschel knitting machines* (see Chapter 29) are used to make patterned laces that look like those made on a Leavers machine. They can be made at much higher speeds and thus are less expensive to produce. Filament

Fig. 27–3 *Bobbin lace: hand-made lace* (left)*; machine-made lace* (right).

Fig. 27–4 Leavers lace machine.

yarns of nylon, polyester, or acetate are commonly used to make coarser laces that are suitable for tablecloths and curtain fabrics. Raschel machines are also used to make crocheted fabric (Figure 27–7). In these machines, needles are set in the machine horizontally instead of vertically.

Quality. Quality in laces is determined by the following: (1) fineness of yarns, (2) number of yarns per square inch or closeness of background net, and (3) intricacy of the design.

Care of Lace. Because lace has open spaces, it is likely to snag and tear easily. Fragile laces should be washed by hand-squeezing suds through the fabric rather than rubbing. Some laces can be put into a cloth bag or pillowcase and machine washed.

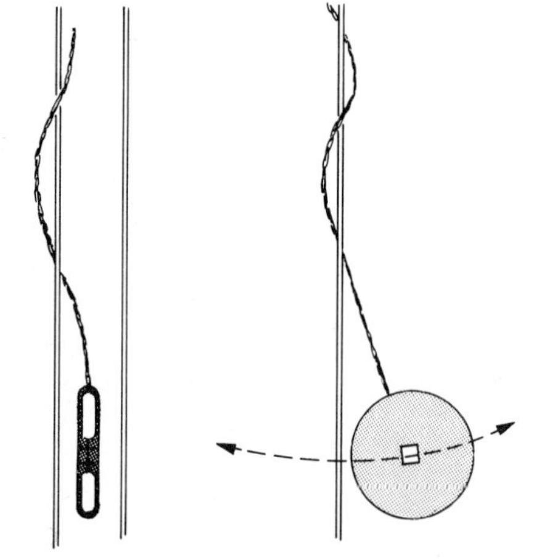

Fig. 27–5 *Brass bobbins. (Reproduced from* Textiles, *1973, Vol. 2, No. 1, a periodical of the Shirley Institute, Didsbury, United Kingdom.)*

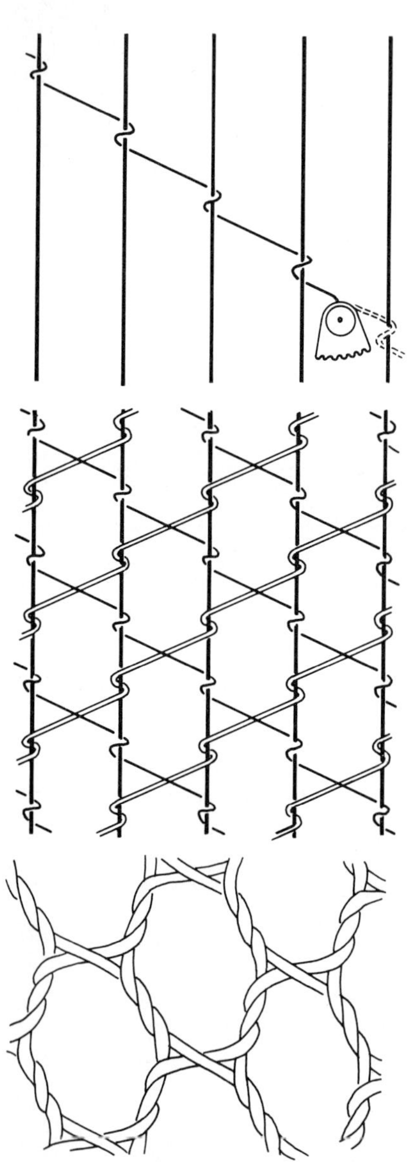

Fig. 27–6 *Brass bobbins, carrying thread, twist around warp yarns. (Reproduced from* Textiles, *1973, Vol. 2, No. 1, a periodical of the Shirley Institute, Didsbury, United Kingdom.)*

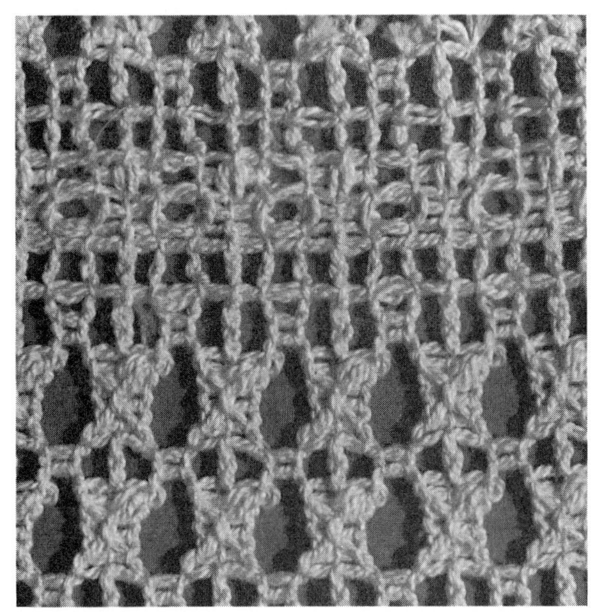

Fig. 27–7 *Raschel crochet* (top)*; raschel lace* (bottom).

Fig. 27–8 *Leno weave.*

Leno Weave

Leno is a weave in which the warp yarns do not lie parallel to each other because one yarn of each pair is *crossed* over the other before the filling yarn is inserted, as shown in Figure 27–8.

Leno is made with a *doup attachment,* which may be used with a plain or a dobby loom. The attachment consists of a thin hairpin-like needle supported by two heddles. One yarn of each pair is threaded through an eye at the upper end of the needle, and the other yarn is drawn between the two heddles. Both yarns are drawn through the same dent in the reed. During weaving, when one of the two heddles is raised, the doup-warp yarn that is threaded through the doup needle is drawn across to the left. When the other heddle is raised, the same doup-warp yarn is drawn across to the right.

By glancing at a leno fabric, one might think that the yarns were twisted fully around each other, but this is not true. Careful examination shows that they are *crossed* and that one yarn of the pair is always above the other. The fabrics made with leno weave are lace-like in character. The word *leno* comes from the French word *linon,* which means *flax.* At one time this weave was called *gauze* weave, meaning fine peculiar weave originating in Gaza, Asia. Today the word *gauze* refers to a low-count plain weave used for bandages, and a wrinkled-looking fashion fabric. Leno refers to the lace-like weave.

Fabrics made by leno weave include *marquisette* (Figure 27–9); mosquito netting; and some bags for laundry, fruit, and vegetables. Polyester marquisettes are widely used for sheer curtains. Casement draperies are frequently made with leno weave and novelty yarns. Ther-

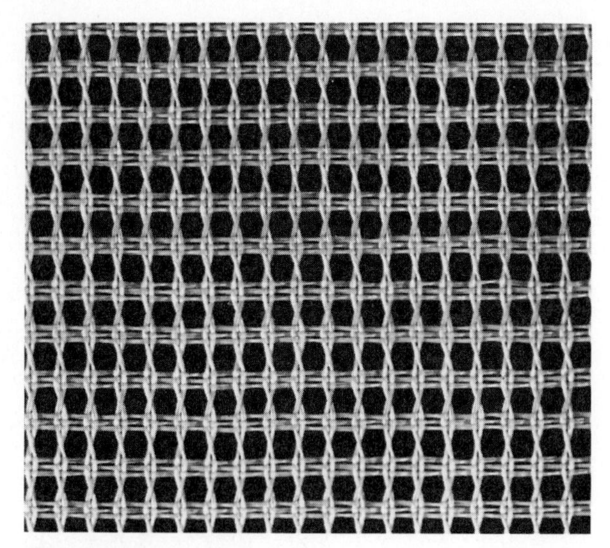

Fig. 27–9 Marquisette.

mal blankets are sometimes made of leno weave. All these fabrics are characterized by sheerness or open spaces between the yarns. The crossed-yarn arrangement gives greater firmness and strength than plain-weave fabrics of the same low count and also gives resistance to slippage of yarns. Care is determined by the fiber content, but one needs to be sure that the fabric will not be snagged during care.

Narrow Fabrics

Narrow fabrics encompass a diverse range of products that are up to 12 inches wide and made by braiding, knitting, or weaving. Narrow fabrics include ribbons of all sorts, elastics, zipper tapes, Venetian-blind tapes, couturier's labels, hook and loop tapes such as Velcro, pipings, carpet-edge tapes, trims, braids, safety belts, and harnesses (Figure 27–10 and 27–11). Webbings are an important group of narrow fabrics that will be discussed at the end of this section. Narrow fabrics are only a minor end use for fibers—1.3 percent of fibers are used in their production. Industrial uses of narrow fabrics are twice as important as apparel uses and both are much more important than home-furnishing uses: In 1984, 110.9 million pounds of fiber were used in narrow-fabric industrial uses, compared with 47.9 million pounds in apparel uses and only 4.3 million pounds in home-furnishing uses.

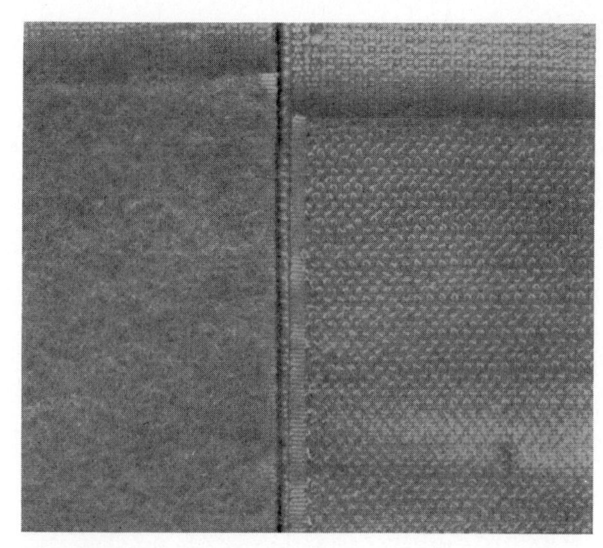

Fig. 27–10 Velcro closure: looped fabric (left)*; hooked fabric* (right).

Braids. *Braids* are narrow fabrics in which yarns are interlaced lengthwise and diagonally (see Figure 20–4). They are very pliable, curve around edges nicely, and are used primarily for trims and shoelaces.

Knitted Fabrics. *Knitted fabrics* are made on a few needles on a multiknit machine. One of the more important type of knit narrow fabrics are knit elastics. Knit elastics account for 35–40 percent of the narrow-elastic market and are used in men's underwear, running shorts, slacks, fleece products, and hosiery.

Both woven and knitted narrow fabrics (of thermoplastic fiber) are made on regular machines in wide widths and slit with hot knives to seal the edges. These are much cheaper to produce and are satisfactory if the heat sealing

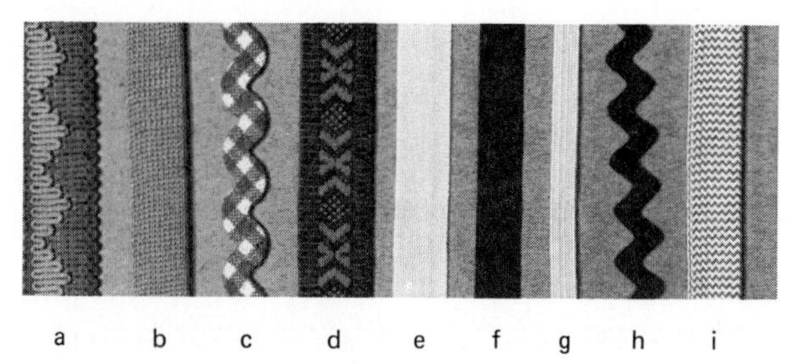

Fig. 27–11 Narrow fabrics: (a) raschel knit; (b) flat filling knit; (c) circular knit; (d) woven fancy; (e) twill tape; (f) woven ribbon; (g) woven elastic; (h) woven rick-rack; (i) braid.

is properly done. Disposeable ribbon of rayon or acetate is also available.

Woven Fabrics. Narrow-fabric looms weave many fabrics side by side. Each fabric has its own shuttle but shares all other loom mechanisms. Plain, twill, satin, pile, and jacquard are the kinds of weaves made.

Bias tape is similar in appearance to braid and has much the same characteristics, but it is made from a plain-weave fabric by cutting strips on the bias and then folding in the edges.

Woven elastics predominate in the narrow-elastics field, accounting for 60–65 percent of the market. They are used for women's bras and underwear. They have better stability and rigidity than knit elastics and are less prone to riding up.

High-Strength Woven Webbing. One very important use for high-strength woven webbing is for seat-belt fabrics. This end use accounts for 0.5 billion yards of 840-denier, 70-filament woven fabric a year! Polyester predominates now because, in comparison to nylon, it is lower priced, elongates only half as much, and can be the same strength although 10–20 percent thinner. Nylon webs are still very important in military webs and cargo handling where the additional elongation is useful. Olefin webbing is used in packaging and cargo handling. Jute webbing is used by the furniture industry.

Other uses for high-strength webbing include the following: children's belts, safety harnesses, truck ties, aircraft-arrester gear, cargo-control and lifting gear, parachute tapes, and webs for animal control (especially for dogs, horses, and show cattle).

28

Weft-Knit Fabrics

Knitting

Knitting is the formation of a fabric by the interlooping of one or more sets of yarns. Knitting has traditionally been a standard construction for some apparel—such as sweaters, underwear, and hosiery—but, for many years, knits represented only a small part of the apparel market.

The technique of knitting is probably not as old as that of weaving. Remnants of knit fabrics dating back to A.D. 250 were found near the borders of ancient Palestine. Knitting was a hand process until 1589, when the Reverend William Lee of England invented a flat-bed machine for kniting cloth for hosiery. This machine could produce cloth at 10 times the rate of hand knitting. The circular-knitting machine and the warp-knitting machine came about 200 years later. Other devices invented about that time include the ribbing device and latch needle.

A unique advantage of knitting is that it can produce—or "fashion"—a completed garment directly on the knitting machine. Sweaters and hosiery are good examples. The knitting of a completed garment was made possible in 1863 when William Cotton invented a machine that could shape garment parts by adding or dropping stitches.

The rate of production of knitting machines is relatively high—about four times as many square yards or meters per hour as for looms. A wide knitting machine will run as fast as a narrow one, whereas wide looms run slower than narrow ones. This speed should be an economic factor in favor of knitting as a method of fabrication, but the increased cost of the yarn more than offsets any savings in the cost of manufacture. There are several reasons for this. First, because the looped position of the yarn imparts bulk, more yarn is required to produce a knit cloth than to produce a comparable woven cloth. Second, the looped structure is porous—has holes or spaces—and, as a result, provides less cover than a woven fabric in which yarns lie side by side. So, in order to achieve an equal amount of cover, the knitter must use smaller stitches (finer gauge) and finer yarns that are more expensive. Also, all knitting yarns are more expensive to make because they must be much more uniform to prevent the formation of thick-and-thin places in the fabric.

Knitting is a very efficient and versatile method of making fabric. This versatility has resulted from the use of computer-aided design systems wherein electronic-patterning mechanisms permit rapid adjustment to fashion changes. Fabric-pattern designs can be changed quickly, so profitable items can be made as soon as their fashion appeal becomes evident. There is now a knitted counterpart for almost every woven fabric—knitted seersucker, piqué, denim, crepe, satin, terrycloth, velour, and fur-like fabrics are examples.

Other knitting-machine technological developments helped broaden the range of knit end uses. The double knits used in women's wear, for example, lacked the lighter weight, finer gauge, and stability needed for men's wear, so knitting machines had to be modified to make the kind of fabrics needed for men's wear. These modifications led to the development of attachments that made the knitting machine capable of producing a fabric with stability more like that of wovens. The *weft-insertion knitting machine* introduced filling yarn for more crosswise stability and the *warp-insertion knitting machine* added warp yarns for greater lengthwise stability.

Knits, like weaves, can be made from any kind of fiber or yarn and can have many textures—soft as cashmere or boardy as felt, loose or tight, inert or elastic, and rough or smooth. They can also be opaque or transparent.

Any fiber can be processed into a knit fabric for apparel, home-furnishing, and industrial uses. For example, cotton is used in shirting fabrics where it is frequently blended with polyester. Wool is used in sweaters and in better dresses and suits. Single knits and medium-weight interlocks gain popularity where styles are soft and drapey. When fashion calls for fabrics with more body, heavier double knits or bonded single knits are more popular.

Prior to 1920, most knitted items, except silk hosiery, were made of spun yarns. Man-made filament yarns created interest in knit goods of silk-like nature and they soon became widely used in women's lingerie. However, men objected to the filament-yarn knits as being too soft, slippery, and cool. Filament-tricot shirts and sheets appeared on the market in the 1950s but were unsuccessful because of poor hand, static, and a clammy feel in hot weather. (Both items were improved and reintroduced, static free and with a luxury feel, in the 1970s.)

The invention of the yarn-texturing process brought about expanded use of filament yarns in knit goods. The textured yarns were first used in stretch-nylon hosiery, leotards, and skiwear. The textured-polyester double knits, however, sparked spectacular growth in the knitting industry. By the mid 1970s, knit's share of the apparel, home-furnishing, and industrial market had risen to about 50 percent and for the first time, knitting became a serious competitor of weaving. Knitted fabrics have maintained a fairly constant share of the market. In apparel, they are very popular in active sportswear. Stabilized knits have increased the use of knit fabrics in furnishing and industrial products.

The major advantages of knitted garments to the wearer are comfort and appearance retention. Comfort in clothing is based on the ability of the garment to adapt to body movement without binding or inhibiting the wearer. The loop structure provides the fabric with outstanding elasticity (stretch/recovery) that is distinct from any elastic properties of the fibers and yarns that are used. The loop can change shape by lengthening or widening to give stretch in either direction of the cloth (Figure 28–1).

The elasticity can be controlled from a minimum to a maximum by means of the stitch construction. Unfortunately, all knits are not constructed properly for the end use and may sag, bag, or snag, thus disappointing the consumer. Knitted fabrics have higher potential shrinkage than woven fabrics. The accepted standard is 5 percent for knits, whereas 2 percent is standard for wovens. However, the performance specifications developed by the American Society for Testing and Materials list a recommended maximum shrinkage for both woven and knit products of 3 percent in both vertical and horizontal directions.

Warmth and coolness are also factors of comfort. The bulky structure of a knit provides many dead-air cells for good insulation in still air but a wind-repellent outer layer is needed to prevent chill winds from penetrating. On a warm, humid day, knits may be too warm because they tend to fit snugly and keep warm air close to the body. Knits of 100 percent polyester, however, make rather cool fabrics for winter wear and need to be blended with wool, rayon, or acrylic to alleviate this problem.

Appearance retention means lack of wrinkles during wear, care, and packing or storage. Wrinkle recovery is based somewhat on the loop structure, but it is also strongly influenced by fiber content and kind of yarn. A combination of polyester fiber, textured-filament-yarn structure, and loop stitch will produce an easy-care fabric, such as a double knit, that rarely wrinkles under any circumstances.

The following chart (page 252) summarizes some of the major differences between the processes of knitting and weaving and the fabrics made by these processes.

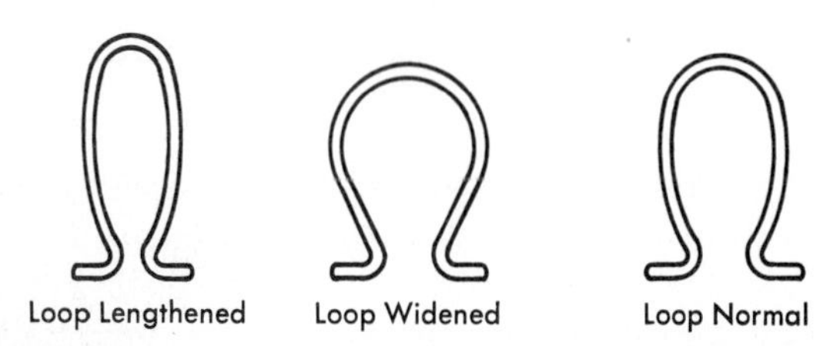

Fig. 28–1 *Loop can change its shape to give stretch.*

KNITTING TERMINOLOGY

Definition. *Knitting* is a fabrication process in which needles are used to form a series of interlocking loops from one or more yarns or from a set of yarns.

Methods of Knitting. *Filling,* or *weft, knitting* is a process in which one yarn is carried back and forth (or around) and under needles to form a fabric. Yarns run horizontally in the fabric. *Warp knitting* is a process in which a warp beam is set into a machine and yarns are interlooped to form a fabric. Yarns run vertically in the fabric. These names were borrowed from the weaving techniques and refer to the way the yarns move in the fabric.

Needles. Knitting is done by *spring-beard needles, latch needles,* or *compound needles,* which are shown in Figure 28–2. Most filling knits are formed with the latch needle. The spring-beard, or bearded, needle may be used to produce fully fashioned garments and knit-fleece fabrics. Spring-beard needles are usually used with fine yarns, whereas latch needles may be used in making coarse fabrics. A double-latch needle is used to make purl loops. The compound needle is used primarily in warp knitting.

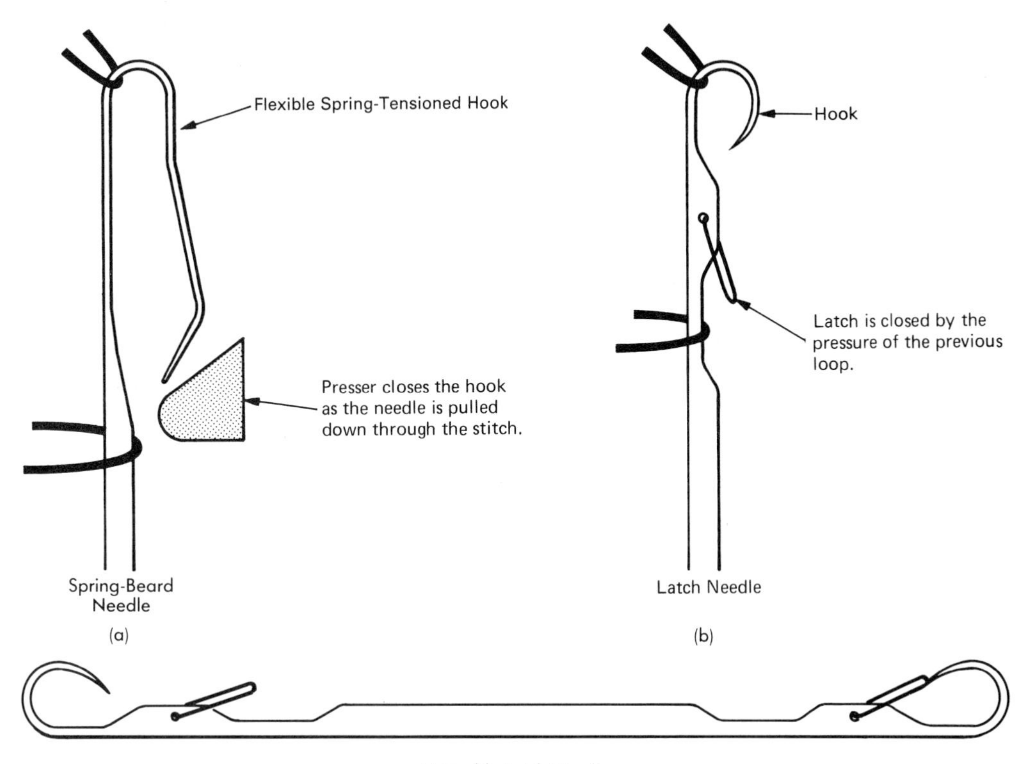

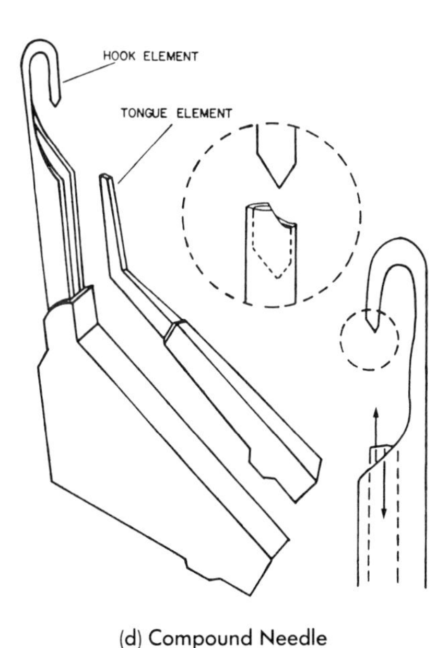

Fig. 28–2 *Knitting needles: (a) spring-beard needle; (b) latch needle; (c) double-latch needle; (d) compound needle.*

Stitches. *Stitches* are made by needles. The stitches are named based on the way they are made. Stitches may be open or closed, depending on how the stitch is formed. Open stitches are most common in filling knitting. In warp knitting, either kind may be found, depending on the design of the knit. Open or closed stitches are useful primarily in identifying the way the fabric was made. Open or closed stitches have little relationship to performance characteristics.

Comparison of Knit and Woven Fabrics

Knitting	*Weaving*
Comfort and Appearance Retention	
Mobile, elastic fabric. Adapts easily to body movement. Good recovery from wrinkles.	Rigid to stress (unless made with stretch yarns). Varies with the weave.
Cover	
Porous, less opaque. More open spaces between yarns let winds penetrate.	Provides maximum hiding power. Maximum cover per weight of yarn. Less air permeable, especially if count is high.
Fabric Stability	
Less stable in wear and care. Many shrink more than 5 percent unless synthetic fibers have been heat set.	More stable in wear and care. Many shrink less than 2 percent.
Versatility	
Sheer to heavyweight fabrics. Plain and fancy knits. Can be made to look like weaves, lace, and other fabrics.	Sheer to heavyweight fabrics. Many different textures and designs.
Economics	
Design patterns can be changed quickly to meet fashion needs. Process is less expensive but is offset by expensive raw material costs. Speedier regardless of fabric width.	Machinery less adaptable to rapid changes in fashion. Most economical method of producing a unit of cover. Wider looms weave slower.

Wales and Courses. *Wales* are vertical rows of loop stitches in the knit fabric. *Courses* are horizontal rows of loops. In machine knitting, each wale is formed by a single needle. Wales and courses show clearly on filling-knit jersey (see Figures 28–3 and 28–4).

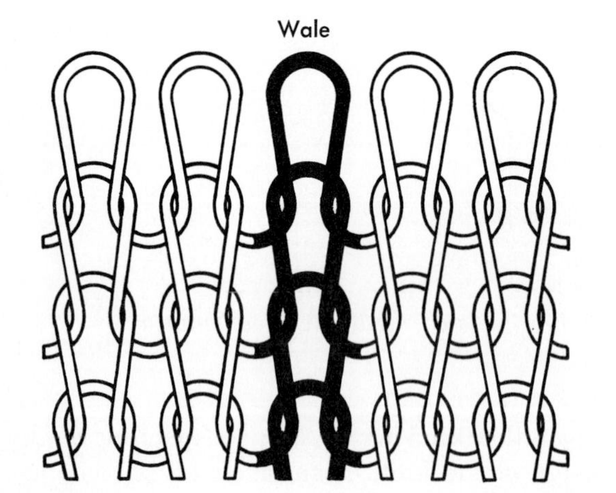

Fig. 28–3 *Wale from the technical face of filling-knit jersey fabric.*

Gauge. *Gauge,* or cut, indicates the fineness of the stitch; it is measured as the number of needles in a specific space on the needle bar and often expressed as needles per inch (npi).

The higher the gauge, or cut, the finer the fabric. The finished fabric may not have the same gauge as the machine that made it because of shrinkage or stretching during finishing. A fine-gauge, weft-knitting machine may have 28 npi or more.

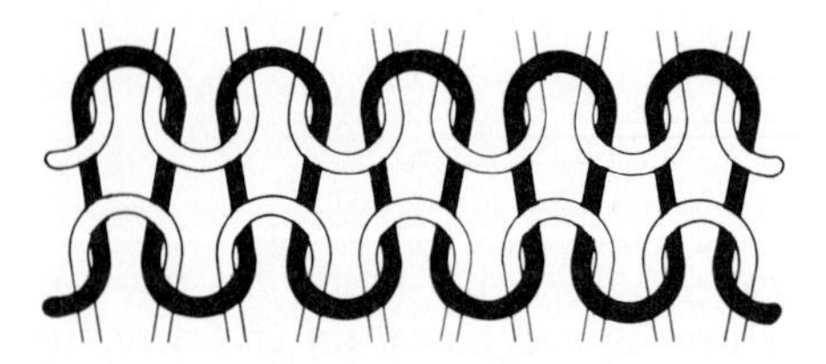

Fig. 28–4 *Course from the technical back of jersey.*

IDENTIFYING THE TECHNICAL FACE, OR RIGHT SIDE, OF A FABRIC

It may be difficult to identify the technical face of the fabric. The following list identifies some characteristics for which to look in determining the technical face of a knitted fabric (see Figure 28–3 and 28–4). *Technical face* refers to right side of fabric as knitted. This may not be the side used as the fashion side in a garment.

1. The technical face side has a better finish.
2. If two kinds of yarn or fiber are used, the more expensive one is used on the face side.
3. If floats are present, the least snaggable ones are on the face.
4. Finer yarns are on the face.
5. If the two sides differ, the design is on the face side.
6. If the fabric curls, it curls to the technical back, parallel to the wales.

Filling (or Weft) Knitting

Knitting is essentially two different industries—one is the production of finished garments and the other is the production of piece goods for cut and sewn garments.

Filling knitting can be either a hand or a machine process. In *hand knitting,* a yarn is cast (looped) onto one needle, another needle is inserted into the first loop, the yarn is thrown around the needle, and, by manipulating the needle, the new loop is taken off onto the second needle. The process is repeated with all the loops being taken off from one needle to the other. In *machine knitting,* many needles (one for each wale) are set into a machine and the loop is made in a series of steps. In the *running position,* the loop begins to move down the needle. In *clearing,* the old loop is moved down to the stem of the needle. During the *yarn-feed step,* the new yarn is positioned in front of the needle. In the *knock-over step,* the old loop is removed from the needle. The final step is the *loop-pulling step,* when the new loop is formed at the hook of the needle. These five steps are repeated in one continuous motion to form a knit (see Figure 28–5).

Both hand and machine knitting can be flat, in which the yarn is carried back and forth, or circular, in which the yarn is carried around helically like the threads in a screw. In hand knitting, many kinds of stitches can be made by varying the way the yarn is placed around the needle (in front or behind) and by knitting stitches together, dropping stitches, or transfer-

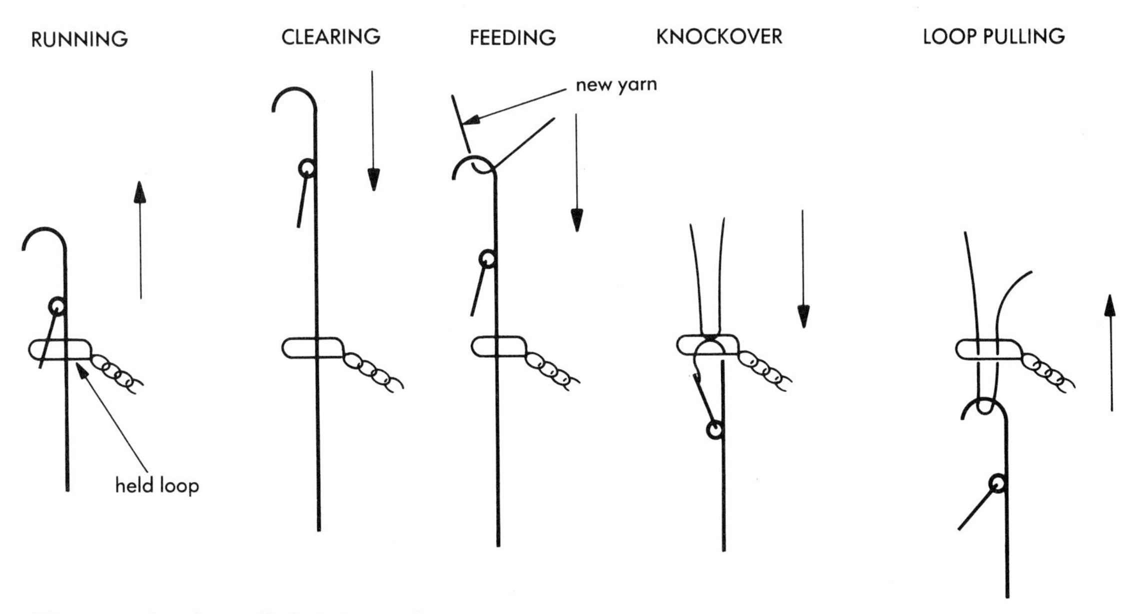

Fig. 28–5 *Latch-needle knitting action.*

ring stitches. Special mechanisms have to be used to obtain these variations in machine knitting. Fabrics may be categorized by the machine used to produce them or by the number of yarn sets in the fabric. Categorizing filling knits by the number of yarn sets is a carry-over from handknitting. Some fabrics, such as ribknits, are made with one set of yarns on a machine with two needle beds. Thus, ribknits could be categorized as single or double knits. In this book, knits are categorized by the machine used to produce them.

MACHINES USED IN FILLING KNITTING

Machine knitting is done on single- and double-knit circular and flat-bed machines. The *circular machines* are faster in production. They make yardage primarily, but are also used to make sweater bodies, pantyhose, and socks. The *flat-bed machines* knit full-fashioned garment parts and have much slower operating speeds.

FILLING-KNIT STRUCTURES—STITCHES

Filling-knit fabrics are also classified according to the stitch type used. There are four possible stitch types. The type of stitch is controlled by selection of *cams,* or guides, that control the motion of the needle. The first stitch is the *knit stitch.* This is the basic stitch used to produce the majority of filling knit fabrics (see Figure 28–6).

Fig. 28–6 *Close-up of a plain-jersey stitch: technical face* (left)*; technical back* (right).

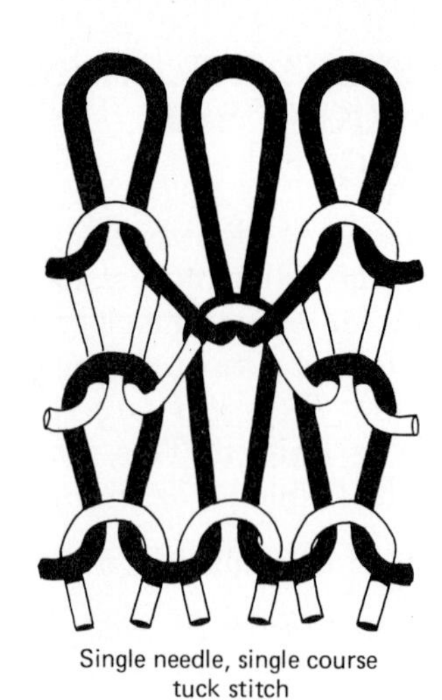

Fig. 28–7 *Tuck stitch. (Courtesy of* Knitting Times, *the official publication of National Knitwear and Sportswear Association.)*

The *tuck stitch* is used to create a pattern in the fabric. In the tuck stitch, the old stitch is not cleared from the needle. Thus there are two stitches on the needle. Figures 28–7 and 28–8 show how the tuck stitch looks in a fabric. In a knit fabric with tuck stitches, the fabric is thicker and slightly less likely to stretch crosswise than a basic-knit fabric with the same number of stitches.

The *float stitch* is also used to create a pattern in the fabric. In the float stitch, no new stitch is formed at the needle while adjacent needles form new stitches. The float stitch can be used when yarns of different colors are used to create patterns. Figure 28–9 shows how the float stitch looks in a fabric. A knit fabric with float stitches is thinner and much less likely to stretch cross-

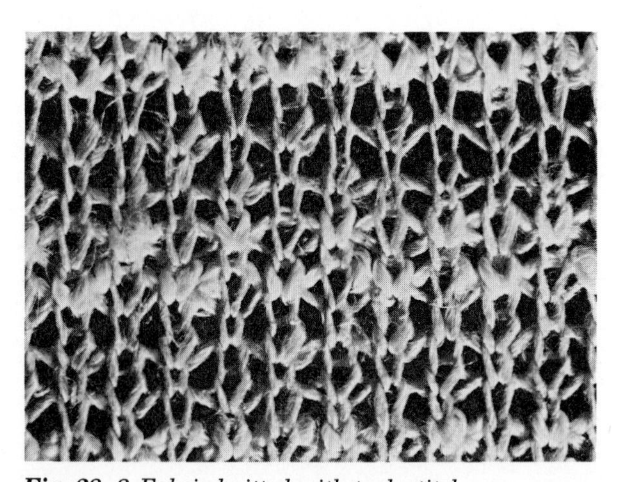

Fig. 28–8 *Fabric knitted with tuck stitch.*

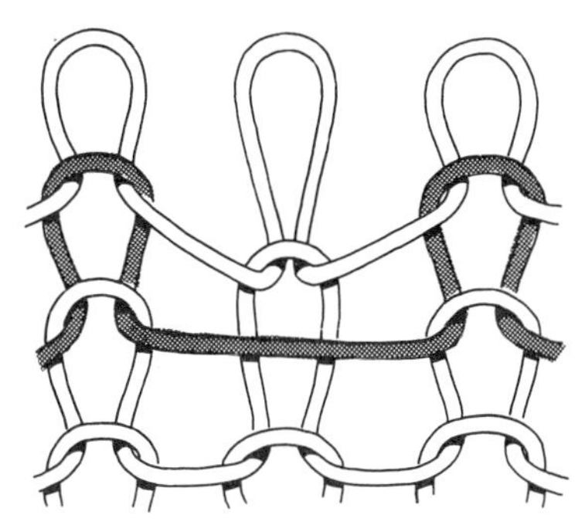

Fig. 28–9 Float, or miss, stitch. (Courtesy of Knitting Times, *The official publication of National Knitwear and Sportswear Association.)*

wise than a basic-knit fabric with the same number of stitches.

The *purl,* or *reverse,* stitch forms a fabric that looks like the technical back of a basic-knit fabric on both sides. The fabric is reversible (see Figures 28–10 and 28–11).

Filling-Knit Classifications

SINGLE-FILLING KNITS

Single-filling knits are made using a machine with one set of needles. This machine is usually a circular one, but it may also be a flat-bed one.

Single knits can be plain color or printed, striped or patterned, gossamer thin or heavy

Fig. 28–10 Purl stitch looks the same on both sides.

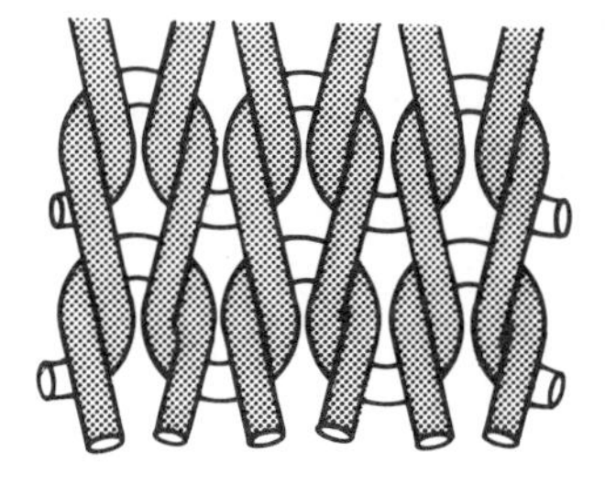

(a) Technical Face

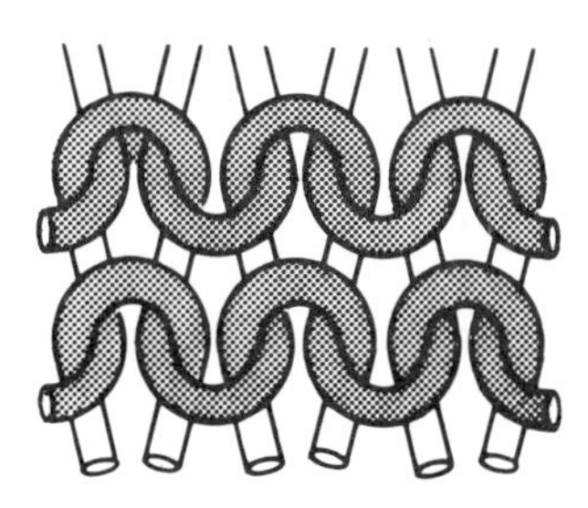

(b) Technical Back

Fig. 28–11 Single jersey: (a) technical face; (b) technical back.

enough for use in winter sweaters. They are less stable than double knits, tend to curl at the edges, and run readily if made of filaments. Fabric names for knits are not as specific as for woven fabrics.

Plain or Single Jersey. *Single jersey fabric* is the simplest of all filling-knit structures. The face side has prominent wales—columns of loops running lengthwise. The back has prominent courses—rows of stitches running crosswise (see Figure 28–6). Stretch a swatch of jersey crosswise and it will curl to the wrong side at the lengthwise edges. The ends will curl toward the face. Ravel out a yarn. It will ravel crosswise because the yarns run horizontally in the fabric. A run can be created by pulling on a cut edge of the fabric or by cutting or breaking a yarn. The run then forms vertically when the broken or cut loop drops loops above and below it. Single jerseys made of staple-fiber yarns tend to resist running because of fiber cohesiveness and tend to be flatter than other knits. They stretch more widthwise than lengthwise. Figure 28–11 shows a circular jersey-knitting machine.

The single-jersey structure, or plain knit, is widely used because it is the fastest method of filling knitting and is made on the least complicated knitting machine.

Jersey is a light-to-heavy weight fabric usually knitted on a circular-jersey machine and sold in tubular form or cut and sold as flat goods. When tubular fabrics are pressed at the factory in the finishing operation, the creases are seldom parallel to the wales of the fabric—they are off-grain. The tubular cloth does not need to be cut and opened out, when cutting out the garment, unless there is a specific reason for doing so.

Figure 28–12 shows a child's top that was cut from tubular cotton jersey with crosswise stripes (wales are crosswise in the garment instead of lengthwise as knitted in the fabric). When the garment was purchased the stripes were vertical and the side seam was perpendicular to the lower edge. After washing, the fabric assumed its normal position causing the side seams to twist and the stripes to spiral.

End uses for plain-knit structures include hosiery, underwear of cotton or blends, shirts, T-shirts, dresses, and sweaters.

Variations in plain knit are made by programming the machines to knit stitches together, to drop stitches, and to use colored yarns to form patterns or vertical stripes. Extra yarns or slivers are used to make terrycloth, velour, and fake-fur fabrics.

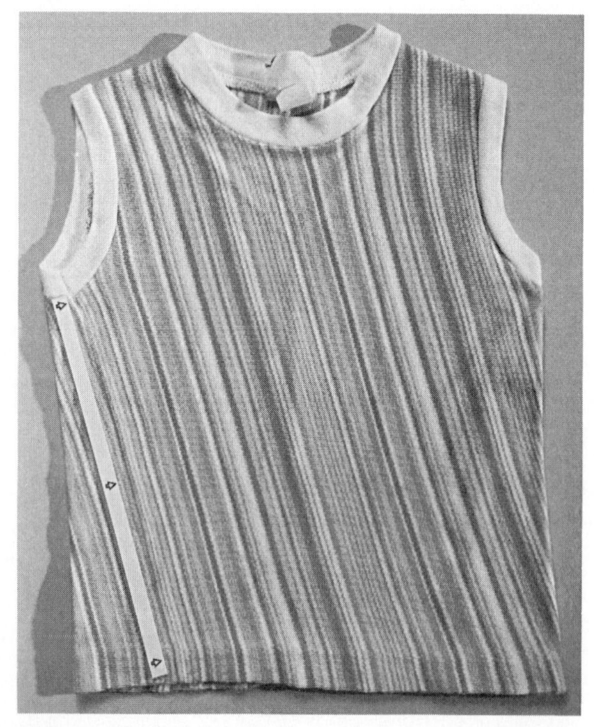

Fig. 28–12 *Child's knitted top. Side seam is shown by white strip with arrows.*

Two stitches commonly used to make jersey variations are the tuck stitch and the float stitch. *Tuck stitches* are used to create blisters or special effects and to secure laid-in yarns or long floats of yarns on the wrong side of the fabric. Figure 28–8 shows tuck stitches in fabric. *Float stiches* are used to carry colored yarn on the back of fabric for knitted-in designs.

Stockinette is another name for jersey. Wool, acrylic, polyester-spun yarns, cotton, and cotton/polyester blends are widely used in jersey.

Single-figured jerseys are made by a jacquard mechanism on circular-jersey machines. Printed jerseys are more commonly used.

Intarsia designs in jersey are made by laying in colored yarns. True intarsia designs have a clear pattern on both the right and wrong side of the cloth with no bird's-eye backing that is characteristic of jacquard designs. Fabrics have no extra weight, and the stretch is not impaired. Mock intarsia designs are made by knitting and float-knitting (float or miss stitch), which results in a heavier-weight fabric with floating yarns on the reverse side. These floating yarns reduce the elasticity of the fabric and may snag readily.

Pile-knit jerseys are made on a modified circular jersey machine. The fabrics look like woven pile but are more pliable and stretchy. The pile surface may consist of (1) cut or uncut loops of yarn or (2) fibers (see the following discussion of sliver knits). *Knitted terrycloth* is a loop pile used for beachwear, robes and babies' towels and washcloths. It is softer and more absorbent than woven terry but does not hold its shape as well. *Velour* is a cut-pile fashion fabric used in men's wear, in women's pant suits, and in robes. Velour is knit with loops that are cut evenly. Then the yarn twist is uncurled to give better coverage and the fabric is dyed, tentered, and steamed. The fabric is 60 inches wide and about 17½ ounces/yard.

Sliver-pile knits are made on a special weft-knit, circular, sliver-knitting machine and are called fur-like, high-pile, or deep-pile fabrics. They have been available since 1955. Examine Figure 28–13 and notice that yarns are used for the ground; the *sliver* furnishes the fibers for the pile. Sliver is an untwisted rope of fiber and is the product of either carding, drawing, or combing (Chapter 19).

Fibers from the sliver are picked up by the knitting needles—along with the ground yarns—and are locked into place as the stitch is

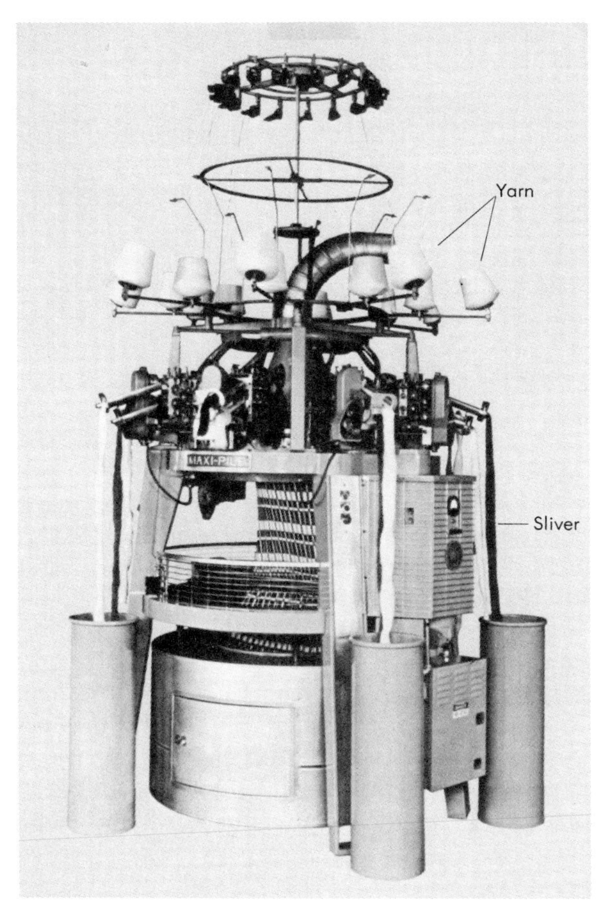

Fig. 28–13 Sliver-knitting machine. (Courtesy of Rockwell International Corp.)

formed. A denser pile can be obtained with sliver than with yarn because the amount of face fiber is not limited by yarn size or by the distance between yarns (Figure 28–14).

The following are the steps used in finishing fur-like fabrics:

- *Heat setting,* which shrinks the ground fabric and expands the diameter of the individual face fibers.
- *Tigaring,* a brushing operation that removes surplus fiber from the face of the fabric.
- *Electrifying,* also known as *polishing,* is a brushing operation. The fibers are combed first in one direction and then in the other by grooved, heated cylinders that rotate at high speed. This may be repeated several times to develop the required finish. The process gives high luster.

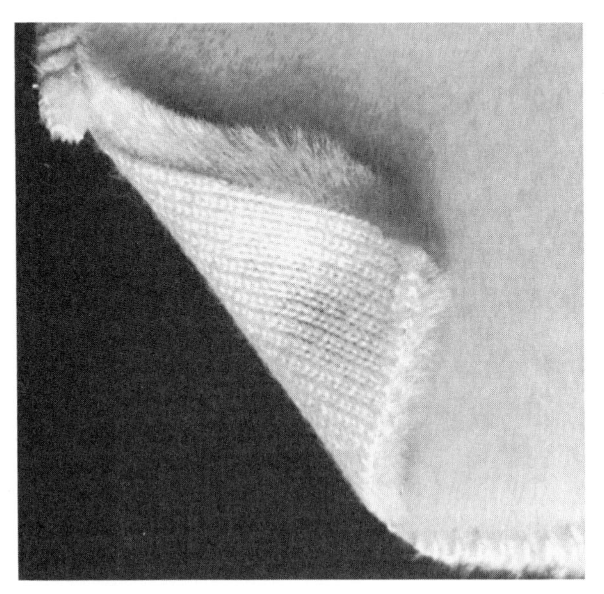

Fig. 28–14 Sliver knit fur-like fabric.

The fur-like knits are usually made from acrylic, modacrylic, polyester, and olefin fibers or from blends or combinations of these fibers. A modacrylic fiber was originally used for the ground yarns of the pile because its high shrinkage when heated could be used to advantage to make a much more compact pile. Cotton and olefin are also used.

The surface pile can be made with guard hairs to resemble mink, for example; printed to resemble pony, jaguar, or leopard; or printed in other designs for fun furs. Fibers are usually solution dyed because piece dyeing distorts the pile very badly. Prints are made by screen printing.

Fur-like fabrics are used for the shells (the outer surface) and for linings (the inner surface) of coats and jackets. The difference is mainly one of weight, the shells being heavier than the linings. In actual use the dividing line is less distinct because shell fabrics are used as linings in expensive garments and as shells in low-priced garments.

Fur-like fabrics are much lighter in weight, are much more pliable, and have better comfort characteristics than real fur. They require no special storage and can be successfully dry cleaned by using a cold tumble dryer and combing the pile rather than steam pressing it.

Weft-Insertion Jersey. In *weft insertion,* a yarn of any type is laid in a course as that course is being knit. The yarn is not knit into stitches, but is laid in the loops of the stitches as they are

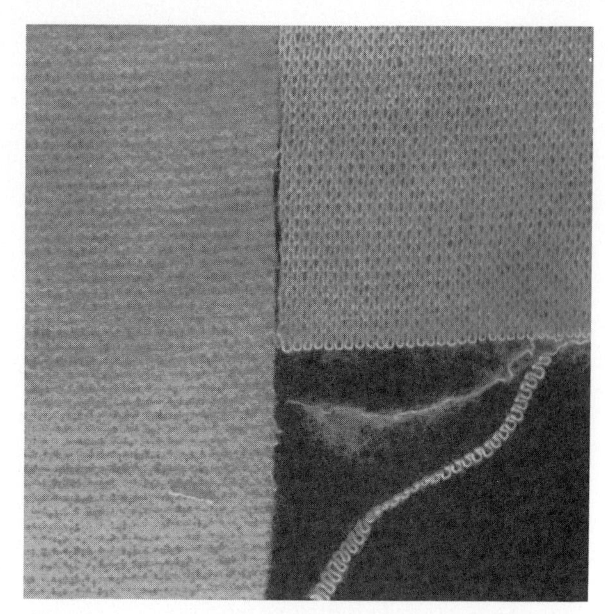

Fig. 28–15 Weft-insertion weft knit: napped side (left) *and technical face with knitting and in-laid yarns* (right).

being formed. The yarn may be novelty or too large or too irregular for normal knitting. The laid-in yarn increases the crosswise stability of the fabric. The yarn may be used for decorative, strength, stability, or elasticity reasons. The yarn also may be designed to give a pile effect on napping or brushing during finishing. In-laid yarns for pile-effect fabrics such as napped jerseys may have so little twist that they are too weak to withstand the stress of knitting (see Figure 28–15).

DOUBLE-FILLING KNITS

Double-filling knits are made using a machine with two sets of needles, with the second bed or set of needles located at a right angle to the first bed of needles. Most double-knitting machines have the two needle beds arranged in an inverted V and are called vee-bed machines. The double-knit fabric may be made with one or more sets of yarns. Double knits are categorized based on the arrangement of the needles in the double-knitting machine, or the gait of the machine. In a *rib-gait machine,* the two beds of needles are positioned so that both needles can be knitting at the same time. The needles of one bed are located in the spaces between the needles of the other bed. In *interlock gaiting,* the needles are positioned so that only one needle bed can be knitting at a time. The needles of one

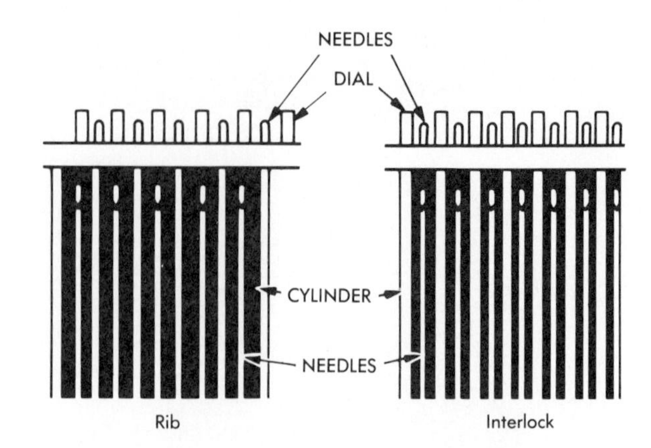

Fig. 28–16 Gaiting: rib (left); *interlock* (right).

bed are located directly across from the needles of the other bed (see Figure 28–16 for rib and interlock gaiting).

Double-knit fabrics are made on circular or flat-bed machines with two sets of needles and may be made with any combination of the four stitches. In the flat-bed machine, the needles from one bed pull the loops to the back and those in the other bed pull the loops to the front (Figure 28–17). In the circular machine, the loops

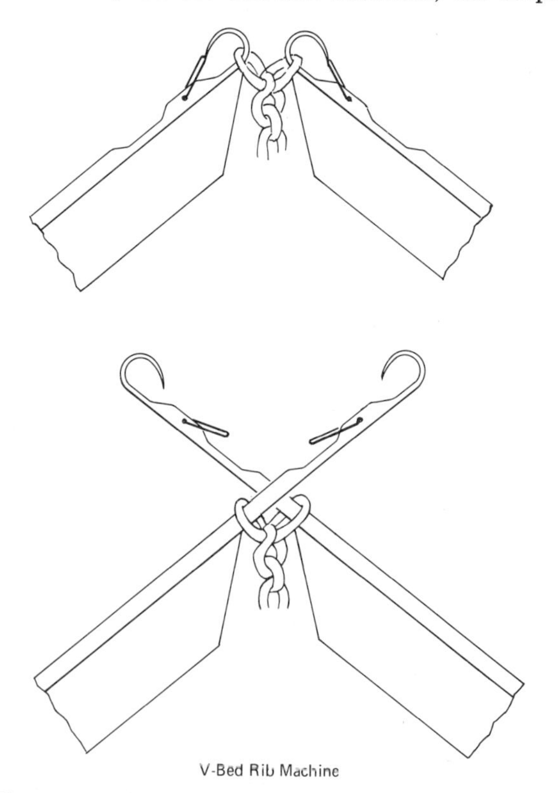

Fig. 28–17 Needle action in flat-bed machine. (Courtesy of Knitting Times, *the official publication of National Knitwear and Sportswear Association.)*

are pulled to the face and back by setting one set of needles vertically in a cylinder and the other set of needles horizontally in a dial (Figure 28–18).

Fabrics have two-way stretch and relatively high-dimensional stability—especially the polyester double knits. They do not curl at the edges, and are less apt to stretch out than single knits. They do not run. Double knits can be made to look like any woven structure and are often given the woven fabric name—denim, seersucker, double piqué, and the like.

A technique used to illustrate the production of double knits is a diagram based on the two needle beds and the type of gaiting. A center line represents the space between the beds. A short vertical line represents a needle. In *interlock gaiting,* the short vertical lines are directly opposite each other (Figure 28–19). In *rib gaiting,* the short vertical lines stop at the horizontal line and needle lines on one side of the line are staggered with needle lines on the opposite side

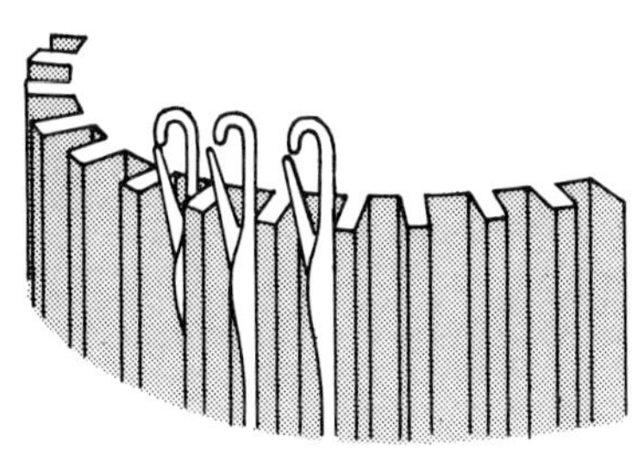

(a) Section of Cylinder

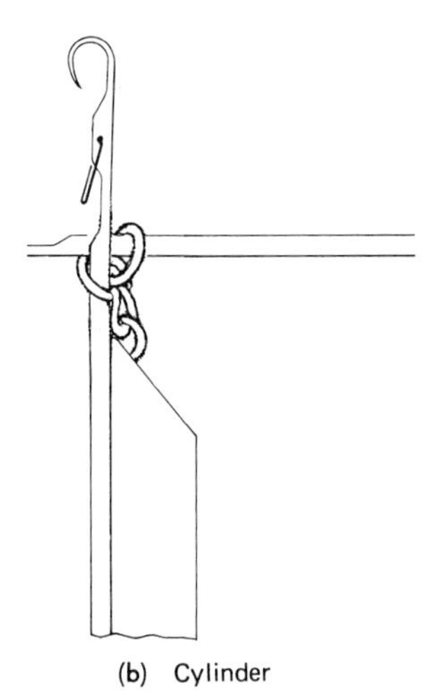

(b) Cylinder

Fig. 28–18 *(a) Needle beds and (b) knitting action in circular-knitting machine. (Courtesy of* Knitting Times, *the official publication of National Knitwear and Sportswear Association.)*

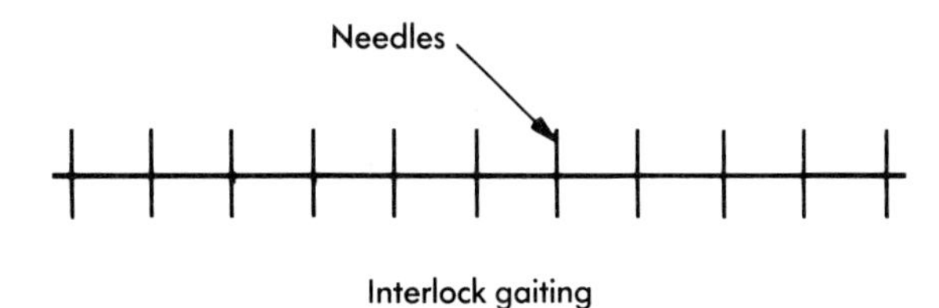

Fig. 28–19 *Interlock gaiting.*

(Figure 28–20). In the diagram, loops represent a knit stitch, an inverted V represents a tuck stitch, and a — represents a float or miss stitch. Thus a 1 × 1 rib would be diagrammed on rib gaiting, as shown in Figure 28–21. A simple interlock would be diagrammed in two steps on interlock gaiting because two steps are required to create the interlock fabric (see Figure 28–22). A double knit on interlock gaiting might be a ponte de roma, which requires four steps to create the fabric (see Figure 28–23). A double knit on rib gaiting might be la coste, which requires eight steps in knitting (see Figure 28–24). An easy way to identify a double knit is to look at the edge of the fabric parallel to a course. If all loops point in one direction it is a single knit. If some of the loops point toward the front and some toward the back, it is a double knit. If the loops are directly opposite each other, it is made with interlock gaiting. If the loops are not directly opposite each other, it is made with rib gaiting.

Double-knit jersey, like *interlock jersey,* looks the same on both sides. It differs in that it is made on rib gaiting and needles from the cam and cylinder are not opposite each other but are positioned so that the needles from one bed work between the needles from the other bed. They knit a 1 × 1 rib. To distinguish between an interlock and a rib-gaiting fabric, cut along a

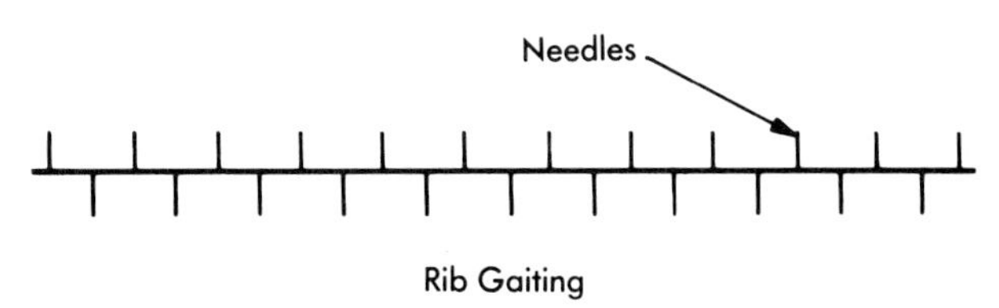

Fig. 28–20 *Rib gaiting.*

Fig. 28–21 *1 × 1 rib (rib gaiting).*

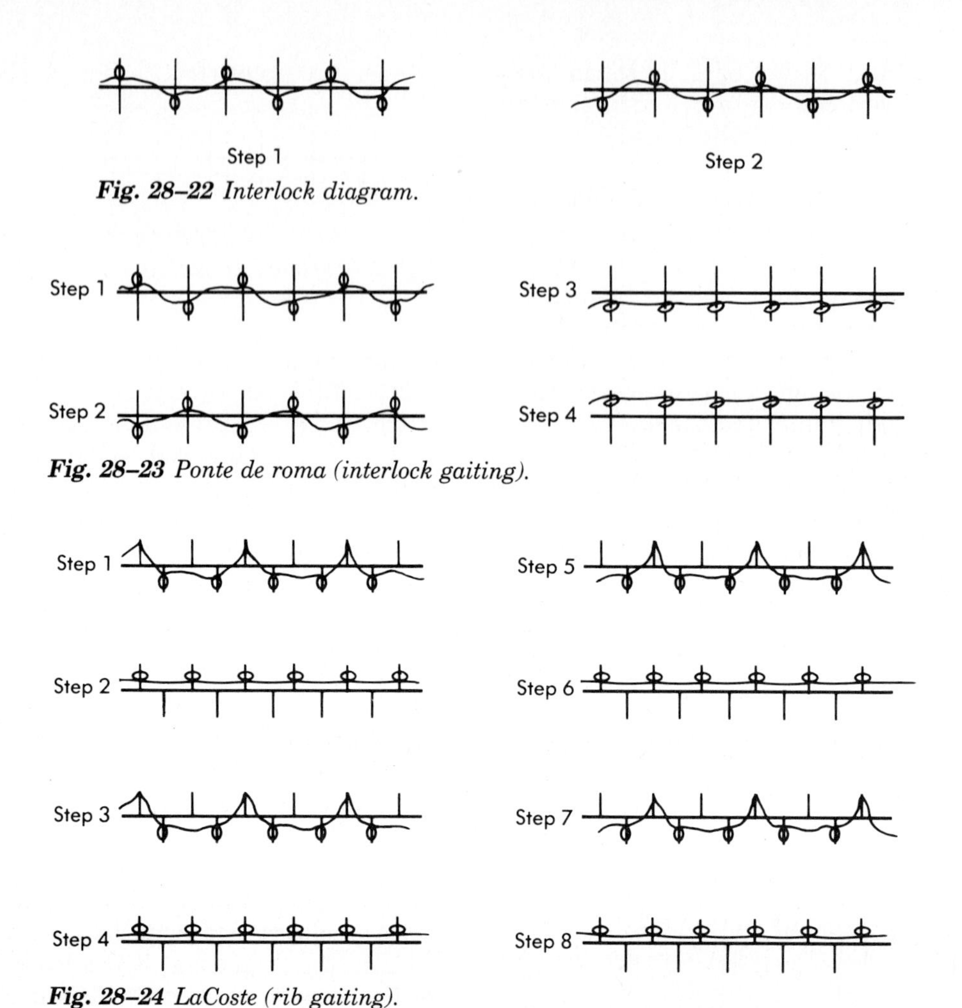

Fig. 28–22 *Interlock diagram.*

Fig. 28–23 *Ponte de roma (interlock gaiting).*

Fig. 28–24 *LaCoste (rib gaiting).*

course and stretch the edge widthwise. Examine the edge. If it is interlock, there will be a back loop opposite each front loop; if it is rib double knit, the back loops will alternate between the front loops.

Rib Structure. A *rib structure* is made of face wales and back wales. The lengthwise ridges are formed on both sides of the fabric by pulling loops first to the face and next to the back of the cloth. In hand knitting, ribs are made by alternating knitting and purling. These may be in various combinations 1 × 1, 2 × 2, 2 × 3, and so on (Figure 28–25). Figure 28–26 shows a T-shirt fabric in rib knit. The 1 × 1 rib is the simplest double-knit fabric produced using rib

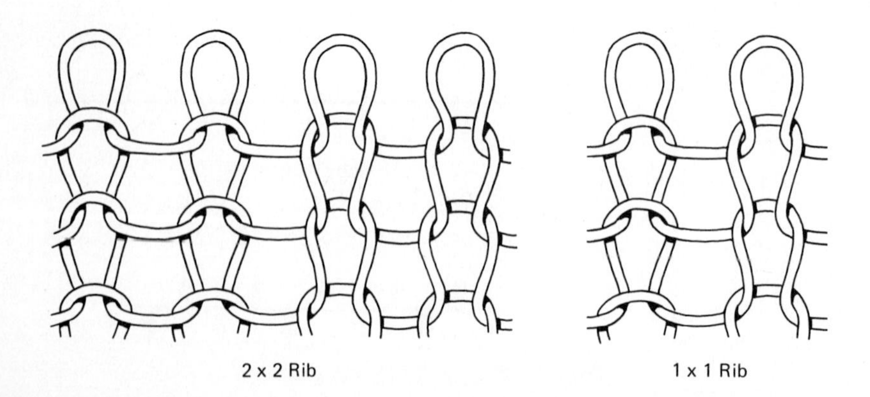

Fig. 28–25 *Rib stitches: 2 × 2 rib* (left); *1 × 1 rib* (right).

Fig. 28–26 Rib-knit fabric: fabric relaxed, left side 2 × 2 rib, technical face (top)*; fabric stretched to show difference in stitches* (bottom).

gaiting. It usually consists of one set of yarns.

Rib knits have the following properties: (1) they have the same appearance on the face and back, (2) they have twice the extensibility crosswise as that of single jersey, (3) they do not curl at the edges, (4) they run, (5) they unravel from the end knit last, and (6) they are twice as thick as single jersey.

Rib-Jacquard Double Knits. These fabrics have almost limitless design possibilities. The intermeshing of the two yarns is the same as the double-knit jersey but have added needle-selecting mechanisms—pattern wheels, pattern drums, punched tapes or cards, or photographic electronic film (Figure 28–27).

Bourdelet is a ripple stitch, or corded, fabric produced by knitting and tucking. *Double piqué* has a fine, diamond-shaped pattern on both sides of the fabric. Names of woven fabrics are often used for double knits that resemble them but

Fig. 28–27 Tape-punching machine prepares pattern tape (top)*; circular knitting machine using pattern tapes* (bottom). *(Courtesy of Rockwell International Corp.)*

more often the term polyester double knit is the only name needed to identify this type of fabric.

Purl Structure. *Purl knits* usually are made on machines with two needle beds and double-hooked needles with either interlock or rib gaiting. Purl is generally the slowest form of knitting, but it is also the most versatile. The purl is the only weft-knitting machine that can produce all three types of weft-knit fabrics—plain, rib, and purl—although as a rule not as economically. Thus a knitter can use a purl machine to make a garment that is part plain, part rib, and part purl.

Fabrics produced by the purl stitch are thick, wide, and short as compared to single jersey with the same number of plain stitches. Fabrics do not curl, but they do run and may be made to unravel from both ends. See Figure 28–28.

The two major end uses for purl structures are children's and infants' wear and sweaters. Purl stitches are often used at the shoulder seams of sweaters to stabilize the garment since they have less crosswise stretch than plain knit.

Fancy purl knits are made by knitting groups of face loops and back loops to form a pattern. The areas curl in opposite directions giving very puffy designs.

The purl machine always moves to the left so it is sometimes called a links-links machine.

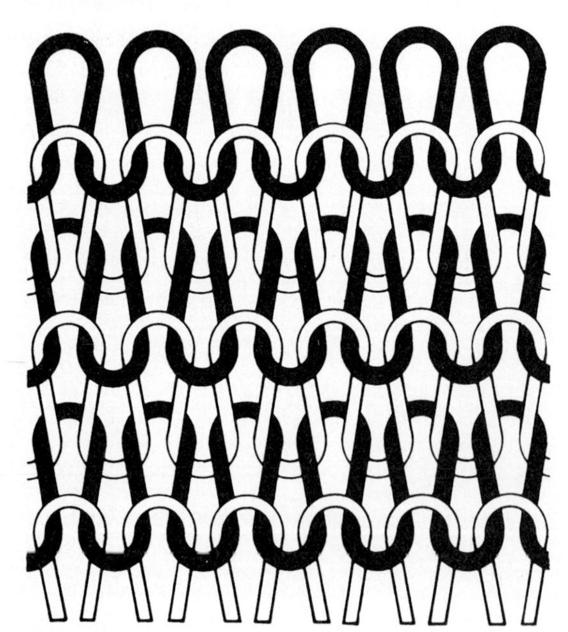

Fig. 28–28 Purl fabric. (Courtesy of Knitting Times, *the official publication of National Knitwear and Sportswear Association.)*

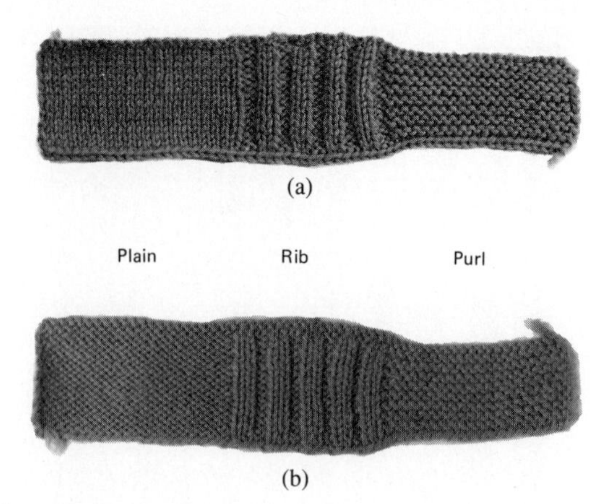

Fig. 28–29 Hand-knit structure: (a) technical face; (b) technical back.

The term links is a German word meaning leftward.

Figure 28–29 is a hand-knit sample in which the same number of stitches was used to knit each section. Notice that the sections differ in length and width as well as in appearance. The plain or single jersey has a different appearance on face and back but the rib and purl look the same on both sides.

Interlock Structure. The *interlock* is the simplest double-knit fabric produced using interlock gaiting (see Figure 28–22). Interlock fabrics are composed of two 1 × 1 rib stitches intermeshed. (Figure 28–30 is offset for clarity).

Both sides of the fabric are alike and resemble the face side of single jersey. Interlock stretches like plain jersey, but the fabric is firmer. Inter-

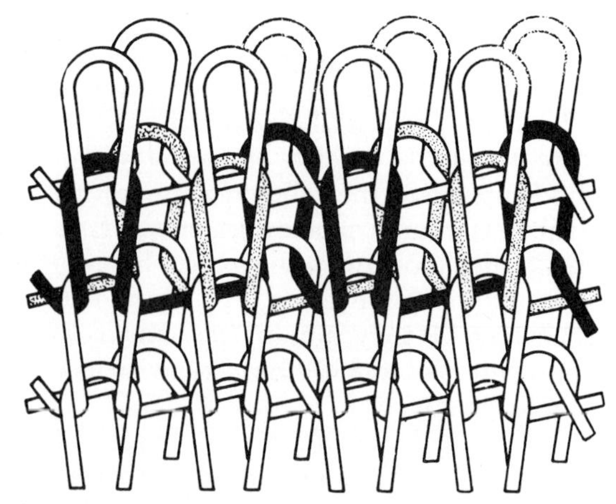

Fig. 28–30 Interlock (diagram offset for clarity). (Courtesy of Knitting Times, *the official publication of National Knitwear and Sportswear Association.)*

locks do not curl and fabrics run and unravel from one end only.

A run occurs when the stitches in a wale collapse or pull out. A run occurs in a stepwise fashion when one stitch in a wale after another collapses due to stress on the loop and when a yarn is severed. Most interlock fabrics are plain or printed. Colored yarns can be knitted to give spot effects or horizontal or vertical stripes.

Polyester Double Knits. Polyester double knits have had a tremendous impact on the knitting industry as well as on home sewing and the ready-to-wear (RTW) industry. The development of textured-set yarns, proper dyeing techniques, and electronic patterning devices helped the "overnight success" (1967) of polyester double knits.

The first textured "set" polyester double knits were made for women's wear and were of fall or winter weight—standard weight was 14½ ounces in a 16–18 gauge because this was what the existing machinery could produce. Development of finer gauge machinery made it possible to make double knits that were compact and that more closely resembled woven fabrics.

Snagging is probably the single most serious problem encountered in the use of knit fabric. When a yarn is snagged so it pulls out and stands away from the surface of the fabric, "shiners" or tight yarns are formed on either side of the snag. If snags are cut off (rather than being worked back into original position) a run may start in some knits—particularly in weft knits. Incorrect amount of twist and improper knitting are major reasons for cloth snagging.

The *mace test* is a particularly tough test for snagging. It consists of running a round, spiked, iron ball a specified number of times along the surface of the cloth to test its resistance to snags (Figure 28–31). Finer yarns, smaller stitches, and higher twist all contribute to snag resistance. This test has been found to give good results in terms of correlation with what happens in actual wear. Antisnag, antistatic, and soil release finishes can be applied to knit fabrics. If these finishes are present, they should be identified on the label.

KNITTING GARMENT PARTS

Garment parts—sweater bodies, fronts, backs, sleeves, skirts, socks, seamed hosiery, and col-

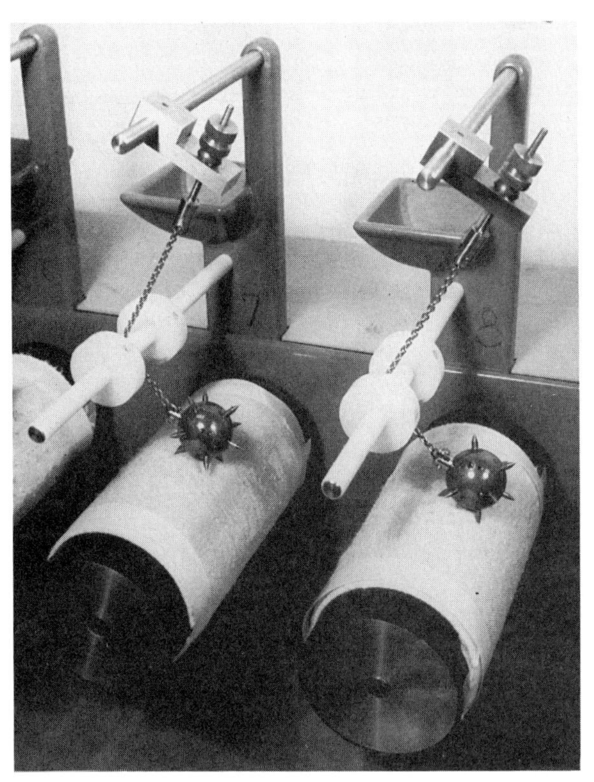

Fig. 28–31 *Mace test.*

lars—can be knitted to shape on flat-bed machines. The stitch used for shaping is called *loop transfer*. A knit loop is transfered from one needle to another, usually near the end of a course, so that the width of the fabric is decreased. The process is called *fashioning* and done to make armholes, neckline curves, collar points, and the like. Garment parts have finished edges.

A *looping machine* is used to join the shoulders and sleeves of the shaped parts with an effect of continuous knitting rather than of seams. This machine is also used to join collars to cut-and-sewn knit garments.

To identify fashioned garments, look for "fashion marks" accompanied by an increase or decrease in the number of wales. (Figure 28–32). Mock fashion marks are sometimes put in the garment but they are not accompanied by an increase or decrease in the number of wales, so no shaping is done by the mock fashion marks (Figure 28–33). Full-fashioned sweaters are almost always made with a jersey stitch. Circular jersey sweaters are cut and sewn.

Full-fashioned garments do not necessarily fit better than cut-and-sewn garments because this

Fig. 28–32 Raglan sleeve portion of full-fashioned sweater. Notice that stitches are dropped.

depends on the size and shape of the pieces. But full-fashioned garments are always on-grain, look better to the discerning eye, and they should not become misshapen during washing—no twisted seams.

The following chart summarizes the differences and similarities between flat-bed and circular machines from a single- and double-knit perspective.

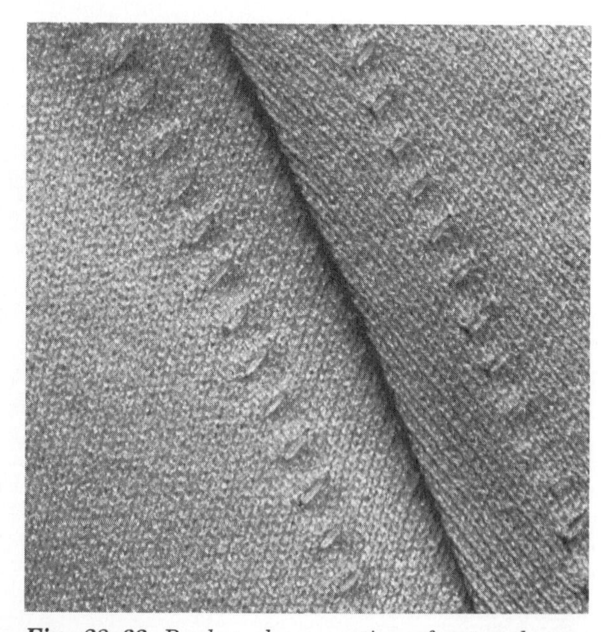

Fig. 28–33 Raglan sleeve portion of cut-and-sewn sweater. Notice mock-fashion marks.

Hosiery

Before the advent of knitting, people wore ill-fitting stockings that were cut and sewn.

Fashion and man-made fibers have been responsible for many of the developments in hosiery. Women's legs were shown in public for the first time in the twentieth century. Dress lengths gradually went from the floor in 1900 to far above the knee in 1970, creating a need for sheer, well-fitting women's hose and finally panty hose. Stretch nylon caused a change in construction methods.

YARNS USED FOR HOSIERY

Spun Yarns. Spun yarns are used for socks. They may be of any fiber content, acrylic and cotton blends are most commonly used. Two-ply mercerized cotton, called *lisle,* is stronger and more durable than regular cotton. Spandex is used in the tops of socks. Nylon is commonly used as reinforcement in the heels and toes of socks.

Filament Nylon Yarns. Filament nylon yarns are used in women's hosiery and lighter weight socks. They may be monofilament or multifilament.

When nylon was introduced in 1939, women were wearing silk hosiery. The first nylons were 30 denier, a service-weight stocking that was extremely durable. As nylon hose became more sheer, durability decreased because finer yarns do not wear as well as heavier yarns. Most nylon hosiery is now 15 denier. Nylon stockings were first made of conventional nylon fiber (Figure 28–34), and elasticity and fit were dependent on the knit loop and the fiber's elasticity. In 1954, textured stretch yarns were introduced. Textured stretch yarns give better fit and better wear and make it possible for retailers to carry fewer sizes.

Bicomponent, self-crimping filaments are also used in hosiery. Cantrece II made by du Pont is advertised as fitting like a second skin (Figure 28–35).

Bicomponent fibers, spandex, and rubber are used in support and surgical hose. *Support hose* are worn by both men and women for comfort. People who work in jobs that require them to be

Filling-Knitting Machines and Fabrics

	Single Knit		Double Knit	
	Jersey—Flat	*Jersey—Circular*	*Flat (Rib/Interlock)*	*Circular (Rib/Interlock)*
Description	Straight bar holding one set of latch needles	See Figure 28–11. One set of latch needles	Two flat needle beds formed in ∧ position, see Figure 28–17	See Figure 28–18. One set of needles mounted on dial, one set on cylinder
	Yarn carried back and forth	Yarn carried around	Yarns carried back and forth	Multiple feed-yarn carried around needle selection mechanism
	Purpose is to shape garments	Electronic control patterns make range of designs	Stitch-transfer carriage can switch from one bed to another to make variety of stitches	
Kinds of knits and end uses	Basic knit stitch	Workhorse of knitting industry		Double knits—plain and jacquard double knits
	Fabric has different appearance on face and back	Fabric has different appearance face and back	Same appearance face and back	
	Full-fashioned garments	High-volume production Seamless hose Jersey, velour, terry	Used when fabric must have finished edge Collars, trims	
Advantages	Economical use of yarn Garments always on grain Design variations possible	Fastest method	Less waste than circular rib	High-speed production Excellent design flexibility Versatile in yarn usage
Limitations	Quite slow in production Higher priced end product Single-feed system	Variety of pattern possibilities available	Slow speed	Complex machine Downtime can be a problem

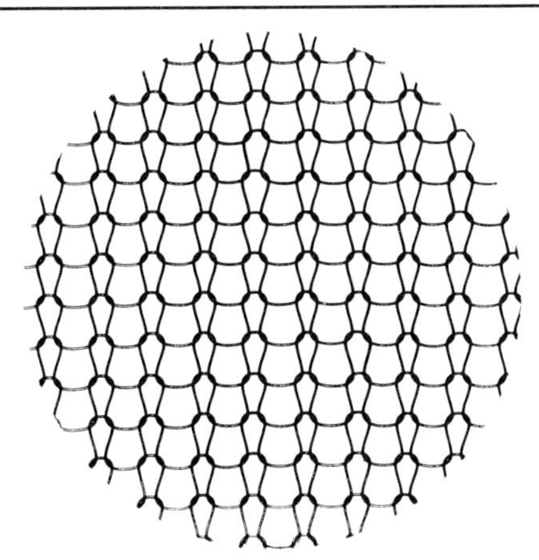

Fig. 28–34 Monofilament hosiery yarn. (Courtesy of E. I. du Pont de Nemours & Company.)

on their feet most of the time wear support hose to prevent muscular fatigue. Support hose are also beneficial to pregnant women. They are available in nylon and spandex. See Figure 28–36.

Surgical hose, prescribed by doctors for leg disorders such as varicose veins, are worn by hospital patients to prevent blood clots after an operation. They may be purchased singly or in pairs, with or without heels and toes, and in various lengths—ankle, knee, or over the knee. It is very important to choose the correct size of surgical hose so that the support from the stocking is placed to give support where it is needed.

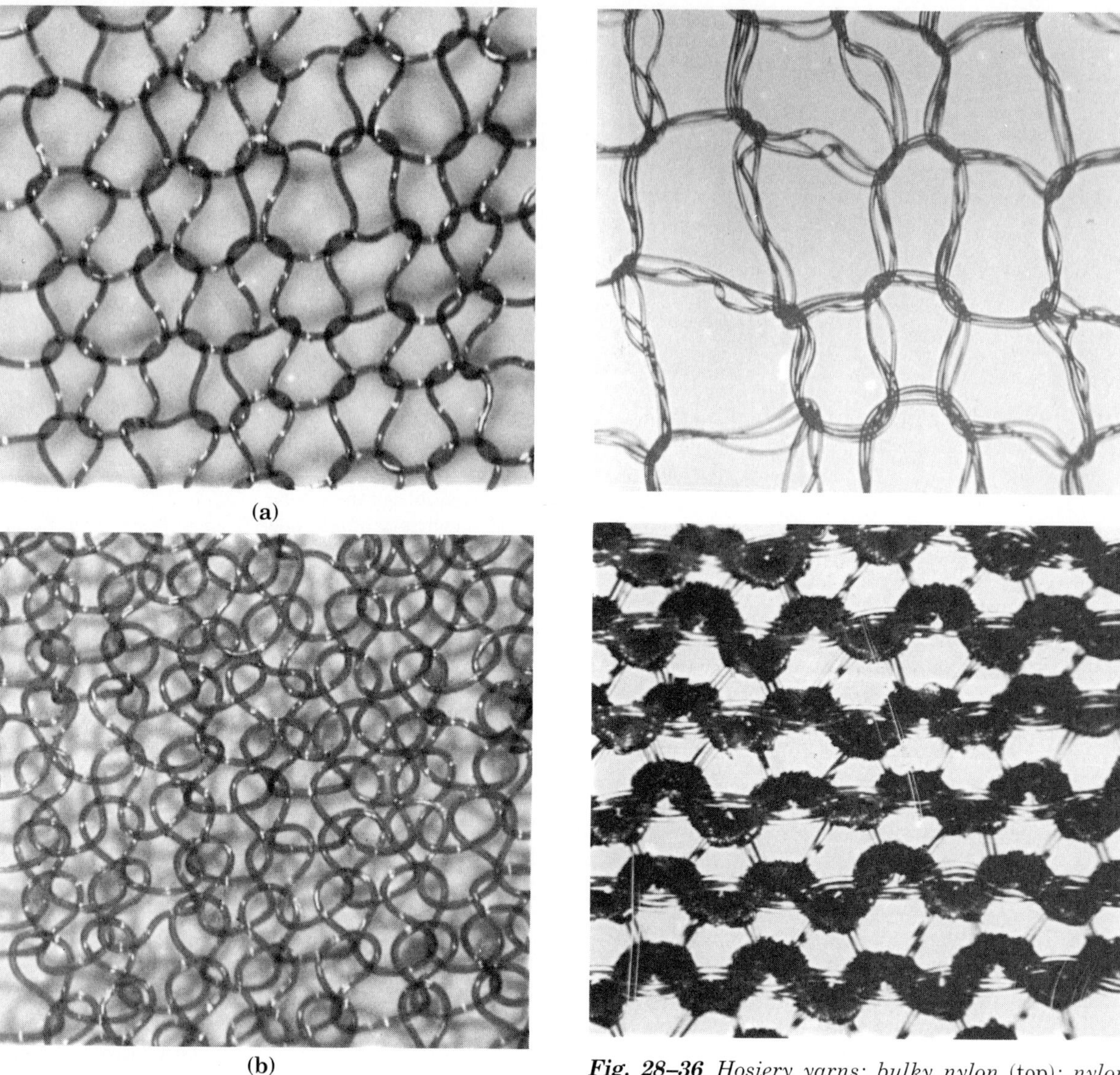

(a)

(b)

Fig. 28–35 *Nylon Cantrece: (a) stretched; (b) relaxed. (Courtesy of E. I. du Pont de Nemours & Company.)*

Fig. 28–36 *Hosiery yarns: bulky nylon* (top)*; nylon and spandex* (bottom). *(Courtesy of the Fibers Division of Monsanto Chemical Co., a unit of Monsanto Co.)*

The yarns are spandex, or rubber for the elastic core, and cotton and nylon are used as the warp. Surgical hose are quite expensive and can be made to order. Figure 28–35(b) shows a wrapped spandex and nylon construction.

CONSTRUCTION

All hosiery is a filling knit. The types include plain (or jersey), rib, mesh, and micromesh. The plain knit has stretch in both directions, and hose can be very sheer if made of fine denier monofilaments. Plain knit has the disadvantage of running readily when a loop is broken. Mesh hose are lacelike knits that do not run, but they snag and holes will develop. Micromesh has loops knitted so that a run goes up only. Mesh and micromesh stockings are not as elastic or as smooth as plain jersey. Rib stitches and jersey are used in socks. Fancy knits such as cables and argyles are often used. Some hose and socks may be heat transfer printed (see Chapter 36).

Seamless. *Seamless hose* have the same number of stitches from top to toe and are knit on circular machines. Shaping may be done by decreasing the size of the loop gradually from top to toe. If shaping is done at toe and heel, a circular fashioning mechanism that drops stitches is used. Heavier yarn can be knit into toe and heel to give greater comfort and durability. Seamless hosiery is knit in one piece as a continuous operation. When knitting is finished, the toe is closed and the stocking is turned right side out.

Long hose of regular nylon and socks with shaped heels are *preboarded,* a process in which stockings are placed on metal leg forms of the correct size and shape and then steamed to press. Tube socks and stockings do not have shaped heels. They are seamed across the toe end.

Panty hose are usually made from textured stretch nylon. The panty portion is heavier than the stocking portion. Panty hose are knitted in tube shape with a guide for slitting. After slitting the panty section, two tubes are stitched together in a U-shaped crotch seam with a firm, serged stitch. A separate crotch section is often inserted for better fit. The crotch section is of a double layer of nylon or nylon and cotton. Control top panty hose may have spandex or bicomponent nylon/spandex in the panty portion, or a separate panty of cut and sewn power net (see Chapter 29) may be stitched to stockings.

29

Warp-Knit Fabrics

Warp knitting is unique in that it developed as a machine technique without ever having been a hand technique. Warp knitting started about 1775 with the invention of the *tricot machine* by Crane of England. The tricot (pronounced *tree-ko'*) machine is sometimes called a warp loom because it uses one or more sets of yarns that are wound on warp beams and mounted on the knitting machine (Figure 29–6). The first tricot machine made fabric 16 inches wide for silk-stocking cloths. In 1880, Kayser established the first warp-knitting mill in the United States.

Warp knitting provides the fastest means of making cloth from yarns. It has been said that warp knits combine the best qualities of both double knits and wovens. Warp knits can duplicate wovens in many respects while, at the same time, offering the performance and easy care of knits. Warp-knit fabrics tend to be less resilient and lighter weight than weft knits. They can have stability in both directions of the cloth or exhibit a degree of stretch, as determined by the control of the knitting stitch.

Warp knitting produces a vertical-loop construction, as shown by the line of loops in Figure 29–1. It is a machine process of making fabric in flat or sheet form using one or more *sets* of warp yarns that are fed from warp beams to a row of knitting needles extending across the width of the machine. Each set of yarns is controlled by yarn guides mounted in a guide bar that also extends across the width of the machine. Warp-knitting machines are wider than looms with a width of 170 inches. If there is one set of yarns, the machine will have one warp beam and one guide bar; if there are two sets of yarns, there will be two warp beams and two guide bars, and so on: hence the terms *one-bar tricot* and *two-bar tricot.* All guide bars feed yarn to the same set of needles. Each yarn guide on the bar guides one yarn to the hook of one knitting needle. Chains with links of various heights control the movement of the guide bars. More guide bars give greater design flexibility. The loops of one course are all made simultaneously when the guide bar raises and moves sideways, laying the warp around the needles to form the loops, which are then pulled down through the loops of the preceding course.

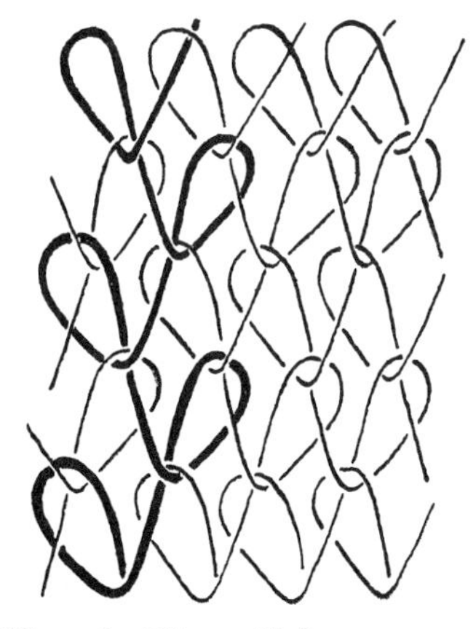

Fig. 29–1 *Warp-knitting stitch.*

Warp knits are usually diagrammed using a *point-paper notation.* In this notation, each point in a horizontal row represents a needle. In warp knitting, all the needles knit at one time. Thus each row of points represents the needles used to produce one course. In the point-paper notation, the movement of a guide bar is diagrammed for each course. Because the guide bar controls the movement of the yarn, the guide-bar diagram shows yarn movement. Thus the diagram for each row of points represents the movement of the guide bar to create the yarn loops for that course. The next row of points would represent the next course, and so on until one complete repeat has been represented. The diagram starts at the bottom row of points and moves up the paper from course to course as time progresses. Figure 29–1 is a diagram for a warp knit. In Figure 29–2, steps (a) through (g) show the movement of the guide bar as it directs the yarn's movement to create that warp-knit stitch. In (a) of Figure 29–3, the seven steps are combined in one diagram; in (b), the resulting yarn loop is shown; in (c), the two repeats of the pattern are shown. In Figure 29–4, the knitting action of a tricot machine is diagrammed.

Yarn from the front bar usually predominates on the surface, whereas yarn from the back bars provides run resistance, elasticity, and weight.

The warp-knitting industry is expanding into nonapparel uses. It is also focusing on short runs rather than on volume production and lead times of weeks rather than months. Apparel end uses for warp knits include lingerie, underwear, sportswear, and outerwear. Industrial end uses are the most rapidly expanding area and include fabrics for sun and light protection, for controlling rock falls, for grass collection, for snow barriers, and for dam reinforcement. Warp-knit fabrics are also used as medical implants such as

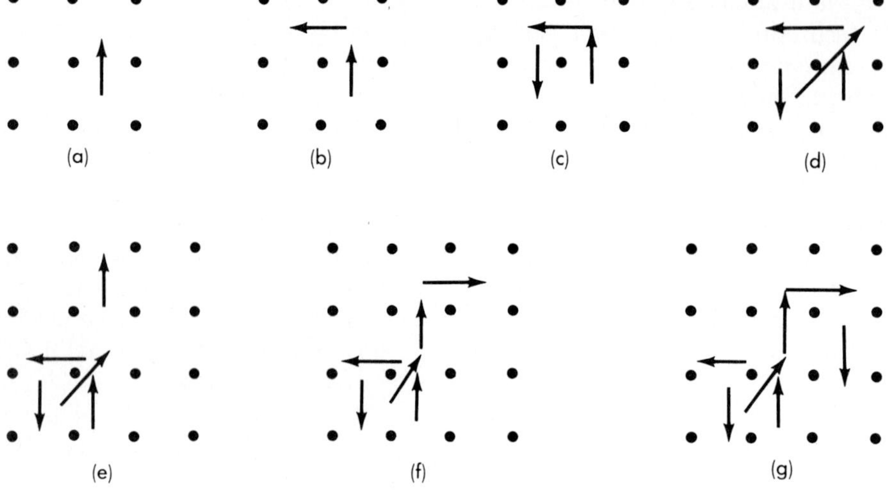

Fig. 29–2 Guide-bar movements, step by step, for warp-knit stitch.

artificial veins and tissue-support fabrics.

Even though warp knitting is fast, warp knits are not inexpensive because the process requires very regular yarns. Thus the cost of the yarns offsets the fast speed of the process.

MACHINES USED IN WARP KNITTING

Warp knits are classified by the equipment type used to produce the fabric and the characteristics of the resulting fabric. Tricot machines use a single set of spring-beard or compound needles. Tricot-knitting machines with computer-controlled guide bars, electronic beam control, and computerized take-up are able to knit 2,000 courses per minute. Raschel machines use one or two sets of vertically mounted latch needles. Jacquard Raschel knitting machines with as many as 78 computer-controlled guide bars are available. The differences between the fabrics produced by these machines have become less distinct in the past several years. Several types of warp knitting machine are listed in the chart on page 272. Tricot and Raschel machines, however, account for the manufacture of about 95 percent of all warp-knit goods.

WARP KNITS VERSUS FILLING KNITS

The two kinds of knits differ because of the different knitting techniques and the machines used in their manufacture. The major differences are summarized in the following chart (on page 272).

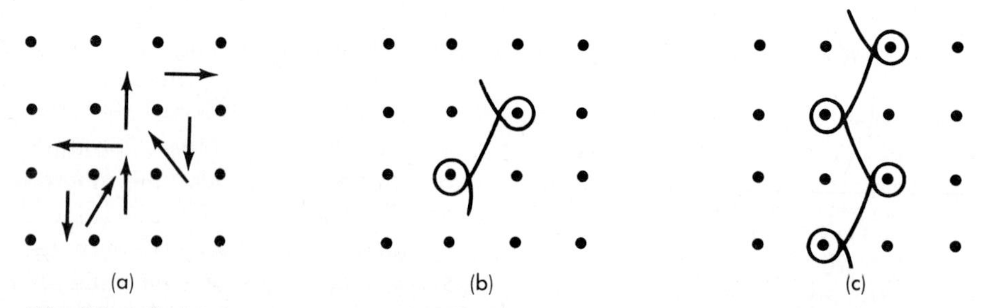

Fig. 29–3 Guide-bar movements: (a) all steps of the guide-bar movement; (b) yarn movement following the guide-bar movement; (c) series of yarn loops creating warp-knit fabric.

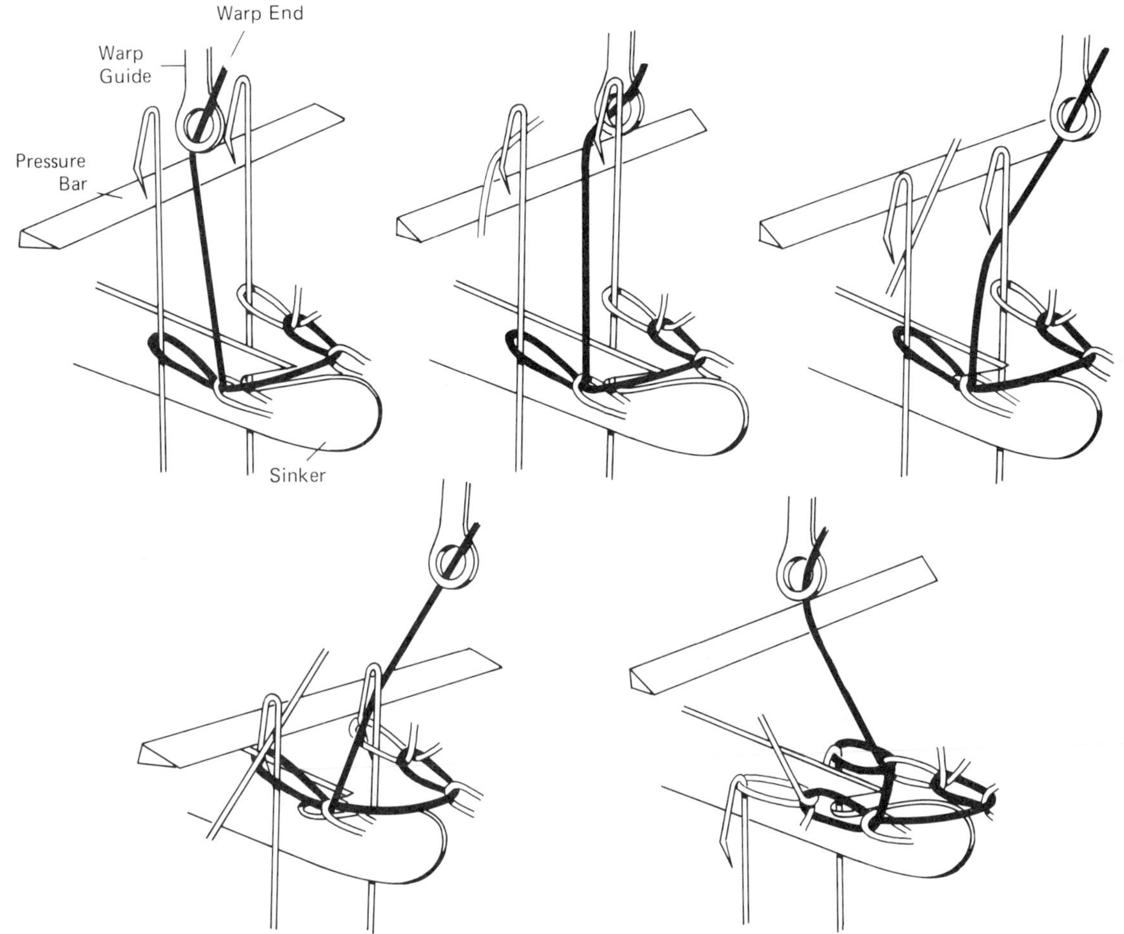

Fig. 29–4 Knitting action of tricot machine. (Courtesy of Knitting Times, *the official publication of National Knitwear and Sportswear Association.)*

Tricot-Warp Knits

The name *tricot* has been used as a generic name for all warp-knit fabric; specifically, it is the fabric produced on the tricot machine using the *plain stitch*. This fabric is called tricot. Tricot comes from the French word *tricoter,* meaning *to knit.*

The plain stitch is shown in Figure 29–5. The face of the fabric (top photo) is formed of vertical portion of loops and the back (lower photo) has horizontal portion of loops. The face has a much finer appearance than the back. Tricot is runproof and nonraveling. The fabric will curl just as weft-knit jersey does. Tricot fabric has high tear strength, high resiliency, and elasticity, which is greater in the crosswise direction. Some

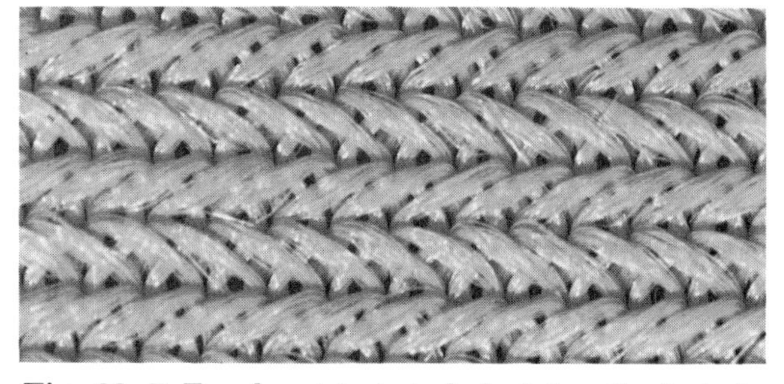

Fig. 29–5 *Two-bar tricot: technical face* (top); *technical back* (bottom).

Warp-Knitting Machines

Tricot	*Raschel*	*Simplex*	*Milanese*
Single bed	One or two needle beds	2 sets of needles	Flat—spring-beard needles
Spring-beard needles or compound needles	Latch needles Coarse gauge	Spring-beard needles	Circular–latch needles
2–3–4 bars indicate number of sets of warp yarns	May have as many as 65 guide bars		Yarn travels diagonally from one side of material to the other
Simple fabric	Complex fabric		
High-speed, high-volume	Great design possibilities	Rare	Rare
Usually filament yarns	Usually spun or spun and filament		
Wide fabric 170 inches	Narrow fabric (100 inches)		
	End Uses		
Plain, patterned, striped, brushed fabric Underwear Outerwear	Sheer laces and nets Draperies Power net Thermal cloth Outerwear	Warp double-knits Gloves	Underwear Outerwear

important end uses of tricot fabrics include the following:

- Lingerie
- Sleepwear
- Loungewear
- Men's shirts
- Uniforms for nurses, waitresses and the like
- Jersey dresses and blouses
- Other outerwear
- Automotive fabrics (upholstery)

Comparison of Filling and Warp Knits

Filling Knits	*Warp Knits*
Yarns run horizontally	Yarns run vertically
Loops joined one to another in the same course	Loops joined one to another in adjoining course
Connections are horizontal	Connections are diagonal
More design possibilities	Higher productivity
More compact fabric	More compact fabric
Two-way stretch	Crosswise stretch, little lengthwise stretch
Run, most ravel	Do not run or ravel
Hand or machine process	Machine process
Flat or circular	Flat
Can have finished edges	
Can knit shaped garments, garment pieces, or yardage	Produced as yardage

The Tricot Machine. The first tricot machines which used spring-beard needles, were designed to use the finer yarns for blouses, underwear, and loungewear. Spun yarns were not knit on these machines, because the increased friction from the fuzzy ends and the irregularities that are typical of spun yarns caused excessive yarn breakage. Until the early 1970s, only 2 percent of the warp knits were made of spun yarns. Then tricot machines, such as the Reading Spunwarp machine that used a latch needle, were developed for knitting spun yarns and novelty yarns as well as textured yarns.

The modern tricot machine is the mainstay of the warp-knitting industry (Figure 29–6). It is a high-speed machine that can knit flat fabric up to 170 inches wide. The machine makes a plain-jersey stitch or can be modified to make tuck stitch, clipped dot, jacquard, and other designs. Another modification is the attachment for laying in yarn in a tricot structure.

Tricot Fabrics. *Plain tricot* is made on a machine employing one set of needles and two guide bars. Filament yarns are used in either smooth or textured form. Acetate is an inexpensive yarn that is used for tricot backing of bonded fabrics.

Few, if any, methods of cloth manufacture can produce anything comparable to nylon tricot in the standard ranges of 15–40 denier. Nylon tricot is light weight (17.5–6.5 yards/pound), has exceptional strength and durability, and can be heat set so that it is dimensionally stable. One of the unique features of nylon tricot is that the same piece of gray goods can be finished under different tensions to different widths that have different appearances; for example, 168-inch gray goods can be finished at 98, 108, 120, 180, or 200 inches wide.

Brushed tricots are a common application of warp-knit fabrics. The velvet-like surface consists of loops raised from the surface. Heat-set

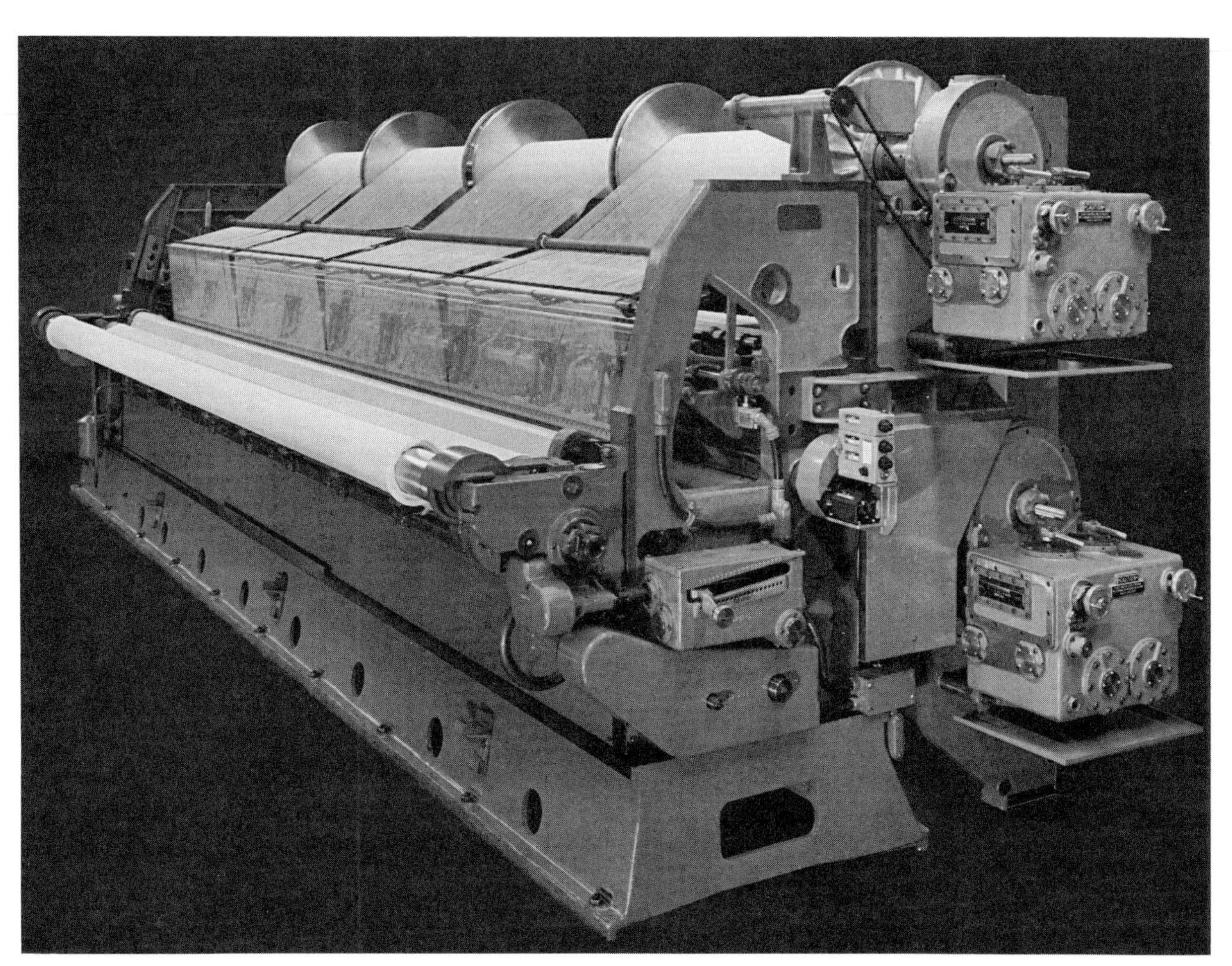

Fig. 29–6 *Warp-knitting tricot machine. (Courtesy of Textile Machine Works.)*

nylon must be used, so it is possible to raise the loops without breaking them. The fabric is versatile and has several end uses such as sleepwear, evening gowns, shoes, slacks, upholstery, and draperies. Velcro closures with loops acting as catches for hooks are another end use.

Napped velours can be dyed, dried, heat set, and then napped on a double-action napper, heat set again to proper width, and then sheared.

The knit stitches have long underlaps. One set of yarns is carried over 3–5 yarns to form floats; the second set of yarns interloops with adjacent yarns. Nylon is used for the adjacent looping to provide strength and durability. The floats are broken when the fabric is run through the napper. The napped side is used as the right side of the fabric even though it is the technical back (Figure 29–7).

Satin-like tricots are made in the same way as napped tricots except that the fabric is Schreinered (see Chapter 33) instead of napped. These fabrics are usually 100 percent nylon or polyester and the floats are longer.

Tricot-net fabric can be made by skipping every other needle so only half as much yarn is used and open spaces are created in the fabric.

Tricot strips for trim can be made by cutting on a slitter or by knitting in a yarn that is later dissolved.

Striped tricot is made by using acetate yarn on the second and/or third bars and pulling the heavier yarn to the face in alternating horizontal rows. Cross dyeing gives the difference in color.

Tuck effects use the same yarns as in the striped fabrics, but a change in the yarn arrangement forms a "lip," or tuck. The tucks may be straight, wavy, irregular, intermittent, wide, or narrow.

Automotive tricot upholstery—a double-knit velvet—has been introduced (see Figure 29–8).

Tricot Finishing. Nylon and polyester require heat setting to stabilize the fabric against shrinkage, to improve the hand, and to give good stitch definition and desirable cutting properties. Most heat setting is done on the pin-tenter frame. Either radiant heat or superheated steam is used. Shrinkage can be controlled to within 1 percent. If the tricot is dry heat set on the tenter frame with 15 percent extension in width, improved crease resistance during wear is achieved.

The Schreiner finish (see Chapter 33) was first used on tricot in 1957. Called Satinette, the finish causes a permanent flattening of the yarns, thus filling in the open spaces of the knit structure to give more cover and make the fabric more opaque. The fabric has a satiny smoothness.

Pleating is a natural for nylon tricot and is done with a pleating machine, which folds the fabric with pleating paper in the form desired. The fabric is steamed and the paper is removed. (A pleated effect can be achieved during the knitting process by leaving out yarns in alternate areas.)

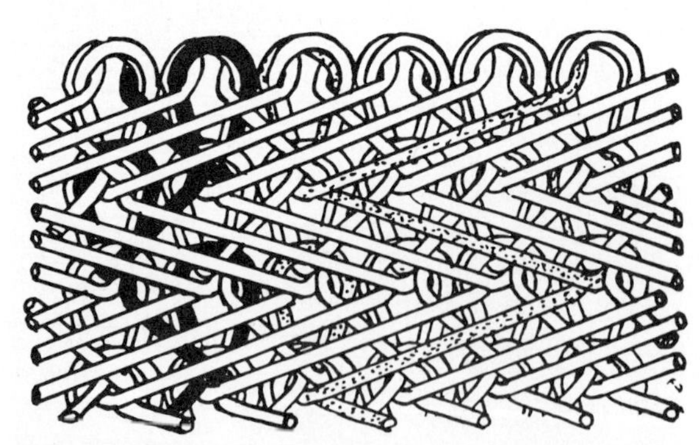

Fig. 29–7 *Warp knit velour before napping. Black yarn is nylon; white and gray yarns will be broken during napping. (Courtesy of* Knitting Times, *the official publication of National Knitwear and Sportswear Association.)*

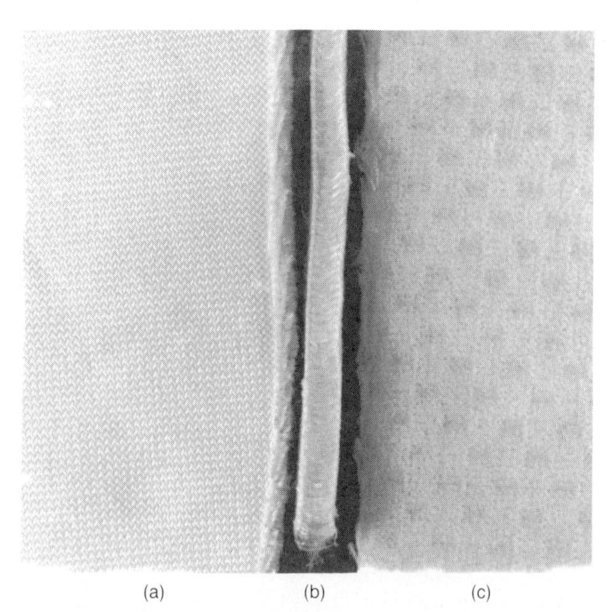

Fig. 29–8 *Knit-velvet car upholstery: (a) as knit; (b) cross-sectional view; (c) after separation.*

Bonding is usually done with single-tricot jerseys. Acetate or nylon tricot is commonly used for the backing of bonded woven goods. Bonding provides greater stability and gives body to the fabric.

Antistatic, antisnag, and flame-retardant finishes are also used for tricot fabrics.

Raschel-Warp Knits

The Raschel-warp-knitting machine has one or two needle beds with latch needles set in a vertical position and as many as 78 guide bars. The fabric comes from the knitting frame almost vertically instead of horizontally as in the tricot machine. The various Raschel machines knit a wide variety of fabrics from gossamer-sheer nets and veilings to very heavy carpets.

Raschel knits are used industrially for laundry bags, fish nets, dye nets, safety nets, and covers for swimming pools.

Raschel Fabrics. *Crochet-type fabrics* have rows of chain-like loops, called *pillars,* with laid-in yarns in various lapping configurations (Figures 29–9 and 29–10). These fabrics can be identified by raveling the laid-in yarn and noticing that the fabric splits or comes apart lengthwise.

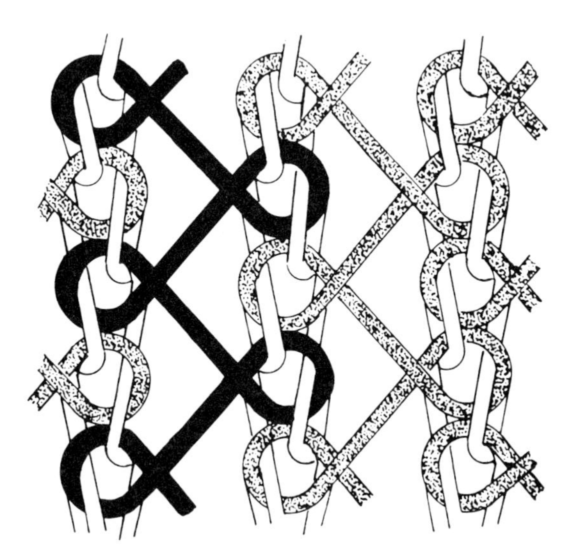

Fig. 29–9 *Raschel knit. (Courtesy of* Knitting Times, *the official publication of National Knitwear and Sportswear Association.)*

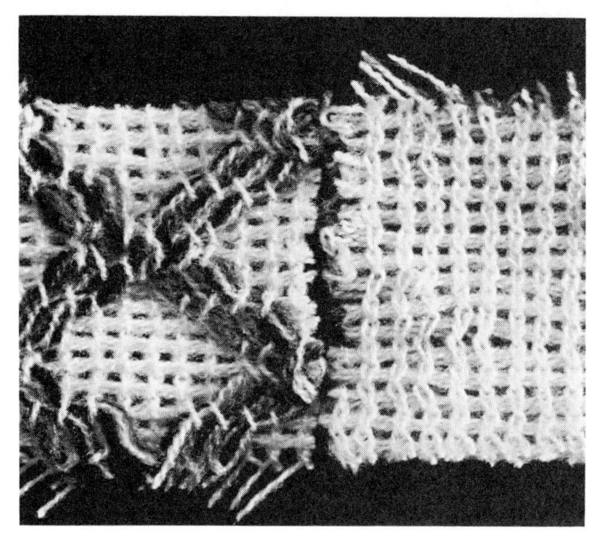

Fig. 29–10 *Raschel knit with extra yarn: technical face of fabric* (left)*; technical back of fabric* (right).

Window-treatment fabrics and outerwear fabrics are knitted on this standard-type machine.

Carpets have been knitted since the early 1950s. Since their production is faster, knitted carpets are cheaper to make than woven carpets. Tufted carpets, being still much cheaper to produce, have captured most of the carpet market. Knitted carpets have two- or three-ply warps for lengthwise stability; laid-in crosswise yarns for body and crosswise stability; and pile yarns of nylon, wool, or acrylic. Knitted carpets can be identified by looking for chains of stitches on the underside. They seldom have a secondary backing. These carpets are usually commercial carpets.

Lace and *curtain nets* of the kind made on Leavers lace machines (see page 241) can be made at much higher speeds on a Raschel machine. The Raschel machine has a single needle bed and up to 78 guide bars. Window-treatment nets, which can have square, diamond, or hexagonal meshes, are made on tricot machines. Laces are usually made of nylon or polyester (Figure 29–11).

Thermal cloth has pockets to trap heat from the body knitted in; it looks like woven waffle cloth and is used mainly for winter underwear. Brynje is the name given to this fabric, which is much like the Norwegian fisherman's vest. This knit is also used for some thermal blankets.

Power net is an elasticized fabric used for foundation garments and bathing suits. Nylon is used for the two-bar ground construction and

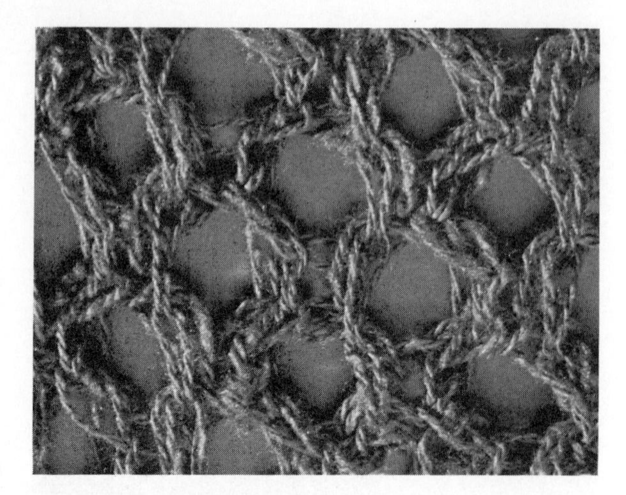

Fig. 29–11 Raschel fabric.

spandex is laid in by two other guide bars (Figure 29–12).

WEFT INSERTION

Weft insertion is done by a warp knitting machine with a weft-laying attachment. Several models have been developed. The simplest attachment carries a single filling yarn to and fro across the warp knitter, and this yarn is then fed steadily into the needle zone of the machine. A firm selvage is formed on each side.

More-complex attachments supply a sheet of filling yarns to a conveyor that travels to and fro across the machine. Figure 29–13 shows the Weft-Loc machine, which carries as many as 24 filling yarns in 6-inch bands. The yarns are then fed into the stitching area of the machine. A cutting device trims filling yarn "tails" from the selvages and a vacuum removes the tailings.

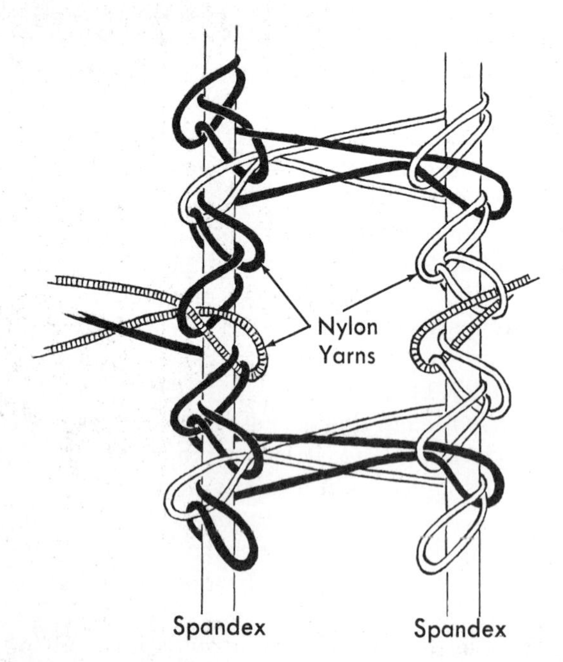

Fig. 29–12 Raschel power-net stitch.

Weft-insertion fabrics offer the best properties of both woven and knitted cloth: namely, strength, comfort, cover without bulk, and weight. They are lighter weight than double knits but have more covering power. They have increased crosswise stability of weaves but retain the comfort of knits. The fabric in Figure 29–14, which was made on a Weftamatic modified tricot machine, resembles a ticot jersey. It has a 20-denier-nylon knitting yarn and a 150-denier-polyester insertion yarn. The polyester is of several colors, making a horizontal stripe. Warp knits usually have vertical stripes unless printed because by nature the fabric contains only vertical yarns.

In weft-insertion warp knits, the inserted yarn is caught in a vertical chain of stitching. These fabrics are used for hospital curtains and table linens, as well as for other home-furnishing and industrial uses.

WARP INSERTION

The insertion of warp yarn into a knit structure gives the fabric the vertical stability of woven cloth while retaining the horizontal stretch of knit fabric (Figure 29–15). The cloth is basically a single tricot.

WARP AND WEFT INSERTION

Warp and weft insertion fabrics have characteristics very similar to woven fabrics. These fabrics can be much less expensive than woven fabrics and are available in wide width. They are frequently used as window treatment fabrics (see Figure 29–16).

Minor Warp Knits

SIMPLEX

The *simplex machine,* which is similar to the tricot machine, uses spring-beard needles, two

Fig. 29–13 *Weft-Loc machine. (Courtesy of the Crompton & Knowles Corp.)*

needle bars, and two guide bars. It produces a two-faced fabric somewhat like circular double knits. End uses are gloves (traditional), swimwear, and dresses.

One company, Blue Ridge Winkler Textiles of Bangor, Pennsylvania, produces most of the simplex knits. Glove fabrics are given a suede finish by passing them two or three times over

Fig. 29–14 *Weftamatic fabric. The raveled yarns are the polyester filling that are inserted in the warp-knit stitches.*

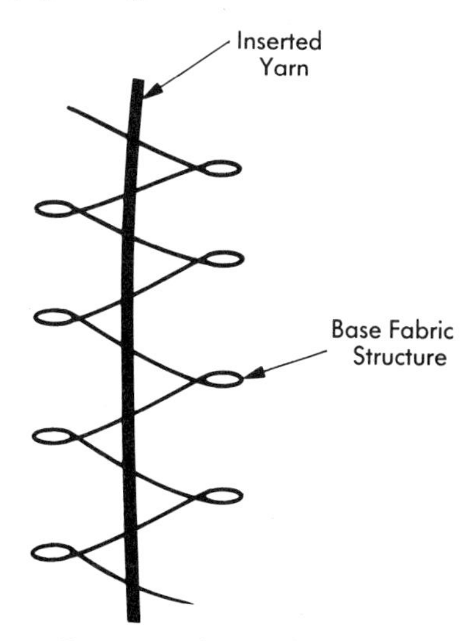

Fig. 29–15 *Diagram of warp-knit warp insertion.*

Fig. 29–16 Warp and weft insertion. Warp knit casement fabric for windows.

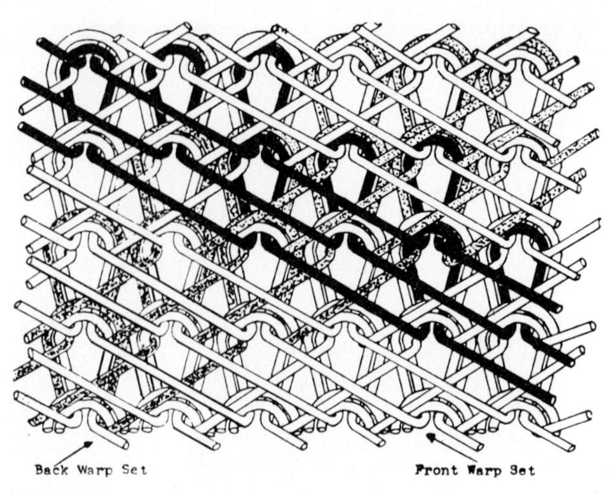

Fig. 29–17 Milanese. (Courtesy of Knitting Times, *the official publication of National Knitwear and Sportswear Association.)*

revolving sandpaper-covered rollers. The number of times the fabric is passed through the machine is determined by the degree of sueding desired. Dyeing follows.

MILANESE

The *Milanese machine* is especially constructed to produce superior warp-knit fabrics. The machine can use both kinds of needles. It is equivalent to a two-bar tricot fabric and made from two sets of warp yarns with one needle bar and one guide bar. But the lapping movements are arranged so each warp thread moves across the full width of the fabric, one set knitting from right to left and the other from left to right. This results in a diagonal formation (Figure 29–17), which shows up on the back of the cloth. The face has a very-fine rib. The fabric is runproof and is used for gloves and lingerie.

30

Film, Foam, Coated Fabrics, Leather, and Fur

Film, foam, leather, and fur usually are not classified as fabrics. However, they are used for apparel and home furnishings and can have a fabric-like hand and drape. Many films are given a leather-like grain and widely used where the more-expensive leathers were formerly used. Because suede is such an attractive, desirable, and expensive material, many fabrics are finished to resemble it. Fur, like leather, is a natural product, highly prized for its beauty and durability and widely imitated in various fabric constructions.

Films

Films are made directly from solution by melt-extrusion or by casting the solution onto a hot drum. The solutions are similar to the spinning solutions for fibers.

Apparel and home-furnishing textile films are made from vinyl or polyurethane solutions. They are similar in appearance but vary in the care required. Vinyl films become stiff in dry-cleaning solvents. They are washable but not dry cleanable. Urethane films are both washable and dry cleanable. Urethane films remain soft in cold weather, whereas vinyl films stiffen.

There are several kinds of films. *Plain films* are firm, dense, and uniform, and may also be referred to as *nonreinforced films.* These films are usually impermeable to air and water. They have excellent soil and stain resistance and recover well from deformation.

Expanded films are spongier, softer, and plumper as a result of a blowing agent that incorporates tiny air cells into the compound. They are neither as strong nor as abrasion resistant as plain films. Expanded films also may not be as impermeable to air and water. To increase the comfort characteristics of plain and expanded films, thousands of tiny pin holes called micropores may be punched in the fabric to permit air and water vapor, but not liquid water, to pass through the fabric.

Because plain films and expanded films are seldom durable enough to withstand normal use, these products are usually attached to a woven, knit, or fiberweb support fabric or substrate. The final product is a *supported,* or *reinforced, film.* Supported films are multiplex fabrics, but will be discussed in this chapter because many are designed to imitate the appearance of leathers and suedes. Supported films are more durable, easier to sew, and less likely to crack and split than nonreinforced films.

Plastic films and coated fabrics are better than any other material for waterproof items. They can be made to look like most any other textile by printing, embossing, or flocking. They can vary in thickness from very thin transparent film used to make a sandwich bag to heavy leatherette used to cover a dentist's chair. They have the advantage of being uniform in appearance and quality, they can be obtained by the yard or meter in wide widths, and they are much cheaper and easier to cut into apparel compared to leather.

The following chart lists some polymers that are used as films and fibers.

Foam

Foams are made by incorporating air into an elastic-like substance. Rubber and polyurethane are the most commonly used foams. The outstanding characteristics of foams are their bulk and sponginess. They are used as carpet backings and underlays, furniture padding, pillow

Fig. 30–1 *Foam carpet pad: surface view* (top) *and cross-sectional view* (bottom).

Films Are Made from Solutions

Solution	*Fiber*	*Film*	*End Uses for Film*
Acetate	Acetate	Acetate	Photographic film; projection film
Polyamide	Nylon	Nylon	Cooking Magic Bags*
Polyester	Polyester	Mylar*	Packaging; metallic yarns
Polypropylene	Olefin		
Polyethylene		Polyethylene	Packaging, garment and shopping bags, squeeze bottles
Polyurethane	Spandex	Polyurethane	Leather-like fabrics
Polyvinyl chloride	Vinal	Vinyl	Packaging, garment bags, leather-like fabrics for apparel and upholstery, seed tapes, water-soluble bags
Vinylidene chloride	Saran	Saran Wrap*	Packaging
Viscose (regenerated cellulose)	Rayon	Cellophane	Glitter weaving yarns—mostly in hand-woven textiles

*Trade names.

forms, and foam laminated to fabric for apparel and home-furnishing textiles (Figure 30–1). Shredded foam is used to stuff accent pillows and toys.

Polyurethane foam can be obtained in a wide range of physical properties from very stiff to rubbery. The size of the cells can be controlled. Foams will yellow on exposure to sunlight, but this neither causes a chemical change in the urethane foam nor affects its usefulness and durability. Exposure to sunlight does, however, cause rubber to disintegrate. Polyurethane is prepared by the reaction of diisocyanate with a compound containing two or more hydroxyl groups in the presence of a suitable catalyst. Chemicals and foaming agents are mixed thoroughly. After the foam is formed, it is cut into blocks 200–300 yards long, and strips of the desired thickness are cut from these blocks.

Coated Fabrics

A *coated fabric* combines the best characteristics of a textile fabric with a polymer film. The woven, knit, or fiberweb fabric provides such characteristics as strength and elongation control. The coating or film provides protection from environmental factors such as water, chemicals, oil, abrasion, and so forth. Common films include rubber and synthetic elastomers such as polyvinyl chloride (PVC) and polyurethane. PVC-coated fabrics are the most common, and they are used in window shades, book covers, upholstery, wall coverings, apparel, shoe liners, and shoe uppers. Most polyurethane-coated fabrics are used in women's shoe uppers and apparel. Heavier-weight polyurethane coated fabrics are used in industrial tarpaulins.

The coating is added to the fabric by several methods. The first and most common method is *lamination,* in which a prepared film is adhered to fabric with adhesive or heated to slightly melt the film. The second of these is calendering. In *calendering,* the polymer is mixed with a filler, stabilizing agent, pigment, and plasticizer, and the mixture is calendered directly onto a preheated fabric by passing the fabric and the mixture between two large metal cylinders spaced close together. The final method is *coating,* in which a more-fluid compound is required. The fluid mixture is applied by knife or roll. Degree of penetration of the mixture is controlled by allowing the mixture to solidify or gel slightly before it comes in contact with the fabric.

End Uses for Films and Coated Fabrics

Air-supported roofings	Shower curtains
Draperies	Tablecloths
Hospital-bed coverings	Umbrellas
Hose containers for fuel and water	Upholstery
Inflatable flood gates for water control	Waterproof apparel such as raincoats, boots, and mittens
Leather-like coats, jackets	

Coated fabrics can be printed or embossed. They are used for apparel, shoe uppers and liners, upholstery, vinyl car tops, floor and wall coverings, window shades, bandages, acoustical barriers, filters, soft-sided luggage, awnings, pond and ditch liners, and air-supported structures.

Bion II by Biotex Industries is a monolithic, or solid, polyurethane coating that is waterproof, breathable, and flame retardant. It is used in active sportswear such as running suits, outerwear such as parkas, diaper covers, mattress covers, and incontinent products. It can be applied to most fiber contents in woven or knit forms.

A polyester-monolithic coating is Sympatex by BASF. Sympatex can be laminated to an outer shell, a lining fabric, or a lightweight insert fabric such as tricot or fiberweb for use in skiwear. Sympatex is washable or dry cleanable.

Poromeric Fabrics

Poromeric, or *microporous, fabrics* are coated fabrics, but they are classified in a separate category because the coating is very fine and microporous. These two distinctions are major ones because they determine many characteristics of the resulting fabric. The micropores of the fabric are so small that they allow the passage of water vapor, but not liquid water; hence, they are water-vapor permeable. This factor greatly enhances comfort in apparel items. Poromeric films can be made from polytetrafluoroethylene (Gore-Tex) or polyurethane (Figure 30–2). These products are waterproof, windproof, and breathable. They can be applied to a wide variety of

Fig. 30–2 *Gore-Tex fabric: face fabric, polytetrafluoethylene film, backing* (left)*; face and back of fabric* (right).

fabrics and fiber contents. They are used for active sportswear; rugged outdoor wear, such as hunting clothes; tents; sleeping bags; medical products; filters; and coatings for wires and cables.

Suede-Like Fabrics

Because of the beautiful texture and hand of suede and the problems encountered in care, *suede-like fabrics* have been produced. These fabrics are needle-punched fabrics made from extremely fine denier fibers combined with a resin coating and nonfibrous polyurethane. The microfine-denier fibers are arranged in a manner that reproduces the microscopic structure of natural suedes. The fabric is dyed and finished. The process was developed by Toray Industries, Inc., of Japan. Facile® and Ultrasuede® are made of 60 percent microfine polyester and 40 percent polyurethane foam. Facile® and Ultrasuede® are registered trade names of Springs Industries. Facile® is a lighter-weight fabric, while Ultrasuede® weighs approximately 9 ounces per square yard. Ultrasuede® and Facile® are used in apparel. Ultrasuede® may be backed with a cotton-woven fabric when used for upholstery. Belleseime® is a similar fabric of 65

Suede-Like Fabrics

Construction Technique	*Characteristics*	*Trade Name*
Composite fabric—polyester fibers and polyurethane mixed, cast on drum, napped	Washable, dry cleanable Looks like leather	Ultrasuede
Composite fabric	Easy care Looks like natural leather	Belleseime
Woven cotton/polyester substrate with surface coating of polyurethane	Dry cleanable Washable	
Substrate with polyurethane on both sides		
100 percent polyester-pile fabric with suede-like finish	Dry cleanable Washable	
Flocked cotton	Least expensive. Flock may wear off at edges of garment	
100 percent polyester-warp knit-napped	Washable, dry cleanable	Super-suede

percent polyester/20 percent nylon matrix fiber on a 15 percent polyurethane-foam substrate produced by Kanebo Company of Japan. Suede-like fabrics are made in various ways (see the chart above).

Leather

Leather is processed from the skins and hides of animals, reptiles, fish, and birds. It is an organic substance derived from living animals and therefore varies greatly in uniformity. The hides from different animals differ in size, thickness, and grain. *Grain* is the marking that results from the skin formation and varies not only from animal to animal but also within one hide. Other factors influence the surface of hides. Animals scratch themselves, run into barbed-wire fences, and fight, causing scars that cannot be erased; brand marks or skin diseases mar the hides. Animals are raised primarily for meat or fiber, not for their hides or skins. Leather is a relatively unimportant by-product. Of 100 hides, it is estimated that less than 5 percent are suitable for conversion into smooth top-grain cowhide in aniline finish, 20 percent are suitable for smooth leathers with a pigment finish, and the remaining 75 percent must be embossed, buffed, snuffed, or corrected.

Dried skins and hides are stiff, boardy, nonpliable, and subject to decay. *Tanning* is the process in which skins and hides are treated with a tanning agent to make them pliable and water resistant. *Vegetable tanning,* the most expensive process, is done with an extract leeched out from the bark of various trees. *Chrome tanning* (bichromate of soda, sulfuric acid, and glucose) is used to make soft, pliable leather. *Oil tanning* is used to make chamois. *Alum tanning* is used for white leather.

Skins go through many processes to become leather: salting; cleaning to remove the hair and epidermis; tanning; bleaching; stuffing; coloring or dyeing; staking; and finishing by glazing, boarding, buffing, snuffing, or embossing, depending on the desired end use. These many processes explain why leather is an expensive product.

Leather is a nonseparable-fiber product. As shown in Figure 30–3, the fibers are very dense on the skin side and less dense on the flesh side. Thick hides are often split to make them more pliable and economical (Figure 30–4).

The first layer is called *top grain* and has the typical animal grain. It takes the best finish and wears well. Splits have a looser, more-porous

Fig. 30–3 Cross-sectional drawing of a strip of leather, showing variations in density of fiber.

structure and are cut across the fibers. They are not as smooth as top grain and tend to rough up during wear. Most split leathers are given an embossed finish. Although splits are not identified as such on labeled products, top grain is usually mentioned.

Leather is a durable product with a pleasant odor. It varies greatly in quality—not only from skin to skin but within one skin. Like wool fibers, leather from the backs and sides of the animal is better, whereas that from the belly and legs tends to be thin and stretchy or very coarse. Leather picks up oils and grease readily. It requires special care in cleaning since it is stiffened by solvents. Leather should be cleaned by specialists. Most dry cleaners send leather and suede items to a specialist for cleaning.

Reconstituted leathers have been made by grinding up leather, mixing with urethane, and forming into sheets. This "leather" product is uniform in thickness and quality and is not limited in length and width.

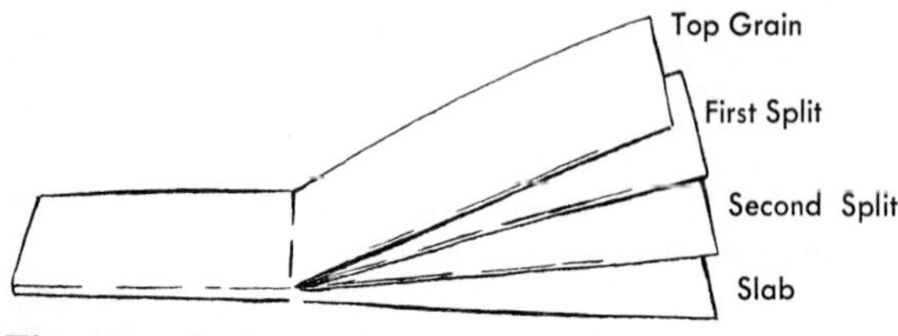

Fig. 30–4 *Split leather.*

Suede. *Suede* is a popular leather for coats, jackets, dresses, and trims. The soft, dull surface is made by napping (running the skin under a coarse emergy board) on the flesh side to pull out the fibers. Suede is a very durable product, but it requires special care. Rain and wet snow will damage suede. Cleaning of suede also should be done by specialists.

Furs

A *fur* is defined as any animal skin or part of an animal skin to which the hair, fleece, or fur fibers are attached. Furs are considered luxury items by the United States government. Consumers usually purchase furs because of their beautiful appearance rather than for their warmth, durability, or easy care. It is important to know about the kinds of fur, how fur garments are made, and how to care for them in order to make wise selections and to maintain their beauty.

Furs are natural products and therefore vary in quality. Good-quality fur has a very dense pile and if it has guard hairs they are long and very lustrous. The fur is usually soft and fluffy. The quality depends on the age and health of the animal and the season of the year in which it is killed. Fur trapping has long been an important industry in all parts of the world. Fur farming, started in 1880, has been a boon to the fur industry because better pelts are produced as a result of scientific breeding, careful feeding and handling of the animals, and slaughtering them when the fur is in prime condition. Silver fox, chinchilla, mink, Persian lamb, and nutria are the animals raised on ranches. By crossbreeding and inbreeding, new and different colored furs have been produced.

The cost of fur garments depends on fashion, the supply and demand of fashionable furs, and the work involved in producing the garment. Chinchilla, mink, sable, platina fox, and ermine have always been very expensive.

Furs go through many processes before they are sold. Dressing of fur is comparable to tanning of leather and the purpose is the same—to keep the skins from putrifying and make them soft and pliable. Dressing must be more-carefully done than with leather so that the fur will not be damaged. After tanning, pelts are combed, brushed, and beaten. The final process is drumming in sawdust to clean and polish the hair and to absorb oil from the leather and fur. The sawdust is removed.

Many furs are dyed. Originally, furs were dyed to make less-expensive furs look like the expensive ones, and this is still done. Muskrat is dyed to resemble seal, rabbit is stenciled to look like leopard or ocelot, and so forth. But today furs are dyed to improve their natural color as well as to give them unnatural colors—red or green, for example. Tip dyeing is brushing the tips of the fur and guard hairs with dye. Furs also are dip dyed, in which the entire skin is dipped in dye. Some furs are bleached and some are bleached and then dyed.

Skins are gathered together from all over the world and sold at public auction. The four largest markets for fur are St. Louis, New York, London, and Montreal.

Furs require care to keep them beautiful. They should not be stored in damp places or in hot humid places and never in plastic bags. Between seasons, if possible, garments should be sent to a furrier for cold storage, where the furs are kept in special vaults in which the temperature and humidity are controlled. To restore luster and clean the garment, it is usually best to send it to a furrier once a year. Furs should never be dry cleaned.

Furs should be protected from abrasion. Avoid sitting on them; hang on a wide, well-constructed hanger and allow plenty of space between garments; shake rather than brush. Driving in furs is very hard on them. Much of automobile upholstery, which is made of plastic or synthetic fibers, breaks off the fur in the shoulder and seat area of the garment.

The Fur Products Labeling Act, which became effective on August 9, 1952, requires that the true English-language name of the animal must be used, and if the fur is dyed, this must be stated. In addition, the country of origin must be indicated. The Act requires that used, damaged, or scrap fur also be identified. The Act has been amended to identify animals by name and expand the list of alterations to the natural fur to include tip dyeing, pointing (coloring the tips of the guard hairs), and other means of artificially altering the color of the fur.

This law protects the consumer from buying furs sold under names resembling expensive furs. For example, prior to the enactment of this law, rabbit was sold as lapin, chinchilctte, ermaline, northern seal, coney, marmink, Australian seal, Belgian beaver, and Baltic leopard. Hudson seal was muskrat plucked and dyed to look like seal. The law does not provide for quality designations; poor-quality mink is still mink. Fur dealers sometimes deplore this law because the consumer may misinterpret it. The consumer may not realize that the price represents the quality of the fur and the quality of construction of the garment.

31

Fiberweb, Net-Like, and Multiplex Structures

The fabrics with which this chapter deals are those that are made directly from fibers, fiber-forming solutions, or other fabrics. Thus there may be no processing of fibers into a yarn. Included in this chapter are both very old processes and very new processes. Felt and tapa cloth have origins lost in antiquity; net-like structures used to bag fruits and vegetables use new technologies; multiplex fabrics are made by combining fabrics. Fabrics in this chapter, especially those made from fibers, are the fastest growing area of the textile industry. These fabrics often have industrial uses, but some are also used in apparel and home-furnishing items. Much research and development work is being focused on industrial fabrics; new products made from existing fibers provide expanded markets for fiber companies.

These fabrics are often referred to as *nonwovens,* meaning that these fabrics are not made from yarn. However, the term nonwoven creates confusion because knits are nonwovens as well. This term nonwoven is used to refer to a wide variety of fabric structures. Because the term nonwoven is so indefinite, a new classification system has been proposed to eliminate the problems of the nonwoven system. The new system includes three categories: fiberweb structures, net-like structures, and multiplex structures.*

Use of fiberwebs, net-like structures, and multiplex structures has increased almost five times from 1974 to 1984 (*American Textiles,* December 1985).

The following are reasons for the increased usage of these fabrics:

1. Increased cost of reconditioning traditional textiles—especially the labor cost.

2. Scarcity and fluctuating cost of low-cost fibers (cotton, rayon, jute).

3. Production and promotion of some man-made fibers.

4. Easier cutting and sewing especially with unskilled labor.

5. New technologies that result in made-to-order inexpensive products.

*S. K. Batra, S. P. Hersh, R. L. Baker, D. R. Buchanan, B. S. Gupta, T. W. George, and M. H. Mohamed, "Neither Woven nor Knit: A New System for Classifying Textiles," *Nonwovens Industry,* September 1985, pp. 28, 30, 32, 34, 35.

Fiberweb Structures

Fiberweb structures include all textile-sheet structures made from fibrous webs, bonded by mechanical entanglement of the fibers or by the use of added resins, thermal fusion, or formation of chemical complexes. In this category, fibers are the fundamental units of structure arranged into a web and bonded so that the distances between fibers are several times greater than the fiber diameter. These fiberweb structures are less paper-like. It is these two properties that distinguish fiberweb structures from paper. Thus fiberweb structures are more flexible than paper structures of similar construction.

Properties of fiberweb structures are controlled by selection of the geometrical arrangements of fibers in the web, the properties of the fibers used in the web, and the properties of any binders that may be used.

PRODUCTION PROCESSES

The following are steps in making fiberwebs:

1. Selecting the fibers.

2. Laying the fibers to make a web.

3. Bonding the web together to make a fabric.

Fibers. Any fiber can be used to make fiberwebs. The inherent characteristics of the fibers are reflected in the fabric. Filaments and strong staple fibers are used where strength and durability are important; rayon and cotton are used for absorbency; thermoplastics are used for spun-bonded webs.

Web Formation. Webs are made in the following ways:

1. Dry laid.

2. Wet laid.

3. Spun bonded.

4. Spun laced.

5. Melt blown.

The first fiberweb—tapa cloth—dates back to biblical times and was made from the fibrous inner bark of the fig tree. It was used chiefly for clothing. The bark was used by primitive people in many areas where the fig tree grows, the Pacific Islands, Central America, and elsewhere. The cloth was made by soaking the inner bark to loosen the fibers, beating them with a mallet, smoothing them out into a paper-like sheet, and decorating them with block prints (Figure 31–1). The finest cloth was made by the Hawaiians.

Dry-laid fiberwebs are made by carding or air laying the fibers. Webs delivered from the carding machine have fibers oriented lengthwise. Webs can be cross-laid by stacking the carded web so that one layer is oriented lengthwise and the next layer is oriented crosswise to give added strength and pliability. *Cross-laid webs* do not have grain and can be cut more economically than woven or knitted fabrics. *Air-laid,* or *random, webs* are made by special machines that disperse the fibers by air. This web is much like the cross-laid web but has more random distribution. Dry-laid fiberwebs can be oriented or random. Oriented webs have the majority of the fibers parallel to the longitudinal axis of the fabric. Oriented webs have good strength in the direction of orientation, but poor cross-orientation strength. Random webs have the fibers oriented in different directions in a random fashion. Random webs have uniform strength in all directions. Typical end uses for dry-laid fiberwebs include wipes, wicks, battery separators, backing for quilted fabrics, interlining, insulation, abrasive base, filters, and base fabric for laminating and coating.

Wet-laid fiberwebs are made from a slurry of short (paper-process length) and textile-length fibers and water. The water is extracted, leaving a fiber web. The advantage of these webs is their exceptional uniformity. They comprise 15 percent of the fiberweb-fabric market. Typical end uses for wet-laid fiberwebs include laminating and coating bases, filters, interlining, insulation, roofing substrates, adhesive carriers, wipes, and battery separators.

Spun-bonded webs are made directly from spinnerets. The continuous filaments are laid down in a random fashion on a fast-moving belt, and, in their semimelted state, they fuse together at their cross points. They may be further bonded by heat and pressure. Spun-bonded fiberwebs have high-tensile and -tear strength, and low bulk (Figure 31–2). Typical end uses for spun-bonded fiberwebs include carpet backings, geotextiles, adhesive carriers, envelopes, tents and tarps, wall coverings, house-wrap vapor barriers, tags and labels, bags, protective apparel, filters, insulation, and roofing substrate. Reemay by the Intertech Group is a spunbonded polyester used in geotextiles and roofing (Figure 31–2). Tyvek is a spun-bonded, high-density fiberweb made from microfine olefin fibers used for protective clothing, sterile packaging, and construction uses. Typar is another spun-bonded

Fig. 31–1 *Tapa cloth from Samoa.*

Fig. 31–2 *Spun-bonded filament: Reemay.*

olefin used for carpet backing. Both Typar and Tyvek are made by du Pont.

Spun-laced webs are similar to spun-bonded webs except that jets of water are forced through the web, separating the filaments into a woven-like structure to produce a looser-bonded fabric (Figure 31–3). They have more elasticity and flexibility than spun-bonded fabrics. End uses of spun-laced fabrics are draperies and bedspreads. Spun-bonded and spun-laced fabrics comprise 25 percent of the market for fiberwebs.

Melt-blown fiberwebs are made by extruding the polymer through a single-extrusion orifice into a high-velocity, heated-air stream that breaks the fiber into short pieces. The fibers are collected in a web form on a moving conveyor belt. Fibers are held together by a combination of fiber interlacing and thermal bonding. Because the fibers are not drawn, the strength of the resulting fiberweb is low. Olefin and polyester are the fibers used commercially with this process to produce hospital/medical products and battery separators.

Bonding. Webs become fabrics by the following processes:

1. Using needling, a mechanical process.
2. Using a chemical substance or adhesive.
3. Using heat.

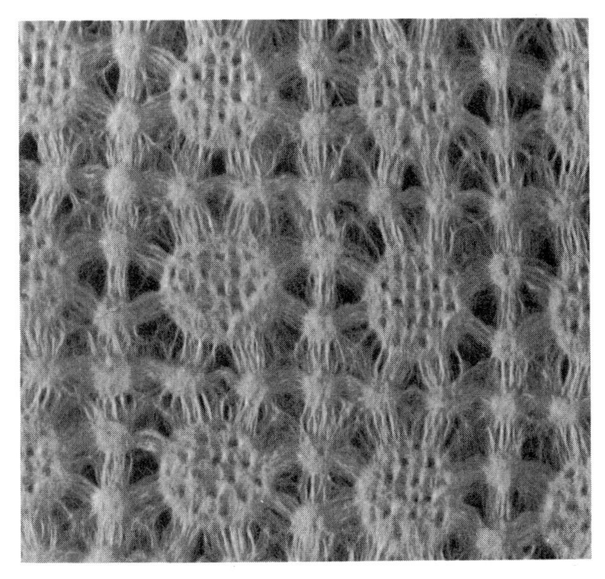

Fig. 31–3 Spun-lace fabric.

Needle-punching consists of passing a properly prepared dry-laid web over a needle loom as many times as is necessary to produce the desired strength and texture. A *needle loom* is a board with barbed needles protruding 2 or 3 inches from the base (Figure 31–4). As the needles push up and down through the web, the barbs catch a few fibers, causing them to inter-

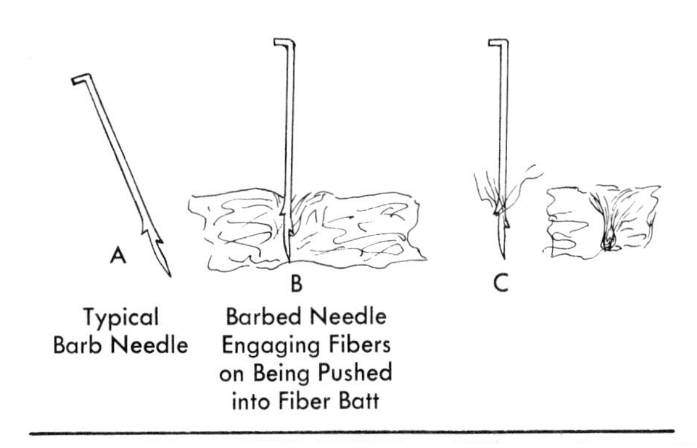

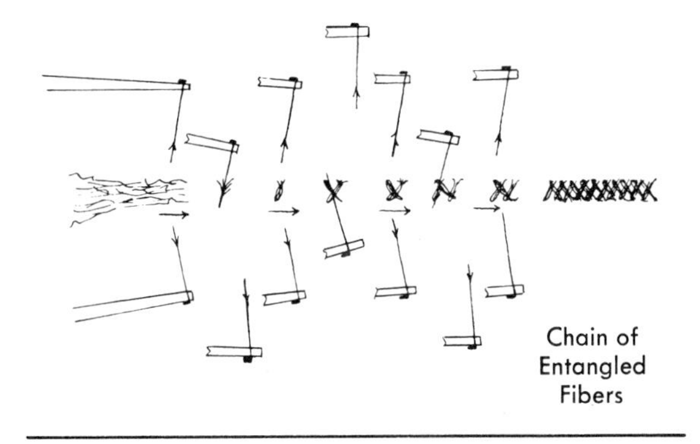

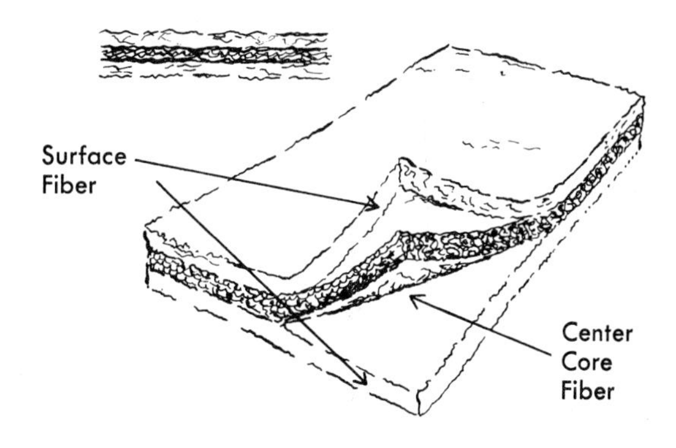

Fig. 31–4 Needle-punch process.

lock mechanically. The construction process is relatively inexpensive.

Attractive blankets and carpeting have been made by needle punching. In the Fiberwoven blanket, 100 percent Acrilan fiber or blends of Acrilan fiber and other fibers have been used. Fiber denier and fiber type may be varied. Blankets may be lofty or compact (see Figure 31–4).

Indoor/outdoor needle-punched carpeting made of olefin fibers is used extensively for patios, porches, pools, and putting greens. The fiber is impervious to moisture. Locktuft, a needled nonwoven made of Marvess olefin, is a carpet backing designed especially for use with tufted carpets.

Needled fabrics can be made of a web consisting of two layers, each with a different color. By pulling colored fibers from the lower layer to the top surface, geometric designs can be made. If the fibers are pulled above the surface, a pile fabric will result. The army has developed a ballistics-protective vest for combat use from needle-punched fabrics. Needle-punched fabrics are finished by pressing, steaming, calendering, dyeing, and embossing. Solution-dyed fibers are often used.

The Arachne and Maliwatt systems of knit-sew can be used without stitching threads to make fiberwebs. In these processes, a closed needle penetrates the web, opens, grabs some fibers, and draws them back as a yarn-like structure that is then chain-stitched through the web.

Chemicals or *adhesives* are used with dry-laid or wet-laid webs. Acrylic emulsions are usually used.

Heat and pressure are used to bond thermoplastic-fiber webs. Figure 31–5 shows Mirafi 140 fabric composed of a random mixture of two different continuous-filament fibers that are heat bonded. The fibers are polypropylene (olefin) homopolymer and bicomponent-polypropylene core-nylon sheath filaments. Mirafi 140 is an industrial fabric used for roadbed and area stabilization.

Batting, wadding, and fiberfill are not fabrics, but they are important components in apparel for snowsuits, ski jackets, quilted robes, and jackets, and in home-furnishing textiles for quilts, comforters, furniture paddings, mattresses, and mattress pads.

Batting is made from new fiber, *wadding* is made from waste fiber, and *fiberfill* is the name given to a man-made staple made especially for these end uses. Carded fibers are laid down to form the desired thickness and are often covered with a sheet of fiberweb fabric.

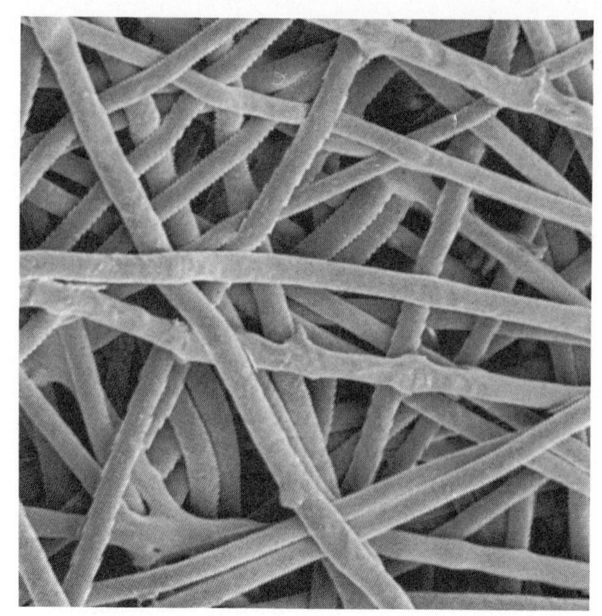

Fig. 31–5 *Mirafi 140 fabric. (Courtesy of Dominion Textiles of Canada.)*

Fiber density relates to the heaviness of a unit volume. Thus fiber density is important for batting, wadding, and fiberfill because consumers desire lightweight fabrics, especially for outerwear items such as ski jackets.

Resiliency is important because fabrics that maintain their loft incorporate more air space. When fibers stay crushed, the fabric becomes thinner and more compact. *Resistance to shifting* is important in maintaining uniformity of thickness in the fabric. For instance, down comforters need to be shaken often because the filling tends to shift to the outer edges. The thermoplastic-fiber batts can be run through a needle-punch machine in which hot needles melt parts of the fibers that they touch, causing them to fuse and form a more-stable batt. The thicker the batt, the warmer the fabric, regardless of fiber content. In apparel, there is a limit to the thickness, however, because too much bulk restricts movement, a limiting factor in styling.

Two recent types of polyester fiberfill by du Pont are Thermoloft, which is a high-loft insulation, and Thermolite, which is a thin insulation made with microfibers.

Comparison of Properties of Commonly Used Battings

Fiber	Density	Resiliency	Resistance to Shifting	Care
Down (costly)	Lightweight	Excellent	Poor	Dry cleanable
Acetate (low cost)	1.30	Fair	Poor	Washable, dries more quickly than cotton
Polyester (medium cost)	1.30–1.38	Good	Good—can be spot welded	Washable, quick drying
Cotton (low cost)	1.52	Poor	Poor	Washable, but slow drying

FUSIBLE FIBERWEBS

Fusible fiberwebs give body and shape to outerwear and are widely used in tailored garments. They also are used as interfacing in shirts, blouses, and dresses, and as interlining in outerwear (Figure 31–6).

A fusible is a fabric that has been coated with a heat-sealable adhesive that is thermoplastic. It also may be a thin, spider web-like of thermoplastic filaments (Figure 31–7). The fusible fabric is applied to a face fabric and bonded to it by heat and pressure.

The adhesives used are polyethylene, hydrolyzed ethylene vinyl acetate, plasticized polyvinyl chloride, and polyamides. Adhesives were first applied to the woven or nonwoven substrate as continuous coatings that resulted in very stiff end products or as discontinuous coatings in which powdered adhesives were spread over the fabric surface to give a softer hand. Today the adhesive is usually printed on the substrate in a precisely positioned manner to give the desired hand to the end product.

The following are the advantages of fusibles in apparel:

1. They eliminate certain areas of stitching, such as zigzag stitches used in coat and suit lapels.

2. They require less skilled labor in garment production.

3. They increase productivity.

4. They improve the appearance of garments at the point of sale and over time.

The disadvantages of fusibles in apparel are:

1. They may separate during care.

2. Adhesives may bleed through to the face fabric.

3. The fusible and the face fabric may shrink differently during care.

4. The change in hand and drape may be significant.

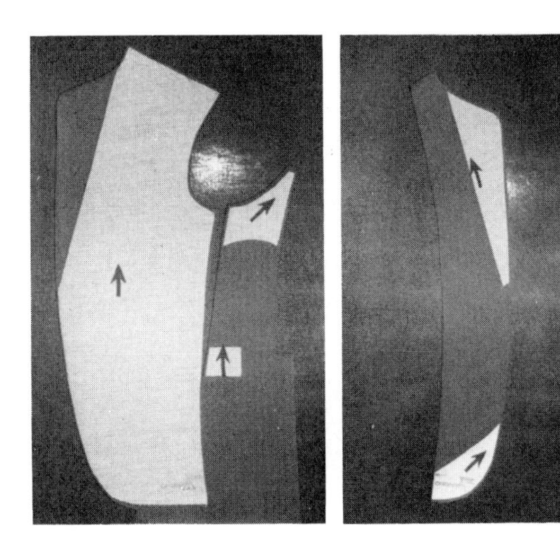

Fig. 31–6 *Fusible interfacings.*

Fig. 31–7 *Fusible interfacing. Tape under web to show sheerness of web.*

Uses of Fiberwebs

Durable	*Type*	*Disposable*	*Type*
Bedding and coated fabrics, mattress ticking, backing for quilting, dust cloth for box springs	Spun bonded	Diapers, underpads, sanitary napkins/ tampons	Dry laid
		Surgical packs and accessories	Dry laid, melt blown
		Wipes and towels	Dry laid, or wet
Carpet backing	Needled, spun bonded dry laid	Packaging	Spun bonded
Filters			
Interfacings	Dry or wet laid		
Interlinings	Needled, dry laid		
Others—draperies, upholstered furniture, backings, facings dustcovers, automotive, shoe parts, geotextiles, labels	Dry and wet laid, spun bonded, spun lace, melt blown		

END USES

Fiberwebs are used for disposable goods, such as diapers and wipes, durable goods that are incorporated into other products or used alone for draperies, mattress pads, and possibly some apparel (see the chart above).

Felt

True felt is a mat or web of wool or part-wool fibers held together by the interlocking of the scales of the wool fibers. Felting is one of the oldest methods of making fabrics. Primitive peoples made felt by washing wool fleece, spreading it out while still wet, and beating it until it had matted and shrunk together in fabric-like form. Figure 31–8 shows a Numdah felt rug. These rugs are still made in India. In the modern factory, layers of fiber webs are built up until the desired thickness is attained and then heat, soap, and vibration are used to mat the fibers together and to shrink or full the cloth. Finishing processes for felt resemble those for woven fabrics.

Felts do not have grain; they are rather stiff and less pliable than other structures; they do not ravel; they are not as strong as other fabrics; they vary in quality depending on the quality of the wool fiber used.

Felt has many industrial and some clothing uses. It is used industrially for padding, soundproofing, insulation, filtering, polishing, and wicking. In the past, felt was used under practically all machinery to absorb sound. Foams, being much cheaper, have replaced felt in this end use. Felt is not used for fitted clothing because it lacks the flexibility and elasticity of fabrics made from yarns. Felt has wide use in such products as hats, house slippers, clothing decorations, and pennants. Because felt does not fray, it needs no seam finish. Colored felt letters or decorations on apparel often fade in washing and should be removed or the garments should be sent to a professional dry cleaner.

Fig. 31–8 *Numdah felt rug.*

Net-Like Structures

Net-like structures include all textile structures formed by extruding one or more fiber-forming polymers in the form of a film or a network of ligaments or strands. In the integral-fibrillated-net process, the extruded film is embossed while molten by being passed through a pair of heated rollers that are engraved to form a pattern on the fabric. When the film is stretched biaxially, slits occur in the fabric creating a net-like structure. In the integral-extruded-net process, the spinneret consists of two rotating dies. When the polymer is extruded, the fibers form as single strands that interconnect when the holes of the two rotating dies coincide. The process produces a tubular net that is used for packaging fruit and vegetables, agricultural nets, bird nets, and plastic fencing (Figure 31–9).

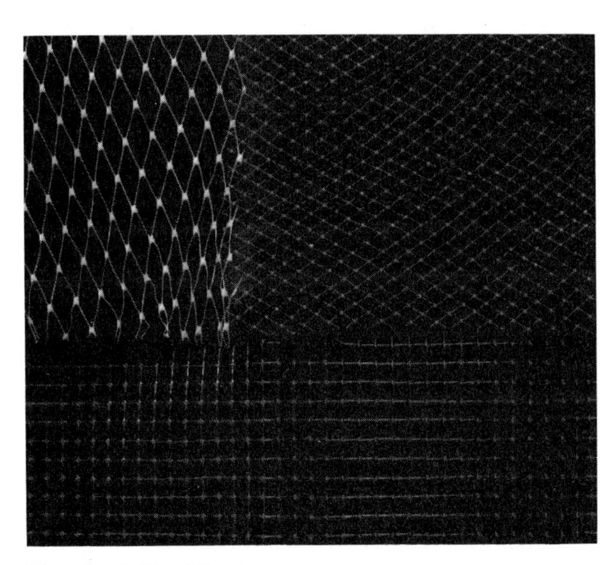

Fig. 31–9 *Net-like structures.*

Multiplex Fabrics

Multiplex fabrics are fabrics that combine several primary and/or secondary structures, at least one of which is a recognized textile structure, into a single structure. This broad category includes such diverse fabrics as stitch-bonded structures, laminates, tufted and flocked structures, scrim structures, and coated fabrics. Coated fabrics are discussed in Chapter 30; tufted and flocked fabrics are discussed in Chapter 33.

STITCH-BONDED FABRICS

Stitch-bonded fabrics include those fabrics that combine textile structures by adhering fabric layers with fiber or yarn loops, chemical adhesives, or fusion of thermoplastic fibers. Stitch-bonded fabrics are divided into knit-through fabrics and quilted fabrics. Stitch-bonded fabrics can be produced from fiberwebs and any woven or knit fabric.

Knit-through fabrics are made in several ways. In the first process, a Raschel warp-knitting machine knits yarns through a fiberweb structure. These fabrics can also be made by knitting fibers or knitting yarns around laid (not woven) warp and filling yarns. To make these knit-through fabrics, needles are used to create interconnected loops from yarns or fibers to stabilize the structure.

Araknit® is a knit-through fiberweb fabric used as a coating substrate. Arachne® and Maliwatt® are trade names for fabrics made by warp-knitting yarns through a fiberweb structure. These knit-through fabrics can be produced at high speeds and are used for home-furnishing items—such as upholstery, blankets, and window-treatment fabrics—and industrial uses—such as insulation and interlining. Malimo® uses warp or filling or warp and filling laid-in yarns with warp-knitting yarns (see Figures 31–10 and 31–11). These fabrics also can be produced at very high speeds and are used for tablecloths, window-treatment fabrics, vegetable bags, dishcloths, and outerwear. A new development is a knit-through fabric with warp, weft, and diagonal insertion.

Some knit-through fabrics utilize split-polymer films from recycled carbonated-beverage bottles. These fabrics are used in the carpet, geotextile, and bale-wrap industries.

QUILTED FABRICS

Quilted fabrics are multiplex fabrics consisting of three layers: face fabric, fiberfill or batting,

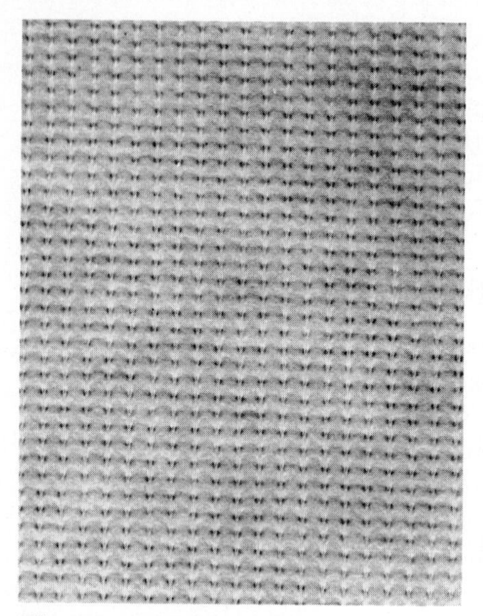
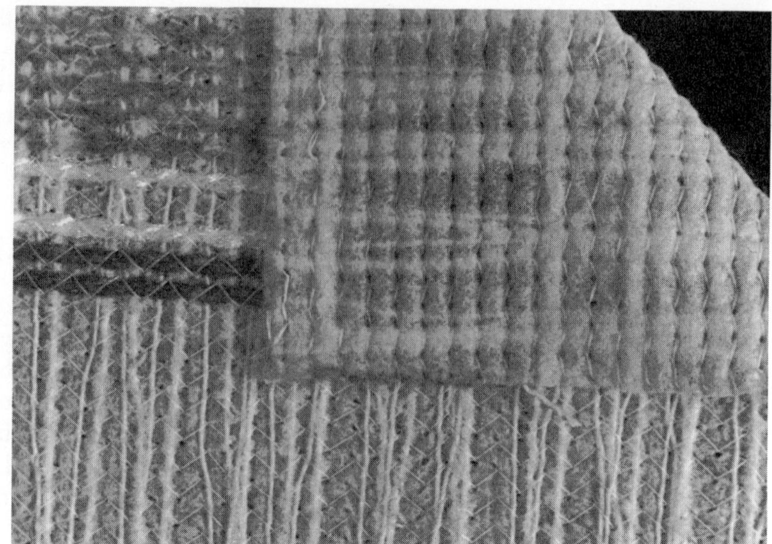

Fig. 31–10 Malimo fabrics.

and backing fabric. The three layers are stitch bonded with thread, chemical adhesive or fusion by ultra-high frequency sound. The bonding connects the layers in a pattern. The actual area physically bonded together is a very small percentage of the fabric's surface in order to get the high loft and bulky appearance desired in quilted fabrics.

Most quilted fabric is made by stitching with thread. The thread and type of stitch used in quilting are good indicators of the quality and durability of the finished fabric. A durable quilt will have a lock-type stitch with a durable thread. Twistless nylon-monofilament thread is often used because of its strength and abrasion resistance and because it is transparent and picks up the colors of the fabric.

Almost any thread can be used for quilting, but those designed for quilting are better than regular sewing thread. The thread should be durable to abrasion and should be appropriate for the desired end use of the quilt. The disadvantage of thread stitches in quilting is that the threads break when one sits in the garment or on the bed, or from abrasion or snagging. Broken threads are unsightly and with cotton, wool, down or acetate fiberfill, the loose fiber is no longer held in place.

Any fabric can be used for the shell or covering. A fashion fabric is always used on one side. If the article is reversible or needs to be durable or beautiful on both sides, two fashion fabrics are used. If the fabric is to be lined or used as a chair covering or bedspread, the under layer is often cheesecloth, tricot, or a fiberweb fabric.

The wadding or batting may be foam, cotton, down, or fiberfill. Fiberfill is a modified-fiber type with crimp designed to help maintain fluffiness and air space. Quilting is usually done in squares, or in straight or wavy lines. In upholstery and expensive quilts and comforters or bedspreads, the stitching is done outlining printed figures. This is a hand process (machine quilting guided by hand) and thus makes the fabric more costly.

Beauty of fabric is important for all end uses. For ski jackets and snowsuits, a closely woven water- and wind-repellent fabric is desirable; for comforters, resistance to slipping off the bed is important; for upholstery, durability and resistance to soil are important.

Chemical adhesives seldom are used at present. Chemstitch is a trade name for a quilted fabric using chemical adhesives applied in a pattern. These fabrics are neither as appealing nor as durable as the other quilting methods.

Heat generated by ultra-high-frequency sound or ultrasonic vibrations will melt thermoplastic fibers, fusing several layers. Ultrasonic quilting requires thermoplastic fibers. Figure 31–12 shows a Pinsonic Thermal Joining machine that "sews" by heat sealing thermoplastic materials by ultrasonic vibrations. The machine quilts seven times as fast as conven-

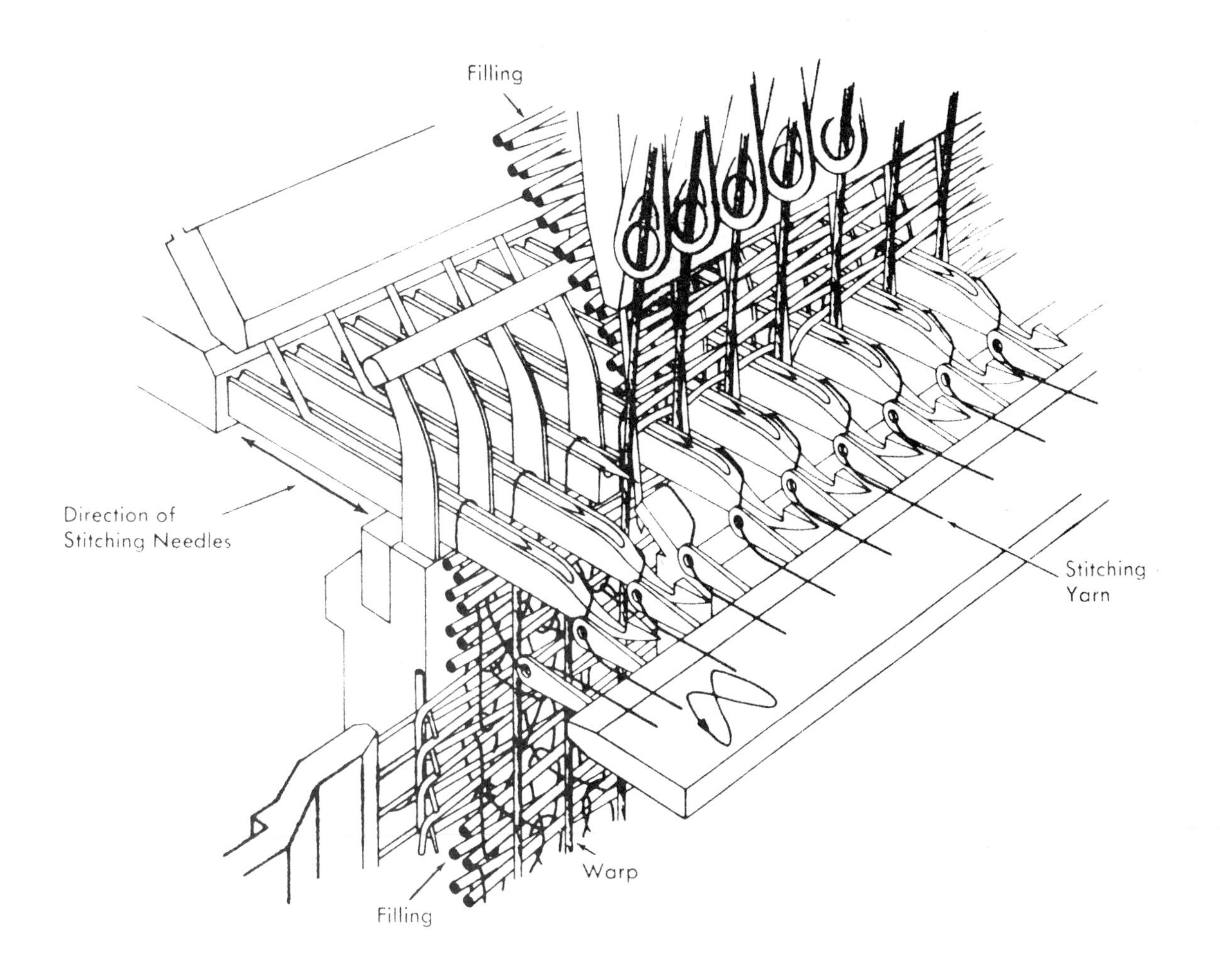

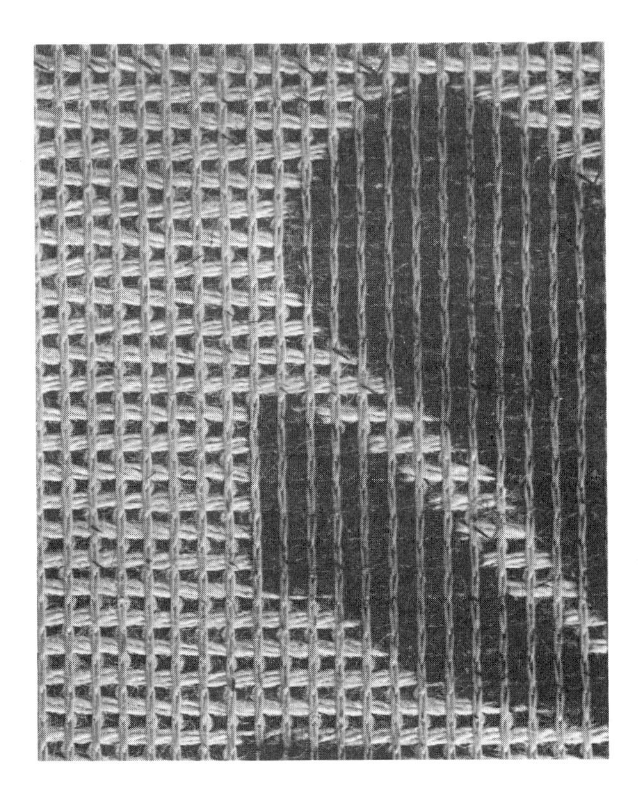

***Fig. 31–11** Malimo textile machine* (top); *Malimo fabric* (bottom).

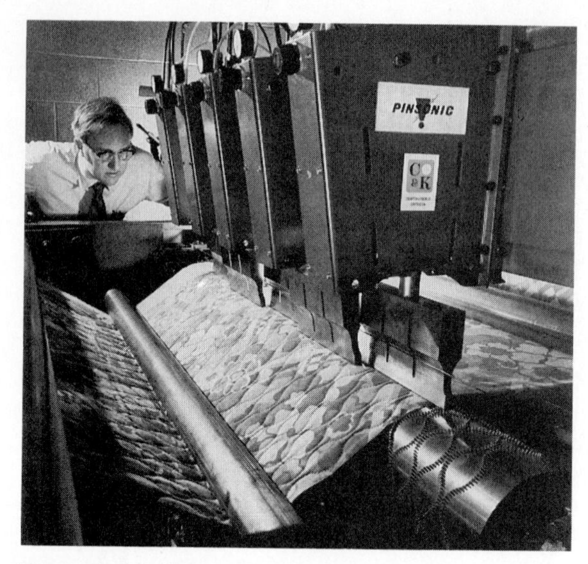

Fig. 31–12 Pinsonic Thermal Joining machine. (Courtesy of Branson Ultrasonics Corp.)

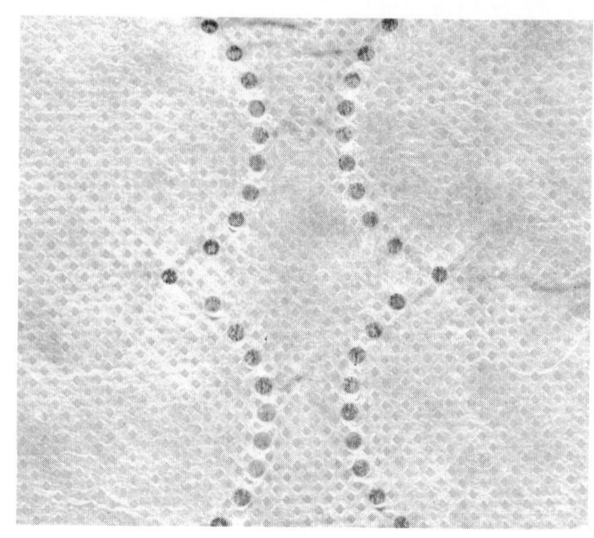

Fig. 31–13 Mattress pad. Two layers of fiberweb fabric and fiberfill batt joined by Pinsonic Thermal Joining machine.

tional quilting machines. This process is widely used on mattress pads and bedspreads. It eliminates the problem of broken threads—a boon for institutional bedding in the hospitality industry. However, the outer layer may tear along the quilting lines. Figure 31–13 shows Pinsonic fabric.

LAMINATES

Laminates include those fabrics in which two layers of fabric are adhered with an adhesive or foam. Laminate may be used to refer to a fabric in which an adhesive was used. Bonded may be used to refer to a fabric in which a foam was used. Both terms are used interchangeably.

The first fabric of this type (1958) was an inexpensive wool flannel bonded to acetate jersey with an adhesive. In the early stages, bonding was a way to deplete inventories of tender or lightweight fabrics. Some converters were marginal operators who were not interested in quality. A bonder could buy two hot rolls discarded by finishers and be in business. The fabric could be stretched as it went through the rollers. Consequently, many problems were associated with these bonded fabrics. The layers separated (delaminated) or shrank unevenly. There were problems with blotchy colors when the adhesive bled through to the technical face.

Because of these problems, laminated fabrics have a poor reputation. However, the textile industry has worked to improve the quality of laminated fabrics and current laminates have greatly improved performance characteristics. Some advantages and limitations are given in the chart. Figure 31–14 shows a laminated fabric with label.

Fig. 31–14 Bonded herringbone fabric and label.

Laminated Fabrics

Advantages	*Limitations*
Less costly fabrics are upgraded.	Top-quality fabrics are not bonded.
Self-lining gives comfort.	Backing does not prevent bagging, so a lining may be needed.
Stabilized if good quality.	Uneven shrinkage* possible.
Reduces time in sewing.	May be bonded "off-grain."
Interfacings may be eliminted.	May delaminate.*
Underlinings, stay-stitching, and seam finishing are not needed.	Hems, darts, and facings are stiff and boardy.
	Do not hold sharp creases.

*These are the two major problems.

Backing. Knits are usually used as the *backing* fabric. Knit will give with the stresses applied to the face fabric. Acetate tricot, the least expensive, is most often used. Other fabrics used are nylon tricots. The color of the backing can be a decorative feature in bonded laces.

Two methods of bonding are used (Figure 31–15):

1. Wet-adhesive method:
   Aqueous acrylic adhesive
   Solvent urethane adhesive
2. Foam-flame method.

In the *wet-adhesive method,* the adhesive is applied to the underside of the face fabric, and

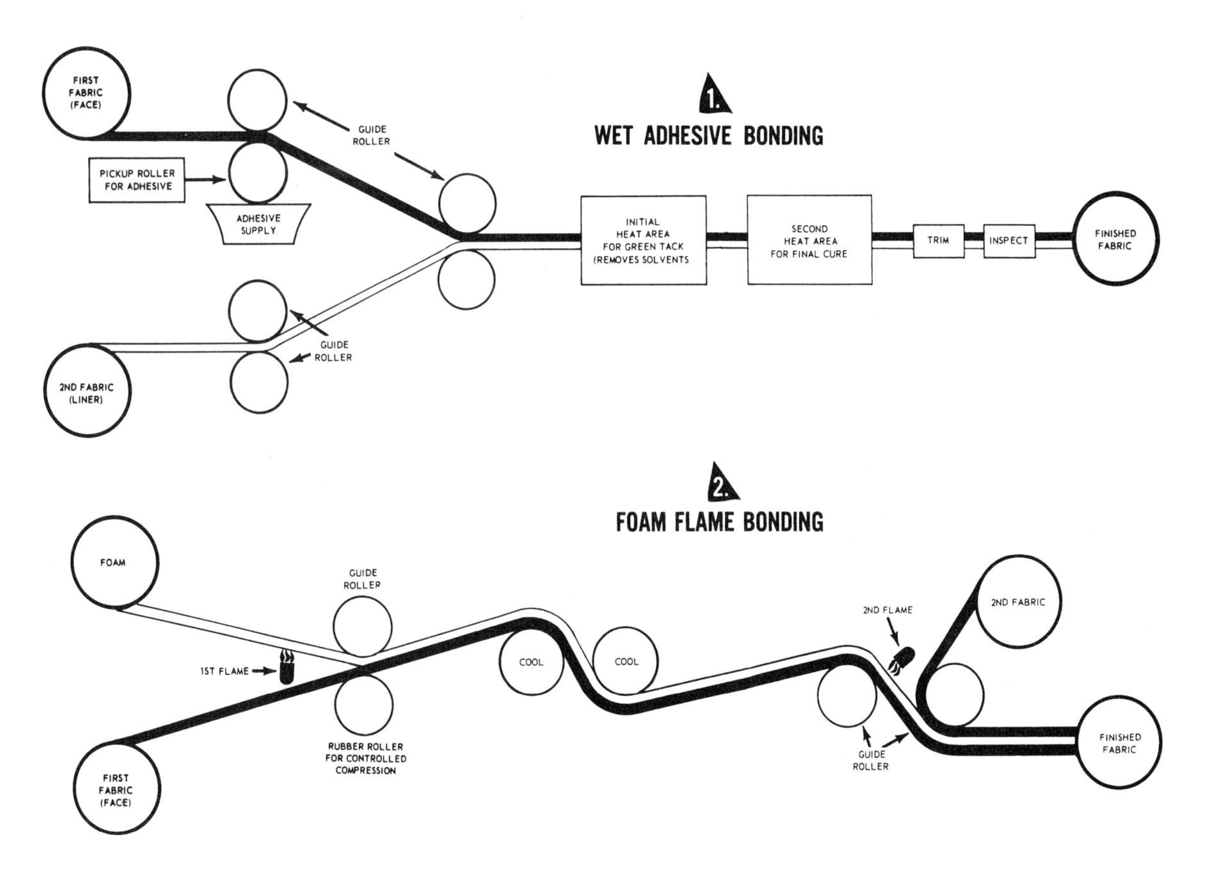

Fig. 31–15 *Two basic methods for producing bonded textiles. (Courtesy of* American Fabrics.*)*

the liner fabric is joined by being passed through rollers. It is heated twice, the first time to drive out solvents and to give a preliminary cure, and the second time to effect a permanent bond.

In the *foam-flame process,* polyurethane foam acts as the adhesive. Foam laminates consist of a layer of foam covered by another fabric or between two fabrics. The foam is made tacky first on one side and then on the other by passing the foam under a gas flame. The final thickness of the foam is about 15/1,000 of an inch. This method gives more body but reduces the drapability of the fabric.

Another foam process generates the foam at the time it is to be applied, flowing it onto the cloth, and curing on the cloth. Foam laminates were first visualized as thermal garments for outdoor workers because they are light weight but warm. Foams were quilted to lining fabrics for a lining, interlining combination. The first fabric to sell in volume was a dress-weight jersey laminated to foam and made up in spring coats. This was so successful that coats were made for all-purpose uses. Foam laminates today are made using all kinds and qualities of fabric with many different thicknesses of foam.

SUPPORTED-SCRIM STRUCTURES

Supported-scrim structures include the foam- and fiber-type blankets currently on the market. These fabrics use a lightweight nylon scrim, a loose warp knit fabric, in the center between two thin layers of polyurethane foam. A short nylon-flock fiber is adhered to the surface. These fabrics are attractive, durable, easy care, and inexpensive. Vellux and Vellux-Plus are two trade names.

Another type of a supported scrim includes those that are needle punched with a scrim between the fiberweb layers. These fabrics are used industrially for things such as roadbed-support fabrics.

32

Finishing: An Overview

A *finish* is defined as anything that is done to fiber, yarn, or fabric either before or after weaving or knitting to change the *appearance* (what is seen), the *hand* (what is felt), or the *performance* (what the fabric does). All finishing adds to the cost of the end product.

The order normally followed in textile processing is an involved one. Often several steps will be repeated. A common sequence is fiber processing followed by yarn processing. Fabrication (producing a fabric) usually follows some preparation steps. In preparation, the yarn or fabric is made ready for additional steps in the sequence. Bleaching is almost always done before dyeing. Coloration is usually done before finishing and reworking (repairing). The following is a diagram of a normal pattern in finishing:

Fiber Processing → Yarn Processing →
Preparation → Fabrication → Bleaching →
Coloration → Finishing → Reworking

Finishing may be done in the mill where the fabric is constructed or it may be done in a separate establishment by a highly specialized group called *converters*. Converters operate in two ways: They perform a service for a mill by finishing goods to order, in which case they are paid for their services and never own the fabric; or they buy the fabric from a mill, finish it according to their own needs, and sell it under their own trade name.

A *permanent finish* lasts the life of the garment. A *durable finish* lasts longer than a temporary finish, but not for the life of the garment. A *temporary finish* lasts until the garment is washed or dry cleaned. A *renewable finish* can be applied with no special equipment, or it may be applied by the dry cleaner.

Some finishes—such as dyeing, printing, or embossing—are easy to recognize because they are visible. Other finishes—such as durable press—are not visible but have an important effect on fabric performance. The consumer needs to recognize the visible finishes and the need for nonvisible finishes. The consumer also needs to know the serviceability of the finish.

Many finishing processes are used on both woven and knitted fabrics. The processes are very similar for either type of fabric. The major differences that exist between finishing woven or knitted fabrics occur in the way the fabric is handled or transported. Woven fabrics have little give or stretch. Knits have a much greater potential for stretch. Hence precautions need to be taken to minimize stretching during finishing of knits.

Gray goods (grey, greige, or loom state) are fabrics, regardless of color, which have been woven on a loom and have received no wet- or dry-finishing operations. Some gray-goods fabrics have names, such as print cloth and soft-filled sheeting, which are used only for the gray goods. Other gray-goods names, such as lawn, broadcloth, and sateen, are also used as names for the finished cloth.

Converted, or *finished, goods* have received wet- or dry-finishing treatments such as bleaching, dyeing, or embossing. Some converted goods retain the gray-goods name. Others, such as madras gingham, are named for the place of origin; and still others, such as silence cloth, are named for the end use. Figure 32–1 shows a print-cloth gray goods, and the fabric after bleaching, printing, and dyeing.

Fig. 32–1 *Print-cloth gray goods: as produced, bleached, printed, piece dyed* (from lower left to upper right).

Mill-finished fabrics can be sold and used without converting, although they may be sized before they are sold.

For years, *water-bath finishing* was standard. In water-bath finishing, the chemical was placed in a water solution and padded onto the fabric by immersing it in the solution and squeezing out any excess. The fabric was heavy with water and required a lot of energy to move it and remove the excess water from it. A great deal of water was used to scour or clean the fabric. Recently, with water pollution concerns and energy costs, *foam finishing* has become an alternative means of adding a finish. Foam finishing is a fairly recent development in textile finishing. This type of finishing uses foam rather than a liquid in applying the finishing chemical to the fabric. A foam is a mixture of air and liquid that is lighter weight than a solution of the liquid. Foam finishing is used because of the low-wet pickup (a much smaller amount of liquid is added to the fabric). In addition, energy is conserved in moving and drying the lighter fabric. The higher production speeds of foam finishing mean that the costs of finishing can be kept low. Foam finishing uses less water in scouring and cleaning, and it is used to add routine, as well as special-purpose, finishes to the fabric.

Another development in finishing is the use of *solvent finishing*. Solvent finishing was developed in response to a need to decrease water pollution and energy costs. In solvent finishing, a solvent other than water is used to mix the solution. Solvent finishing is not as popular as foam finishing because of the cost of solvents and reclaiming processes and environmental, regulatory, and health problems.

In this chapter, routine finishing will be discussed. Routine finishing includes those steps in finishing that are done to most fabrics to prepare them for dyeing and special purpose finishes. These routine finishes are often referred to as *preparation*. The normal order of production of a cotton/polyester, bottom-weight, plain-weave fabric will be discussed. Because many steps of production are discussed in detail elsewhere, these steps will be included in the discussion for continuity, but students are directed to the other chapters for more-detailed information. Routine-finishing steps for other fiber contents or fabric types will be discussed at the end of this chapter.

Routine-Finishing Steps

FIBER PROCESSING

In *fiber processing,* the cotton fibers are processed separately from the polyester fibers because of the differences in the two fiber types. The processing of these two fiber types is discussed in Chapters 4 and 12.

YARN PROCESSING

In *yarn processing,* the fibers are aligned, blended, and twisted. Yarn processing is discussed in Chapter 19.

YARN PREPARATION

Preparation involves several steps. The first steps involve the yarns and will be discussed before the fabrication step.

Slashing. In *slashing,* the warp yarns are coated with a mixture of natural and synthetic resins to enable the yarns to resist the abrasion and tension of weaving. Slashing is adding a protective coating to the yarn to obtain optimum weaving efficiency. It is done before the fabric is woven. The sizing may contain a gum, starch, metal-to-fiber lubricant, preservative, or defoamer. In the case of the cotton/polyester–blend fabric, the sizing is probably a mixture of a gum or starch and a lubricant or poly (vinyl alcohol). The sizing must be removed after weaving in order for the finishes to bond with the fiber.

FABRICATION

Fabrication normally follows the slashing step. In fabrication, the fabric is woven, knitted, or created in some other manner. For the cotton/polyester–blend fabric, the fabric would be woven on a loom set up to create a plain weave. For a more-detailed discussion see Chapters 21 and 22.

FABRIC PREPARATION

Desizing. In *desizing,* the sizing added to the warp yarns in the slashing step is removed.

Physical or chemical desizing may be used depending on the type of sizing used and the fiber content of the warp yarns. For example, in cotton-blend fabrics, physical desizing (agitation) with chemical desizing (an enzyme) may be used.

Singeing. *Singeing* is the burning of free projecting fiber ends from the surface of the cloth. These protruding ends cause roughness, dullness, and pilling, and interfere with finishing. Singeing is the first finishing operation for all smooth-finished cotton fabrics and for clear-finished wool fabrics. Fabrics containing heat-sensitive fibers such as cotton/polyester blends are often singed after dyeing because the melted ends of the fibers may cause unevenness in color. Singeing is one of the best solutions for pilling.

Scouring. *Scouring* is a general term used to refer to removal of foreign matter or soil from the fabric prior to finishing or dyeing. The procedure followed is related to the fiber content of the fabric. The foreign matter involved may be processing oils, starch, natural waxes, and tints or color added to aid in fiber identification during production. Common scouring chemicals are soaps or detergents.

BLEACHING

Most bleaches are oxidizing agents. The actual *bleaching* is done by active oxygen. A few bleaches are reducing agents. These are used to strip color from dyed fabrics. Bleaches may be either acid or alkaline in nature. They are usually unstable, especially in the presence of moisture. Bleaches that are old or have been improperly stored will lose their oxidizing power. In bleaching, the goals are a uniform removal of hydrophobic impurities in the fabric; and a high, uniform degree of whiteness of the fabric in order to get bright uniform colors when dyeing.

Any bleach will cause some damage and, because damage occurs more rapidly at higher temperatures and concentrations, these factors should be carefully controlled.

The same bleach is not suitable to all kinds of fibers. Because fibers vary in their chemical reaction, bleaches must be chosen with regard to fiber content.

The finisher uses bleaches to clean and whiten gray goods. The natural fibers are an off-white color because of the impurities they contain. Because these impurities are easily removed from cotton, most cotton gray goods are bleached without damage. The bleaching step is often omitted with wool because it has good affinity for dyes and other finishes even if not bleached.

Peroxide bleaches are common factory bleaches for cellulose and protein fibers and fabrics. *Hydrogen peroxide* is an oxidizing bleach. A 3 percent solution is relatively stable at room temperature and is safe to use. Peroxide will bleach best at a temperature of 180–200°F in an alkaline solution. These bleaching conditions make it possible to do peroxide bleaching of cellulose gray goods as the final step in the kier boil.

In the peroxide cold bleach procedure, the fabric is soaked overnight or for a period of 8 hours. This procedure is often used on cotton-knit goods and wool to preserve a soft hand. Peroxide is good for removing light scorch stains.

PREPARATION

Mercerization. *Mercerization* is the action of an alkali (caustic soda) on a fabric. Mercerizing was a revolutionary development discovered in 1853 by John Mercer, a calico printer. He noticed that his cotton-filter cloth shrank, and became stronger, more lustrous, and more absorbent after filtering the caustic soda used in the dye process. Little use was made of mercerization at that time because the shrinkage caused a 20–25 percent yardage loss, and the increased durability caused mill men to fear that less fabric would be used. In 1897, Lowe discovered that if the fabric were held under tension, it did not shrink but became very lustrous and silk-like.

Mercerization is used on cotton and linen for many different reasons. It increases the luster and softness, gives greater strength, and improves the affinity for dyes and water-borne finishes. Plissé effects can be achieved in cotton fabrics. "Mercerized cotton" on a label is associated with luster. Cotton is mercerized for luster in both yarn and fabric form.

Yarn mercerization is a continuous process in which the yarn under tension passes from a warp beam through a series of boxes with guide rolls and squeeze rolls, a boil-out wash, and a final wash (Figure 32–2).

Fig. 32–2 *Mercerization of warp yarn. (Courtesy of Coats & Clark, Inc.)*

Fabric mercerization is done on a frame that contains mangles for saturating the cloth; a tenter frame for tensioning the fabric both crosswise and lengthwise while wet; and boxes for washing, neutralizing with dilute sulfuric acid, scouring, and rinsing. Chapter 4 includes a discussion of the changes that occur in cotton.

Greater absorbency results from mercerization because the caustic soda causes a rearrangement of the molecules, making more of the hydroxyl groups available to absorb more water and water-borne substances. Thus dyes can enter the fiber more readily, and when they can be fixed inside the fiber they are faster. (Caustic soda is also used in vat dyeing to keep the vat dye soluble until it penetrates the fiber.) Mercerized cotton and linen take resin finishes better for the same reason.

Increased strength is another important gain. Mercerized cotton fibers are stronger because, in the swollen fiber, the molecules are more-nearly parallel to the fiber axis. When stress is applied, the end-to-end molecular attraction is harder to rupture than in the more spiral-fibril arrangement.

Slack mercerization consists of dipping 100 percent cotton cloth in a 23 percent caustic solution, allowing it to react for 1½ minutes, and then washing and drying. The cloth shrinks and the yarn crimp increases. The straightening of the crimp when stress is applied gives the stretch. One-way or two-way stretch can be obtained by variations in the mercerization conditions. Stretch is achieved in cotton by slack mercerization. It gives comfort in fabrics. Stretch diapers are an end use.

Ammoniating Finishes. *Ammoniating finishes* are used on cotton and rayon yarns and fabrics. Yarns or fabrics are treated with a weak ammonium solution at −33°C and are then passed through hot water, stretched, and dried in hot air. The finish is similar to mercerization but is less expensive and less polluting. It gives more strength to cotton yarns and fabrics, increases luster, gives better affinity for dyes, and improves the hand of fabrics. Duralized and Sanforset are trade names. Ammoniating finishes are an important alternative to mercerization. They are a means of improving easy-care properties of heavy cotton fabrics like denim. Liquid ammonia treatments frequently are used as a substitute for mercerization on most cotton sewing threads.

COLORATION

Color is normally added to the fabric at this stage in the sequence. Dyeing and printing are discussed in detail in Chapter 36.

FINISHING

Special-Purpose Finishes. *Special-purpose finishes* that would be appropriate for the cotton/polyester–blend fabric might include durable-press, soil-release, and a fabric-softening finish. These finishes usually follow dyeing to avoid interfering with the absorption of the dye by the fibers. These finishes are discussed in Chapter 34.

Tentering. *Tentering,* one of the final finishing operations, performs the double process of straightening and drying fabrics. In tentering, the fabric can be fed to the pins or clips at a speed slightly greater than that of the chains of the tentering frame. The result of this is that the amount of lengthwise shrinkage can be reduced to a degree. Tentering is an important finishing step in terms of the fabric's quality. If the filling yarns are not perfectly perpendicular to the warp yarns, the fabric is off-grain. If the two ends of the filling yarn are not directly across from each other, the fabric will exhibit *skew.* If

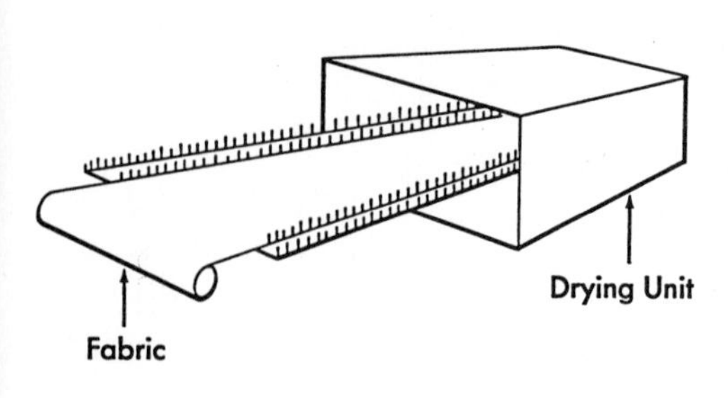

Fig. 32–3 Tenter frames: clip tenter (left); *drawing of pin-tenter frame* (right).

the center of the fabric moves at a slower speed than the two edges, the fabric will exhibit *bow*. Both of these off-grain problems can be eliminated by proper tentering. If a fabric was tentered off-grain, it will be printed off-grain as well. Some tentering frames have electronic sensors that help control the grain. The fabric may go through the tentering frame several times during finishing.

Tenter machines are of two types: the pin tenter and the clip tenter (Figure 32–3). The mechanism on the two sides moves around like a caterpillar tractor wheel, holding the fabric by a series of pins. More tension can be exerted by the clip tenter, but it may also damage some fabrics, in which case the pin tenter is used. The marks of the pins or the clips are often evident along the selvage.

Loop Drying

Fabrics with a soft finish, towels, and stretchy fabrics such as knits are not dried on the tenter frame but are dried on a *loop dryer,* where the drying can be done without tension. Many rayon fabrics are dried on loop dryers.

Heat Setting. In *heat setting,* the fabric is usually placed on a tenter frame and passed through an oven where the time of exposure and the temperature are carefully controlled based on the fiber content and resins added to the fabric. The cotton/polyester fabric would require heat curing if it had been given a durable-press or soil-release finish or if the percentage of polyester is high enough to give a degree of shrinkage control.

Calendering. *Calendering* is a mechanical finishing operation performed by a "stack" of rollers through which the cloth passes. There are several types: the simple calender, the friction calender, the moiré calender, the Schreiner calender and the embossing calender. Each produces a different finish (see Chapter 33).

Most calender machines have three rollers. (Others may have two, five, or seven.) Hard-metal rollers alternate with softer, cloth-wrapped rollers or with solid-paper rollers. Two metal rollers never run against each other.

The *simple calender* corresponds to the iron and gives a smooth, ironed finish to the fabric. The cloth is slightly damp before it enters the calender. The metal roll is heated. The cloth travels through the calender at the surface speed of the rollers so the rollers simply exert pressure to smooth out the wrinkles and give a slight sheen (Figure 32–4).

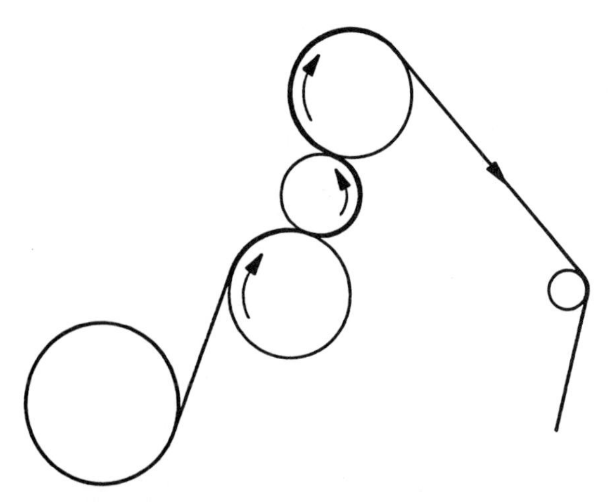

Fig. 32–4 *Calender machine.*

REWORKING

Inspecting. Fabrics are *inspected* by pulling or running them over an inverted frame in good light. Inspectors mark flaws in the fabric.

Repairing. Flaws marked by the inspectors are repaired, if possible. Broken yarns are clipped, snagged yarns are worked back into the cloth, and defects are marked so that adjustments can be made when fabrics are sold. The fabric is then wound on bolts or cylinders ready for shipment.

Other Routine-Finishing Steps

CLEANING

All gray goods must be *cleaned* and made ready for the acceptance of the finish. Gray goods contain a warp sizing, which makes the fabric stiff and interferes with the absorption of liquids. The fabric must be desized before further finishing can be done. Also, fabrics are often soiled during fabrication and must be cleaned for that reason. Warp sizing, dirt, and oil spots have always been removed by a washing process—*degumming* of silk, *kier boiling* of cotton, and *scouring* of wool.

CRABBING

Crabbing is a wool-finishing process used to set wool fabrics. Fabrics are immersed in hot water, then in cold water, and passed between rollers.

DECATING

Decating produces a smooth, wrinkle-free finish and lofty hand on woolen and worsted fabrics and on blends of wool and man-made fibers. The process is comparable to steam ironing. A high degree of luster can be developed by the decating process because of the smoothness of the surface. The dry cloth is wound under tension on a perforated cylinder. Steam is forced through the fabric. Moisture and heat relax tensions and remove wrinkles. The yarns become set and are fixed in this position by cooling, which is done with cold air. For a more-permanent set, dry decating is done in a pressure boiler. Wet decating often precedes napping or other face finishes to remove wrinkles that have been acquired in scouring. Wet decating as a final finish gives a more-permanent set to the yarns than does dry decating.

BEETLING

Beetling is a finish used on linen and a few fabrics resembling linen. As the cloth revolves slowly over a huge wooden drum, it is pounded with wooden-block hammers. This pounding may continue for a period of 30–60 hours. It flattens the yarns and makes the weave appear less open than it really is. The increased surface area gives more luster, greater absorbency, and smoothness to the fabric.

OPTICAL BRIGHTENERS

Optical brighteners are also used to whiten off-white fabrics. They are fluorescent-white compounds, not bleaches. The fluorescent-white compounds are absorbed by the fiber and emit a bluish fluorescence that masks yellow. At the mill, optical brighteners give best results when used in combination with the bleach rather than as a substitute for it. They also are added to the spinning solution of some man-made fibers to optically brighten them.

CARBONIZING

Carbonizing, which is the treatment of wool yarns or fabrics with sulfuric acid, destroys vegetable matter in the fabric, and allows for more-level dyeing. Carbonizing is also done on recycled wool to remove any cellulose that may have been used in the original fabric. Carbonizing gives better texture to all-wool fabrics.

PRESSING

Pressing is the term used with wool or wool blends. In pressing, the fabric is placed between heavy-metal plates that steam and press the fabric.

SIZING

Sizing is similar to slashing in that a starch or resin is added to a textile. The major difference is that slashing is padded on yarns prior to weaving and sizing is added to fabrics to add body. Sizing is temporary if starch is used and permanent if resin is used. Sizing adds weight and body to the fabric.

33

Aesthetic Finishes

Finishes that change the appearance and/or hand of fabrics often create a new or particular fabric; for example, eyelet embroidery, flannel, and organdy are made by special finishes. Figure 33–1 shows several fabrics that were converted from print cloth. Percale is roller printed, chintz is waxed and friction calendered, plissé is printed with caustic soda, and embossed cotton is run through an embossing calender. This fabric also could be flocked, embroidered, or given a surface coating.

Types of Aesthetic Finishes

Finishes discussed in this chapter are grouped by the aesthetic change caused by the finish into the following areas:

1. Luster.
2. Drape.
3. Texture.
4. Hand.

In each of the groupings, the way the finish is applied to the fabric, the effect of the finish, and the relationship of the finish to the fabric name will be explained.

Many of these finishes are additive finishes that give texture (body, stiffness, softness), luster, embossed designs, and abrasion resistance to the fabric. They are held on the surface mechanically and their permanence depends on the efficiency of the finish and the type of finish itself.

The *padding machine,* often called the "workhorse" of the textile industry, is used to apply dyes, chemicals, and additive finishes. It will apply them in either liquid or paste form, on one or both sides (Figure 33–2).

Padding is done by passing the fabric through the finishing solution, under a guide roll, and between two padding rolls. The rolls are metal or rubber, depending on the finish to be applied. The rolls exert tons of pressure on the fabric to squeeze the finish into the fiber or fabric to assure good penetration. Excess liquid is squeezed off. The fabric then travels into the steaming or washing and drying machine.

The *backfilling machine* is a variation of the padding machine. It applies the finish to one side only, usually to the wrong side of the fabric (Figure 33–3).

LUSTER

Luster finishes result in a change in the ability of a fabric to reflect light. Most of the finishes in this group increase the ability of the fabric to reflect light and improve the luster or shine of the fabric. The increase in luster may be over the entire fabric—as in the glazed, ciré, and Schreiner finishes—or it may be a localized increase in luster—as in the moiré and embossed finishes.

Glazed. Glazed chintz and polished cotton are two fabrics named because of the surface *glaze* that results from this finish. The *friction calender* is used to give a highly glazed surface to the cloth. If the fabric is first saturated with starch and waxes, the finish is only temporary; but if resin finishes are used, the glaze will be durable. The cloth is first passed through the finishing solution and then dried to a certain degree of dryness. It is then threaded into the calender. The speed of the metal roller is greater than the speed of the cloth, and the roller polishes the surface just as the sliding motion of the hand iron polishes the fabric.

Ciré. A *ciré finish* is similar to a glazed finish, except that the metal roll is hot, which results in more luster on the surface of the fabric. These fabrics often are made of a thermoplastic-fiber content. Because thermoplastic fibers are heat sensitive, the surface of the fibers that come in contact with the metal roll melts and flattens slightly and gives the highly polished appearance to the fabric. Ciré is a taffeta or satin fabric hot-friction calendered to give a high gloss, or "wet" look.

Moiré. To produce a *moiré pattern* on a fabric, two techniques can be used. In the first method, called *true moiré,* rib fabrics such as unbalanced taffeta or faille are used. True moiré is made by placing two layers of ribbed fabric, one on top of the other, so that the ribs of the top layer are

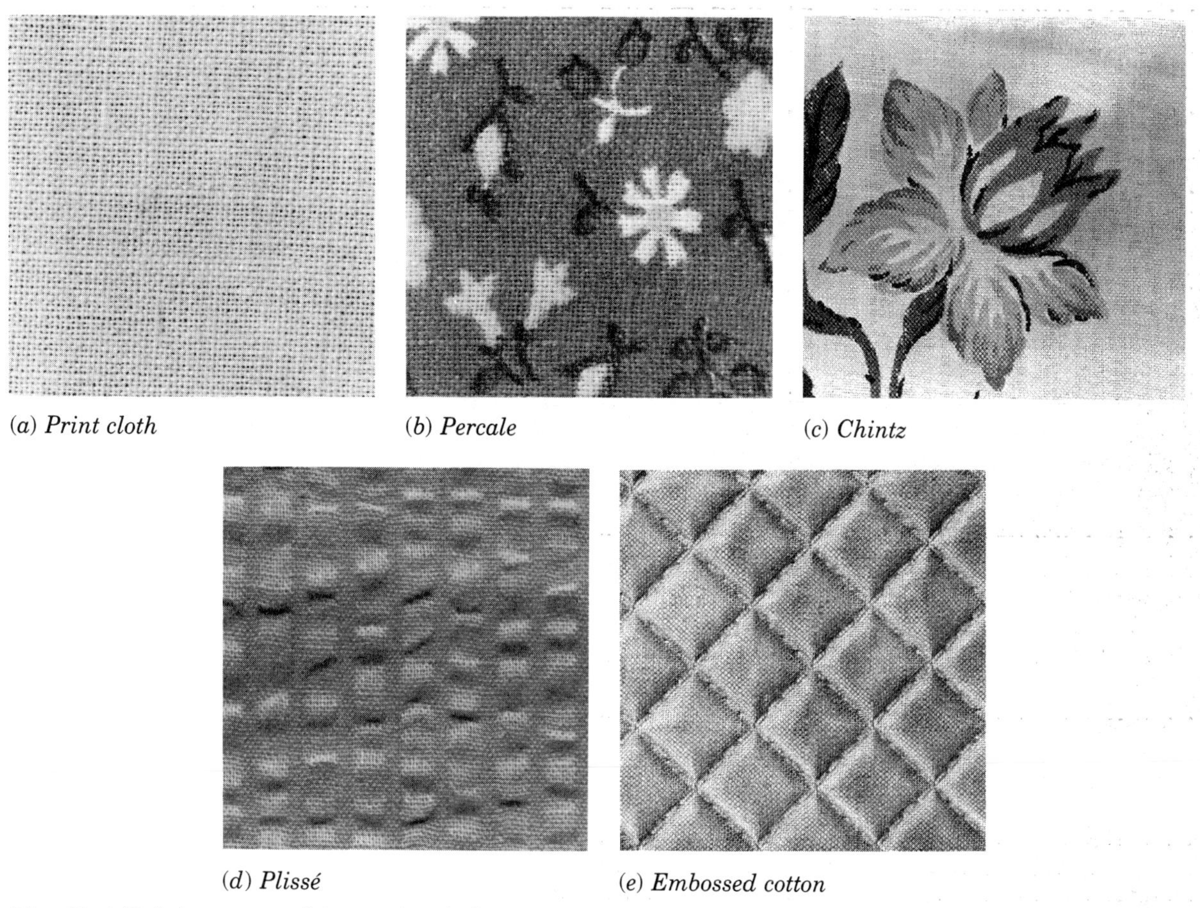

(*a*) *Print cloth* (*b*) *Percale* (*c*) *Chintz*

(*d*) *Plissé* (*e*) *Embossed cotton*

Fig. 33–1 *Fabrics converted from print cloth: (a) gray goods; (b) roller printed; (c) waxed and friction calendered; (d) printed with caustic soda; (e) embossed calendered.*

slightly off-grain in relation to the under layer. The two layers are stitched or held together along the selvage and are then fed into the smooth, heated-metal roll calender. Pressure of 8–10 tons causes the rib pattern of the top layer to be pressed into the bottom layer and vice versa. Flattened areas in the ribs reflect more light and create a contrast to unflattened areas. This procedure can be modified to produce patterned moiré designs other than the watermarked one.

In the second procedure, an embossed-metal roll is used. The embossed roll has a moiré pattern engraved on it. When the roll is passed over a ribbed fabric, the ribs are flattened in areas and a moiré pattern is created. If the fabric is of a thermoplastic-fiber content and the roll is heated, the finish is permanent.

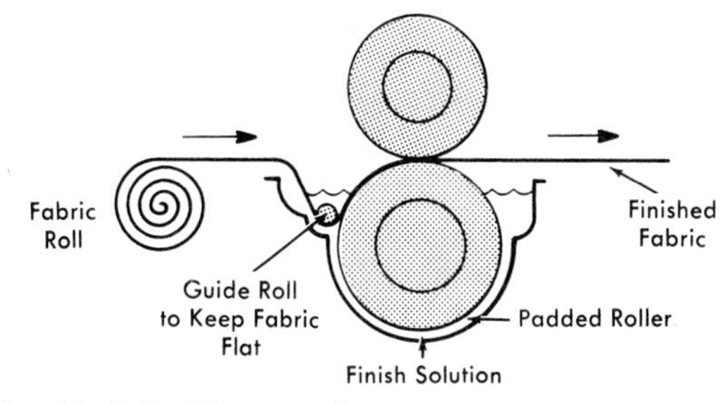

Fig. 33–2 *Padding machine.*

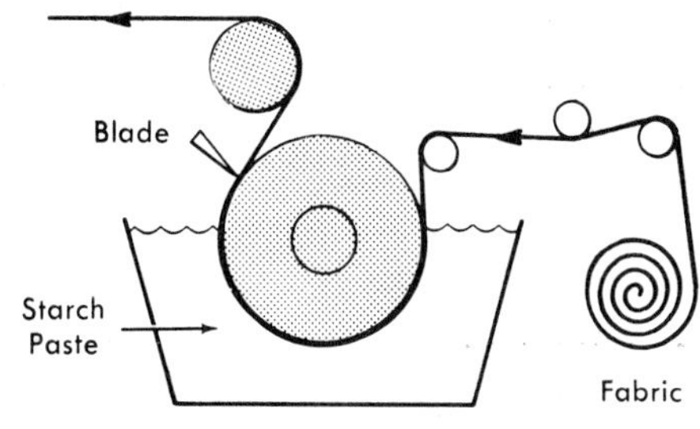

Fig. 33–3 *Backfilling machine.*

Schreiner. Fabrics with a Schreiner finish will have a softer luster than with most of the other luster finishes. The *Schreiner calender* (Figure 33–4) has a metal roller engraved with 200–300 fine diagonal lines that are visible only under a magnifying glass. (The lines should not be confused with yarn twist.) Until the advent of resins and thermoplastic fibers, this finish was temporary and was removed by the first washing. The primary purpose of this finish is to produce a deep-seated luster, rather than a shine, by breaking up reflectance of light rays. It also flattens the yarns to reduce the openness between them and give smoothness and cover. It can upgrade a sleazy or lower quality fabric. This finish was originally used with cotton sateen and table damask to make them more lustrous. It was later used on polished, resinated cottons and sateens as a durable finish.

Embossed. *Embossed designs* are created using an embossing calender that produces either flat or raised designs on the fabric. Embossing became a much more important finish after the heat-sensitive fibers were developed because it was possible to produce a durable, washable, embossed pattern. Nylon, acrylic, acetate, polyester, and fabrics made of nylon and metallic yarns are used. If the fabrics are made of solution-dyed fibers, they can be embossed directly off the loom and are then ready for sale. Embossed satins are used in high-style garments and can be sold for a higher price than the unembossed fabric.

The embossing calender consists of two rolls, one of which is a hollow, engraved-metal roll heated from the inside by a gas flame. The other is a solid-paper roll exactly twice the size of the engraved roll (Figure 33–5). The fabric is drawn between the two rollers and is embossed with the design.

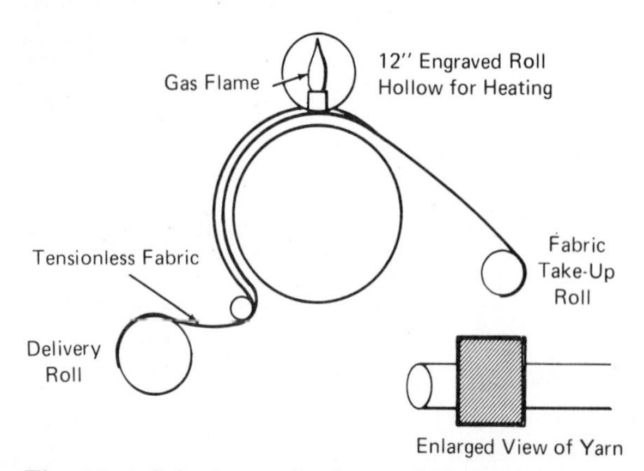

Fig. 33–4 *Schreiner calender machine for tricot.*

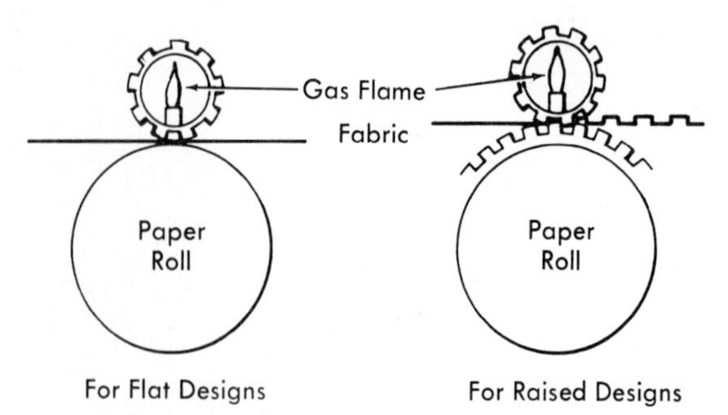

Fig. 33–5 *Embossing process.*

The process differs for the production of flat and raised designs. Raised-embossed designs will be discussed later in this chapter under the section on texture. Flat embossed designs are the simplest to produce. A copper roll, engraved in deep-relief (Figure 33–6) revolves against a smooth-paper roll. The hot, engraved areas of the roll produce a glazed pattern on the fabric. Embossed brocades are an example of this type of design.

DRAPE

Drape finishes change the way a fabric will fall or hang over a three-dimensional shape. These finishes make the fabric stiffer or more flexible.

Fig. 33–6 *Embossing rolls. (Courtesy of Consolidated Engravers Corp.)*

Crisp and Transparent. *Transparent* (or parchment) effects in cotton cloth are produced by treatment with strong sulfuric acid. This fabric is often referred to as parchmentized. One of the oldest finishes is a Swiss or organdy finish produced by the Heberlein process. Since acid damages cotton, the process must be very carefully controlled, and "split-second" (5–6 seconds) timing is necessary to prevent *tendering,* or weakening of the fabric. These effects are possible: all-over parchmentization, localized parchmentization, and plissé effect on either of the first two.

Because *all-over parchmentizing* is for the purpose of producing a transparent effect, a sheer fabric of combed lawn is used. The goods are singed, desized, bleached, and mercerized. Mercerization is such an important part of the process that the fabric is mercerized again after the acid treatment in order to improve the transparency. The fabric is then dyed or printed with colors that will resist acid damage. The cloth is immersed in the acid solution and partial solution of the surface of the cellulose takes place. On drying, this surface rehardens as a cellulosic film and gives permanent crispness and transparency. After the acid treatment, the cloth is neutralized in weak alkali, washed, and then calendered to give more gloss to the surface. This all-over treatment produces *organdy* fabric.

In *localized parchmentizing,* if the design is a small figure with a large transparent area, an acid-resist substance is printed on the figures and the fabric is run through the acid bath. The acid-resistant areas retain their original opacity and contrast sharply with the transparent background (Figure 33–7). If a small transparent design is desired, the acid is printed on and then quickly washed off.

Fig. 33–7 *Localized parchmentizing (acid finish) gives transparent background.*

Burned-Out. *Burned-out,* or *etched, effects* are produced by printing certain chemicals on a fabric made of fibers from different fiber groups: rayon and silk, for example. One fiber will be dissolved, leaving sheer areas. Figure 33–8 shows an etched rayon/silk velvet. The rayon has been dissolved by acid.

Sizing. *Sizing,* or *starching,* at the mill is similar to starching at home except that the starch mixture contains waxes, oils, glycerine, and similar compounds, which act as softeners. For added weight, talc, clay, and chalk are used. Gelatin is used on rayons because it is a clear substance that does not detract from the natural luster of the fibers but enhances it. Sizing adds stiffness to the fabric. The permanence of the sizing is related to the type of sizing and method of application. If the sizing is water soluble, it will be removed during washing. If the sizing is resin based and heat set, the sizing will be permanent.

TEXTURE

Sheared. A *sheared fabric* is a pile or napped fabric in which the pile or nap has been cut to remove loose fiber or yarn ends, knots, and sim-

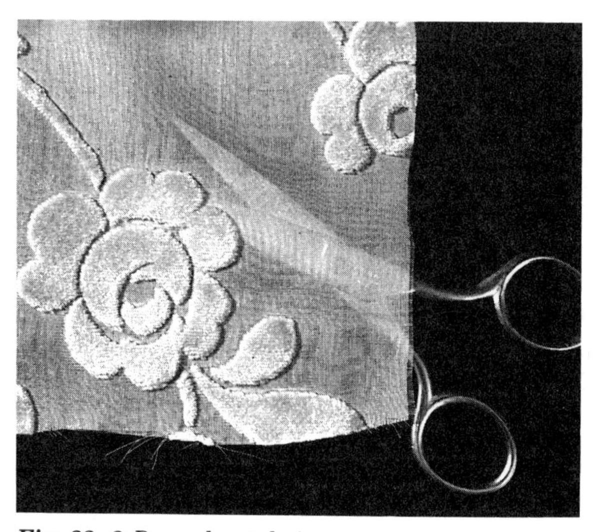

Fig. 33–8 *Burned-out design.*

ilar irregularities or surface flaws. Shearing is a finishing process done by a machine similar to a lawn mower. Shearing controls the length of the pile or nap and may create a patterned or a smooth surface. Sculptured effects are made by flattening portions of the pile with an engraved roller and then shearing off the areas that are still erect. Steaming the fabric raises the flattened portions.

Brushed. A *brushed fabric* is one where the surface of the fabric has been cleaned of fiber ends. Brushing follows shearing to clean the surface of clear-face fabrics. When combined with steaming, it will lay nap or pile in one direction and fix it in that position, thus giving the up-and-down direction of pile and nap fabrics.

Embossed. *Embossed fabrics* also may have a raised design or pattern. The embossed design may be permanent if the fabric is of a thermoplastic-fiber content or if a resin is used and heat set. Raised, or relief, designs require a more-complicated routine than flat-embossed fabrics. The paper roll is soaked in water and then revolved against the steel-engraved roll (without fabric) until the pattern of engraving is pressed into the paper roll. The temperature is adjusted to suit the fabric, which is then passed between the rolls.

Pleated. A *pleated fabric* is made by a special variation of embossing. Pleating is an ancient art that dates back to the Egyptians, who used hot stones to make the pleats. Colonial women in the United States used heavy pleating irons to press in fancy pleats and fluting. Today, pleating methods are highly specialized operations done by either the paper-pattern technique or by the machine process.

The paper-pattern technique is a hand process and is therefore more costly, but it produces a wider variety of pleated designs. Garments in partly completed condition, such as hemmed skirt panels, are placed in a pleated-paper pattern mold. The fabric is placed in the paper mold by hand and another pattern mold is placed on top so that the fabric is pleated between the two pleating papers. The whole thing is rolled into a cone shape, sealed, and put in a large curing oven for heat setting.

The machine-pleating process is less expensive. The machine has two heated rolls. The fabric is inserted between the rolls as high-precision blades put the pleats in place. A paper backing is used under the pleated fabric and the pleats are held in place by paper tape. After leaving the heated roll machine, the pleats are set in an aging unit.

Puckered Surface. *Puckered surfaces* are created by partial solution of the surface of a nylon or polyester fabric. Sculptured and "damasque" effects are made by printing a chemical on the fabric to partially dissolve it. Shrinking occurs as it dries, thus creating a puckered surface.

Plissé. *Plissé* is converted from either lawn or print-cloth gray goods by printing sodium hydroxide (caustic soda) on the cloth in the form of stripes or designs. The chemical causes the fabric to shrink in the treated areas. As the treated stripe shrinks, it causes the untreated stripe to pucker. Shrinkage causes a slight difference in count between the two stripes. The treated or flat stripe increases in count as it shrinks. The upper portion of the cloth in Figure 33–9 shows how the cloth looks before finishing, and the lower portion shows the crinkle produced by the caustic-soda treatment. This piece

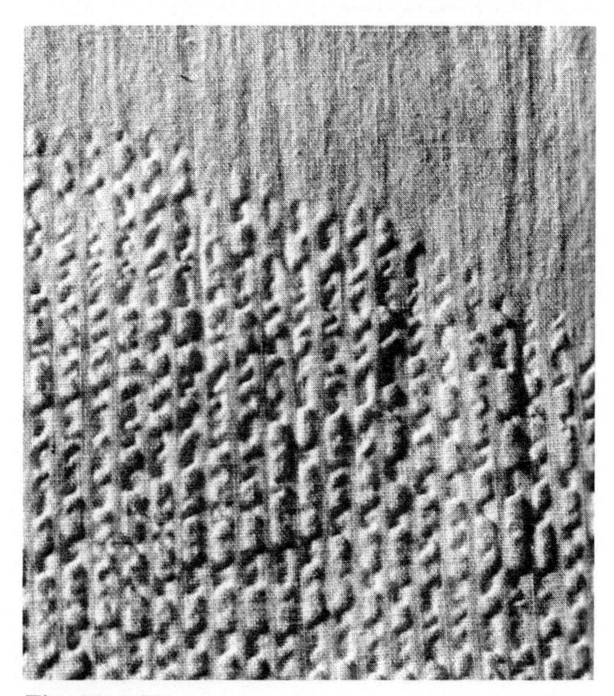

Fig. 33–9 *Plissé.*

of goods is defective because the roller failed to print the chemical in the unpuckered area.

Flocked. *Flocked fabrics* are made to imitate pile fabrics. In a flocked fabric, a surface fiber is applied to the fabric after the base fabric has been produced. See Chapter 24 for details of the process.

Embroidered. *Embroidered fabrics* can be produced either by hand or by machine. These fabrics are decorated with a surface-applied thread. The machine-embroidery operation is similar to making fancy stitches on a sewing machine, with zigzag stitches of various lengths very close together. The Schiffli embroidery machine is a frame 10 or 15 yards long (Figure 33–10) with 684 or 1,026 needles, respectively.

The designer makes a careful sketch of the embroidery pattern, which is enlarged six times and used as a guide for punching holes in a roll of thin, flexible cardboard. The perforated roll guides the placement of each stitch in the automated machine. Some embroidered fabrics are shown in Figures 33–11 and 33–12. Embroidering can be done on any kind of fabric.

Embroidered figures are very durable, often outlasting the ground fabric. It is expensive compared to the same fabric unembroidered. Like other applied designs, the figure may or may not be on-grain.

Fig. 33–10 *Schiffli embroidery machine.*

Fig. 33–11 *Embroidered linen* (right); *printed to look like embroidery* (left).

Eyelet is a widely used embroidery fabric. Small, round holes are cut in the fabric, and stitching is done completely around the hole. The closeness and amount of stitching, as well as the quality of the background fabric, vary tremendously.

Eyelet fabrics are a perennial favorite. In 1978, a lower-cost fabric that looks like eyelet came on the market. The technique used to make this fabric is called *expanded foam.* A colored-foam substance is printed on the fabrics. The substance expands during processing to give a three-dimensional effect. Initially the pattern simulated hand-embroidery stitches

Fig. 33–12 *Embroidery used in purse. (Courtesy of Lesco Lona.)*

Fig. 33–13 *Fabric before and after napping.*

such as the cross stitch. The effect seems to be durable.

TEXTURE

Napped. *Nap* consists of a layer of fiber ends, on the surface of the cloth, that are raised from the ground weave by a mechanical-brushing action. Thus napped fabrics are literally "made" by a finishing process. Figure 33–13 shows a fabric before and after napping.

Napping was originally a hand operation in which the napper tied together several teasels (dried, thistle-like vegetable burs, shown in Figure 33–14) and swept them with a plucking motion across the surface of the cloth to raise fibers from the ground weave. The teasels had a gentle action and the barbs would break off before causing any damage to the cloth. The raised fibers formed a nap that completely changed the appearance and texture.

Teasels may be used in the machine finishing of fine-wool fabrics such as duvetyn. For machine processing (gigging), they are mounted on rollers, and, as the barbs wear off or break off, the worn teasels are replaced by new ones. The fabric may be either wet or dry.

Most napping is now done by rollers covered by a heavy fabric in which bent wires are embedded (Figure 33–15). Napping machines may be single action or double action. Fewer rollers are used in the single-action machine. They are all alike and travel at the same speed. They are called pile rolls, and the bent ends of the wires point in the direction in which the cloth travels, but the rollers are all mounted on a large drum or cylinder that rotates in the same direction as the cloth. The pile rolls must travel faster than the cloth in order to do any napping.

In the double-action napping machine, every

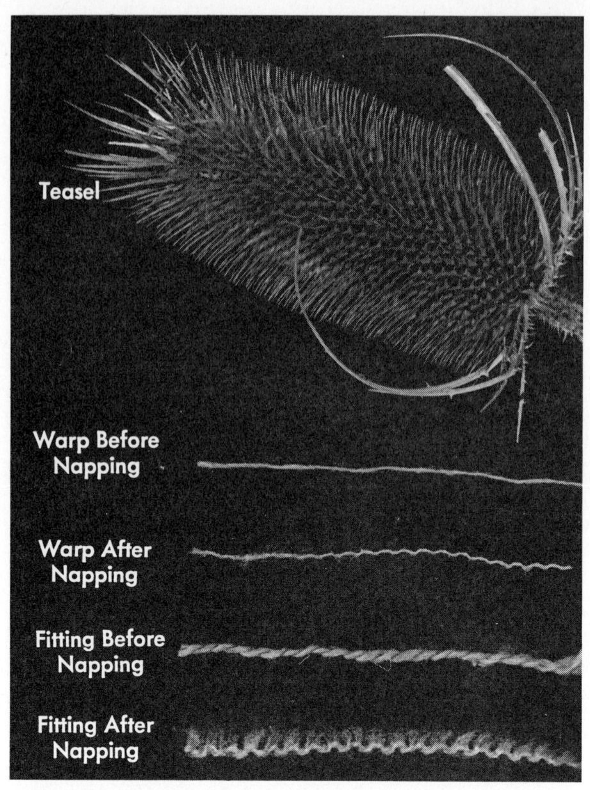

Fig. 33–14 *Teasel and yarn before and after napping.*

other roll is a counter-pile roll. This roll has wires that point in the direction opposite to those of the pile roll. The counter-pile roll must travel slower than the cloth in order to produce a nap. When the speed of the rolls is reversed (pile rolls at slower speed and counter-pile rolls at faster speed), a "tucking" action occurs. Tucking pushes the raised fibers back into the cloth and makes a smooth surface. Reasons for napping include the following:

1. *Warmth.* A napped surface and the soft twist of the filling yarns increase the dead-air space. Still air is one of the best insulators.

2. *Softness.* This characteristic is especially important in baby clothes.

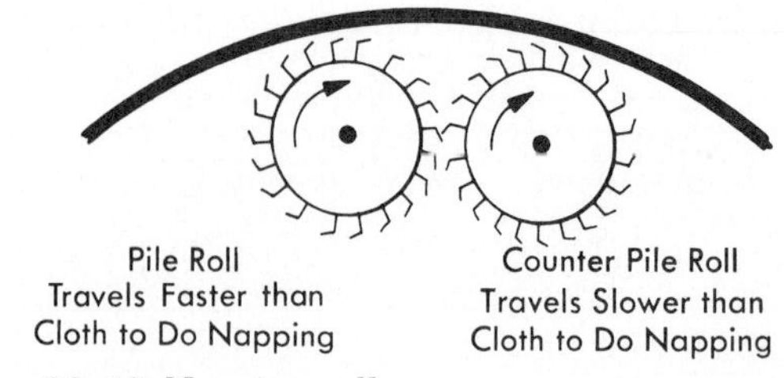

Fig. 33–15 *Napping rolls.*

3. *Beauty.* Napping adds much to fabric attractiveness.

4. *Water and stain repellence.* Fiber ends on the surface cut down on the rapidity with which the fabric gets wet.

The amount of nap does not indicate the quality of the fabric. The amount may vary from the slight fuzz of Viyella flannel to the very thick nap of imitation fur. Short compact nap on a fabric with firm yarns and a closely woven ground will give the best wear. Stick a pin in the nap and lift the fabric. A good, durable nap will hold the weight of the fabric. Hold the fabric up to the light and examine it. Press the nap aside and examine the ground weave. A napped surface may be used to cover defects or a sleazy construction. Rub the fabric between the fingers and then shake it to see if short fibers drop out. Thick nap may contain flock (very short wool fibers). Rub the surface of the nap to see if it is loose and will rub up in little balls (pilling). Notice the extreme pilling in the sweater in Figure 33–16.

Some napped fabrics have an up and down. To test this, brush the surface of the fabric. Brushing against the nap roughs it up and causes it to look darker because more light is absorbed. This is the "up" direction of the fabric. When the nap is smoothed down, the reflection of light from the surface gives a lighter shade of color. Napped fabrics should be made with the nap "down" so that garments will be easier to brush. However, the direction of the nap is not as important as the fact that the same direction of nap is used in all parts of a garment.

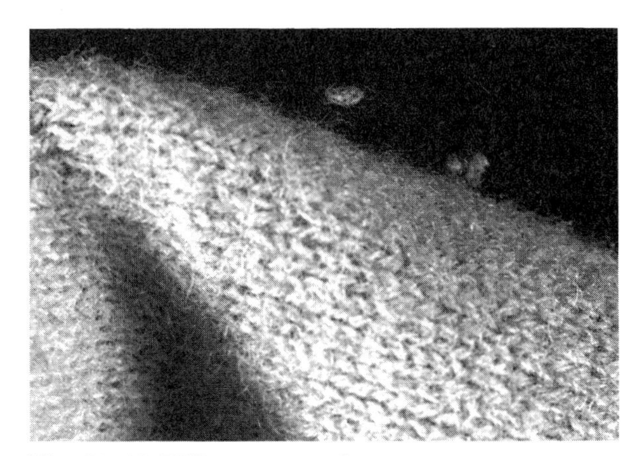

Fig. 33–16 *Pilling on a wool sweater.*

Low-count fabrics usually have low-twist yarns; when strain is applied, the fibers slip past one another and do not return to their original position. Thus garments tend to "bag" in the seat and elbow areas. Tightly twisted yarns are more resistant to bagging.

Wear on the edges of sleeves, collars, buttonholes, and so forth causes an unsightly contrast to unworn areas. Very little can be done to it except that a vigorous brushing will give a fuzzier appearance. Price is no indication of resistance to wear, since more-expensive wool fibers are finer and less resistant to abrasion. Wear will flatten the nap, but, if nap is still present on the fabric, it can be raised by brushing or steaming. Loosely napped fabrics will shed short fibers on other garments or surfaces. They also shed lint in the wash water, so they should be washed separately or after other articles are washed.

Napped fabrics must be made from specially constructed gray goods in which the filling yarns are made of low-twist staple (not filament) fibers. (See Chapter 17 for information about yarn twist.) The difference in yarn structure makes it easy to identify the lengthwise and crosswise grain of the fabric. Figure 33–14 shows the warp and filling yarns from a camel's-hair coat fabric, before and after napping.

Fabrics can be napped on either or both sides. The nap may have an upright position or it may be "laid down" or "brushed." When a heavy nap is raised on the surface, the yarns are sometimes weakened. Wool fabrics are fulled or shrunk to bring the yarns closer together and increase fabric strength.

Yarns of either long- or short-staple fibers may be used in napped fabrics. Worsted flannels, for example, are made of long-staple wool. The short-staple yarns used in woolen flannels have more fiber ends per inch and thus can have a heavier nap. In blankets which are heavily napped for maximum fluffiness, a fine-cotton (core) ply is sometimes used in the yarn to give strength.

Napped fabrics can be made of any staple fiber. Pilling and attraction of lint due to electrostatic properties are problems with the nylons and polyesters. Napping is less expensive than pile weave as a way of producing a three-dimensional fabric.

Napped fabrics may be plain weave, twill weave, or knit. More filling yarn is exposed on the surface in a $\frac{2}{2}$ twill or a filling-faced twill; therefore a heavier nap can be raised on twill fabrics. The knit construction in napped fabrics is often used for articles for the baby.

Flannel is an all-wool napped fabric made in dress, suit, or coat weights. It may be made with either worsted or woolen yarns. They may be yarn dyed. Worsted flannels are important in men's suits and coats and are used to a lesser extent in women's suits and coats. They are firmly woven and have a very short nap. They wear well, are easy to press, and hold a press well. Woolen flannels are fuzzier, less-firmly woven fabrics. Many have been given a shrinkage-control treatment that alters the scale structure of the fiber. Because napping causes some weakening of the fabric, 15–20 percent nylon or polyester is blended with the wool to improve the strength.

Fleece is a coatweight fabric with long brushed nap or a short clipped nap. Quality is very difficult to determine.

Cotton flannels flatten under pressure and give less insulating value than wool because cotton fibers are less resilient. The fibers are also shorter; thus there is more shedding of lint from cotton flannels. The direction of nap (up and down) is relatively unimportant in these fabrics because their chief uses are robes, nightwear, baby clothes, and sweatshirts. Cotton/polyester blends are often used. *Flannelette* is a plain-weave fabric that is converted from a gray-goods fabric called soft-filled sheeting. It is napped on one side only, has a short nap, and often has a printed design. The nap will form small pills and is subject to abrasion. *Suede* and *duvetyn* are also converted from the same gray goods but are sheared close to the ground to make a smooth, flat surface. Of the two, duvetyn is lighter weight. *Outing flannel* is a yarn-dyed (or white) fabric that is similar in fabric weight and nap length to flannelette but is napped on both sides.

As the warp yarns in both flannelette and outing flannel are standard weaving yarns, it is easy to identify the grain of the fabric. Napped, knitted fabrics are often given pile-fabric names such as velvet or velour.

Fulled. *Fulling* is done on wool fabrics to improve the appearance and hand. Fabrics are fulled by moisture, heat, and friction—a very mild felting process. A fabric that has been fulled will be denser and more compact (see Figure 6–8).

34

Special-Purpose Finishes

Special-purpose finishes are also known as *functional finishes.* These are treatments that are applied to fabrics to make them better suited for a specific end use. These finishes usually do not alter the appearance of fabrics, but they improve performance. They help solve some consumer problems with textile products.

Stabilization: Shrinkage Control

A fabric is *stabilized* when it retains its original size and shape during use and care. Unstable fabrics shrink or stretch. Of these, shrinkage is the more-serious problem. *Shrinkage* is the reduction in size of a product.

The shrinkage problem began when spinning, weaving, and finishing were mechanized. Fabrics are under tension on the loom, and, in wet finishing, fabrics are pulled through machines in long continuous pieces and finally set under excessive warpwise tension that leaves the fabric with high residual shrinkage. Shrinkage will take place when tensions are released by laundering or steam pressing.

Shrinkage is used to advantage in the manufacture of some fabrics. For example, *fulling* of wool cloth closes up the weave and makes a firmer fabric, and shrinkage of the high-twist yarns in crepes creates the surface crinkle. Shrinkage is a disadvantage to the consumer when it changes the length or size of a garment.

There are two types of shrinkage: relaxation (or fabric) shrinkage and progressive (or fiber) shrinkage. *Relaxation shrinkage* occurs during washing or dry cleaning. Most relaxation shrinkage occurs during the first care-cycle. However, small amounts of relaxation shrinkage may continue to occur for several additional care cycles; this is *progressive shrinkage.* If the care is mild in the first cycle and more severe in later cycles, more shrinkage may occur during these later cycles than in the first cycle. For example, if the item was line dried in the first cycle and machine dried in later cycles, shrinkage will be more severe during machine drying. Progressive shrinkage will continue to occur during subsequent care cycles. The following list groups fibers by the kind of shrinkage they normally exhibit.

1. *Cotton, linen, and high-wet-modulus rayon:*
Exhibit relaxation shrinkage.
No progressive shrinkage.

2. *Regular rayon:*
Exhibits high relaxation shrinkage.
Moderate progressive shrinkage.

3. *Wool:*
Exhibits moderate relaxation shrinkage.
High progressive shrinkage.

4. *Other properly heat set man-made fibers:*
Exhibit relaxation shrinkage.
No progressive shrinkage.

Mechanical-control methods or heat are used to eliminate relaxation shrinkage, and chemical-control methods are used to prevent progressive shrinkage.

RELAXATION SHRINKAGE AND METHODS OF CONTROL

Knit Fabrics. *Knit fabrics* shrink because the loops are elongated 10–35 percent lengthwise in knitting and wet finishing (Figure 34–1). During laundering, the stitches will reorient themselves to their normal shape, and the garment will become shorter and wider. Length is a critical dimension in knit apparel because width shrinkage is restorable when the clothing is worn.

Minimal shrinkage of knits can be achieved by running the fabric through rollers, but overfeeding the fabric between sets of rollers that results in lengthwise shrinkage. The increased use of polyester in cotton-like knits permits the fabric to be heat set for stabilizing.

Woven Fabrics. All woven fabrics shrink after the strains of weaving, warp-yarn sizing, and wet finishing are released when the fabric gets

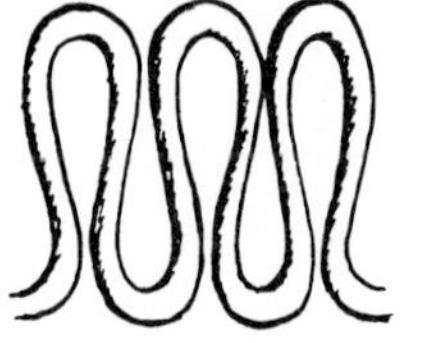

(a) Before Redmanizing ***(b)*** After Redmanizing

Fig. 34–1 *Knit stitches: (a) stretched; (b) relaxed.*

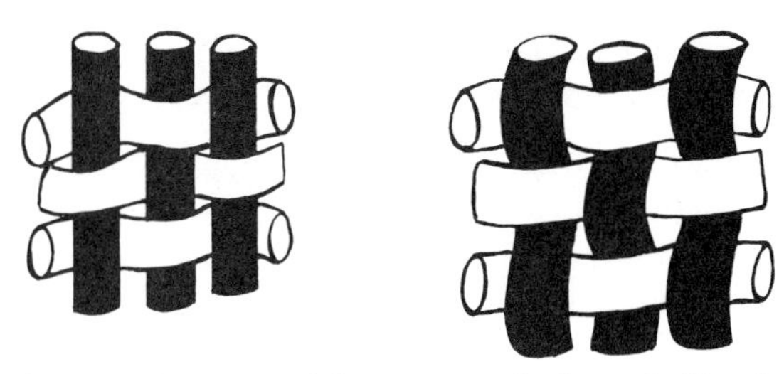

Fig. 34–2 Position of the warp on the loom (left); *after the fabric relaxes when it becomes wet* (right).

wet. The warp yarns are stretched out straight while they are on the loom, and the filling is inserted in a straight line. The filling takes on crimp as it is beaten back into the cloth, but the warp stays straight (Figure 34–2). When the fabric is thoroughly wet and allowed to relax, the yarns readjust themselves and the warp yarns also move to a crimped position (Figure 34–2). This crimp shortens the fabric in the warp direction. With the exception of crepe fabrics, less change occurs in the filling direction.

Compressive shrinkage processes are used on woven fabrics of cotton, linen, and high-wet-modulus rayon. Regular rayons will not hold a compressive-shrinkage treatment because of the high swelling and wet elongation of the rayon fibers.

In the factory process, a thick rubber blanket is the medium that shrinks the cloth. A thick blanket will shrink the cloth more than a thin one. The blanket, with the moist cloth adhering to its surface, is passed around a feed-in roll. In this curved position, the outer surface stretches and the inner surface contracts. The blanket then reverses its direction around a heated drum. The outer curve becomes the shorter, inner surface and the fabric adhering to it is compressed. The fabric, which is now against the drum, is dried and set with a smooth finish. The count will increase, and the cloth will actually be improved after compressing (Figures 34–3 and 34–4).

Research has shown that faulty laundering will cause compressively shrunk fabrics to shrink as much as 6 percent. Tumble drying may also compress the yarns beyond their normal shrinkage.

London shrunk is a 200-year-old relaxation finish for wool fabrics that removes strains caused by spinning, weaving, and finishing. Originally, fabric was laid out in the fields of the city of London and the dew soothed away the

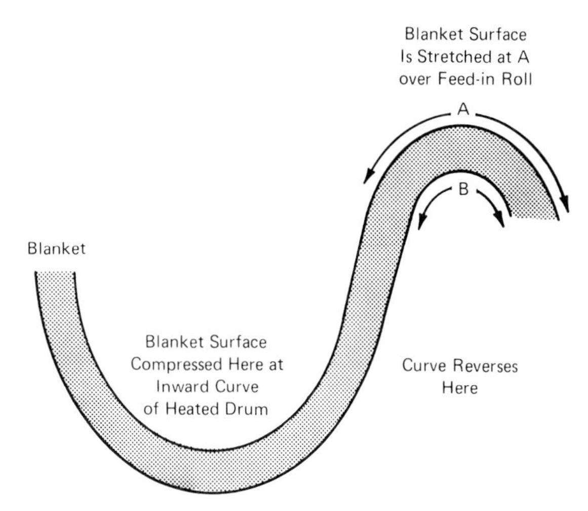

Fig. 34–3 Reversal of curve causes change in size to compress fabric.

stresses and improved the hand of the fabric. Although techniques have been modernized, there is still much hand labor involved. A wet blanket—wool or cotton—is place on a long platform, a layer of cloth is then spread on it, and alternate layers of blankets and cloth are built up. Sufficient weight is placed on top to force the moisture from the blankets into the wool. The cloth is left in the pile for about 12 hours. The cloth is then dried in natural room air by hanging it over sticks. When dry, the cloth is subjected to hydraulic pressing by building up layers of cloth and specially made press boards with

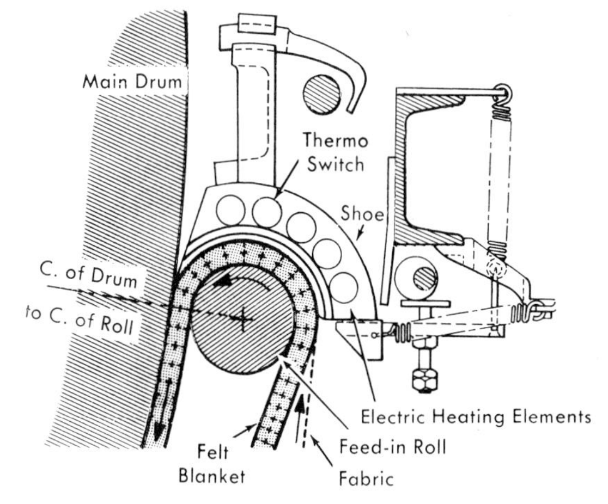

Fig. 34–4 Closeup of compressive-shrinkage process. The electrical-heated shoe holds the fabric firmly on the outside of the blanket so that when the blanket collapses in straightening out, the fabric is shrunk accordingly. (Courtesy of the Sanforized Company.)

a preheated metal plate inserted at intervals. A preheated metal plate is also placed on the top and bottom. This setup of cloth, boards, and plates is kept under 3,000 pounds of pressure for 10–12 hours. London shrinking is done for men's wear fine worsteds—not for woolens or women's wear.

Today the right to use the label "Genuine London Process" or something similar is licensed by the Parrot group of companies, Clothworkers of London, Leeds, and Huddersfield, to garment makers all over the world. The permanent-set finish Si-Ro-set, which produces washable, wrinkle-free wool fabrics, is now applied to some fabrics during London-shrunk processing.

A similar method for home use is that of rolling wool cloth in a wet sheet, allowing it to stand for 6 hours, and then placing it flat on a table or floor to dry. If it is straightened while wet, pressing may be unnecessary. This is the best means for straightening wool that has been tentered or decated off-grain. It should not be used on wool crepe. Fabrics that have a napped surface, such as wool broadcloth or some wool flannels, may be changed in appearance. Wool fabrics should always be tested for shrinkage prior to cutting. A simple method of testing is to draw a right angle on the ironing board, place the warp edge along one side and the filling edge along the other, and hold the steam iron over the fabric. If either edge draws away from the pencil line, the fabric will shrink during steam pressing and it should be shrunk.

PROGRESSIVE SHRINKAGE AND METHODS OF CONTROL

Thermoplastic Fibers. Thermoplastic fibers are stabilized by heat setting, a process in which fabrics are heated at temperatures at or above the glass transition temperature (T_g) and then cooled. The *T_g temperature* is the point at which the amorphous regions of the fiber are easy to distort. It is lower than the melting point of the fiber and differs for various fibers. If properly heat set, fabrics will exhibit no progressive shrinkage and relaxation shrinkage will also be controlled (Figure 34–5).

Wool Fibers. Washable wool is important in clothing and in blends with washable fibers. If wool fabrics are to improve their position in the competitive market with fabrics made from wool-like fibers that have easy-care characteristics, they must be finished to keep their original size and surface texture with laundering. It might be assumed that people who can afford professional care will not be interested in washing wools. Another assumption might be that washable wools (those given a felting shrinkage-control treatment) are the poor- to medium-quality wools. Whether or not these assumptions are true, felting shrinkage is important today, as evidenced by the fact that patents for feltproofing wool continue to be issued. Figure 34–6 shows shrinkage of a wool sock after washing and drying.

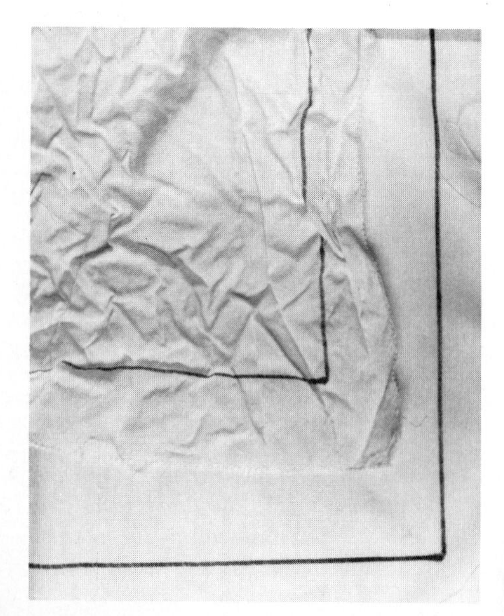

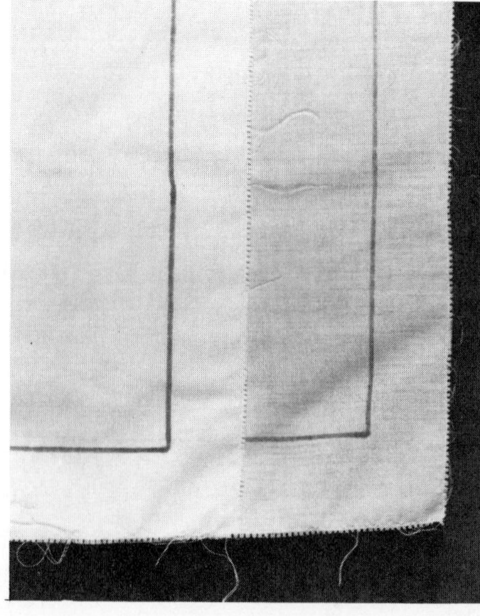

Fig. 34–5 *Comparison of thermoplastic fiber fabrics. Bottom left fabric has been heat set. Note wrinkling* (left); *note shrinkage* (right).

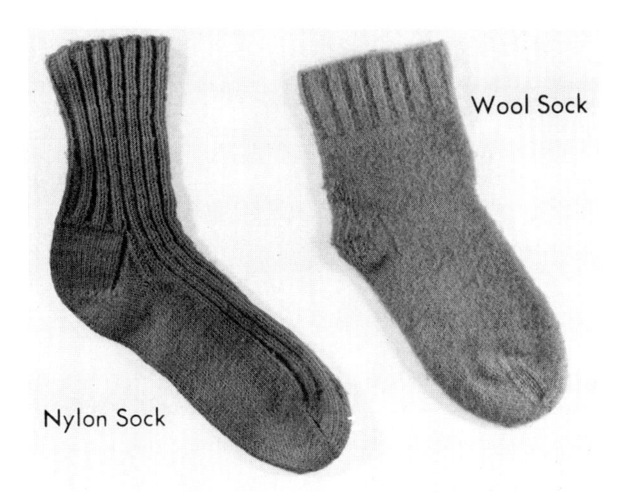

Fig. 34–6 *These socks were the same size when purchased. Nylon* (left); *wool* (right).

To prevent felting shrinkage, the finish alters the scale structure by "smoothing off" the free edges and thus reduces the differential-friction effect that prevents wool fibers from returning to their original position in the cloth. The effectiveness of felting-shrinkage treatments depends on the kind and amount of finish used, and on the yarn and fabric construction. Worsteds need less finish than woolens. Low-count fabrics and low-twist yarns need more finish to give good washability. Treated-wool fabrics are usually considered machine washable, but care should be taken to use warm, not hot, water and a short agitation period. Hand washing is preferable, because soil is easy to remove from the fiber and the hand-washing process ensures lower temperature and less agitation. Machine washing may cause more loosening of fibers, which results in a fuzzy or slightly pilled surface.

Two methods are used to smooth off the free edges of the scales: surface coatings and halogenation treatments.

Surface coatings of a polyamide-type solution are applied to mask the scales. This is a very thin, microscopic film on the fiber. In addition to controlling shrinkage, the coating tends to minimize pilling and fuzzing (one of the greatest problems in wash-and-wear wools), gives the fabrics better wash-and-wear properties, and increases resistance to abrasion. This process, which carries the trade name *Wurlan,* was developed at the Department of Agriculture Laboratory in Albany, California.

Halogenation treatments primarily with chlorine are the most widely used. They are low cost, can be applied to large batches of small items such as wool socks, do not require padding or curing equipment, and are fairly effective. Since the scales are partially dissolved, felting shrinkage is lessened. The processes are likely to damage the fibers, if not carefuly done. The scales are more resistant to damage than the interior of the fiber and should not be completely removed or there will be considerable reduction in wearing properties, weight, and hand. The fabric will feel harsh and rough. To maintain the strength of the fabric, 18 percent nylon fiber is blended with the wool before weaving. A process combining chlorination and resin makes wool knits machine washable and dryer dryable. Shrinkage is less than 3 percent in length and 1 percent in width, and goods retain their loft and resiliency.

Rayon Fibers. The shrinkage of *regular rayon* varies with the handling of the fabric when wet. While it is wet, the fabric can be stretched, and it is difficult to keep from overstretching it during processing. If it is dried in this stretched condition, the fabric will have high potential shrinkage and shrink when wet again and dried without tension because the moisture in the fabric adds enough weight to stretch it.

Shrinkage-control treatments for rayon reduce the swelling property of the fiber and make it resistant to distortion. Resins are used to form crosslinks that prevent swelling and keep the fiber from stretching. The resin also fills up spaces in the amorphous areas of the fiber, making it less absorbent. The nonnitrogenous resins (the aldehydes) are superior to other resins because they do not weaken the fabric, are nonchlorine retentive, and have excellent wash fastness. Treated rayons are machine washable, but the wash cycle should be short. High-wet-modulus rayon is also resin treated, mainly for durable-press purposes, because its shrinkage can be controlled by relaxation shrinkage-control methods.

Shape-Retention Finishes

Before 1940, creases were pressed in and wrinkles were pressed out after garments were washed or dry cleaned. Most people sent silks and wools to the dry cleaner and washed cottons and linens at home. Cottons were starched to prevent mussing. Significant changes have taken place in clothing care. Today, most people want to wash everything in the automatic washer, dry everything in the dryer, and wear the garments with no other care required. With thermoplastic fibers and special shape-retention finishes, wash, dry, and wear is possible. It took nearly 20 years to change the care requirements for textiles.

THEORY OF WRINKLE RECOVERY

Fiber recovery is dependent on crosslinks that hold adjacent molecular chains together and pull them back into position after the fiber is bent, thus preventing the formation of a wrinkle.

The cellulosic fibers do not have natural crosslinks. Molecular chains are held together by hydrogen bonds that operate like the attraction of a magnet for a nail. The hydrogen bonds of cellulose break with the stress of bending and new bonds form to hold the fiber in this bent position, thus forming a wrinkle. Resin crosslinks give fibers a "memory" and will prevent this giving the fiber good wrinkle recovery (Figure 34–7).

Urea formaldehyde was the first resin used to prevent wrinkles; other resins and improved resin combinations were developed later. Although fabrics treated with these resins were smooth and flat and wrinkle resistant, they had poor abrasion resistance, lessened tear strength, and a fishy odor.

WRINKLE-RESISTANT FINISHES

The first step was the development of *wrinkle-resistant finishes* for spun rayon. Resin finishes were first used in England in 1920 and in the United States in 1940 for this purpose. The resin finishes were found to be equally good on cotton and linen fabrics.

Wrinkles, caused by crushing fabrics during wearing and washing, are usually undesirable; creases and pleats made by pressing are desirable style features. Fibers that have strong intermolecular bonds (good molecular memory) resist wrinkling and creasing, whereas those with weak bonds wrinkle and crease readily.

WASH-AND-WEAR

Wash-and-wear finishes were the next development. More resin was used to produce fabrics for drip-dry garments; about 7 percent resin compared with 3 percent for wrinkle resistance. Garments made of these fabrics could be washed and rinsed but needed to be removed from the washer before the spin cycle. They could be hung on hangers, buttoned or zipped, straightened out, and allowed to drip dry. Touch-up pressing might be needed if there were puckers at the seams or in the zipper area.

DURABLE PRESS

Durable press is a descriptive term applied to garments that retain their shape and their pressed appearance even after many washings, wearings, and tumble dryings. The terms durable press and permanent press are used inter-

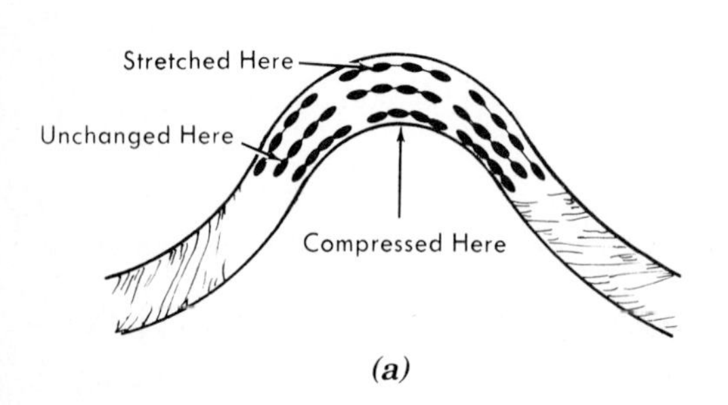

H OH
OH H
OH H
H
H O O
CH_2O
R
Resin or
Reactant
O
H
OH H
OH H
H
H O O
CH_2OH
(b)

Fig. 34–7 *(a) Effect on internal structure when fiber is bent; (b) resin crosslink.*

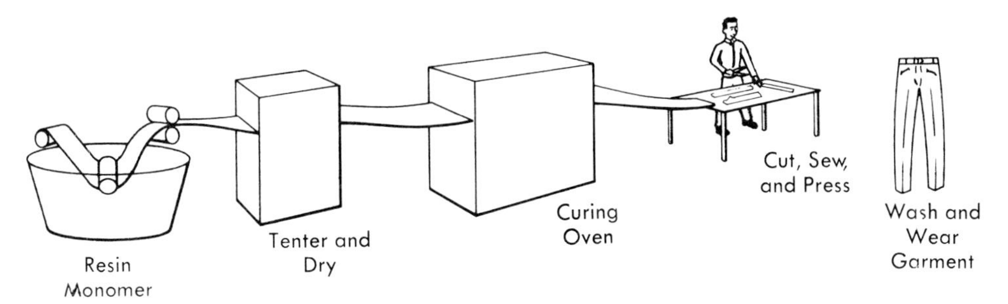

Fig. 34–8 *Precured process.*

changeably, but durable is a more realistic description.

The concept of permanent shape for garments—durable press—was introduced in the early 1960s when Koratron's "Oven-Baked" pants for men were marketed with tremendous and instantaneous success, even though the higher-resin content (10 percent) needed for durable press caused the pants to split at the creases and cuff edges after two or three washings.

The two processes, *precured* and *postcured,* for durable-press garments and fabrics are outlined. The major difference in these two processes is the stage at which cutting, sewing, and pressing take place (Figures 34–8 and 34–9).

The Precured Process

1. Saturate the fabric with the resin-crosslinking solution and dry.

2. Cure in a curing oven (crosslinks form between molecular chains).

3. Cut and sew garment. Press with iron. (All yard goods for home sewing are made this way).

The Postcured Process

1. Saturate the cloth with a resin-crosslinking solution and dry.

2. Cut and sew garment and press shape with hot-head press.

3. *Cure* by putting pressed garment into a curing oven at 300°–400°F.

4. Curing gives shape to the cotton component. The polyester component (or any other thermoplastic fiber) was *set* by the hot-head pressing.

Problems associated with resin finishes—other than reduced tensile strength and abrasion resistance—are listed.

1. Fabric stiffness and poor hand.

2. Chlorine absorption, which causes yellowing and loss of strength.

3. Offensive odors—fishy or formaldehyde-like smell.

4. Color problems: "frosting," or loss of color on abraded edges; migration of color from the polyester fibers to the cotton component as a result of the high curing temperature.

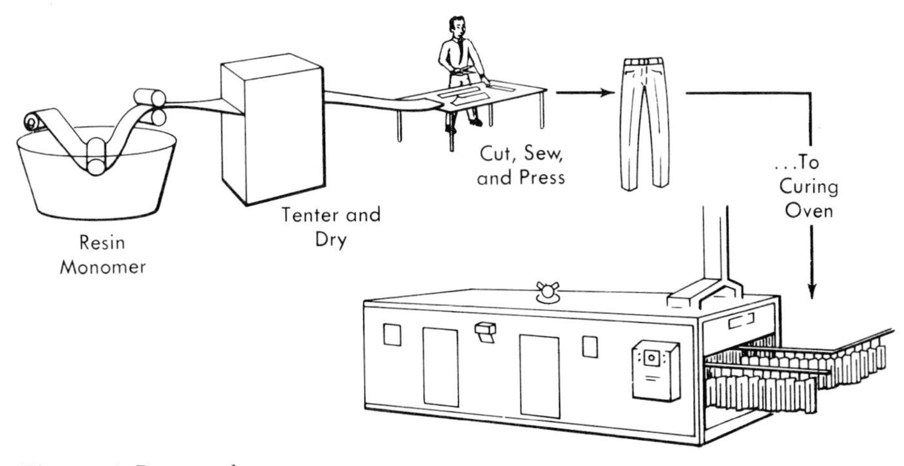

Fig. 34–9 *Postcured process.*

5. Soiling—especially the affinity of resins for oily soils. Soil-release finishes now help with this (see page 325).

6. Static pilling.

7. Garment construction problems—seam puckering, pressing-in or removing creases when altering garments.

Many of these problems have been solved or minimized.

By using blends of cotton/polyester instead of 100 percent cotton, less resin is needed. The high strength and abrasion resistance of polyester make these fabrics much more durable. Special pressing equipment has been developed for use on precured fabrics. Pretreatment of cotton with liquid ammonia or mercerizing cotton under tension adds strength to fabrics so they are not weakened as much from the finish. Polymer sizing on the yarns before curing gives the fabric greater abrasion resistance.

Coneprest, by Cone Mills, is a process in which precured fabrics are made into garments. Where creases are desired, the garment is sprayed with a substance that temporarily inactivates the wrinkle-resistant finish. The garments are then pressed under pressure to recure the finish.

Ameriset, developed by American Laundry Machinery Industries, is a finish in which untreated fabrics are cut, sewed, and pressed, and the garments are placed in a gas chamber where formaldehyde and sulfur-dioxide vapor enters the fabric, causing crosslinks in the cellulose fibers.

Creaset is a silicone-based finish by Creaset, Inc., for all-wool and all-cotton fabrics.

Durable-Press Wool. Wool has good resiliency when it is dry, but it does not have durable-press characteristics when it is wet. *Durable-press wool* is achieved with resin treatments, but this must be accompanied by a treatment with shrink-resist resins in order to control wool's tendency to excessive shrinkage. Several procedures are used, but the one described here is typical.

1. Flat fabric is treated with 1–2 percent of the durable-press resin and steamed (semidecated) for 3–5 minutes.

2. The garment is made up, sprayed with more durable-press resin, and pressed. This gives the permanent-crease effect.

3. Shrink-resist resin is mixed with a dry-cleaning solvent and the garment is dry cleaned. The resin is then allowed to cure in the garment for 3–7 days before the garment can be laundered.

QUALITY STANDARDS AND CARE

Quality-control standards had not been developed during the wash-and-wear era and there was wide variation, dependent on the economic objectives of the individual converter. To avoid problems, the industry developed standards for durable press, and quality has been much more dependable. Registered trade names have been adopted by many producers. These trade names are an indication to the consumer that the product has met certain performance tests. The consumer can also check the fabric for objectionable odor or excessive stiffness—both of these indicate poor processing. If the fabric is a blend, there should be an adequate amount of the polyester or other thermoplastic fiber for the garment to meet performance standards.

The following are some suggestions for care:

- Wash frequently, and do not allow soil to build up. Resins have a special affinity for oil and grease, which should be removed quickly before they can penetrate.
- Pretreat stains, collars, and cuffs. Use a spot-removal agent on grease spots.
- Keep wash loads small. Crowding makes wrinkles.
- Heat sets wrinkles, so avoid heat as much as possible in the laundering process. Avoid wringing and squeezing. Use an automatic dryer if possible, but remove clothes promptly.

Other Finishes

WATER-REPELLENT FINISHES

Waterproof fabrics are compared with water-repellent fabrics in the following chart.

Waterproof Fabrics	*Water-Repellent Fabrics*
Fabrics are films or low-count fabrics with a film coating.	High-count fabrics with a finish that coats the yarn but does not fill up the interstices of the fabric.
Characteristics	
No water can penetrate.	Heavy rain will penetrate.
Most plastic fabrics stiffen in cold weather.	Fabric is pliable and little different from untreated fabric.
Cheaper to produce.	Fabric can "breathe" and is comfortable for raincoats.
Permanent.	Durable or renewable finish.

A *water-repellent* fabric is resistant to wetting, but if the water comes with enough force, it will penetrate the fabric.

The Federal Trade Commission has suggested the use of the terms *durable* and *renewable* in describing water-repellent fabrics.

Water repellency is dependent on surface tension and fabric penetrability and is achieved by (1) finish and (2) fabric construction.

Finishes that can be applied to fabric to make it repellent are wax emulsions, metallic soaps, and surface-active agents. They are applied to fabrics that have a very high warp count and are made with fine yarns.

Wax emulsions and *metallic soaps* coat the yarns but do not fill the interstices between the yarns. These finishes are not permanent but tend to come out when the fabric is washed or dry cleaned. They can be renewed.

Surface-active agents have molecules with one end that is water repellent and one end that will react with the hydroxyl (OH) groups of cellulose. After they are applied, heat is used to seal the finish to the fabric. This finish is permanent to washing and dry cleaning.

It is more difficult to select a water-repellent coat than a waterproof coat because the finish is not obvious and one must depend on the label for information. However, the consumer can recognize some guides for buying. The cloth construction is far more important than the finish. The closer the weave, the greater the resistance to water penetration. The kind of finish used is important in selection because it influences the cost of care. The use of two layers of fabric across the shoulders gives increased protection, but the inner layer must also have a water-repellent finish or it will act as a blotter and cause more water to penetrate. Care is important in water-repellent fabrics. The greater the soil on the coat, the less water repellent it is.

Water-repellent finishes render fabrics spot and stain resistant. Some of the finishes are resistant to water-borne stains, some to oil-borne stains and some to both. Durable water-repellent finishes often hold greasy stains more tenaciously than untreated fabrics. Unisec, Scotchgard, and Zepel are trade names for finishes that give resistance to both oil- and water-borne stains. Hydro-Pruf and Syl-mer are silicone finishes that resist water-borne stains. Zepel, Scotchgard, and Fybrite are trade names for fluorocarbon finishes.

SOIL- AND STAIN-RELEASE FINISHES

Soil-release finishes function by reducing the degree of soiling of the fabric by repelling the soil or preventing a bond between the soil and the fabric. Thus these fabrics are easier to clean than those without soil-release finishes. Fluorochemicals are common, durable, and effective soil-resistant finishes.

A large quantity of the washable material on the market now has a soil-release finish that definitely improves the fabric performance in resisting soil, releasing soil, and retaining fabric whiteness by resisting redeposition of soil from the wash water. The big problem is that these finishes do not last the life of the garment. Some are durable enough to last through 20–30 washings. Lack of permanence results from the fact that the finish is applied to the surface.

Soil-release finishes were developed for use on durable-press garments because of their tendency to pick up and hold oily stains and spots. The resin content of durable press is much higher than that of the older wash-and-wear, so the problem of soiling is much more acute. The tendency to pick up oil (oil affinity) means that oil is absorbed into the resin or the fiber. Soil-release finishes either attract water and permit the soil to be lifted off the fabric or coat the fibers and prevent the soil from penetrating the coating.

Most durable-press garments are blends of cotton/polyester. Untreated cotton is hydrophilic, and hydrophilic surfaces give the best soil-release performance, so cotton releases soil when it is laundered. The resin finish, however, is hydrophobic and does not release the oily soil. Polyester is hydrophobic and has an affinity for oil (oleophilic). It must be spot treated to remove oily soil from contact areas of a garment, such as the collar. When the polyester is coated with resin as it is in durable press, its oil affinity is increased. Finer fibers soil more readily than coarse fibers, and soil can penetrate low-twist yarns more easily than high-twist yarns.

Oily soil is not removed from resinated cottons or polyester in washing, so they must be dry cleaned or spot cleaned with a special solution. Soil-release finishes make the surface less attractive to oil and more easily wetted—more hydrophilic. Many finishing materials fall into two general classes: They are mechanically or chemically bonded to the surface. Many soil release finishes are organosilicon substances. A new soil-release finish by 3M is Scotch Release. Other soil-release finishes include Scotchgard, Visa, and Zepel.

ABSORBENT FINISHES

Absorbent finishes are designed to increase the moisture absorbency of the fabric. These finishes increase the time needed to dry the fabric. They may aid in the dyeing of the fabric. Absorbent finishes are fair in durability. Absorbent finishes are used on towels, diapers, underwear, and sportswear. They are applied as surface coatings for fibers and yarns. On nylon, a solution of nylon 8 is used; on polyesters, the finish changes the molecular structure of the fiber surface so that moisture is broken up into smaller particles that wick more readily; on cellulosics, the finish makes them absorb more moisture. Fiber modifications and different fabric structures are more effective than finishes. Fantessa, Visa, and Zelcon are trade names.

ANTISTATIC FINISHES

Antistatic finishes are important in both the production and the utilization of fabrics. Static charges that develop on fabrics cause them to cling to machinery in the factory and to people, attract dust and lint, and produce sparks and shocks.

Control of static buildup on natural-fiber fabrics was done by increasing humidity and using lubricants but these controls were not adequate with the thermoplastic fibers.

Antistatic finishes were developed to (1) improve the surface conductivity so that excess electrons move to the atmosphere or ground; (2) attract water molecules, thus increasing the conductivity of the fiber; or (3) develop a charge opposite to that on the fiber, thus neutralizing the electrostatic charge. The most effective finishes combine all three effects. Most antistatic finishes are not durable and must be replaced during care. Most finishes are quaternary ammonium compounds. Washing aids such as fabric softeners help to control static.

Incorporating antistatic substances into the fibers gives the best static control. Most manmade fibers are produced in antistatic form especially for rugs, carpets, lingerie, and uniforms (see Chapter 8). Some trade names of antistatic fibers variants are Ultron nylon, Antron nylon, Staticgard nylon, and Anso nylon.

FABRIC SOFTENERS

Fabric softeners were developed to improve the hand of harsh textiles, which may develop as a result of resin finishes or heat setting of synthetics. Types of softeners include anionic softeners, cationic softeners, and nonionic softeners. *Anionic softeners* are usually sulfonated, negatively charged fatty acids and oils. These softeners are padded onto the fibers because of a lack of affinity for the fiber. Anionic softeners are often used commercially on cellulosic fibers and silk. *Cationic softeners* are most often used in domestic washing. These softeners have an affinity for the fiber. They tend to yellow with age and may build up on the fiber if used frequently, reducing the absorbency of the fabric. Cationic softeners may contain quaternary ammonium compounds, and these compounds may confer some incidental antibacterial properties. *Nonionic softeners* must be padded onto the fabric. These commercial softeners are usually a fatty acid.

ABRASION-RESISTANT FINISHES

Abrasion-resistant finishes are used on lining fabrics, especially for pockets. Thermoplastic resins seem to fix fibers more firmly into the yarns so they do not break off as readily. Blending nylon or polyester with cotton or rayon gives better resistance to abrasion than using finishes. In abrasion-resistant finishes, an acrylic resin is often used. The resin may increase the wet soiling of the fabric. These resins are used in areas that receive high degrees of abrasion, such as pockets and waistbands.

ANTISLIP FINISHES

Antislip finishes are used on low-count, smooth-surfaced fabrics. Fabrics are treated with resins, stretched, and dried under tension, causing the yarns to be somewhat bonded at their interlacing points. Antislip finishes are used to reduce seam slippage and fraying. Seam slippage occurs when the yarns in the seam slide toward the seam allowance. This results in an area next to the seam where only one set of yarns can be seen. Areas that have exhibited seam slippage have poor abrasion resistance and an unacceptable appearance. In some cases, the seam can ravel completely. Antislip finishes are also called *slip-resistant,* or *nonslip, finishes.* The most effective and durable finishes are resins of urea or melamine formaldehyde.

METALLIC AND PLASTIC COATINGS

Metallic and plastic coatings are used on the back of fabrics to reduce heat transfer through the fabric, alter the appearance of the fabric, lock yarns in place, and minimize air and water permeability. These coatings often include aluminum as the metallic coating. These coatings are used on apparel and window-treatment fabrics. Plastic coatings reduce the soiling of the fabric.

Plastic coatings also are used to give a smooth, leather-like look to fabrics. Problems of metallic and plastic coatings include cracking and peeling of the finish. In order to increase the life of these coatings, be sure to read and follow the care labels.

Acrylic-foam coatings are common on drapery fabrics. These back coatings are used to minimize air movement through the draperies, give a greater comfort factor by increasing the thickness of the fabric, and minimize the need to have a separate lining fabric. Draperies with the foam-back coating often are sold as self-lined draperies (Figure 34–10).

MOTH CONTROL

Moths and carpet beetles are likely to damage fibers containing protein, such as wool. In addition, insects are likely to cause damage to other fibers if soil is present. Both moths and carpet beetles attack not only 100 percent wool but also blends of wool and other fibers. Although they can digest only the wool, moths and carpet beetles eat through the other fibers. The damage is done by the larvae and not the adult moth. Clothes moths are about ¼-inch long; they are not the large moths that we may occasionally see about the house. Larvae shun bright sunlight and do their work in the dark. For this reason, it is necessary to clean often under sofas, under sofa cushions, in creases of chairs and garments, and in dark closets.

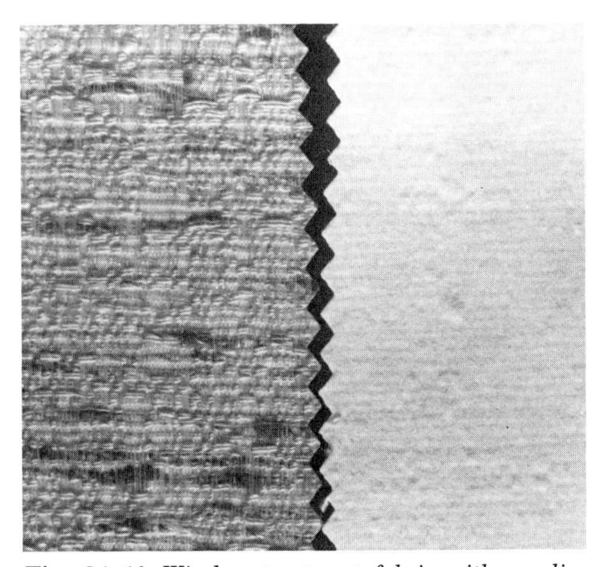

Fig. 34–10 *Window treatment fabric with acrylic-foam coating: face* (left) *and back* (right).

Means of controlling insect damage are as follows:

1. Cold storage.
2. Odors that repel. Paradichlorobenzene and naphthalene (moth balls) used during storage.
3. Stomach poisons. Fluorides and silicofluorides are finishes for dry-cleanable wool.
4. Contact poisons. DDT is very effective but has been banned in the United States.
5. Chemicals added to the dye bath permanently change the fiber, making it unpalatable to the larvae.

MOLD AND MILDEW CONTROL

Molds and *mildew* will grow on and damage both cellulosic and protein textiles. They will grow on, but not damage, the thermoplastic fibers.

Prevention is the best solution to the problem in apparel because cures are often impossible. To prevent mold or mildew, keep textiles clean and dry. Soiled clothes should be kept dry and washed as soon as possible. Sunning and airing should be done frequently during periods of high humidity. An electric light can be used in dark, humid storage places. Dehumidifiers are very helpful.

If mildew occurs, wash the article immediately. Mild stains can be removed by bleaching. Mold and mildew growth is prevented by many compounds. Salicylanilide is often used on cellulosic fibers and wool under the trade names of Shirlan and Shirlan NA.

ANTIPESTICIDE PROTECTIVE FINISHES

Antipesticide protective finishes are designed to protect the wearer of the clothing from pesticides. The finish is designed to prevent penetration of the chemical through the fabric and to aid in removal of the chemical during washing. It is expected that research in this area will continue.

ANTIMICROBIAL FINISHES

Antimicrobial finishes are used to inhibit the growth of bacteria and other odor-causing germs, prevent decay and damage from perspiration, control the spread of disease, and reduce the risk of infection following injury. Antimicrobial finishes are also known as *antibacterial, bacteriostatic,* or *antiseptic, finishes.*

These finishes are important in skin-contact clothing, shoe linings, and hospital linens. The chemicals used are surface reactants, mostly quaternary ammonia compounds. Zirconium peroxides can be formed on the surface of cotton fabrics to give antimicrobial properties. Those substances can be added to the spinning solution of rayon and acetate fibers. Most diaper-service establishments add the finish during each laundering. Eversan and Sanitized are two trade names.

FUME FADING–RESISTANT FINISHES

Fume fading–resistant finishes are available for use on those fibers dyed with dyes susceptible to fading when exposed to atmospheric fumes. The most common of these are acetate fibers dyed with disperse dyes. Of course, this problem was decreased by a significant amount with mass pigmentation. However, there are some cases where mass pigmentation is not economically practical. In these cases, fume fading–resistant finishes of tertiary amines and borax are used to minimize fume fading.

35

Flame-Retardant Finishes and Flammability

Children and the elderly are the most frequent victims of fire because they are unable to protect themselves. Each year, a large number of fatalities, as well as nonfatal injuries, result from fires associated with flammable fabrics. Financial loss from such fires is estimated in the millions of dollars. Five common causes of these fires are smoking in bed, starting fires with flammable liquids, playing with matches and lighters (by children), burning trash, and being caught in a burning structure. Smoke inhalation is as great a danger as burns, and it is regarded as the major cause of death in fires.

Construction of a fabric or garment determines the degree to which oxygen is made available to the fibers. Fabrics that burn quickly are sheer or lightweight fabrics and napped, pile, or tufted surfaces. Some apparel made from these constructions ignite quickly, burn with great intensity, and are difficult to extinguish. "Torch" sweaters, fringed cowboy chaps, and chenille berets are examples of some apparel items that caused tragic accidents. When chenille berets were tested in the laboratory, test strips taken from the berets would begin to burn while the flame was several inches away, indicating that it would be extremely dangerous to wear the beret while lighting a cigarette. Some style features of apparel also present a fire hazard. Long, full sleeves; flared skirts; ruffles; frills; and flowing robes are examples.

There are many terms that are used when discussing the ability of a fabric to resist ignition, burn more slowly than normal, or self-extinguish once the source of ignition has been removed from the fabric. The following is a list of terms with their 1985 American Society for Testing and Materials definitions:

- *Fire:* As related to textile flammability, the destruction of materials by burning, in which the associated flames are not constant in size and shape and that results in a relatively high-heat flux.
- *Fire retardance:* The resistance to combustion of a material when tested under specified conditions.
- *Flame resistance:* The property of a material whereby flaming combustion is prevented, terminated, or inhibited following application of a flaming or nonflaming source of ignition, with or without subsequent removal of the ignition source.

Flammability: Those characteristics of a material that pertain to its relative ease of ignition and relative ability to sustain combustion.

FLAMMABLE FABRICS ACT

Congress enacted the first national law dealing with flammable fabrics in 1953, following several apparel-fire deaths. The Flammable Fabrics Act prohibited the marketing of dangerously flammable material, including all wearing apparel, regardless of their fiber content or construction. The act covered items that were imported or were in interstate commerce. One purpose of the law was to develop standards and tests to separate dangerously flammable fabrics from normally combustible ones.

The act was amended in 1967 to cover a broader range of apparel and interior furnishings with responsibilities for its implementation divided among the secretary of Commerce, the secretary of Health, Education, and Welfare, and the Federal Trade Commission. In 1972, the Consumer Product Safety Act was passed, and the Consumer Product Safety Commission (CPSC), which has broad jurisdiction over consumer safety, was established. The responsibilities and functions, as stipulated in the Flammable Fabrics Act, were transferred to the CPSC. Federal standards were established under the direction of the Department of Commerce and later under the CPSC as shown in the following chart. These standards and/or test methods may be modified in the future depending on further research and evaluation.

It takes considerable time to develop a standard. First, facts must be collected to indicate a need. A notice is published in the Federal Register that there is a need for a standard. Interested persons are requested to respond. Test methods are developed and published in a second notice. A final notice, which includes details of the standard and test method, is published with the effective date of compliance. One year is usually allowed so that merchandise that does not meet the standard can be sold or otherwise disposed of, and new merchandise can be altered (if necessary) to meet the standard.

Mandatory standards have been issued for children's sleepwear, sizes 0–6X and 7–14, large and small carpets and rugs, and mattresses and mattress pads. The Upholstered Furniture Action Council (UFAC) has issued voluntary standards for upholstered furniture. Figure 35–1 shows a photograph of a UFAC label.

Fig. 35–1 UFAC label.

Some cities and states have established standards for textile items. Various sectors of industry have adopted voluntary standards for such items as tents, blankets, and career clothing for people who work near fire.

Flame-Resistant Fibers and Finishes

Fabrics may be made flame resistant in three ways:

1. Use inherently flame resistant fibers.
2. Use fiber variants that have been made flame resistant by adding flame retardants to the spinning solution.
3. Apply flame-retardant finishes to the fabrics.

The burning characteristics of fibers are given in a chart in Chapter 2. Fibers that are inherently flame resistant are aramid, modacrylic, novoloid, saran, PBI, sulfar, and vinal/vinyon matrix fibers. Fibers in which flame-retardant chemicals have been added to the spinning solution are some acetates, nylons, polyesters, and rayons.

Flame-retardant finishes are designed to block one leg of the fire triangle. The fire triangle consists of fuel, ignition, and oxygen. Flame-retardant finishes function in a variety of ways. The finish may form a glassy interface during burning. The glassy interface blocks the flame of the fuel and hinders further flame propagation. A foam-containing, flame-extinguishing gas that blocks oxygen may be produced. The solid may be modified so that the products of combustion are not volatile or require excess heat to continue the fire.

Flame-retardant finishes are used on cotton, rayon, nylon, and polyester fabrics. Flame-retardant finishes must be durable (withstand 50 washings), nontoxic, and noncarcinogenic. They should not change the hand and texture of fabrics or have unpleasant odors. Most finishes are not visible and they add to the cost of the garment, so the consumer is asked to pay for something that cannot be seen.

Flame-retardant finishes can be classified as durable and nondurable. These durable finishes are specific to fiber type and are usually phosphate compounds or salts, halogenated organic compounds, or inorganic salts. Examples of *durable finishes* for polyester and cellulosics include the trade names of Antiblaze and Pyroset. There are many other flame-retardant finishes sold by trade names in addition to these mentioned here. However, because most consumer products that are flame retardant are those mandated by federal law, few products are sold with the finish identified by trade name.

Flame-retardant finishes are less expensive than flame-resistant fibers or fiber variants. Knitting or weaving gray goods that can be given a topical flame-retardant finish or not, depending on its end use, is a more economical

Federal Standards Implementing the Flammable Fabrics Act

Effective Date	*Item*	*Requirements*	*Test Method*
1954	Flammability of clothing Title 16 CRF 1610 (formerly CS 191–53)	Articles of wearing apparel except interlining fabrics, certain hats, gloves, footwear.	A 2 × 6 inch fabric placed in a holder at a 45° angle exposed to flame for 1 second will not ignite and spread flame up the length of the sample in less than 3.5 seconds for smooth fabrics or 4.0 seconds for napped.
1954	Flammability of vinyl plastic film Title 16 CRF 1611 (formerly CS 192–53)	Vinyl-plastic film for wearing apparel.	A piece of film, placed in a holder at an angle of 45° will not burn at a rate exceeding 1.2 inches per second.
1971	Large carpets and rugs Title CFR 1630 (formerly DOC FF1–70)	Carpets that have one dimension greater than 6 feet and a surface area greater than 24 square feet. Excludes vinyl tile, asphalt tile, and linoleum. All items must meet standards.	"Pill" test. 9 × 9 inch specimens exposed to methenamine tablet placed in center of each specimen does not char more than 3 inches in any direction.
1971	Small carpets and rugs Title 16 CFR 1631 (formerly DOC FF2–70)	Carpets that have no dimension greater than 6 feet and a surface area no greater than 24 square feet. May be sold if they do not meet standard if labeled: Flammable. (Fails U.S. Department of Commerce Standard FF 2–70).	Same as for large carpets and rugs.

procedure for fabric producers than weaving or knitting fabrics for children's sleepwear, for example. In the early 1970s, satisfactory flame-retardant finishes were prepared with variable characteristics of hand and strength loss. One of the most efficient was Tris (2,3 dibromopropyl phosphate). Tris was thought to be carcinogenic (according to the CPSC) and, in 1977, all sleepwear garments treated with Tris were removed from the market. Other flame-retardant finishes were satisfactory (or had not been tested) and still others have been prepared. Not all of the flame-retardant finishes have been tested. Some of them as well as chemicals used for other finishes may prove to be unsafe. Other substances will, no doubt, be found to be satisfactory.

Problems

Cost and care are the greatest problems for the consumer. The high cost of research and development of fibers and finishes, testing of fabrics and garments, and liability insurance result in a higher cost of apparel and home-furnishing items. Because the items look no different, the consumer often thinks the item is overpriced. Because of government standards, the consumer has no choice; for example, people who do not smoke in bed must pay a higher price for mattresses, because only those mattresses that pass

Federal Standards Implementing the Flammable Fabrics Act *(Cont.)*

Effective Date	*Item*	*Requirements*	*Test Method*
1972	Children's sleepwear, Sizes 0–6X Title 16 CFR 1615 (DOC FF3–71)	Any product of wearing apparel up to and including size 6X, such as nightgowns, pajamas, or other items intended to be worn for sleeping. Excludes diapers and underwear. Items must meet requirements as produced and after 50 washings and dryings. All items must meet standard.	"Vertical Forced Ignition" Test. Each of five 3.5 inch × 10 inch specimens is suspended vertically in holders in a cabinet and exposed to a gas flame along the bottom edge for 3 seconds. Specimens cannot have average char length of more than 7 inches.
1973	Mattresses (and mattress pads) Title 16 CFR 1632 (DOC FF4–72)	Ticking filled with a resilient material intended for sleeping upon, including mattress pads. Excludes pillows, box springs, sleeping bags, and upholstered furniture. All items must meet standard.	"Cigarette" test. A minimum of 9 cigarettes allowed to burn on smooth top, edge, and quilted locations of bare mattress. Char length must not be more than 2 inches in any direction from any cigarette. Tests are also conducted with 9 cigarettes placed between two sheets on the mattress surfaces.
1975	Children's sleepwear Sizes 7–14 Title 16 CFR 1632 (DOC FF5–74)	Same as preceding. All items must meet standard.	Same as preceding.

flammability standards can be sold in interstate commerce.

Most of the topical finishes require special care in laundering to preserve the flame resistancy of the garments. Labels on garments are very good and should be followed. Most labels indicate the following care: Use phosphate detergents, do not bleach, do not use soap, do not use hot water. In cities where phosphates are banned, soft water and heavy-duty liquid detergents should be used.

For the producer, the problems are much greater. The Flammable Fabrics Act is protective legislation. The producer not only must meet government standards for fabrics and garments but also must be protected from damage suits resulting from disfiguring burns or death caused by burn damage.

Flame-retardant-treated fabrics may exhibit some problems for consumers. The hand may be harsh. The fabric may be less abrasion resistant than it would be without the finish. The finish may give the wearer a false sense of security. Remember, the finish is designed to make the fabric flame retardant. The finish will not prevent the fabric from igniting or burning, although ignition and rate of burning will be slower with the finish.

Using flame-retardant fibers would seem to be the best solution to making flame-resistant articles. Flame resistance is only one of the many properties of fibers needed for good fabrics. The flame-resistant fibers, glass, modacrylic, and vinyon, are lacking in many desirable properties; aramids and novoloids are too expensive for general use at the present time. The modified fibers, rayon, acetate, polyester, and nylon, may be questionable in the future, depending on the additive used, or they may not be available in the desired denier or fiber length. Topical finishes for nylon and polyester fabrics have been satisfactory and more widely used than modified fibers.

The flammability of apparel is related to three characteristics: the fiber content of the apparel items, the weight of the fabric or fabrics involved, and the style and fit of the clothing. Flame-retardant fibers are safer than those that are not flame retardant. Heavier-weight fabrics are safer than lighter weight fabrics. Garments that fit close to the body are safer than those that are free flowing.

Protective fibers and finishes do not sell merchandise. Color, texture, and fashion features are more important to most consumers so the producer must provide these features in flame-resistant items.

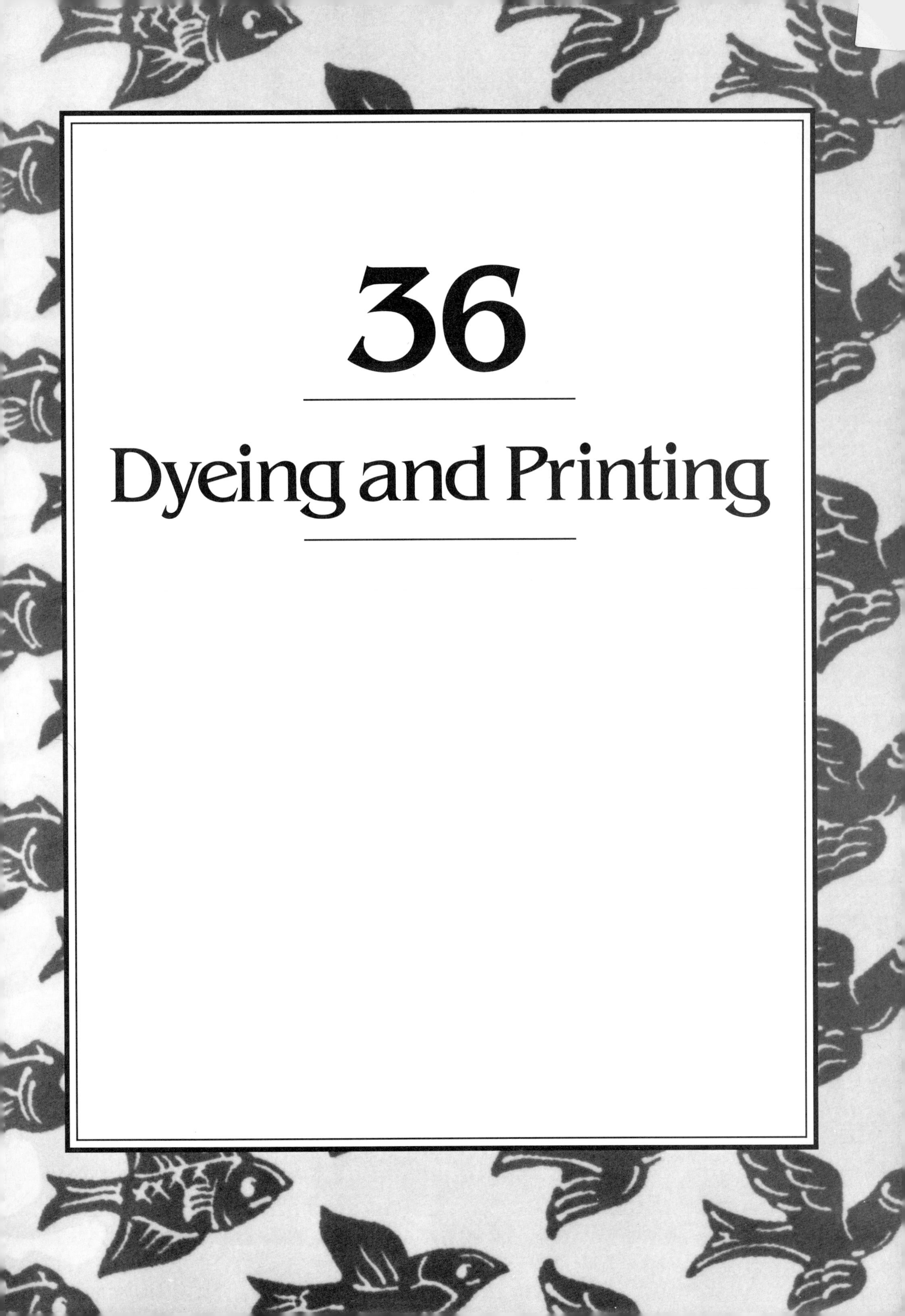

36

Dyeing and Printing

Color is one of the most important characteristics of apparel and home-furnishing items. It is the first feature for which many consumers look. In this chapter, the characteristics of color will be discussed from an identification perspective (when color was added to the product), a process perspective (how color was added to the product), a serviceability perspective (how color will affect the serviceability of the product), and a problem-solving perspective (what kind of problems can develop due to color).

Consumers are not aware of the complex problems involved in achieving a particular color in a uniform manner on a textile product. Consumers expect that the color will remain vivid and uniform throughout the life of the product. It is remarkable that color creates as few complaints from consumers as it does.

In order to understand the complexities of adding color to a textile product, we should begin with the fiber. As was discussed in earlier chapters, fibers differ in their chemical composition. This difference can be seen in various properties and performance characteristics. In addition, any colored textile product may be exposed to a wide variety of potential color degradants, such as detergent, perspiration, drycleaning solvents, sunlight, and makeup. In order to achieve a permanent or fast color, the dye must be permanently attached to or trapped within the fiber by using a combination of heat, pressure, and chemical assistants.

Color has always been important in textiles. Until 1856, natural dyes and pigments were used as coloring agents. These dyes and pigments were obtained from plants, insects, and minerals. When William Henry Perkin discovered mauve, the first synthetic dye, a whole new industry came into being. Europe became the foremost center for synthetic dyes, and it was not until World War I, when trade with Germany was cut off, that a dye industry was developed in the United States. Since that time, many dyes and pigments have been developed, so that today there are hundreds of colors from which to choose.

Color can be added to textile objects by dyes or pigments. Because there are major differences between these coloring substances and the ways they are added to fabrics, the next few paragraphs differentiate between pigments and dyes.

Pigments are insoluble color particles that are held on the surface of a fabric by a binding agent. Their application is quick, simple, and economical. Any color can be used on any fiber, because the pigments are held on mechanically. Stiffening of the fabrics, crocking, and fading are some of the problems encountered. Pigments also may be mixed with the spinning solution for man-made fibers.

A *dye* is an organic compound composed of a chromophore, which is the color-producing portion of the dye molecule, and an auxochrome, which augments or alters the color slightly. The auxochrome also functions by adding solubility to the dye and a possible site for bonding to the fiber. Figure 36–1 shows examples of dye molecules. Dye must be small particles that can be thoroughly dissolved in water or some other carrier in order to penetrate the fiber. Undissolved particles stay on the outside and have poor fastness to crocking and bleeding.

A *fluorescent dye* absorbs light at one wavelength and re-emits that energy at another wavelength. Fluorescent dyes are used to make whites appear whiter. Fluorescent dyes are also used in safety clothing to increase visibility of the wearer at night and some furnishing items to give a bright, intense glow to the color. Fluorescent dyes also are used in apparel, especially in Halloween costumes for the safety and glow-in-the-dark effects and in fashion apparel for bright, neon-like colors.

A *dye process* is the environment created for the introduction of dye by hot water, steam, or dry heat. Accelerants and regulators are used to regulate penetration of the dye. A knowledge of fiber-dye affinity, methods of dyeing, and equipment will give the consumer a better understanding of color behavior.

The stage at which color is applied has little to do with fastness but has a great deal to do with dye penetration, and it is governed by fabric design. In order for a fabric to be colored, the dye must penetrate the fiber and either be combined chemically with it or locked inside it. Fibers that dye easily are those that are absorbent and have dye sites in their molecules, which will react with the dye molecules. The dye reacts with the surface molecules first. Moisture and heat swell the fibers, causing the molecular chains to move farther apart so that more reactive groups are exposed to react with the dye. During drying, the chains move back together, trapping the dye in the fiber. Dyeing of wool

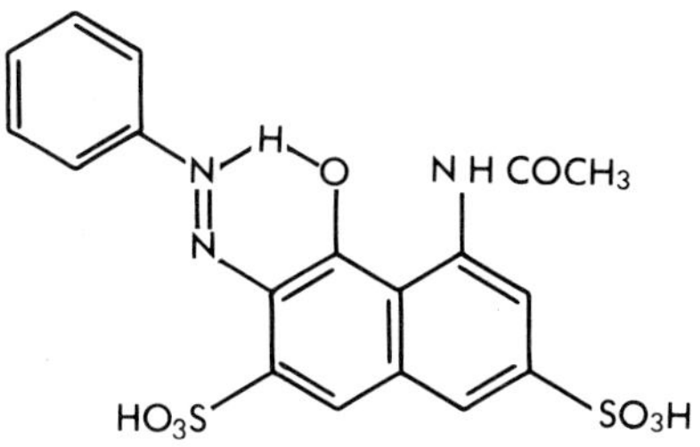

C.I. Acid Red 1

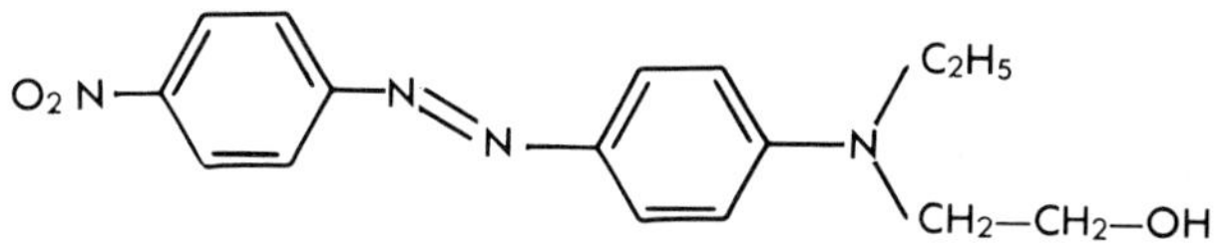

C.I. Disperse Red 1

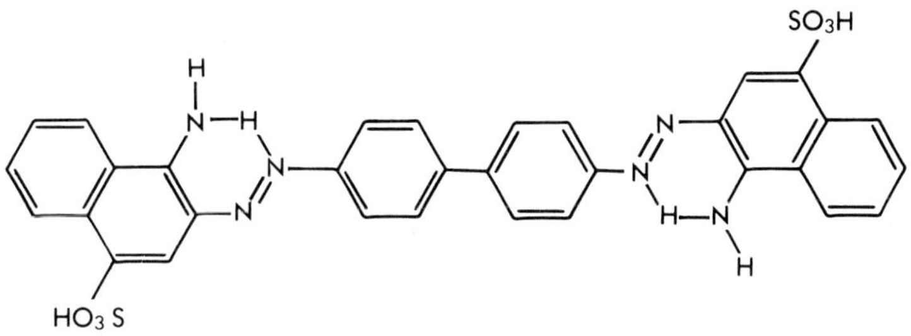

Congo Red, a direct dye

Fig. 36–1 *Dye molecules: (a) C.I. Acid Red 1; (b) C.I. Disperse Red 1; (c) Direct Dye Congo Red.*

with acid dyes is a good example of dyeing fibers that are both absorbent and have many dye sites.

The thermoplastic fibers are difficult to dye because their absorbency is low. Most of the man-made fibers are modified to accept different classes of dyes. This makes it possible to achieve different color effects or a good solid color in blends of unlike fibers by piece dyeing.

No one dye is fast to everything, and the dyes within a class are not equally fast. A complete range of shades is not available in each of the dye classes. The dyer chooses a dye suited to the fiber content and the end use of the fabric. The dyer must apply the color so that it penetrates and is held in the fiber. Occasionally the garment manufacturer or the consumer selects fabrics for uses that are different than the fabric manufacturer intended. For example, an apparel fabric used for draperies may not be fast to sunlight. The consumer should know what to expect from the textile and report to the retailer if fabrics do not give satisfactory performance.

Dyes are classified by chemical composition or method of application. The following chart (page 338) lists the major dyes, along with some of their characteristics and end uses.

Dyeing

Textiles may be dyed during the fiber, yarn, fabric, or product stage, depending on the color effects desired and perhaps on the quality or end use of the fabric. Better dye penetration is achieved with fiber dyeing than with yarn dyeing, with yarn dyeing than with piece

Classification of Dyes

Dyes	*End Uses*	*Characteristics*
Cationic (basic). Used with mordant on fibers other than silk and wool and acrylic. Complete color range.	Used primarily on acrylics. Direct prints on acetate. Discharge prints on cotton. Used on modified polyester and nylon.	Fast colors on acrylics. On natural fibers—poor fastness to light, washing, perspiration. Tends to bleed and crock.
Acid (anionic). Complete color range.	Wool, silk, nylon, modified rayon. Modified acrylic and polyester.	Bright colors. Vary in fastness to light. May have poor fastness to washing.
Azoic (naphthol and rapidogens). Complete color range. Moderate cost.	Cotton primarily. May be used on man-made fibers such as polyester.	Good to excellent light fastness and washing. Bright shades. Poor resistance to crocking.
Developed. Dyes developed in the fiber. Complete color range. Duller colors than acid or basic.	Cellulose fibers primarily. Discharge prints.	Good to excellent light fastness. Fair wash fastness.
Direct (substantive). Largest and most commercially significant dye class. Complete color range.	Used on cellulosic fibers.	Good color fastness to light. May have poor wash fastness.
Disperse. Dye particles disperse in water and dissolve in fibers. Good color range.	Developed for acetate but used on most synthetic fibers.	Fair to excellent light and wash fastness. Blues and violets on acetate fume fade.
Mordant (chrome). Fair color range. Duller than acid dyes.	Used on same fibers as listed for acid dyes.	Good to excellent light and wash fastness. Dull colors.
Reactive. Combines chemically with fiber. Produces brightest shades.	Primarily used on cotton. Some are used on other cellulosics and wool.	Good light and wash fastness. Sensitive to chlorine bleach.
Sulfur. Insoluble in water. Complete color range except for red. Dull colors.	Primarily for cotton. Heavy work clothes. Most widely used black dye.	Poor to excellent light and wash fastness. Sensitive to chlorine bleach. Stored goods may become tender.
Vat. Insoluble in water. Incomplete color range.	Primarily for cotton work clothes, sportswear, prints, drapery fabrics.	Good to excellent light and wash fastness.

dyeing, and with piece dyeing than with product dyeing.

FIBER DYEING

In the *fiber-dyeing process,* the fiber is dyed before yarn spinning.

1. *Mass pigmentation* is also known as *solution, spun, dope dyeing,* or *mass coloration.* It consists of adding colored pigments or dyes to the spinning solution; thus each fiber is colored as it is spun. The color is an integral part of the fiber and fast to most color degradants. This method is preferred for fibers that are difficult to dye by other methods or where it is difficult to get a certain depth of shade. A common fiber that is mass pigmented is olefin. Black polyesters are often mass pigmented.

Another type of dyeing similar to mass pigmentation is *gel dyeing,* in which the color is incorporated in the fiber while it is still in the soft, gel stage.

2. *Stock,* or *fiber, dyeing* is used when mottled or heather effects are desired. Dye is added to loose fibers before yarn spinning. Good dye

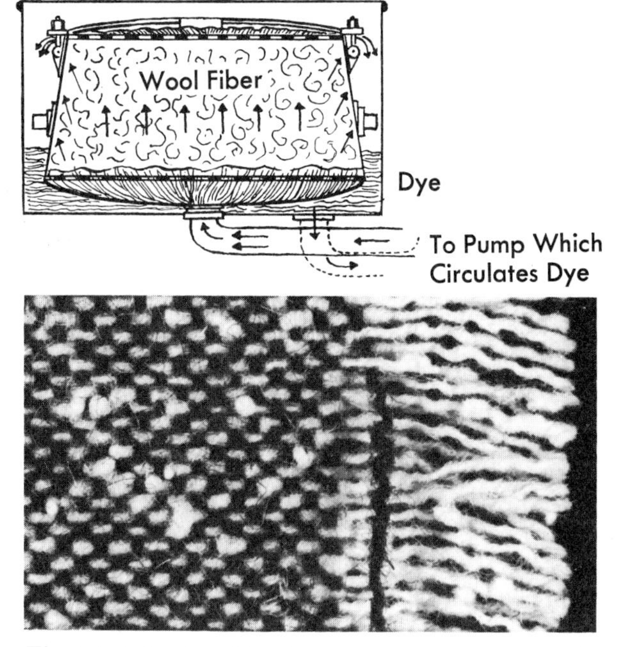

Fig. 36–2 Stock or fiber dye: process and tweed-fabric example (cross dyeing).

penetration is obtained, but the process is fairly expensive (Figure 36–2).

3. *Top dyeing* gives results similar to stock dye and is more commonly used. Tops, the loose ropes of wool from the combing machine, are wound into balls, placed on perforated spindles, and enclosed in a tank. The dye is pumped back and forth through the wool. Continuous processes on loose fiber and wool tops are also used using a pad-steam technique.

YARN DYEING

Yarn dyeing can be done with the yarn in skeins, called *skein dyeing;* with the yarns wrapped on cones or packages, called *package dyeing;* or with the yarn wound on warp beams, called *beam dyeing*. Yarn dyeing is less costly than fiber dyeing but more costly than piece dyeing and printing. Yarn-dyed designs are more limited and larger inventories are involved (Figure 36–3).

PIECE DYEING

When the bolt or roll of fabric is dyed, the process is referred to as *piece dyeing*. Piece dyeing usually produces solid-color fabrics. It generally costs less to dye fabric than to dye loose fiber or yarns. One other advantage is that

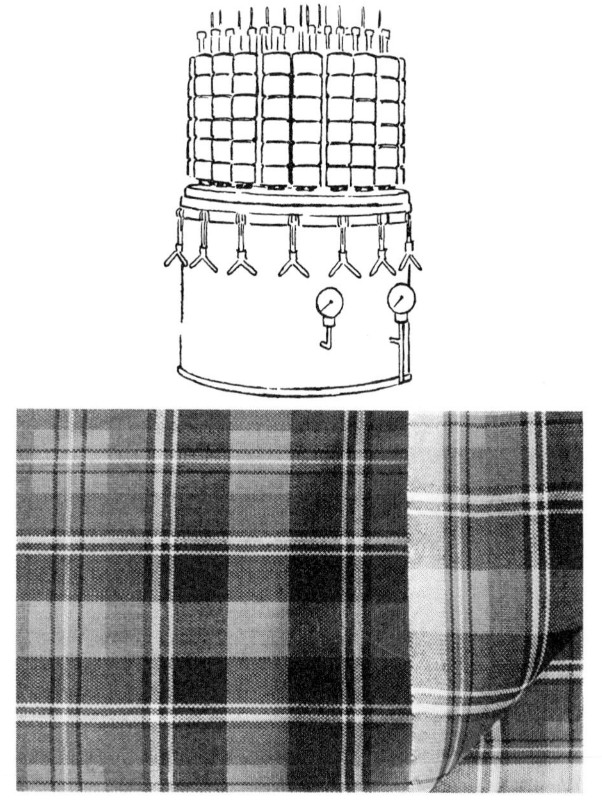

Fig. 36–3 Yarn dye: process and fabric.

decisions on color can be delayed so that fashion trends can be followed more closely.

Cross dyeing is piece dyeing of fabrics (Figure 36–4) made of fibers from different generic groups—such as protein and cellulose—or by combining acid-dyeable and basic-dyeable fibers

Fig. 36–4 Cross-dyed fabric.

of the same generic group. It results in two or more colors in the fabric. An example is a fabric made of wool yarns and cotton yarns dyed with a red acid dye and a blue direct dye, respectively. If the fabric was made with wool warp and cotton filling, the warp would be red and the filling blue.

Union dyeing is piece dyeing of fabrics made of fibers from different groups, but, unlike cross dyeing, the finished fabric is a solid color. Dyes of the same hue, but of composition suited to the fibers to be dyed, are mixed together in the same dye bath. Piece dyeing is done with various kinds of equipment.

PRODUCT DYEING

Before *product dyeing,* the fabric is cut and sewn into the finished product. Once the color need has been determined, the product is dyed. Great care must be taken in the dyeing process to get a level, uniform color throughout the product. Careful selection of components also is required, or buttons, thread, and trim may be a different color because of differences in dye absorption between the various product parts. Product dyeing is becoming more important in the apparel and home-furnishing industries with the emphasis on quick response to retail and consumer demands.

METHODS

The method chosen for piece dyeing depends on fiber content, weight of the fabric, dyestuff, and degree of penetration required in the finished product. Time is money in mass production so that processes in which the goods travel quickly through a machine are used whenever possible. Dyeing and afterwashing require a great deal of water, and waste water causes stream pollution. For this reason, dyers and finishers are always searching for new methods.

Jig Dyeing. *Jig dyeing* consists of a stationary dye bath with two rolls above the bath. The cloth is carried around the rolls in open width and rolled back and forth through the dye bath once every 20 minutes or so. It is on rollers the remaining time. There are some problems of level dyeing. Acetate, rayon, and nylon are usually jig dyed (Figure 36–5).

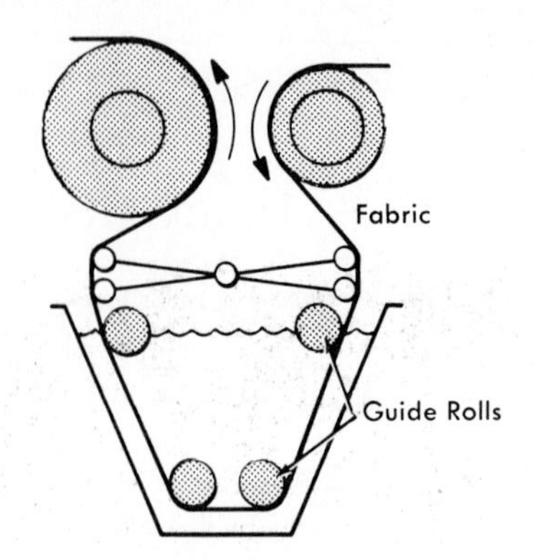

Fig. 36–5 *Jig dyeing.*

Pad Dyeing. *Pad dyeing* is a method in which the fabric is run through the dye bath in open width and then between squeeze rollers that force the dye into the fabric. Notice in Figure 36–6 that the pad box holds only a very small amount of dye liquor, making this an economical method of piece dyeing. The cloth runs through the machine at a rapid rate, 30–300 yards a minute. Pad-steam processes are widely used.

Winch, Reel, or Beck Dyeing. *Winch, reel,* or *beck dyeing* is the oldest type of piece dyeing (Figure 36–7). The fabric, in a loose rope sewed together at the ends, is lifted in and out of the dye bath by a reel. The fabric is kept immersed in the dye bath except for the few yards around the reel. Penetration of dye is obtained by continued immersion in slack condition rather than by pressure on the wet goods under tension. This method is used on lightweight fabrics that cannot withstand the tension of the other methods, and on heavy goods, especially woolens. Reels are of various shapes—oval, round, octagonal.

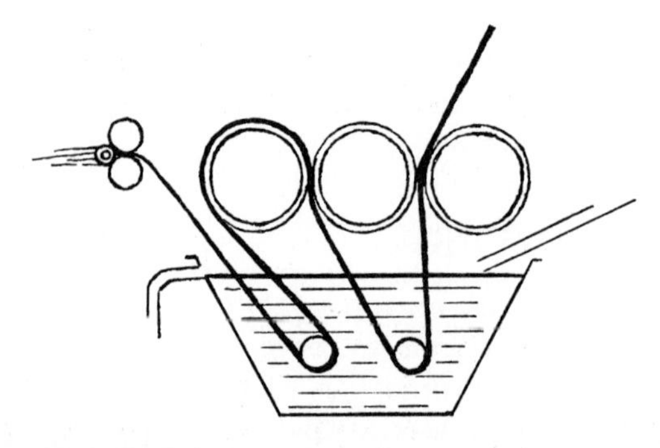

Fig. 36–6 *Pad dyeing.*

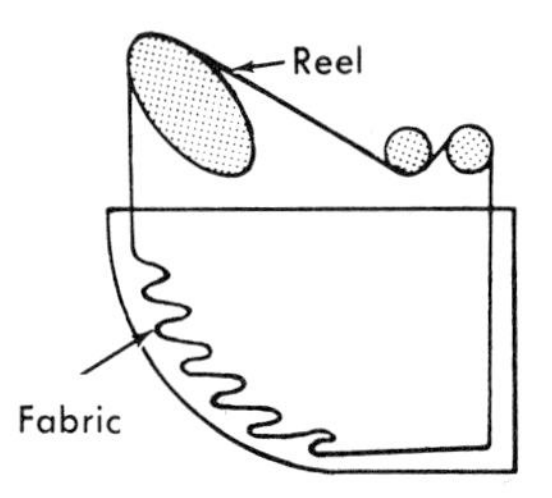

Fig. 36–7 *Winch dyeing.*

Continuous Machines. *Continuous machines,* called *ranges,* are used for large lots of goods. They consist of compartments for wetting-out, dyeing, aftertreatment, washing, and rinsing.

Printing

Color designs are produced on fabrics by *printing* with dyes in paste form or by positioning dyes on the fabric with specially designed machines. Printed fabrics usually have clear-cut edges in the design portion on the right side and the color seldom penetrates completely to the wrong side of the fabric. Yarns raveled from printed fabrics will have color unevenly positioned on them.

Use of foams in printing is becoming more important. In *foam printing,* the color is dispersed in the foam. The foam is applied to the fabric and the foam collapses. The small amount of liquid in the foam limits color migration. At present, designs tend to be limited to simple geometric ones. However, as foam technology is improved, more-complex patterns are expected. It is also expected that foam printing will become as important in printing as foam finishing is in finishing. Printed designs are done by various processes, as listed on the following chart.

DIRECT PRINTING

In *direct printing,* color is applied directly to the fabric in the pattern desired in the finished fabric. Direct printing is one of the most common methods of printing a design on a fabric.

Block Printing. *Block printing* is a hand process and the oldest technique for decorating textiles. It is seldom done commercially because it is costly and slow. A design is carved on a block. The block is dipped in a shallow pan of dye and stamped on the fabric (Figure 36–8). Slight irregularities in color register or positioning are clues to block prints but these can be duplicated in roller printing made to resemble them.

Direct-Roller Printing. *Direct-roller printing* was developed in 1783, about the time all textile operations were becoming mechanized. Figure 36–9 shows the essential parts of the printing machine. The fabric is drawn around a metal cylinder during printing (1 in Figure 36–9). A

Fig. 36–8 *Carved wooden block and design made from it.*

Printing Processes

Direct	*Discharge*	*Resist*	*Jet*	*Other*
Block	Discharge	Batik	TAK	Heat transfer
Direct Roller		Tie-Dye	Microjet	Electrostatic
Warp		Ikat	Polychromatic	Differential
		Screen: Flat, Rotary		
		Stencil		

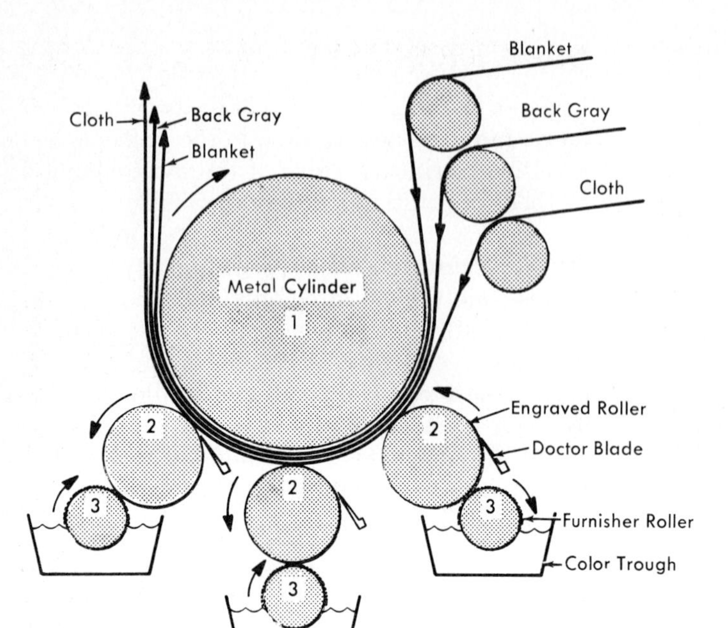

Fig. 36–9 Direct-roller printing.

different printing roller applies each color. The copper printing roller (2 in Figure 36–9) is etched with the design. There are as many different rollers as there are colors in the fabric. In the diagram, three engraved rollers are used. Furnisher rollers are covered with hard rubber or brushes made of nylon, or hard-rubber bristles. They revolve in a small color trough, pick up the color, and deposit it on the copper rollers. A doctor blade scrapes off excess color so that only the engraved portions of the copper roller are filled with dye when it comes in contact with the cloth. The cloth to be printed, a rubberized blanket, and a back gray cloth pass between the cylinder and the engraved rollers. The blanket gives a good surface for sharp printing; the gray goods protects the blanket and absorbs excess dye.

Rayon and knitted fabrics are usually lightly coated with a gum sizing on the back to keep them from stretching or swelling as they go through the printing machine. After printing, the cloth is dried, steamed, or treated to set the dye.

Screen printing and heat-transfer printing are challenging roller printing's position as the most common means of printing a design on a fabric.

Duplex printing is roller printing that prints on both sides of the fabric with the same or different patterns. In duplex prints, both sides of the fabric may be printed at the same time.

Warp Printing. *Warp printing* is done on the warp yarns prior to weaving. This technique gives an interesting, rather hazy pattern, softer than other prints. To identify, ravel adjacent sides. Color in the form of the design will be on the warp yarns. Filling yarns are white or solid color. Imitations have splotchy color on both warp and filling yarns. Warp printing is usually done on taffeta, satin ribbons, or cotton fabric, and on upholstery or drapery fabric (Figure 36–10).

DISCHARGE PRINTING

Discharge prints are piece-dyed fabrics in which the design is made by removing color.

Discharge printing is usually done on dark backgrounds. The fabric is first piece dyed in any of the usual methods. A discharge paste, which contains chemicals to remove the color, is then printed on the fabric. Dyes that are not harmed by the discharging materials can be mixed with printing solution if color is desired in the discharge areas. The fabric is then steamed to develop the design, as either a white or a colored area. Better dye penetration is obtained with piece dyeing than with printing, and it is difficult to get good dark colors except by piece dyeing.

Discharge prints can be detected by looking at the wrong side of the fabric. In the design area, the color is often not completely removed

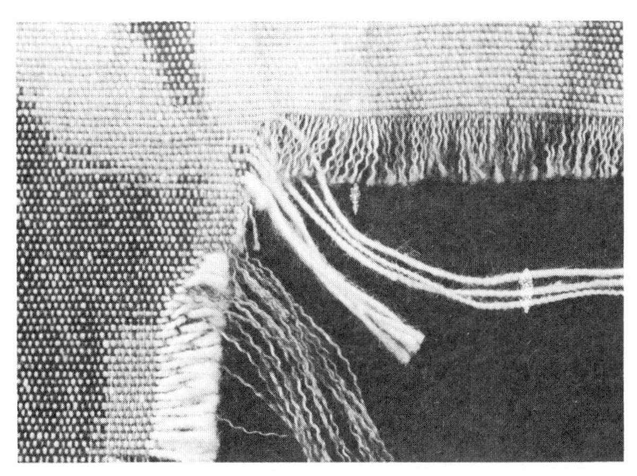

Fig. 36–10 *Warp-printed cretonne. Notice color is on the warp yarns only.*

Fig. 36–12 *Tie-dye. This fabric was rolled on the bias, tied, and piece dyed. A second dyeing was done with the fabric rolled in the opposite direction.*

and one can see evidences of the background colors, especially around the edges of the design. Background colors must be colors that can be removed by strong alkali. Discharge prints are usually satisfactory (Figure 36–11). Discharge prints may cause tendering or weakening of the fabric in the areas where the color was discharged.

RESIST PRINTING

Resist prints are piece-dyed fabrics in which color is prevented from entering the fabric.

Batik. *Batik* is a hand process in which hot wax is poured on a fabric in the form of a design. When the wax is set, the fabric is piece dyed. The wax prevents penetration of color into the wax-covered portions. Colors are built up by piece dyeing light colors first, covering portions, and redyeing until the design is complete. The wax is later removed by a solvent.

Tie-Dye. *Tie-dye* is a hand process in which yarn or fabric is wrapped in certain areas with fine thread or string. The yarn or fabric is then piece dyed and the string is removed, leaving undyed areas (Figure 36–12 and 36–13).

Ikat. *Ikat* is an ancient form of resist printing. In ikat, the yarn is tied for dyeing and weaving. In ikat, the technique can be applied to only the warp yarns (warp ikat), only the filling yarns (filling ikat), or both sets of yarns (double ikat).

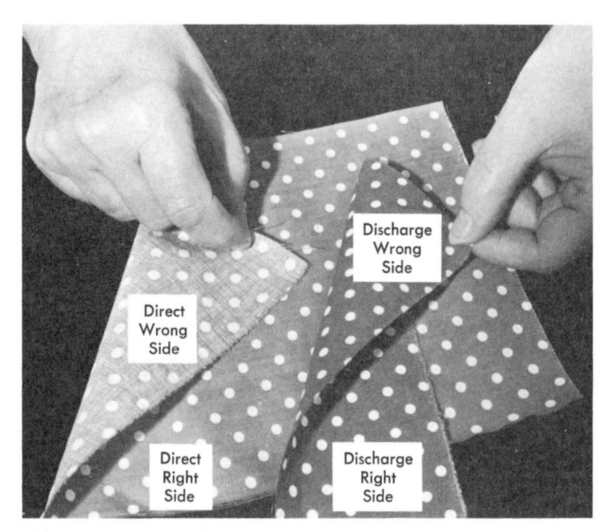

Fig. 36–11 *Discharge print versus direct print.*

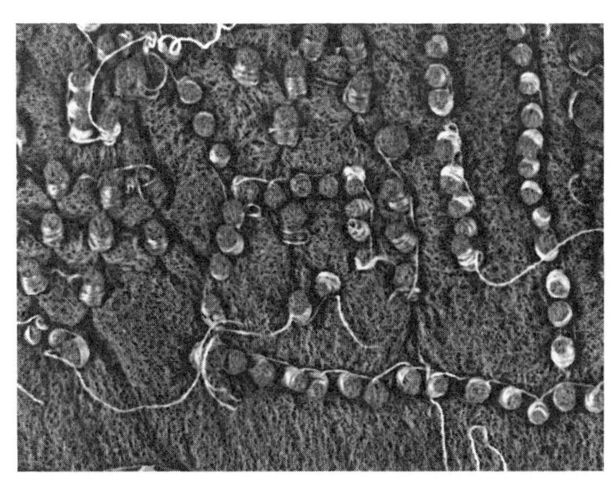

Fig. 36–13 *Tie-dye fabric showing thread used to make the design.*

Ikat designs do not have precise edges. Ikat requires great skill in determining the placement of the design in the finished fabric (see Figure 36–14).

Screen Printing. *Flat-screen printing* is done commercially for small yardages, 50–5,000 yards, and often is used for designs larger than the circumference of the rolls used for roller printing.

The design is applied to the screen so that all but the figure is covered by a resist material. One screen is used for each color. The color is forced through the screen by a squeegee.

In the *hand process,* the fabric to be printed is placed on a long table. Two people position the screen on the fabric, apply the color, move the screen to a new position, and repeat the process until all the fabric is printed.

In the *automatic-screen process,* the fabric to be printed is placed on a conveyer belt. A series of flat screens are positioned above and are lowered automatically. Color is applied automatically, and the fabric is moved automatically and fed continuously into ovens to be dried.

Rotary-screen printing is done with cylindrical metal screens that operate in much the same way as the flat screens, except that the operation is continuous rather than started and stopped as the screens are raised and lowered in the flat process (Figure 36–15). The rotary screens are cheaper than the copper rollers used in roller printing. Rotary screen printing is more common than flat bed screen printing.

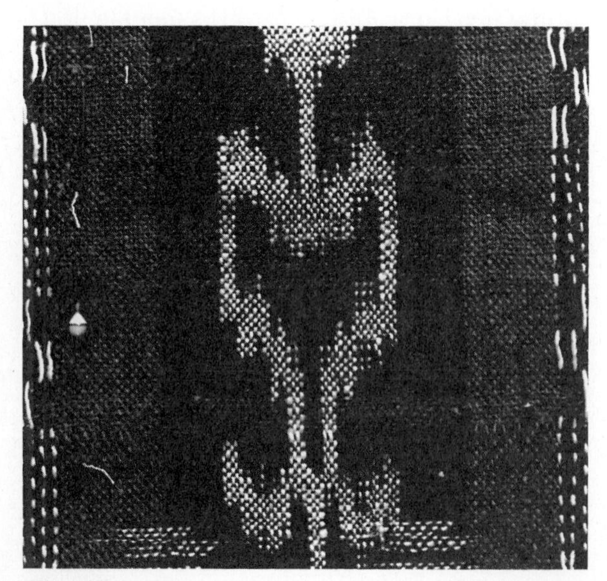

Fig. 36–14 *Ikat.*

Fig. 36–15 *Rotary-screen printing.*

Screen printing is useful for printing large designs on fabrics. Screen printing is offering stiff competition for roller printing.

Stencil Printing. *Stencil printing* is the precursor of screen printing. The pattern is cut from a special wax paper or thin metal sheets. As in screen printing, a separate stencil is cut for each color. Color in a thick solution or paste is applied by hand with a brush or sprayed with an air gun. Stencilling is done on limited yardage.

JET PRINTING

Jet printing is a process that uses continuous streams of dye forced from jets. Jet printing is most common in carpeting. The following are several types of jet printing.

TAK. *TAK* are the initials of a German rug manufacturer and a machinery supplier who invented a machine to print carpets. The machine can be used for other hard-to-print fabrics such as terrycloth, velvet, and upholstery fabrics. Dye is dropped on the fabric from individual channels fed by a trough. A special mechanism deposits drops of dye in a desired design. The fabric is then run through a padder to force the color into the cloth.

Microjet Injection. Millitron is a computer-controlled *microjet system* developed by the Milliken Company. It is used on carpets and upholstery to produce jacquard-like designs. The machine consists of a series of horizontal bars containing the dye, which is fed through small dye jets (10 per inch in a carpet machine, 16 per inch in an upholstery machine). The prepared cloth passes under the bars, and by the use of an electronically controlled tape, dye is deposited in the proper place from the jets on the bars. Dye penetrates to the backing and patterns can be easily changed (see Figure 36–16).

Polychromatic Printing. *Polychromatic printing* is an economical process for use with thick fabrics. Designs are multicolored stripes, abstract splashes, or tie-dye effects. Several colors are applied in one operation from jets set in bars. The fabrics move over an inclined plane, and dye comes out of jets onto the fabric. The cloth then goes through heavy rollers that press the dye completely through the fabric.

OTHER PRINTING METHODS

Heat-Transfer Printing. *Heat-transfer printing* is a process in which designs are transferred to fabric from specially printed paper by heat and pressure (see Figure 36–17). The paper is printed by gravure, flexograph, offset, or converted rotary-screen processes. The fabric or garment is placed on a plastic frame and padded with a special solution. The paper is placed over the fabric and then covered with a silicone-rubber sheet. The sandwich is then subjected to high pressure at a temperature of 200°C for a few seconds during which the print sublimes and migrates from the paper to the fabric.

The advantages of heat-transfer printing are better penetration and clarity of design, lower production costs, and elimination of pollution problems. Transfer printing can be done on circular knits, without splitting them first, and on garments.

Print papers using disperse dyes were developed for polyester-fiber fabrics and have been successful on high polyester/cotton blends and on nylon. For cotton fabrics and 50/50 blends of cotton/polyester, the fabric is treated with a resin that has an affinity for disperse dyes. Print papers with acid dyes for nylon, silk, and wool and with cationic dyes for acrylics are available.

Electrostatic Printing. *Electrostatic printing* is similar to electrostatic flocking. A screen that has the design on it is covered with powdered dye mixed with a carrier that has dielectric properties. The screen is about ½ inch above the fabric and, when passed through an electric field, the dye-resin is pulled onto the material, where it is fixed by heat.

Differential Printing. *Differential printing* is a printing technique using screen printing on carpets tufted with yarns that have different dye affinities.

Fig. 36–16 *Jet dyed fabrics.*

Fig. 36–17 *Heat-transfer printing: design on paper* (left) *is transferred by heat to fabric* (center). *Design on paper is lighter after printing* (right).

Color Problems

Good color fastness is expected, but it is not always achieved. When one considers all the variables connected with dyeing and printing and the hostile environment in which fabrics are used, one can appreciate how good most colored fabrics are.

The following are the factors that influence color fastness:

1. Chemical nature of fibers.
2. Chemical nature of dyes and pigments.
3. Penetration of dyes into the cloth.
4. Fixing the dyes or pigments on or in the fabrics.

The coloring agents must resist washing, dry cleaning, bleaching, and spot and stain removing with all of the variables of time, temperature, and substances used. They also must be resistant to light, perspiration, abrasion, fumes, and other factors.

If the color is not fast in the fabric as purchased, it is not possible to make it fast. Salt and vinegar are used as exhausting agents for household dyes, but there is no available research to support the theory that they will "set" color. Color loss occurs through bleeding, crocking, and migration or through chemical changes in the dye. *Bleeding* is color loss in water. *Crocking* is color loss from rubbing or abrasion. *Migration* is shifting of color to the surrounding area or to an adjacent surface. Atmospheric gases (fume fading), perspiration, and sunlight may cause fading as a result of a chemical change in the dye.

Certain vat and sulfur dyes will *tender,* or destroy, cotton cloth. Green, red, blue, and yellow vat dyes and black, yellow, and orange sulfur dyes are the chief offenders. Manufacturers know which dyes cause the trouble and can correct it by thoroughly oxidizing the dyestuff within the fiber. Damage is increased by moisture and sunlight. The problem is sometimes critical in draperies. Damage may not be evident until the draperies are cleaned and then slits or holes occur (Figures 36–18 and 36–19). Sunlight, smog, and acidic atmospheric gases, as well as dyes, will cause fabric damage.

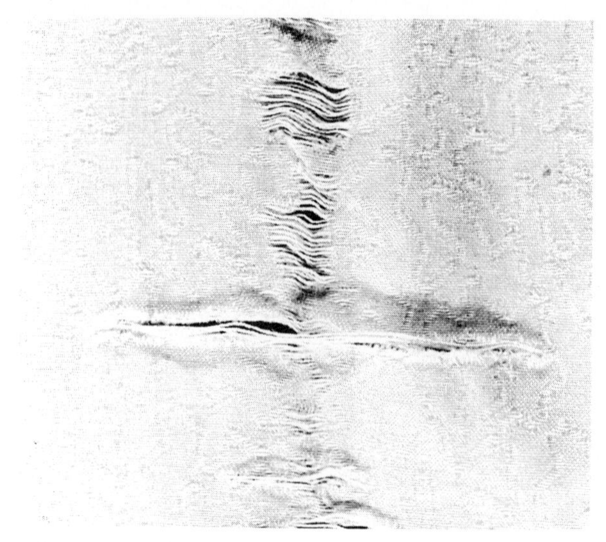

Fig. 36–18 *Cotton-and-rayon drapery fabric. After dry cleaning, yellowish streaks were obvious; after washing, splits occurred.*

Wear may remove the surface color of heavy fabrics (Figure 36–20). Movement of yarns in bending causes undyed fibers to work out to the surface. Color streaks may result from uneven removal of sizing before the dye is applied because the portions of the fabric that did not come in contact with the dye were not even dyed. Some resin-treated fabrics show this sort of color change because either the dye was applied with the resin and did not penetrate sufficiently or the fabric was dyed after being resin treated, in which case there were not enough places for the dye to be anchored. The best way to check the

Fig. 36–19 *Tendering of cotton draperies caused by sulfur dye, atmospheric moisture, and heat.*

Fig. 36–20 *Denim jeans; loss of surface color.*

Fig. 36–21 *One color is printed on a fabric at a time. When colors are not properly aligned, they are out-of-register.*

dye penetration in heavy fabrics is to examine the fabric. In yard goods, ravel off a yarn to see if it is the same color throughout. In ready-to-wear, look at the edge of seams. In heavy prints, look at the reverse side. The more color on the wrong side, the better the dye penetration.

A defect in printed fabrics occurs when two colors of a print overlap each other or where the edges are not clear. This defect is referred to as *out of register* (see Figure 36–21).

Printed fabrics are often printed off-grain. Off-grain prints create problems because the fabric cannot be both straight with the print and cut on-grain. If cut off-grain, the fabric tends to assume its normal position when washed, causing twisted seams and uneven hemlines. If cut on-grain, the print will not be straight. In an all-over design, this is not important, but in large checks and plaids or designs with crosswise lines, matching at seam lines is impossible and slanting lines across the fabric are seldom desirable. The reason for off-grain prints is that the fabric is started into the machine crooked or the mechanism for moving the fabric does not work properly. It is a problem that can easily be corrected at the mill. If consumers would refuse to buy off-grain prints and let retailers know why they are not buying them, better prints would be on the market (see page 196).

A *"frosting" effect* often results from abrasion in cotton/polyester durable-press garments that have been dyed with two different, but color-matched, dyes (union dyed). During wear, the cotton is abraded and becomes lighter in color, while the unabraded polyester keeps its color.

Other problems related to dyeing and printing are problems of concern to the producer. These are problems related to the consistency of the color throughout the width and length of the fabric or from dye lot to dye lot. Producers of apparel and home-furnishing items need to have a fabric that is consistent in color. The color needs to be the same from side-to-side, side-to-center, and end-to-end. If the color is not consistent, then the producer will have problems with product parts not matching in color. Textile product producers need to check shade variations in fabrics they receive in order to make sure the color is consistent. When several rolls of the same color fabric are required, it is important that all the rolls are consistent in color.

The fastness of the dye often determines the method of care that should be used. The consumer must depend on the label, but some knowledge of color problems that occur in use and care will allow for more intelligent choices.

37

Care of Textile Products

The term *care* refers to cleaning procedures or techniques necessary to remove soil from a textile product and return the product to its new or nearly new condition. Care can also be used to refer to storage conditions. This chapter will deal primarily with the Care Labeling Rule and commonly used cleaning procedures. However, a few brief comments will relate to storage.

Care Labeling Rule

In 1971, the Federal Trade Commssion issued the Care Labeling Rule. Because of some problems with the 1971 regulation, an amended version became effective in 1984. The rule requires manufacturers or importers of textile wearing apparel and certain piece goods to provide an accurate, permanent label or tag that contains regular-care information and instructions (relative to washing, drying, bleaching, warnings, and dry cleaning) and that is permanently attached and legible.

The regulation was developed because of consumer complaints regarding care instructions. The 1984 revision of the rule requires more-specific, detailed information concerning only one care method for a product. The label should use common terms that have a standard meaning (see chart, pp. 350–351). The instructions must be in words, not just symbols, although symbols may also be used. When products are produced offshore and sold in the United States, they must meet U.S. care-labeling requirements. When a label identifies washing, it must also state the washing method, water temperature, drying method, drying temperature, and ironing temperature when ironing is necessary. Procedures to be avoided must be identified, such as "Only non-chlorine bleach, when necessary." It multiple care methods are appropriate for that product, the manufacturer is not required to list them on the label. If the care-label instructions are followed and some problem develops during care, the manufacturer is liable. However, if the care-label instructions are not followed, the manufacturer is not liable for any problems due to improper care. Figure 37–1 shows several care labels for items of wearing apparel.

Fig. 37–1 *Care labels for garments.*

The rule applies to most items of wearing apparel. It does not apply to leather, suede, fur garments, ties, belts, and other apparel not used to cover or protect a part of the body. Certain other apparel items such as reversible garments are only required to have removeable, not permanent, care labels. For piece goods, the information must be supplied on the end of the bolt, but neither the manufacturer nor the retailer are required to provide a label to be sewn to the finished product.

Laundering

In *laundering,* it is important to understand the nature of soil and soiling, the manner in which a detergent and water work, and the additives that are used to improve the removal of soil or the appearance of the laundered item.

SOIL AND SOIL REMOVAL

Soil can be divided into several categories based on the soil type and how it is held on the fabric. Soil, such as gum, mud, or wax, can be held on the fabric mechanically. These soils can be removed mechanically by scraping or agitation. Soil, such as lint and dust, can be held on the

Standard Terms for Care Labels

1. Washing, Machine Methods

- **a.** *Machine wash*—A process by which soil may be removed from products or specimens through the use of water, detergent or soap, agitation and a machine designed for this purpose. When no temperature is given, e.g., "warm" or "cold," hot water up to 150°F (66°C) can be regularly used.
- **b.** *Warm*—Initial water temperature setting 90°–110°F (32°–43°C) (hand comfortable).
- **c.** *Cold*—Initial water temperature setting same as cold water tap up to 85°F (29°C).
- **d.** *Do not have commercially laundered*—do not employ a laundry that uses special formulations, sour rinses, extremely large loads or extremely high temperatures or that otherwise is employed for commercial, industrial, or institutional use. Employ laundering methods designed for residential use or use in a self-service establishment.
- **e.** *Small load*—Smaller than normal washing load.
- **f.** *Delicate cycle or gentle cycle*—Slow agitation and reduced time.
- **g.** *Durable press cycle* or *permanent press cycle*—Cool-down rinse or cold rinse before reduced spinning.
- **h.** *Separately*—alone.
- **i.** *With like colors*—With colors of similar hue and intensity.
- **j.** *Wash inside out*—Turn product inside out to protect face of fabric.
- **k.** *Warm rinse*—Initial water temperature setting 90°–110°F (32°–43°C).
- **l.** *Cold rinse*—Initial water temperature setting same as cold water tap up to 85°F (29°C).
- **m.** *Rinse thoroughly*—Rinse several times to remove detergent, soap, and bleach.
- **n.** *No spin* or *Do not spin*—Remove material at start of final spin cycle.
- **o.** *No wring* or *Do not wring*—Do not use roller wringer, nor wring by hand.

2. Washing, Hand Methods

- **a.** *Hand wash*—A process by which soil may be manually removed from products or specimens through the use of water, detergent or soap, and gentle squeezing action. When no temperature is given, e.g., "warm" or "cold," hot water up to 150°F (66°C) can be regularly used.
- **b.** *Warm*—Initial water temperature 90°–110°F (32°–43°C) (hand comfortable).
- **c.** *Cold*—Initial water temperature same as cold water tap up to 85°F (29°C).
- **d.** *Separately*—alone.
- **e.** *With like colors*—With colors of similar hue and intensity.
- **f.** *No wring or twist*—Handle to avoid wrinkles and distortion.
- **g.** *Rinse thoroughly*—Rinse several times to remove detergent, soap, and bleach.
- **h.** *Damp wipe only*—Surface clean with damp cloth or sponge.

3. Drying, All Methods

- **a.** *Tumble dry*—Use machine dryer. When no temperature setting is given, machine drying at a hot setting may be regularly used.
- **b.** *Medium*—Set dryer at medium heat.
- **c.** *Low*—Set dryer at low heat.
- **d.** *Durable press* or *permanent press*—Set dryer at permanent-press setting.
- **e.** *No heat*—Set dryer to operate without heat.
- **f.** *Remove promptly*—When items are dry, remove immediately to prevent wrinkling.
- **g.** *Drip dry*—Hang dripping wet with or without hand shaping and smoothing.
- **h.** *Line dry*—Hang damp from line or bar in or out of doors.
- **i.** *Line dry in shade*—Dry away from sun.
- **j.** *Line dry away from heat*—Dry away from heat.
- **k.** *Dry flat*—Lay out horizontally for drying.
- **l.** *Block to dry*—Reshape to original dimensions while drying.

fabric by electrostatic forces. If the electrostatic force is neutralized, the soil can be removed. Because water is such an excellent conductor of electricity, immersing the fabric in water will neutralize any static charge on the surface of the fabric. Water-soluble soils—such as coffee, sodas, and sugar water—are absorbed into the fiber. When the fabric is immersed in water, the water can dissolve the soil. Organic soils—such as grease, oil, and gravy—require the assistance of the chemical action of a detergent and heat in order to be removed. Of course, many soils are mixtures of these categories and removed by a combination of thermal, mechanical, and chemical actions. If one aspect of removal is decreased, another aspect must be increased in order to maintain the degree of soil removal. For example, if the water temperature is decreased,

Standard Terms for Care Labels (*Cont.*)

m. *Smooth by hand*—by hand, while wet, remove wrinkles, straighten seams and facings.

4. **Ironing and Pressing**
   a. *Iron*—Ironing is needed. When no temperature is given, iron at the highest temperature setting may be regularly used.
   b. *Warm iron*—Medium temperature setting.
   c. *Cool iron*—Lowest temperature setting.
   d. *Do not iron*—Item not to be smoothed or finished with an iron.
   e. *Iron wrong side only*—Article turned inside out for ironing or pressing.
   f. *No steam* or *Do not steam*—Steam in any form not to be used.
   g. *Steam only*—Steaming without contact pressure.
   h. *Steam press* or *Steam iron*—Use iron at steam setting.
   i. *Iron damp*—Articles to be ironed should feel moist.
   j. *Use press cloth*—Use a dry or a damp cloth between iron and fabric.
5. **Bleaching**
   a. *Bleach when needed*—All bleaches may be used when necessary.
   b. *No bleach* or *Do not bleach*—No bleaches may be used.
   c. *Only nonchlorine bleach, when needed*—Only the bleach specified may be used when necessary. Chlorine bleach may not be used.
6. **Washing or Drycleaning**
   a. *Wash or dry clean, any normal method*—Can be machine washed in hot water, can be machine dried at a high setting, can be ironed at a hot setting, can be bleached with all commercially available bleaches and can be dry cleaned with all commercially available solvents.
7. **Dry cleaning, All Procedures**
   a. *Dry clean*—A process by which soil may be removed from products or specimens in a machine that uses any common organic solvent (for example, petroleum, perchlorethylene, fluorocarbon) located in any commercial establishment. The process may include moisture addition to solvent up to 75 percent relative humidity, hot tumble drying up to 160°F (71°C) and restoration by steam-press or steam-air finishing.
   b. *Professionally dry clean*—Use the dry-cleaning process, but modified to ensure optimum results either by a dry-cleaning attendant or through the use of a dry cleaning machine that permits such modifications or both. Such modifications or special warnings must be included in the care instruction.
   c. *Petroleum, Fluorocarbon,* or *Perchlorethylene*—Employ solvent(s) specified to dry clean the item.
   d. *Short cycle*—Reduced or minimum cleaning time, depending upon solvent used.
   e. *Minimum extraction*—Least possible extraction time.
   f. *Reduced moisture* or *Low moisture*—Decreased relative humidity.
   g. *No tumble* or *Do not tumble*—Do not tumble dry.
   h. *Tumble warm*—Tumble dry up to 120°F (49°C).
   i. *Tumble cool*—Tumble dry at room temperature.
   j. *Cabinet dry warm*—Cabinet dry up to 120°F (49°C).
   k. *Cabinet dry cool*—Cabinet dry at room temperature.
   l. *Steam only*—Employ no contact pressure when steaming.
   m. *No steam* or *Do not steam*—Do not use steam in pressing, finishing, steam cabinets, or wands.
8. **Leather and Suede Cleaning**
   a. *Leather clean*—Have cleaned only by a professional cleaner who uses special leather- or suede-care methods.

Source: Federal Trade Commission (1984). *Writing a Care Label.* Washington, D.C.: U.S. Government Printing Office.

either more detergent or more agitation will be required to be as effective.

DETERGENCY

Detergency refers to the manner in which the soap or detergent removes soil. By adding soap or synthetic detergent to water, the surface tension of the water is lowered; thus the water will wet things faster. Water will not bead up on surfaces, but will spread over the surface, wetting the surface. A *soap* or *detergent* molecule consists of an organic "tail" that has an affinity for organic soils and a polar "head" that has an affinity for water. Thus the two parts of the soap or detergent molecule literally dislodge the soil.

Agitation breaks the soil into very tiny globules that are held in suspension until they are rinsed away (Figure 37–2). By using hot water, the oily soils are softer and more likely to break into small globules. Because of the many functions of the ingredients in detergents, it is very important to use the appropriate amount of detergent when doing the laundry. Instructions for the proper amount to use are included on the label. Much research has been done to determine the correct amount to use for optimum results. Do not guess as to the amount to use, use a measuring cup. If too much detergent is used, detergent will build up on textiles in the wash. If too little detergent is used, soil will remain on the textiles.

WATER

Water is used as the solvent because it is cheap, readily available, nontoxic, and does not require special equipment for its use. Water has several aspects of importance in laundering: hardness, temperature, and volume. *Water hardness* refers to the kind and amount of mineral contaminants present. Water that has mineral salts in it is referred to as *hard water.* The more mineral salts dissolved in the water, the harder it is. Hard water makes cleaning more difficult. In order to soften the water, the minerals must be removed, or *sequestered* (bonded to another molecule). The most common procedure for softening water is adding a water-softening agent, such as sodium hexametaphosphate, to the water or using an ion-exchange resin in a water softener.

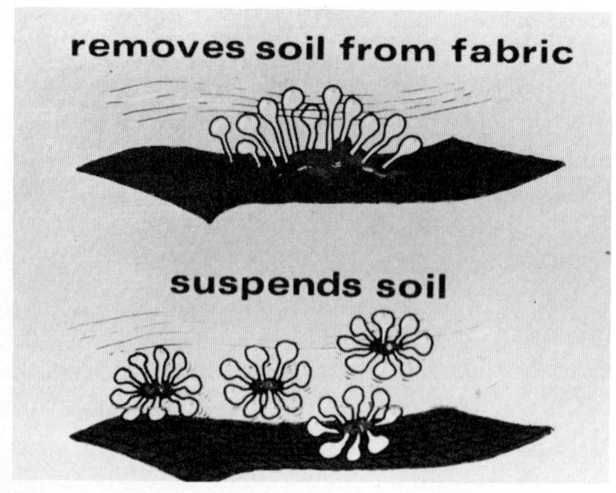

Fig. 37–2 *Mechanism of soiling: detergent surrounds soil and lifts it off the fabric.*

Water temperature is important in determining the effectiveness of the laundry additives used. Some additives are more effective at certain temperatures. Water temperature is also important in removing certain soils. The following are water-temperature ranges as identified by the Federal Trade Commission in the Care Labeling Rule: cold water is 85°F, or the initial water temperature from a cold water tap; warm water is 90°F–110°F, or hand comfortable; and hot water is up to 150°F.

Water volume is important in order to allow for agitation, remove soil and keep soil suspended, and avoid wrinkling items in the load. Water volume is related to the amount of fabric present in the machine.

SOAP AND SYNTHETIC DETERGENT

Soaps and synthetic detergents are used to remove and suspend soils, minimize the effects of hard water, and alter the surface tension of the water.

Soaps. *Soaps* are salts of long, linear-chain fatty acids produced from naturally occurring animals or vegetable oils or fats. Soaps can react with hard water and produce insoluble curds that form a greasy, gray film on textiles and a ring on tubs. Soaps are effective in removing oily or greasy stains, but they are not vigorous soil-removal agents.

Synthetic Detergents. Synthetic detergents are really mixtures of several ingredients: surfactant, builder, filler, antiredeposition agent, perfume, dye, and fluorescent-whitening agent (Figure 37–3). Detergent will be used to refer to the box or bottle of cleaning compound called a *detergent.* Surfactants are sulfonated organic compounds that are soluble in hard water and do not form an insoluble curd. Surfactants are vigorous soil-removal agents and are frequently sulfonated, long, linear-chain fatty acids. There are several types of surfactants: nonionic, anionic, and cationic. *Nonionic surfactants,* such as ethers of ethylene oxide, are used in liquids and recommended for use in cold or warm water because they become less soluble at high temperatures. *Anionic surfactants* are used in powders

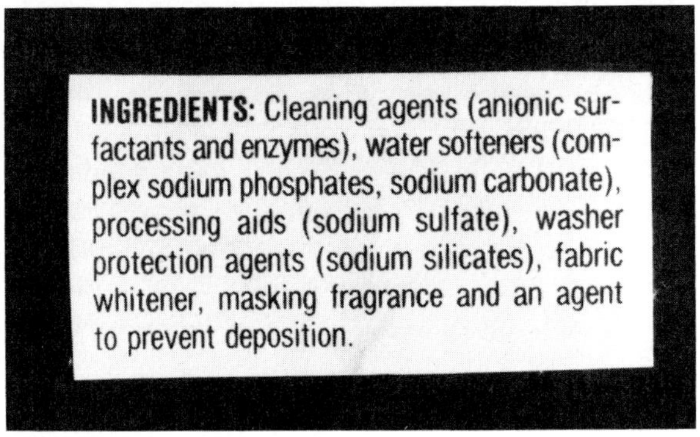

INGREDIENTS: Cleaning agents (anionic surfactants and enzymes), water softeners (complex sodium phosphates, sodium carbonate), processing aids (sodium sulfate), washer protection agents (sodium silicates), fabric whitener, masking fragrance and an agent to prevent deposition.

Fig. 37–3 *Detergent label.*

and good for oily soils and clay-soil suspension. It is this category that is the most common. They are usually linear alkyl sulfonates (LAS) and are biodegradable. Anionic surfactants are most effective in warm and hot water. *Cationic surfactants* are used primarily in disinfectants and fabric softeners.

Builders are used to soften the water, add alkalinity to the solution since a pH of 8–10 is best for maximum cleaning efficiency, emulsify oils and greases, and minimize soil redeposition. Builders include phosphates (usually sodium tripolyphosphate), carbonates (sodium carbonate), citrates (sodium citrate), and silicates (sodium silicate). Of these, phosphate builders offer the best performance over the widest range of laundering conditions. Phosphate builders have been banned or restricted in some parts of the United States due to their role in water pollution. Carbonate builders do not contribute to water pollution. However, they combine with hard-water minerals to form water-insoluble precipitates that may harm the machine, fabric, and zippers. Citrate builders are much weaker at softening hard water and used in heavy-duty liquid detergents. Sodium silicate functions as a builder when present in large concentrations. However, sodium silicate is often present in small concentrations because it also functions as a corrosion inhibitor.

Fillers, such as sodium sulfate in powder and water and alcohol in liquid detergents, are used to add bulk to the detergent, allow for a uniform mix of the ingredients, increase the size of the micelle (the grouping of soap or detergent molecules that remove soil from fabric), protect washer parts, and minimize caking in powders. *Antiredeposition agents*, such as sodium carboxymethylcellulose, are used to minimize the soil redepositing from the wash water on the fabric. *Perfumes* are designed to mask the chemical smell of detergents and to add a "clean" smell to the wash. *Dyes* make the detergent look better and function as bluing .

Fluorescent-whitening agents are also known as *fluorescent-brightening agents, optical-whitening agents,* and *optical-brightening agents.* These compounds are low-grade dyes that fluoresce, or absorb, light at one wavelength and reemit the energy at another wavelength. Thus it is possible to have whites that are "whiter than white." These ingredients do not contribute to soil removal; they mask soil and make yellow or dingy fabrics look white.

Other ingredients commonly found in detergents include fabric softeners and bleaches (which will be discussed later in this chapter), suds-control agents, and foam-control agents. Liquid detergents may contain alcohol to dissolve some ingredients of the detergent, assist in stain removal, and act as an antifreeze during shipping; hydrotopes to assist in keeping ingredients in solution; and opacifiers to give a rich, creamy appearance to the detergent.

OTHER LAUNDRY ADDITIVES

Other *laundry additives* include bleaches, fabric softeners, water softeners, disinfectants, presoaks, pretreatments, starches or sizing, and bluing. Some of these things are seldom used.

Bleach. Most *bleaches* are oxidizing agents. The actual bleaching is done by *active oxygen.* A few bleaches are reducing agents that are used to strip color from dyed fabrics. Bleaches may be either acid or alkaline in nature. They are usually unstable, especially in the presence of moisture. Bleaches that are old or have been improperly stored will lose their oxidizing power.

Any bleach will cause damage, and, because damage occurs more rapidly at higher temperatures and concentrations, these factors should be carefully controlled.

The same bleach is not suitable to all kinds of fibers. Because fibers vary in their chemical reaction, bleaches must be chosen with regard to their fiber content. The sock in Figure 37–4 had been all white, but, when bleached with a chlorine bleach, the wool-ribbed cuff section became discolored while the cotton foot remained white.

Liquid chlorine bleaches were, for many

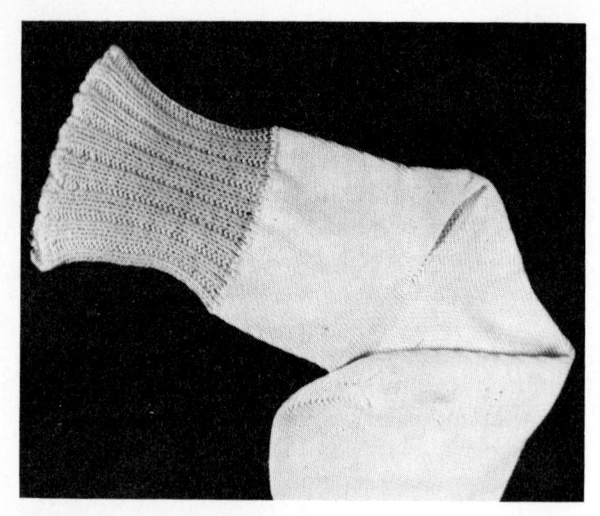

Fig. 37–4 Cotton-and-wool sock after bleaching. Chlorine bleach caused wool in ribbed top to yellow and stiffen.

years, the common household bleaches. They are efficient bactericidal agents and, as such, can be used for sterilizing fabrics. They are cheap and efficient bleaches for cellulosic fibers. The bleaching is done by hypochlorous acid liberated during the bleaching process. Because this tenders cellulosic fibers, the bleach must be thoroughly rinsed out. Chlorine bleaches are of no value on protein and thermoplastic fibers and, if used, will cause yellowing.

Powdered-oxygen bleaches, also called all-fabric bleaches, may be used safely on all fibers and colored fabrics. Their bleaching effect is much milder than chlorine bleaches.

Sodium perborate is a powder bleach that becomes hydrogen perioxide when it combines with water. It is a safe bleach for home use with all kinds of fibers. Powder bleaches are recommended for regular use in the wash water to maintain the original whiteness of the fabric rather than as a whitener for discolored fabrics.

Acid bleaches, such as oxalic acid and potassium permanganate, have limited use. Citric acid and lemon juice are also acid bleaches that are good rust-spot removers.

Fabric Softener. *Fabric softeners* coat the fabric to increase the electrical conductivity of the fabric, minimize static charges, and decrease fabric stiffness. The types of fabric softeners include the following: those added in the final rinse, those present in detergent, and those added in the dryer. Fabric softeners have become much more convenient to use in the past several years with the last two categories. The instructions for use of the fabric softener need to be followed or problems may result. For example, dryer fabric softener sheets should be added to a cold dryer. If they are added to a warm or hot dryer, oil from the fabric softener may spot synthetic items. Fabric softeners have a tendency to build up on fabrics in a greasy layer resulting in less absorbent fabrics. Hence it is not recommended that a fabric softener be used every time a product is laundered. Every other time or every third time is recommended, if necessary.

Water Softener. *Water-softening agents* are found as builders in detergent or as separate ingredients that can be added to increase the efficiency of the detergent if the water is especially hard. If a water-softening additive is used, a nonprecipitating type is recommended to avoid buildup of precipitates on washer parts and items in the wash.

Disinfectant. *Disinfectants* include pine oil, phenolics, chlorine bleach, and coal-tar derivatives. These items are used occasionally to disinfect sickroom garments and bed and bath linens.

Presoak. *Enzymatic presoaks* are used to remove tough stains. These additives contain enzymes—such as protease (for protein stains), lipase (for fat stains), and amylase (for carbohydrate stains)—that act as a catalyst and aid in removal of these soils. Enzymatic presoaks need more time to work than most other additives so a long presoak of one-half hour or more is recommended. Some people prefer to use the presoak overnight. Presoaks often include a builder and a surfactant to improve the efficiency of the presoak. These additives are not safe for use with protein fibers, such as silk and wool, and other specialty wool fibers.

Pretreatment. *Pretreatments* are another means of removing difficult stains, and they are usually added directly to the stain shortly before the item is laundered. Pretreatment products often contain a solvent, surfactant, and builder.

Starch or Sizing. *Starch* or *sizing* is used after washing to add body and stiffness to fabrics. Starch is seldom used at present.

Bluing. *Bluing* is a weak blue dye that masks yellowing in fabrics. Bluing is seldom used by itself today because it is incorporated in detergents as dyes. In addition, the use of fluorescent-whitening agents in detergents supersedes the need to add bluing.

SORTING

Before laundering, it is important to *sort* the items to be washed in order to minimize problems and remove soil as efficiently as possible. Sorting is often done by color, type of garment (for example work garments separate from delicate items), type of soil, recommended-care method, and propensity of fabrics to lint. During sorting, it is a good idea to close zippers and buttons so they do not snag other items in the wash. It is also a good idea to check pockets for pens, tissues, and other items that may create problems during washing. In addition, this is the time to check items for stains, holes, or tears.

The table "Suggested Care of Textile Products by Fiber Group" summarizes the care required, based on fiber content. However, it is important

Suggested Care of Textile Products by Fiber Group

Fiber Group	*Cleaning Method*	*Water Temperature*	*Safe to Use Chlorine Bleach*	*Dryer Temperature*	*Iron Temperature*	*Special Storage Considerations*
Acetate	Dry clean*	Warm (100–110°F)	Yes	Low	Very low	Avoid contact with nail-polish remover
Acrylic	Launder	Warm	Yes	Warm	Medium	—
Cotton	Launder	Hot (120–140°F)	Yes	Hot	High	Store dry to prevent mildew
Polyester/Cotton DP	Launder	Hot	Yes	Warm	Medium	—
Flax	Launder	Hot	Yes	Hot	High For longest wear, do not press in sharp creases	
Glass	Hand wash only	Hot	Yes	Line dry	Do not iron	Prevent fiber breakage by storing as flat as possible
Modacrylic	Launder	Warm	Yes	Low	Very low	—
Nylon	Launder	Hot	Yes	Warm	Low	—
Olefin	Launder	Warm	Yes	Warm	Very low	—
Polyester	Launder	Hot	Yes	Warm	Low	—
Rayon	Launder	Hot	Yes	Hot	High	Store dry to prevent mildew
Silk	Dry clean*	Warm	No	Warm	Medium	—
Spandex	Launder	Warm	No	Warm	Very low	—
Wool	Dry clean*	Warm	No	Warm	Medium, with steam	Protect from moths, do not store in plastic bags

*Or hand wash, avoiding excessive agitation and stretching.

to remember that care is dependent not only on fiber content, but also on other things such as dye, fabrication, finish, product construction, type of soil, and extent of soiling.

DRYING

The *drying* procedure is usually specified on the care label. Machine drying is considered the most severe method because of the abrasion and agitation. Line drying may also be too severe for some items because wet fabrics are extremely heavy. Fibers that weaken when wet, such as wool and rayon, may be under too much stress if the item is hung to dry. Drying flat is the least severe method because the fabric is under little stress.

DRY CLEANING

In *dry cleaning,* the solvents include the following: perchloroethylene (perc), a petroleum solvent (Stoddard's solvent), or a fluorocarbon solvent (Valclene). Of these three, perc is most common. You may need to check with your dry cleaner as to which solvent is used in that system. Dry cleaning is referred to as dry because the solvent does not feel wet like water. Along with items labeled for dry cleaning, many machine-washable items may be dry cleaned.

A professional organization, the International Fabricare Institute (IFI), provides training and updates for dry cleaners, establishes a fair-claims adjustment guide for use in consumer complaints, and provides an evaluation service to members when problems develop. Members of IFI will display an IFI plaque in their business.

In dry cleaning, the items are brought to the cleaners and identified with a tag that includes the special instructions, the owner's identification number, and the number of pieces in the group. Items are first inspected and treated at a spot board. Because a solvent is used, stains that are water soluble and other hard-to-remove spots must be treated at the spot board. Customers who identify stains for the dry cleaner make the cleaning task easier and ultimately improve their satisfaction with the cleaned product.

After treatment at the spotting board, items are placed in the dry cleaning unit to be tumbled with a charged solvent (solvent plus detergent plus a small percentage of water) (Figure 37–5). After tumbling, the solvent is reclaimed in the same unit or a separate unit called a *reclaimer.* The reclaimer serves the same function as a dryer in laundering, except that the solvent is condensed and filtered to be used again. Solvents must be reclaimed because of the cost. Filtering and distilling remove soil, color, odor, and other residue.

Fig. 37–5 *Dry-cleaning unit.*

After the items are removed from the reclaimer, they go to the pressing area, where steam and special steam-air forms are used to give a finished appearance to the item. For example, pants are pressed with a topper that finishes the top part of the pants. Each leg is pressed separately with a press. Jackets, shirts, and blouses are finished with a suzie, a steam-body torso form (see Figure 37–6).

Additional treatments that many dry cleaners are equipped to do include replacing buttons; doing minor repairs to items; replacing sizing, water repellency, and other finishes; adding permanent creases to pants; and fur and leather cleaning. Some dry cleaners can also clean and sanitize feather pillows and clean and press draperies.

STORAGE

When products are not in use, they need to be stored. Many problems develop because the storage was not appropriate for the product. Short-term storage is not as likely to develop problems as long-term storage. All items should be stored clean and as free from wrinkles as possible. Care should be taken that items are protected from

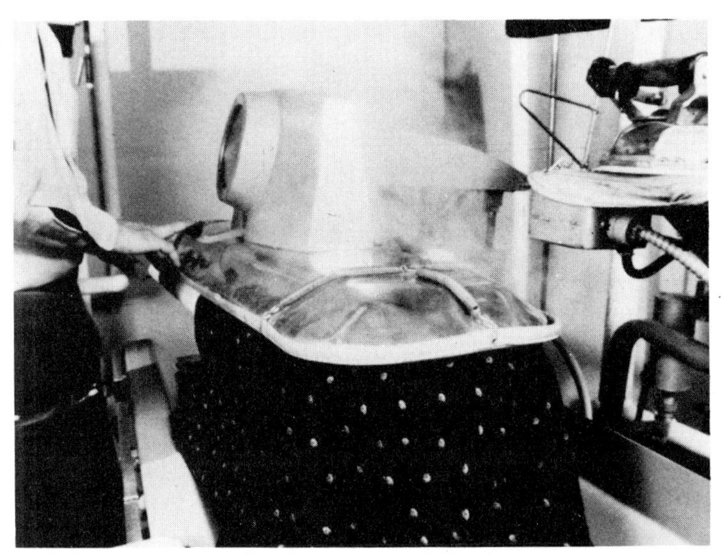

Fig. 37–6 *Pressing equipment: pants press* (left); *inflatable "Suzie" for steaming blouses and shirts* (right).

insects, dust or dirt, excessive moisture, and light during storage. Most products should never be stored in direct contact with raw wood or wood finishes. Raw wood produces acid as it ages. Cellulosic fibers are degraded by acid, and brown or yellow stains may develop as a result of exposure to the wood (see Figure 37–7). Plastic bags from dry cleaners are provided as a service to avoid soiling freshly cleaned items during transport. These bags are not intended for storage and should be discarded immediately after the product is brought into the house. Items stored in dry cleaning bags may discolor due to the acids in the bag; build up static-attracting dust; or trap moisture, creating an ideal environment for mildew. For more information regarding storage, see the appropriate fiber chapter.

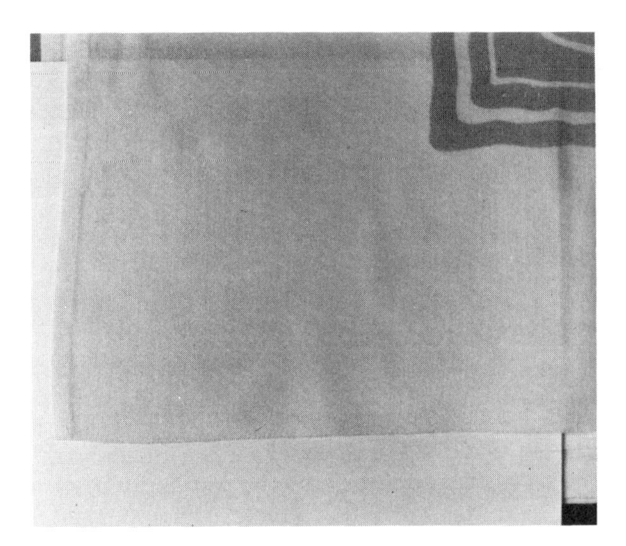

Fig. 37–7 *Yellowed-cotton tablecloth.*

Fabric Glossary

Antique satin is a reversible satin-weave fabric with satin floats on the technical face and surface slubs on the technical back created by using slub-filling yarns. It is usually used with the technical back as the right side for drapery fabrics and often made of a blend of fibers.

Batiste is an opaque, lightweight, spun-yarn, plain-weave fabric with a smooth surface. When made of cotton or cotton/polyester, the yarns are usually combed. It can be made of all wool, silk, or rayon.

Bedford cord is a heavy, warp-faced, unbalanced pique-weave fabric with wide warp cords created by extra filling yarns floating across the back to give a raised effect.

Bengaline is a lustrous, durable, warp-faced fabric with heavy filling cords completely covered by the warp.

Bouclé is a woven or knit fabric with bouclé yarns. The loops of the novelty yarns create a mock-pile surface.

Broadcloth is a close plain-weave fabric made of cotton, rayon, or a blend of either cotton or rayon with polyester. It has a fine rib in the filling direction caused by slightly larger filling yarns, filling yarns with a lower twist, or a higher warp-yarn count. High-quality broadcloth is made with plied warp and filling yarns. The fabric may be mercerized. It has a soft, firm hand. The term *broadcloth* is also used to refer to a plain- or twill-weave lustrous wool or wool-blend fabric that is highly napped and then pressed flat.

Brocade is a jacquard-woven fabric with a pattern that is created with different colors or with patterns in twill or satin weaves on a ground of plain, twill, or satin weave. It is available in a variety of fiber contents and qualities.

Brocatelle is similar to brocade, but the pattern is raised and often padded with stuffer yarns. The pattern is warp faced and the ground is filling faced. Brocatelle is often a double cloth. It is mainly used in home furnishings.

Buckram is a heavy, very stiff, spun-yarn fabric converted from cheesecloth gray goods with adhesives and fillers. It is used as an interlining to stiffen pinch-pleated window-treatment fabrics.

Bunting (*See* Cheesecloth)

Burlap is a coarse, heavy, loosely woven plain-weave fabric often made of single irregular yarns of jute. It is used in its natural color for carpet backing, bagging, and furniture webbing. It is also dyed and printed for home-furnishing uses.

Butcher cloth is a coarse-rayon or rayon-blend fabric. It is made in a variety of weights. A Federal Trade Commission ruling prohibits the use of the word *linen* for this type of fabric.

Cambric is a fine, firm, plain-weave balanced fabric with starch, and has a slight luster on one side. It is difficult to distinguish from percale.

Canvas is a heavy, firm, strong fabric often made of cotton or acrylic and used for awnings, slipcovers, and covers for boats. It is produced in many grades and qualities. It may have a soft or firm hand. It is made in plain or basket weave.

Casement cloth is a general term for any open-weave fabric used for drapery or curtain fabrics. It is usually sheer.

Cavalry twill is a steep, pronounced, double-wale line, smooth-surfaced twill fabric.

Challis (shal'i) is a lightweight, spun-yarn, plain-weave balanced fabric with a soft finish. It can be made of any staple fiber or blend of fibers.

Chambray is a plain-weave fabric usually of cotton, rayon, or a blend of these with polyester. Usually chambray has white yarns in the filling direction and yarn-dyed yarns in the warp direction. Iridescent chambray is made with one color in the warp and a second color in the filling. It can also be made in striped patterns.

Cheesecloth is a lightweight, sheer, plain-woven fabric with a very soft texture. It may be natural colored, bleached, or dyed. It usually has a very low count. If dyed, it may be called *bunting* and could be used for flags or banners.

Chiffon is a sheer, very lightweight, plain-weave fabric with fine crepe twist yarns of approximately the same size and twist used in warp and filling. The fabric is balanced.

China silk is a soft, lightweight, opaque, plain-weave fabric made from fine-filament yarns and used for apparel.

Chino is a steep-twill fabric with a slight sheen, often made in a bottom-weight fabric of cotton or cotton/polyester. Often it is made of combed two-ply yarns in both warp and filling and vat-dyed in khaki.

Chintz is a medium- to heavyweight, plain-weave, spun-yarn fabric finished with a glaze. Chintz may be piece dyed or printed.

Clip spot refers to a fabric in which a structural design is created with an additional yarn that interlaces with the ground fabric in spots and floats along the technical back of the fabric. The floats are removed by shearing.

Cloque fabric is a general term used to refer to any fabric with a puckered or blistered effect.

Corduroy is a filling-yarn pile fabric where the pile is created by long-filling floats that are cut and brushed in the finishing process. The ground weave may be either a plain or twill weave.

Covert was first made in England, where there was a demand for a fabric that would not catch on brambles or branches during fox hunts. To make this tightly woven fabric, a two-ply yarn, one cotton and one wool, was used. Because the cotton and wool did not take the same dye, the fabric had a mottled appearance.

Cotton covert is always mottled, and it may be made with ply yarns, one ply white and the other colored, or it may be fiber dyed white and a color. It is a $\frac{2}{1}$ twill, of the same weight as denim, and used primarily for work pants, overalls, and service coats.

Wool covert is made from woolen or worsted yarns. It may be mottled or solid color and may be suit or coat weight. It may be slightly napped or have a clear finish. The mottled effect is obtained by using two different-colored plys or by blending different-colored fibers.

Crash is a medium- to heavyweight, plain-weave fabric made from slub or irregular yarns to create an irregular surface.

Crepe refers to any fabric with a puckered, crinkled, or grainy surface. It can be made with crepe yarns, a crepe or momie weave, or a finish such as embossed or plissé. Examples of crepe fabrics include chiffon, crepe-backed satin, georgette, and crepe de Chine. For more information, see these fabric names.

Crepe-backed satin is a reversible satin-weave fabric in which the filling yarns have a crepe twist. The technical face has satin floats and the technical back looks like a crepe fabric. It is also known as *satin-back crepe.*

Crepe de Chine is a lightweight, opaque, plain-weave, filament-yarn fabric. It has a medium luster. Silk crepe de Chine usually is made with crepe yarns.

Cretonne is a plain-weave fabric similar to chintz, except that the finish is dull and the fabric is more likely to be printed with large-scale floral designs.

Damask is a reversible, flat, jaquard-woven fabric with a satin weave in both the pattern and the plain-weave ground. It can be one color or two. In two-color damasks, the color reverses on the opposite side. It is used in apparel and home furnishings.

Denim is a cotton or cotton/polyester blend, twill-weave, yarn dyed fabric. Usually the warp is colored and the filling is white. It is usually a left-hand twill that is commonly available with a blue warp and white filling for use in apparel. It is available in a variety of weights.

Dimity is a sheer, lightweight fabric with warp cords created by using heavier-warp yarns at a regular distance, grouping warp yarns together, or using a basket variation where two or more warp yarns are woven as one. It may be printed or piece dyed. It may be made of combed-cotton yarns. *Barred dimity* has heavier or double yarns periodically in both the warp and filling.

Dotted swiss is a sheer, light- or medium-weight, plain-weave fabric with small dots created at regular intervals with extra yarns, either through a swivel weave or a clip-spot weave. Look-alike fabrics are made by flocking, printing, or using an expanded foam print.

Double cloth is a fabric made by weaving two fabrics with five sets of yarns: two sets of warp, two sets of filling, and one set that connects the two fabrics.

Double knit is a general term used to refer to any filling-knit fabric made on two needle beds.

Double weave is a fabric made by weaving two fabrics with four sets of yarns (two sets of warp and two sets of filling yarns) on the same loom. The two fabrics are connected by periodically reversing the positions of the two fabrics from top to bottom. Double weave is also known as *pocket cloth* or *pocket weave.*

Drill is a strong, medium- to heavyweight, warp-faced, twill-weave fabric. It is usually a $\frac{2}{1}$ left-handed twill and piece dyed.

Duck is a strong, heavy, plain- or basket-weave fabric. Duck comes in a variety of weights and qualities. It is similar to canvas.

Duvetyn is similar to suede, but is lighter weight and more drapable. It has a soft, velvet-like surface made by napping, shearing, and brushing.

Embossed fabrics are created by embossing a fabric with a design with heated, engraved calendars. Often print cloths are embossed to imitate seersucker, crepe, or other structural-design fabrics.

Faille (file) is a medium- to heavyweight, unbalanced, plain-weave fabric with filament yarns, warp-faced, flat ribs created by using heavier filling yarns. It has a light luster.

Felt is a fiberweb fabric of at least 70 percent wool made by interlocking the scales of the wool fibers through the use of heat, moisture, and agitation.

Flannel is a light- to heavyweight, plain- or twill-weave fabric with a napped surface.

Flannelette is a light- to medium-weight, plain-weave cotton or cotton-blend fabric lightly napped on one side.

Foulard. (*See* Surah)

Friezé is a strong, durable, heavy-warp-yarn pile fabric. The pile is made by the over-wire method to create a closed-loop pile.

Gabardine (gaberdine) is a tightly woven, medium- to heavyweight, steep- or regular-angle, twill-weave fabric with a pronounced wale. The fabric can be wool, a wool-blend, or a synthetic-fiber content designed to look like wool. Gabardine can also be 100 percent texturized polyester or a cotton/polyester blend.

Gauze is a sheer, lightweight, low-count, plain- or leno-weave balanced fabric made of spun yarns. It is often cotton, rayon, or a blend of these fibers. *Indian gauze* has a crinkled look and is available in a variety of fabric weights.

Georgette is a sheer, lightweight, plain-weave fabric made with fine-crepe yarns. It is crepier and less lustrous than chiffon.

Gingham is a yarn-dyed, plain-weave fabric that is available in a variety of weights and qualities. It may be balanced or unbalanced. It may be made of combed or carded yarns. If two colors of yarn are used, the fabric is called a *check* or a *checked gingham.* If three or more colors are used, the fabric is referred to as a *plaid gingham.*

Gray goods (grey goods or **greige goods)** is a general term used to describe any unfinished woven or knitted fabric.

Grosgrain (grow'grain) is a tightly woven, firm, warp-faced fabric with heavy, round filling ribs created by a high-warp count and coarse filling yarns. Grosgrain can be woven as a narrow-ribbon or a full-width fabric.

Habutai is a soft, lightweight silk fabric. It is heavier than China silk.

Handkerchief linen is similar in luster and count to batiste, but it is linen or linen-look with slub yarns and a little more body.

Herringbone is a broken twill-weave fabric created by changing the direction of the twill wale from right to left and back again. This creates a chevron pattern of stripes that may be or may not be equally prominent. Herringbone fabrics are made in a variety of weights, patterns, and fiber contents.

Homespun is a coarse, plain-weave fabric with a hand-woven look.

Honan was originally of Chinese silk. Now it is made of any filament fiber. It is similar to pongee, but it has slub yarns in both warp and filling.

Hopsacking is a coarse, loosely woven suiting- or bottom-weight, basket-weave fabric often made of low-grade cotton.

Houndstooth check is a medium- to heavyweight, yarn-dyed, twill-weave fabric in which the interlacing and color pattern creates a unique pointed-check or houndstooth shape.

Huck or **huck-a-back toweling** is a medium- to heavyweight fabric made on a dobby loom to create a honeycomb or bird's-eye pattern. Often the filling yarns are more-loosely twisted to increase the absorbency of the fabric.

Interlock is a firm, double-filling knit where the two needle beds knit two interlocked 1 × 1 rib fabrics. Both sides of the fabric look like the face side of jersey.

Jean is a warp-faced twill of carded yarns. It is lighter weight than drill, and it has finer yarns but a higher warp-yarn count.

Jersey is a filling-knit fabric with no distinct rib. Jersey can be any fiber content and knit flat or circular.

Kersey is a very heavy, thick, boardy, wool-coating fabric that has been heavily fulled and felted. In kersey, it is difficult to see the twill weave because of the fulling and the short, lustrous nap. Kersey is heavier than melton. It may be either a single or a double cloth.

Lace is an open-work fabric with yarns that are twisted around each other to form complex patterns or figures. Lace may be hand or machine made or made by a variety of fabrication methods including weaving, knitting, crocheting, and knotting.

Lamé is any fabric containing metal or metallic yarns as a conspicuous feature.

La Coste is a double-knit fabric made with a combination of knit and tuck stitches to create a mesh-like appearance. It is often a cotton or cotton/polyester blend.

Lawn is a fine, opaque, lightweight, plain-weave fabric usually made of combed-cotton or cotton-blend yarns. The fabric may be bleached, dyed, or printed.

Leno refers to any leno-weave fabric in which two warp yarns are crossed over each other and held in place by a filling yarn. Leno weaves require a doup attachment on the loom.

Lining twill is an opaque, lightweight, warp-faced twill of filament yarns. It may be printed.

Madras shirting is a light- to medium-weight, dobby-weave fabric in which the pattern is usually confined to vertical stripes.

Marquisette is a sheer, lightweight, leno-weave fabric usually made of filament yarns.

Matelassé is a double-cloth fabric woven to create a three-dimensional texture with a puckered or almost-quilted look. Matelassés are made on jacquard or dobby looms often with crepe yarns or very coarse cotton yarns. When finished, the shrinkage of the crepe yarn or the coarse cotton yarn creates the puckered appearance. It is used in apparel as well as in home furnishings.

Melton is a heavyweight, plain- or twill-weave coating fabric made from wool. It is lighter than kersey and has a smooth surface that is napped, then closely sheared. It may be either a single or double cloth.

Moleskin is a napped, heavy, strong fabric often made in a satin weave. The nap is suede-like.

Monk's cloth is a heavyweight, coarse, loosely woven, basket-weave fabric usually in a 2 × 2 or 4 × 4 arrangement. Although it can be made in a 6 × 6 or 8 × 8 arrangement, it seldom is because of the low durability. It is often made of softly spun, two-ply yarns of an oatmeal color.

Muslin is a firm, medium- to heavyweight, plain-weave cotton fabric made in a variety of qualities. Muslin made with low-grade cotton fiber with small pieces from the cotton plant is often used in apparel design.

Net is a general term used to refer to any open-construction fabric whether it is created by weaving, knitting, knotting, or another method.

Ninon is a sheer, slightly crisp, lightweight, plain-weave fabric made of filament yarns. The warp yarns are grouped in pairs, but ninon is not a basket-weave fabric.

Organdy is a transparent, crisp, lightweight, plain-weave fabric made of cotton-spun yarns. The fabric has been parchmentized to create the crisp, wiry hand.

Organza is a transparent, crisp, lightweight, plain-weave fabric made of filament yarns.

Osnaburg (osnaberg) is a coarse, bottom-weight, low-count cotton fabric characterized by uneven yarns that have bits of cellulosic waste.

Ottoman is a firm, plain-weave, unbalanced fabric with large and small ribs made by adjacent filling yarns of different size that are completely covered by the warp.

Outing flannel is a medium-weight, napped, plain- or twill-weave, spun-yarn fabric. It may be napped on one or both sides. It is heavier and stiffer than flannelette.

Oxford chambray is an oxford cloth made with yarn, dyed-warp yarns, and white filling yarns. Sometimes a second color is used for the filling yarns.

Oxford cloth is a light- to medium-weight fabric with a 2 × 1 weave.

Peau de soie is a very smooth, heavy, semi-dull, satin-weave fabric. It often has satin floats on both sides of the fabric. It can be made of silk, acetate, or other man-made fibers.

Percale is a plain-weave, medium-weight, piece-dyed or printed fabric finished from print cloths of better quality. Percale is usually a firm, balanced fabric.

Piqué is a fabric made in a variety of patterns. It can be made on a dobby or jacquard loom with carded or combed yarns. Some piqués have filling cords. Most piqués have three or more sets of yarns.

Plissé is a fabric finished from cotton-print cloth by printing with a caustic-soda (sodium-hydroxide) paste. The paste causes the fabric to shrink, thus creating a three-dimensional effect. The strip that was printed usually is darker in piece-dyed goods because the sodium hydroxide increases the dye absorbancy.

Polished cotton is a medium-weight, plain-weave fabric that has been given a glazed-calendar finish.

Pongee is a medium-weight, balanced, plain-weave fabric with a fine regular warp and an irregular filling. It was originally a tussah or wild-silk fabric, but now pongee is used to describe a fabric that has the general appearance of fine warp yarns and irregular filling yarns.

Poplin is a medium- to heavyweight, unbalanced, plain-weave, spun-yarn fabric that is usually piece dyed. The filling yarns are coarser than the warp yarns. Poplin has a more pronounced rib than broadcloth.

Power net is a Raschel-warp knit in which an inlaid spandex fiber or yarn is used to give high elongation and elasticity.

Print cloth is a general term used to describe unfinished, medium-weight, plain-weave, cotton or cotton-blend fabrics. These fabrics can be finished as percale, embossed, plissé, chintz, cretonne, or polished cotton.

Raschel knit is a general term for patterned, warp-knit fabric made with coarser yarns than other warp-knit fabrics.

Sailcloth is a bottom-weight, half-basket-weave (2 × 1), unbalanced fabric. It may be made of spun- or textured-filament yarns. It can be piece dyed or printed.

Sateen is a strong, lustrous, medium- to heavyweight, spun-yarn, satin-weave fabric that is either warp faced or filling faced. A warp faced, spun yarn fabric with a satin weave may be called cotton satin.

Satin is a strong, lustrous, medium- to heavyweight, filament-yarn, satin-weave fabric.

Seersucker is a light- to heavyweight, slack-tension weave fabric. It can be made with a variety of interlacing patterns. Seersucker always has vertical crinkled or puckered stripes made by two sets of warp yarns. One set is under normal tension for weaving. The other set has a much looser tension.

Serge is a general term used to refer to twill-weave fabrics with a flat, right-hand wale. The interlacing pattern is $\frac{2}{2}$. The fabric is often wool or wool-like.

Shagbark is usually a gingham with an occasional warp yarn under slack tension. During weaving, the slack-tension yarns create a loop at intervals giving the fabric a unique surface appearance.

Shantung is a rough-texture, plain-weave, filament-warp yarn and irregular-spun filling-yarn fabric. Shantung is heavier than pongee.

Sharkskin is a wool or wool-like $\frac{2}{2}$ twill made with alternating warp and filling yarns of two different colors and having a smooth, flat appearance. The twill line, unlike most of the wool-twill fabrics, is left-handed. Occasionally a plain-weave or basket-weave fabric is called sharkskin.

Silence cloth is a white double-faced fabric used under table linens to minimize noise during dining.

Suede cloth is a plain-weave, twill-weave, or knitted fabric that is napped and sheared to resemble suede leather. Suede cloth can be napped on both sides. Any suitable fiber can be used. Suede also is used to refer to brushed leather.

Suiting is a general term for heavyweight fabrics. Suiting can be any fiber content or fabric construction that works well in men's or women's suits.

Surah is a soft, lightweight, filament-yarn, twill-weave fabric. It is woven in a $\frac{2}{2}$ twill weave. It can be piece dyed or printed.

Taffeta is a general term that refers to any plain-weave fabric with a fine, smooth, crisp hand made with filament yarns. The unbalanced taffetas have a fine rib made by heavier filling yarns and more warp yarns. Faille taffeta has a crosswise rib made by using many more warp yarns than filling yarns. Balanced taffetas have warp and filling yarns the same size. Moiré taffetas have an embossed water-mark design.

Tapestry is a firm, heavy, stiff, jacquard-weave fabric made with several warp and filling yarn sets. Tapestry is also the term used for fabric made by hand in which the filling yarns are discontinuous. In hand-made tapestries, the filling yarn is used only in those areas where that color is desired.

Terrycloth (terry) is a slack-tension, warp-yarn pile fabric. Terrycloth may have loops on one or both sides of the fabric. Terrycloth may have a jacquard pattern and may be made with plied yarns for durability. There are also weft- or filling-knit terrycloths.

Ticking is a general term used for fabrics of any weave used for mattress covers, slip covers, and upholstery. It may also be used in apparel.

Tricot is a warp-knit fabric made with filament yarns with one or more bars. Tricot has fine, vertical wales on the technical face and horizontal ribs on the technical back.

Tweed is a general term used to refer to wool or wool-like fabrics made of flock or flake novelty yarns. Tweeds are most often made in plain, twill, or twill-variation weaves.

Velour is a general term used to describe pile fabrics. Velours tend to have dense, long, or deep pile. Velours can be woven or knitted.

Velvet is a warp-pile fabric most often made as a double cloth with five sets of yarns. One pair of ground warp and filling create one side of the fabric and a second pair of ground warp and filling create the other side. A fifth set of yarns (pile warp) interlace between the two sets of ground fabrics. The woven fabric is separated into two complete fabrics when the pile warp is cut. Velvet is usually a filament-yarn fabric.

Velveteen is a filling-pile fabric made with long floats that are cut in the finishing process. The ground fabric can have a plain or twill weave. The pile in velveteen is short. Velveteen is usually a spun-yarn fabric.

Viyella™ is a medium-weight, twill-weave fabric made of an intimate blend of 55 percent wool and 45 percent cotton.

Voile is a sheer, lightweight, low-count, plain-weave, spun-yarn fabric in which the yarns have a high, hard, or voile twist to give the fabric a crisp hand. It has a lower count than lawn.

Waffle cloth is a dobby-weave fabric in which the interlacing pattern creates a three-dimensional honeycomb.

Index

C

D

E

F

G

H

I

J

K

L

M

N

O

P

Q

R

S

T

U

V

W

Y

Z